The Gestation of German Biology

The Gestation of German Biology

Philosophy and Physiology from Stahl to Schelling

JOHN H. ZAMMITO

The University of Chicago Press

CHICAGO AND LONDON

The University of Chicago Press, Chicago 60637
The University of Chicago Press, Ltd., London

Published 2018
Printed in the United States of America

27 26 25 24 23 22 21 20 19 18   1 2 3 4 5

ISBN-13: 978-0-226-52079-7 (cloth)
ISBN-13: 978-0-226-52082-7 (e-book)
DOI: 10.7208/chicago/9780226520827.001.0001

Library of Congress Cataloging-in-Publication Data

Names: Zammito, John H., 1948- author.
Title: The gestation of German biology : philosophy and physiology from Stahl to Schelling / John H. Zammito.
Description: Chicago ; London : The University of Chicago Press, 2018. | Includes bibliographical references and index.
Identifiers: LCCN 2017023339 | ISBN 9780226520797 (cloth: alk. paper) | ISBN 9780226520827 (e-book)
Subjects: LCSH: Biology—Study and teaching—Germany. | Physiology—Study and teaching—Germany.
Classification: LCC QH320.G3 Z355 2018 | DDC 570.760943—dc23
LC record available at https://lccn.loc.gov/2017023339

♾ This paper meets the requirements of ANSI/NISO Z39.48-1992 (Permanence of Paper).

To Katie

Contents

INTRODUCTION

The Gestation of German Biology

> Historians want to write histories of biology in the eighteenth century, but they do not realize that biology did not exist then, and that the pattern of knowledge that has been familiar to us for a hundred and fifty years is not valid for a previous period. And that, if biology was unknown, there was a very simple reason for it: that life itself did not exist. All that existed was living beings, which were viewed through the grid of knowledge constituted by *natural history*.
>
> FOUCAULT, *The Order of Things*[1]

This study traces the gestation of German biology from the debate about organism between Georg Ernst Stahl and Gottfried Leibniz at the beginning of the eighteenth century to the formulation of developmental morphology in the era of Carl Friedrich Kielmeyer and Friedrich Schelling at its close. Developments across the eighteenth century in Germany culminated, in the decades around 1800, in the assertion of a new research program. As Schelling famously put it in 1798, with Kielmeyer "a whole new epoch of natural history" took shape.[2] That gestation needs to be reconstructed.

In the eighteenth century "biology" certainly did not exist as a disciplinary rubric.[3] The very term "scientist" was not invented until the nineteenth century.[4] Prior to that, those who pursued inquiry into the natural world went perforce by other rubrics. The set of categories through which I propose to orient my study all derive, quite unsurprisingly, from the protean term "nature." I term "naturalists" the protagonists in this study.[5] They operated in a conceptual field with two distinct poles of orientation. "Natural philosophy" *explained* the physical world in terms of general principles.[6] "Natural history," by contrast, *described* all the plants, animals, and minerals encountered in the material environment.[7] By the mid-eighteenth century, natural history came to a crossroads in its self-definition and articulation vis-à-vis natural philosophy.[8] That coincided with some fundamental crises within natural philosophy itself concerning the possibility and importance of a "nonmathematical *physique*," setting the stage for a shift in the "semantic field" of natural inquiry, a paradigm shift that has been conceptualized by historians of science as "vital materialism."[9] My thesis is that, over the eighteenth century, naturalists undertook to reformulate some domains of natural *history* (living things) into a distinct branch of natural *philosophy* (ultimately, the science of biology).

Those who studied the domain of living things before 1800 were quite serious and systematic, but what they understood themselves to be doing may well have differed substantially from what the discipline of biology later saw as its project. Carl Linnaeus (1707–78), Georges-Louis Leclerc de Buffon (1707–88), or Albrecht von Haller (1708–77), the most eminent naturalists of the eighteenth century, might well have found themselves in a "different world," in a Kuhnian sense, by the mid-nineteenth century.[10] But that does not signify that their project did not have vital connections to what came after. Like the work of Dirk Stemerding, this study is animated by "fascination with an historical period in which those who studied plants and animals were called 'naturalists' and in which 'biology' was only just becoming a catchword, introduced by those who dreamed about a real science of 'life.'"[11] But my interest is driven by concerns for our own philosophy of biology and naturalism more generally.[12] Accordingly, this study is devoted to the *historicist* project of reconstructing the progress over the eighteenth century that opened the way for a "special science" in a *presentist* sense.[13]

The word "biology" came to be evoked by a number of theorists independently around 1800.[14] However, in his classic *The Growth of Biological Thought*, Ernst Mayr proclaimed that "the coining of the *word* 'biology' did not create a *science* of biology"; rather, "what existed was natural history and medical physiology. The unification of biology had to wait for the establishment of evolutionary biology and for the development of such disciplines as cytology."[15] Dissenting from Mayr, Trevor DeJager aptly observes, "the coining of new terms indicated something significant was happening conceptually as well as culturally."[16] Thomas Bach makes a similar point: "The appearance of the term *Biology* is an indication of a shift in perception in the domain of the natural sciences in which the necessity of the elaboration of a new science that concerned itself exclusively with the phenomena of life became manifest."[17] I contend that the embrace of the new term around 1800 signaled a theoretical and methodological *convergence* of natural history with medical physiology in comparative (i.e., *zoological*) physiology that resulted in the field of developmental morphology.[18] The research program (*Fachgebiet*) of *Naturgeschichte*, as it made the transition toward historicization, found it increasingly necessary to move from "external" traits (taxonomic description) to "internal" organs, structures, and processes (comparative anatomy and physiology) to explain and generalize its findings.[19] Conversely, the *Fachgebiet* of physiology found it increasingly important to create developmental and genetic accounts—not only ontogenetically (e.g., in embryology) but even phylogenetically for varieties and species.[20] The emergent research program of morphology was, for all its equivocations, drawn toward *actual* historical development alongside "ideal" typological sequencing.[21] In that sense, it reached out toward the new historicizing *Naturgeschichte*. Developments in each respective *Fachgebiet* drew them, in the apt summation of Thomas Bach, to the "same result: descent explains the similarities in organization."[22] Kielmeyer was the pioneer of this convergence. His "physiological theory of forces" and his "temporalized natural history" achieved the connection that offered a systematic basis for the emergent science.

WHY GERMANY?

It makes perfect sense, of course, to maintain that the development of life science in the eighteenth century was a "transnational" affair and that a national focus can be misleading.[23] Nonetheless, there are reasons for a focus on Germany. First, the Germans were operating within a distinctive cultural context, especially in religion and philosophy.[24] Second, and in some measure as a result of the first, historians of biology have looked askance at the trajectory of German life science over the eighteenth century, considering it foredoomed to the metaphysical errancies they associate with *Naturphilosophie*, ostensibly anathema for any sound constitution of the life sciences.[25] This study aims to rebut that way of thinking.[26] I propose to link the gestation of biology in Germany with that most despised phenomenon in the history and philosophy of science, Idealist *Naturphilosophie*.

As Frederick Beiser observes, "*Naturphilosophie* has been ignored or spurned for decades, by historians of philosophy and science alike. Its reputation suffered greatly under the shadow of neo-Kantianism and positivism, which had dismissed it as a form of pseudoscience. . . . For many philosophers and scientists, *Naturphilosophie* became the very model of how *not* to do science."[27] His assessment is apt, and it is shared by Robert Richards.[28] I take my stand with Beiser and Richards for a new revisionism.[29] Finally, as Daniel Steuer observes, "there seems to be a growing consensus amongst historians of science that the division between empirical science, based on experience and experiments, and speculative Romantic *Naturphil-*

osophie, based on ideas, is an invention of the later nineteenth century."[30] It is well-past time to put this prejudice to rest along with all the other complacent dogmas of the positivist epoch.[31] The effort to consolidate biology found positive reinforcement in German Idealist philosophy, and instead of viewing *Naturphilosophie* as a contamination, we might view it as historical evidence that something essential to the character of biology as a special science was at stake, and thus, this episode in the history of biology might reopen issues in our *own* philosophy of biology.[32] That makes a specific account of German developments indispensable.

Three contexts seem pertinent in reconstructing what was distinctive about the German Enlightenment insofar as it bore upon the gestation of biology. First, in Germany the religious tenor was distinctly stronger than in western Europe.[33] More concretely, the tension between Pietism and "philosophical rationalism" (Wolffian "school philosophy") informed the trajectory of philosophy (and medicine) in the early Enlightenment.[34] Consequently, my point of departure will be the relationship between Enlightenment and Pietism in the *medical* faculty at the University of Halle. Second, ensconcement within academia was a pervasive feature of the German Enlightenment, given the relative "backwardness" of career opportunities in Germany's public sphere.[35] This was true a fortiori for the emergent life sciences. My claim is that *at least in Germany* the eventual discipline of biology gestated primarily in the traditional medical school world.[36] Two university medical faculties played a decisive role: first Halle and then Göttingen. Newly formed and animated by a quite innovative temper, Halle and Göttingen were from the outset *research universities* in the sense that Wilhelm von Humboldt would later make famous at Berlin, and this was true of their medical faculties. Halle had almost half a century of preeminence before Göttingen rose to challenge it in the person of Albrecht von Haller.

A final distinctive characteristic of the German Enlightenment is the tension between the imposition of French cultural models and the assertion of a distinctly German national culture. The nascent German culture's adversarial reaction to the hegemonic imposition of French standards has been a perennial theme, and of course, it remains important; French cultural dominance clearly provoked a reactive nationalism.[37] But not all French influence was unprofitable, even if it aroused resistance, and as regards the gestation of biology, I will argue that the French influence was both substantial and *salutary*. To grasp this creative reception of French "vital materialism" around the midcentury in Germany, we must contextualize more concretely the emergent German Aufklärung, especially in the Royal Prussian Academy of Sciences in Berlin (Berlin Academy).[38]

In tracing out the gestation of biology in Germany, our itinerary will take us through many of these key sites: the University of Halle, the University of Göttingen, the Berlin Academy, and ultimately the University of Jena. The context provided by each of these sites proved crucial to the constitution of the emergent science.

NATURALISTS AND THE MEDICAL FACULTY

As Roger French made clear in his strikingly titled *Medicine before Science*, crisis had beset the traditional identity of the "learned physician" in Europe by the close of the seventeenth century. Physiology had traditionally served as "a bridge between medical theory proper and the larger domain of natural philosophy"; that is, it dressed medicine with academic learning.[39] But traditional physiology (Aristotle and Galen) had come under direct attack by a rival new philosophy, the mechanization of the body, associated especially with Descartes.[40] "By the end of the seventeenth century, . . . the attack on learned physic had succeeded almost entirely."[41] According to Descartes and the other "mechanical philosophers" of the late seventeenth century, a universal physics had no place for a distinct *life* science: animate and inanimate matter needed to be submitted to a single method—*mathesis universalis*—whose principle must derive from physical mechanics.[42] Descartes concluded that animals had to be mechanisms.[43] His impact is vivid in the famous opening argument from the *Leviathan* (1651) of Thomas Hobbes (1588–1679): "Life is but a motion of limbs. . . . For what is the heart, but a spring, and the nerves, but so many strings, and the joints, but so many wheels, giving motion to the whole body?"[44]

The *bête machine* hypothesis became the point of departure for a mechanistic medicine in the seventeenth century. Later still it evoked the eighteenth-century elaboration of vital materialism, calling Descartes's bluff with the notorious proposition that man, too, was a machine.[45] From the outset, Descartes's contemporaries "objected that if the *bête machine* concept were accepted, it would be very difficult in the end to prove that man himself was not a machine."[46] Thus, the animal-machine hypothesis "was not primarily about what animals could do but about the implications for man."[47] The wily skeptic Pierre Bayle (1647–1706) saw all the implications and laid them out in his article "Rorarius."[48] Physiology was inextricably entangled with philosophy, and from this entanglement would spring the concerns that eventually spawned biology. The furious struggles across the entire eighteenth century between the defenders of "physicotheology" and their dread opponents, the Epicurean materialists, proved, in fact, the birth pains of life science.[49]

The Italian Giovanni Borelli (1608–79) composed *De Motu Animalium* (posthumous, 1680–81), and it went through fifteen editions from 1700 to 1723, propagating "iatromechanism."[50] This view proposed the derivation of medical diagnosis and therapy from the physics of motion, a "purely corporeal (*res extensa*) investigation of bodily functions" denying the soul any influence in physiology.[51] Concurrently, an alternative, "iatrochemical" approach proposed to derive medical diagnosis and therapy from the analysis of "mixtures" and "ferments," drawing on the writings of Paracelsus (Philippus von Hohenheim, 1493–1541) and Jan Baptist van Helmont (1579–1644).[52] Neither view could oust the other. Because the theoretical state of the discipline was in stark disorder, there was simply no doctrinal purity to be had.[53]

Adding to the theoretical quandaries was a stark practical challenge. Thomas Sydenham (1624–89) confuted the learned physicians and offered an alternative, relentlessly practice-oriented medicine, invoking the hero of the learned tradition, Hippocrates himself.[54] Sydenham became famous in Europe for contending that physicians had more to learn from natural histories of disease and from case studies of individual patients than from either the book learning of the schools or the elaborate experimental work in anatomy and chemistry that was preoccupying a number of leading physicians.[55]

Confronted by all these challenges, academic medicine desperately sought a new orthodoxy.[56] The figure who achieved a measure of reintegration for learned medicine was Herman Boerhaave (1668–1738).[57] While he was not a major innovator, he was able to achieve a fusion of the mechanistic approach derived from Descartes with a measure of the "chymistry" of the Paracelsians, all the while enthusiastically affirming the Hippocrates-Sydenham emphasis on clinical practice.[58] Making it all cohere for over 1,900 medical students from all over Europe, notably Britain and Germany, at the University of Leiden from 1701 to 1738, Boerhaave emerged as the "teacher of Europe."[59] He enjoyed an unquestioned reputation as the foremost European teacher of medicine.[60] His personality and pedagogy exerted an enormous influence on students.[61] For Albrecht von Haller, Boerhaave was not only his teacher but his "great scientific and human model."[62] "To him I owe eternal affection and everlasting gratitude.... Perhaps future centuries will produce his equal in genius and learning, but I despair of their producing his equal in character."[63] Indeed, Boerhaave served as "Albrecht Haller's grand model for his [entire] life."[64] Given their number and eminence, the students of the Leiden medical school—"Boerhaave's men," as they have been called—had a decisive impact on the whole medical profession in the eighteenth century. They formed a crucial medical network across Europe, in constant and fruitful communication.[65] Moreover, Boerhaave had a transformative impact on the organization of other medical schools. The program at Leiden became the model for the most advanced medical schools of the eighteenth century. Three great medical schools were formed after the image of Boerhaave's Leiden: Edinburgh, Vienna, and Göttingen.[66] The glaring exception was France. Boerhaave had no impact on the Paris medical school, and Montpellier ultimately became a bastion of support for his great critic, Georg Ernst Stahl (1659–1734).[67]

Boerhaave's physiological theories were only moderately mechanist, but he was generally taken to be the eighteenth-century standard-bearer of this approach, especially after his inaugural lecture of 1703.[68] Later, Boerhaave became famous for bedside instruction of medical students. His most prominent students, however, did not report attending any clinical rounds. Notably, Haller's diary contained "no reference . . . concerning this institution," and in the exhaustive manuscripts of Gerard van Swieten (1700–1772), chronicling the whole sweep of Boerhaave's teachings, "only two clinical lectures are mentioned."[69] Nonetheless, on questions of practice and on questions of theory—particularly in physiology—Boerhaave assumed a towering eminence in the ensuing era.

Medicine was the only academic and professional path for a naturalist in early modern Europe, and especially in Germany. As Irmtraut Scheele puts it: "Since the career path of the botanist, the zoologist, or even of the biologist in general did not yet exist, anyone with a special interest in one of these natural sciences was forced to undertake a course of study in medicine for the sake of earning a living later."[70] But within the medical faculty mutations emerged. Central to my account is the "calving" over the eighteenth century of research physiology from the larger clinical-practical structure and orientation of German academic medicine. There widened a decisive division between the clinical-practical direction that the profession of medicine clearly sought to pursue and the pure research into a variety of life-forms (their relevance often quite remote from human therapeutic ends) that a few members of the medical faculty preferred instead.[71] In short, a very important trend—one that would be embodied in the key eighteenth-century Göttingen scholars Albrecht von Haller and Johann Friedrich Blumenbach (1752–1840), who will be central to my account—was the emergence of an increasingly specialized research practice within the traditional medical faculty whose agenda proved somewhat at cross-purposes with what the larger faculty and profession sought to achieve.[72] The new physiology became a special research field (*Fachgebiet*) interested not at all in clinical application but rather in *zoological research* for its own sake.[73]

For this research community to carve out its own institutional space, it would need to create alliances with parallel impulses within academic medicine and beyond it. Paula Findlen writes: "Natural history would continue to be closely associated with medicine through the eighteenth century. But increasingly its leading practitioners studied nature apart from medicine."[74] Thus, neither the greatest of the late seventeenth-century British naturalists, John Ray (1627–1705), nor the "prince of naturalists" in the first half of the eighteenth century in France, René-Antoine Ferchault de Réaumur (1683–1757), was a physician.[75] Key experimental naturalists of the mid-eighteenth century, especially the Genevans Abraham Trembley (1710–84) and Charles Bonnet (1720–93), would also follow that nonmedical path.[76] Perhaps the greatest French naturalist of the whole century, Georges-Louis Leclerc de Buffon, was no physician either.[77] From outside medicine the experimental physiologists were abetted by those who shared their ambitions, both experimental and philosophical. In fact, philosophers and physicians came together as naturalists to create a curious persona—the *médecin philosophe* or *philosophischer Arzt*—through which they could articulate a *common* new research domain exploring body-mind interaction and the place of life in the order of the physical world.[78]

The term *médecin-philosophe* came to prominence in France around the middle of the eighteenth century. The maverick Julien Offray de La Mettrie (1709–51), the most notorious example, insisted that *only* philosophical physicians could penetrate the labyrinth of man.[79] It became a rubric especially for the school of Montpellier. Théophile Bordeu (1722–76) was among the most explicit in identifying himself as a *médecin-philosophe*.[80] The French *médecins-philosophes* adopted the "optimistic attitude that a physiological consideration of man would throw light upon obscure epistemological and moral-legal problem constellations."[81] They believed that even "the most impalpable and spiritual functions of man were to reveal themselves empirically, to exhibit sensible signs, and to permit an empirical analysis."[82] Conversely, the *médecins-philosophes* "accepted the decisive argument that physiological states affected all human behavior, including intellection, acts of will, and moral behavior."[83] This committed them to *influxus physicus* (direct interaction between mind and body) as a *methodological* premise, even if they recognized that they could achieve no *metaphysical* solution to the conundrum of the body-mind interaction (*commercium corporis et mentis*).[84] Accordingly, they dared intrude into spheres sacrosanct to metaphysics, to become *philosophical* physicians.[85] Crucially, this created the opportunity for *philosophers* who were *not* physicians to share this persona. Such a consolidated cadre of *médecins-philosophes*—actual physicians and their philosophical allies—played a conspicuous role in vitalizing nature in the Enlightenment.

THE FRENCH CHALLENGE: VITAL MATERIALISM

The struggle to establish autonomy for the German language and German culture in light of the brilliance and cosmopolitan ubiquity of French culture in the aftermath of the *siècle d'or* was intense.[86] French culture shaped Old Regime courtly-aristocratic culture across all Europe and certainly in the Germanies. Frederick II's Potsdam was a conspicuous instance.[87] There is no question that, in the culture-transfer balance, the Germanies weighed heavily as importers. The impact of the French on fashion, manners, and thought extended well beyond court culture to the urban literate population.[88] For Germany generally, the period around midcentury saw a massive invasion by French ideas. Montesquieu exerted a profound influence.[89] The impact of Rousseau was enormous.[90] Voltaire proved an obstreperous presence not only in text but in person in these years.[91] While no materialist, his sojourn in Prussia was definitely understood as part of the same inundation by the *esprits forts*.

Still, it would be misguided to infer that the Germans were passive in this situation: there can be enormous creativity in (selective) reception, and this they definitely demonstrated across the eighteenth century. Moreover, the energies of indigenous creativity were already stirring. By midcentury, not only had the Swiss German theorists Johann Jakob Bodmer (1698–1783) and Johann Jakob Breitinger (1701–76) challenged Johann Christoph Gottsched (1700–1766) and his French neoclassical gospel of taste, but the Swiss German poet Albrecht von Haller had made a European mark.[92] And Haller was soon joined by Friedrich Klopstock (1724–1803), whose *Messias*, the key literary work of German *Empfindsamkeit* (Sensibility), began appearing in 1748.[93] To be sure, from Berlin, the king of Prussia deemed German a language suitable only for servants, and only an obscure provincial historian of

Osnabrück dared publicly to differ.[94] But by the 1770s, a decisive new generation (the Sturm und Drang) would burst upon the scene, and "a German way and art" would prove monumental.[95]

At midcentury, the "German movement" was still embryonic, but a more general Enlightenment was well under way.[96] The Hochaufklärung (1750–80) responded to the striking growth of large cities, especially Hamburg and Berlin but also Leipzig and Frankfurt am Main.[97] With that growth came distinct changes in the cultural milieu. The periodical press took shape, for instance.[98] In addition to urban newspapers, one of the most important developments was the circulation of a vast number of "moral weeklies," which offered instruction in taste and style along with—as their title suggests—a great deal of moral instruction.[99] With the rise of the so-called *gebildeten Stände* (educated strata) in the cities came a new "public sphere," redefining the sociocultural meaning of Aufklärung.[100] It was no longer a matter merely for *Gelehrten* (scholars). The new *Bürger* needed to achieve independence in judgment: *Selbstdenken*. To be capable of thinking for oneself was to achieve "maturity [*Mündigkeit*]."[101]

The Berlin Enlightenment, led by the "philosophers on the Spree," as Gotthold Efraim Lessing (1729–81) and Moses Mendelssohn (1729–86) have been called, provided an indispensable incubation.[102] Berlin was a European center; 20 percent of its population was Huguenot-French; French was an important language not only for Frederick II's court and for his academy but for the city itself. In the Berlin Aufklärung, French-language philosophical discourse had a significance that is only now coming to be sufficiently appreciated.[103] John Yolton's *Locke and French Materialism* has shown that this Francophone discourse turned upon the so-called "Three Hypotheses" (occasionalism, preestablished harmony, and physical influx) conceived by Leibniz.[104] Avi Lifschitz demonstrates a similar centrality for "Epicurean naturalism," especially regarding the link of language with mind and culture.[105] Part of the revitalized Berlin Academy's mission under Frederick II was to introduce into German culture Western—primarily French—Enlightenment ideas, even at the expense of Pietist and Wolffian domestic traditions.[106] What followed was the influx of French vital materialism and its uptake by German *Freigeister* (free spirits).[107]

This high German Enlightenment began the moment that Lessing chose to abandon his formal university studies in Leipzig and move to Berlin to take up a career as a freelance writer (*freier Schriftsteller*).[108] What made his experience possible? The old-fashioned connection to the reign of Frederick II of Prussia (1740–86) retains its historical plausibility.[109] Three particular features of his new reign are central here: first, his reinstatement of Christian Wolff (1679–1756) at the University of Halle; second, his revitalization of the Berlin Academy, especially through the recruitment of French intellectuals; and third, closely related to this, his interest in fostering "freethinking" in the sphere of religion. The third connection is tightly interwoven with the second because "freethinking," in the German mind, could not be dissociated from French materialism.

In Germany, Berlin was the unquestioned center of such freethinking, with two distinct poles: Frederick II and his court, on the one hand, and Lessing and his circle in the publishing world of Berlin, on the other.[110] As a university student in Leipzig, Lessing had pursued freedom of thought and expression into its most problematic quarters. He was drawn to a consideration of dissenters and their fates, creating a mini-genre for himself, the *Rettung*, or "vindication," to which he devoted a good deal of writing from the late 1740s into the early 1750s.[111] When he moved to Berlin, he brought his commitment to tolerance with him.[112] Lessing believed it was essential to break loose the idea of freedom of thought from the orthodox clerical opprobrium and popular dread it conventionally encountered. That goal animated Lessing's pioneering critical journalism in Berlin: "the penetrating and constantly repeated call to 'thinking for oneself [*Selbstdenken*],' 'judging for oneself [*Selbsturtheilen*],' is the organizing center of all the individual efforts regarding the public even in Lessing's early years."[113] He set out to cultivate—indeed, to create—a new urban reading public by weaning it away from the religious confines of traditional culture, on the one hand, and from acquiescence to representations of courtly and aristocratic preeminence, on the other. As E. Schmidt puts it, "Even before exposure to the dangerous atmosphere of Berlin Lessing was . . . already enlightened enough to consider a freethinker, *esprit fort*, *Freigeist* something more than a puppet with which the moral weeklies like governesses keep their children full of fear of the Lord."[114] The young Lessing relished the notion of *Freigeisterei* and it was central to his personal project.

The terminology of *Freidenker* or *Freigeister* became established only around midcentury, displacing an older reliance on imported French terms, such as *libertin* and *libertinage*. Thus, *Zedlers Universallexicon* (1736) had no entries for *Freidenker* or *Freigeist* but did address their conceptual content under the French derivation *Libertiner*.[115] By 1758 Johann Mehlig's *Historisches Kirchen- und Ketzer Lexikon* had no use for that French term, inserting instead an article on *Freydenker* in which Mehlig reported: "that is how those are called who would earlier generally have been termed libertines."[116] Reiner Wild takes this to be a sign that the latter term was "already by the midcentury an obsolete usage."[117] The next year, 1759, Johann Anton Trinius (1722–84) published his remarkable *Freydenker-Lexikon*.[118] Thus, we have good grounds for believing that the question of "freethinking" was prominent in the German cultural context of midcentury, especially in Berlin.[119] Vernacular translation did not, however, displace cultural meaning: freethinking remained, for Germans, very "French." As Trinius's fascinating lexicon makes clear as well, the connection to Epicureanism and Spinozism was constitutive for *Freigeisterei*. With the opening volumes of the *Encyclopédie* of Diderot and d'Alembert, the "party of the *philosophes*" achieved notoriety in Germany.[120] One of the casualties of the early French battles over the *Encyclopédie*, the abbé de Prades (1720–82), found refuge in Berlin. The most notorious "materialist," La Mettrie, had already taken refuge there. For the indigenous Aufklärung in Berlin, and for Germany more widely, "freethinking" and French, Epicurean-Lucretian, and Spinozist "materialism" came to be inextricably associated. All this would prove fateful for the rise of life science in Germany.

CHAPTER ONE

Animism and Organism: G. E. Stahl and the Halle Medical Faculty

I am convinced that Stahl, who is disposed to explain animal processes in organic terms, was frequently closer to the truth than Hofmann [sic] or Boerhaave, to name but a few.

KANT, 1764[1]

PIETISM AND ENLIGHTENMENT AT THE HALLE MEDICAL FACULTY

Envisioned from about 1688 and starting its first operations already in 1691, the University of Halle celebrated its formal inauguration in 1694. It was created by the rulers of Brandenburg-Prussia to train new civil servants for the ambitiously expansive territorial state and to serve as a cultural symbol of its prominence in Germany by rivaling the nearby Saxon university at Leipzig.[2] Brandenburg-Prussia had to deal internally with the problem of multiple confessions. In 1613, to the dismay of his subjects, the Hohenzollern *Kurfürst* (elector) of Brandenburg converted to the Reformed faith, even though the bulk of the population in his territories remained staunch Lutherans.[3] For their training, most Lutheran clergy in Brandenburg attended the stringently orthodox Lutheran universities of Leipzig and Wittenberg in Saxony. This appeared a double detriment to Hohenzollern dynastic ambitions. Thus, both for bureaucratic staffing but also for religious policy, a new Lutheran university to rival Leipzig (and Wittenberg) became the priority of Frederick III/I (b. 1657; r. 1688–1713).[4]

The Hohenzollern dynastic strategy found powerful convergence with the needs of a new religious movement, Pietism.[5] "[T]he regime in Berlin saw in the Pietists a group within the Lutheran church that would be much more accommodating toward the Huguenot refugees, much less hostile to the Reformed faith in general, and perhaps quite useful to the state as a counterweight to the heavily orthodox Lutheran leadership."[6] A new university at Halle served both Pietist aspirations and those of the dynasty.[7]

Two figures proved central in establishing the character of the new university: the legal and political philosopher Christian Thomasius (1655–1728) and the Pietist theologian and activist August Hermann Francke (1663–1727).[8] When he was named professor of Greek and Oriental languages, Francke was also appointed pastor of the Lutheran congregation in the Halle suburb of Glaucha, where he immediately began active social reform.[9] The famous orphanage (*Waisenhaus*) and its penumbra of attached institutions (the so-called Halle *Anstalten*) made Halle a remarkable venue for experiments in social welfare. The educational institutions that Francke established at the *Anstalten* were as important as his orphanage.[10] From very early on, Francke included medical care for his orphans and the local poor as part of the *Anstalten*. The manager of the highly lucrative *Anstalten* pharmacy, Christian Friedrich Richter (1676–1711), took charge of medical care in 1697, working closely with Francke in institutionalizing "Pietist notions about the meaning of illness and health."[11] By 1708, substantial medical operations were under way at the *Anstalten*, with student volunteers from the Halle medical faculty working with the patients in exchange for free meals and thus gaining important clinical experience. Eventually, this program was taken over by Johann Juncker (1679–1759), who proceeded to make clinical training at the *Anstalten* part of the official curriculum of the Halle medical school after 1716.[12]

Medical enrollment at Halle had remained very low until Juncker's institutionalization of clinical training at the *Anstalten*. By the 1730s over five hundred students were enrolled in the Halle medical faculty.[13] Several cohorts of students, otherwise too impecunious and socially marginal to have had any other opportunity, received their preliminary education via the *Anstalten* and eventually attended the university with substantial assistance (e.g., the famous *Freitische*) from the *Anstalten*, in exchange for services to its charitable institutions of education and health care.[14] Tension between high-living and high-class law students and the more abstemious theological and medical students marked an important cultural divide within the university, leading to a famous allegation about attending Halle: "'*Halam tendis? Aut pietista aut atheista reversurus!*' ('So you're going to Halle? You'll return either a pietist or an atheist!')."[15]

The Halle medical school played a major role in the gestation of German biology in the first half of the eighteenth century. It was a tiny faculty, founded with only two professorships, extended to three in 1718, and not to grow again until 1780. The founding faculty, Friedrich Hoffmann (1660–1742) and Georg Ernst Stahl (1659–1734), presided from the university's founding in 1694 to Stahl's departure for Berlin in 1715. Hoffmann was the first called to the faculty, and he proved decisive in configuring its program, "introducing into medicine at Halle the innovations of the Dutch and the English."[16] Stahl joined the faculty only two years after Hoffmann, and they were friendly rivals until 1715, when Stahl was called to Berlin to become head of the Royal College of Medicine. Notably, in their earlier careers both Hoffmann and Stahl had close relations to the Pietist movement, and evidence suggests Francke played a role in their recruitment to Halle.[17]

All in all, what distinguished the Halle medical school in its heyday was the prominence of its two great theorists. The medical faculty had real weaknesses in other areas, however. Hoffmann and Stahl had "neither resources nor facilities [*weder Hilfsmittel noch Anstalten*] at their disposal to carry out their teaching mission," one historian has noted.[18] Another is even more critical: "For its first decades Halle had no *theatrum anatomicum* or even a *hortus medicus*. . . . Compared to Leiden, 'science' was at a minimum."[19] The medical faculty at Halle was not nearly as advanced as Leiden in basic physics and chemistry and severely lacked facilities for direct experiment in botany and anatomy. In the area of medical theory, "physiology and physiological chemistry are absent as independent disciplines in the curriculum of Halle in the eighteenth century."[20] The botanical garden got a tiny start on a tract of land acquired in 1683; after the medical school was established, this garden was entrusted to Stahl, but he left it "utterly neglected [*völlig verwahrlost*]."[21] Things proved no better under his successor.[22] There was no anatomical theater until 1725, when Georg Daniel Coschwitz (1679–1729) started one at his own expense; it became operational only in 1727.[23] Coschwitz (MD, Halle, 1694) was a student of Hoffmann's and Stahl's who became extraordinary professor in 1716, then third ordinary professor in 1718. When he died in 1729, no ordinary professor of anatomy replaced him (his full professorship went to Juncker, instead), and so extraordinary professors (professors who did not receive salaries) had to take it up. That included buying the anatomy theater from Coschwitz's heirs! These anatomy instructors proved mediocre, leaving the field in limbo at Halle for the better part of two decades.[24] The contrast with Leiden (and eventually Göttingen) could not be more striking.[25]

Jürgen Helm argues compellingly that one cannot separate the history of the medical faculty from the history of the Halle *Anstalten*.[26] In Francke's theology, the source of illness was sin; the affliction was a *test* from God for the already pious and a *warning* from God to the impious. Thus, illness required a "soul-cure" as well as a "body-cure." Only God's intervention would ultimately heal, though the physician should also work on the physical symptoms. This approach to medical practice, which Helm calls "*Waisenhausmedizin*," dominated not only the *Anstalten* but the Halle medical faculty, where it merged fully with Stahl's theoretical approach and with the clinical orientation that Stahl and Hoffmann shared.

Unquestionably, "Stahl's medical theory became a rallying-point for Pietism."[27] By contrast, Hoffmann "did not share the ideological [*weltanschauliche*] opposition of Pietism towards dualism and rationalism," even though he "sincerely supported Pietism's practical concerns in Halle."[28] Hoffmann, as Roger French discerns clearly, saw himself as a moderate, an "enlightened Pietist," who did not wish to let the differences between his theoretical stance and that of his esteemed rival, Stahl, become divisive; rather, he believed that they should serve as complementary stimuli in the education of medical students at Halle.[29] The diversity was, in his view, constructive and contained, especially via the convergence of views concerning practice. In Stahl, by contrast, Pietists perceived an "unequivocal emphasis on the soul and its power to influence the body . . . in relation not only to the physiological but to the pathological."[30] His psychosomatic conception of the origins of illness seemed to have tight affinities to *Waisenhausmedizin*.

Most of Stahl's famous disciples came through the *Anstalten*. They were particularly important in translating and popularizing Stahl's difficult Latin texts. Francke's close associate Christian Friedrich Richter published the most famous popularization of Stahl's system: *Höchst=Nöthige Erkenntnis des Menschen, sonderlich nach dem Leibe und natürlichen Leben* (1710).[31] Another key disciple was Christian Weisbach (1684–1715), who studied first with Stahl before obtaining his MD at Basel in 1711, then published the Stahl-Pietist guidebook *Wahrhaffte und gründliche Cur aller dem menschlichen Leibe zustossende Kranckheiten* in 1712.[32] A mainstay for the learned physicians was Johann Storch's *Praxis Stahliana, das ist Herrn Georg Ernst Stahls . . . Collegium Practicum* (1728; 2nd ed., 1732; 3rd ed., 1745), about 1,500 pages long.[33] Michael Alberti (1682–1757), Johann Juncker, and Johann Samuel Carl (1676–1757), all Stahl students, published compendia of his ideas for classroom use.[34] Carl was Stahl's "most highly-esteemed student [*gepriesener Meisterschüler*]." He received his MD in 1699, then went on to edit Stahl's works and publish a key popularization: *Zeugnüsse von Medicina Morali* (1726).[35] These disciples adopted Stahl's medical doctrines but inflected them in a clearly religious vein, along the lines of Francke's theological approach to illness.

After his departure from the Halle medical faculty, Stahl's acolytes Alberti and Juncker upheld the teachings of their absent mentor and participated actively in the Pietist campaign against the philosopher Christian Wolff (1679–1756), culminating in his 1723 expulsion from Halle and Prussia.[36] Totally rooted in Francke's Pietist *Waisenhausmedizin*, they dominated the medical faculty from 1715 through the 1730s. Both of these figures had begun as theologians, then switched to medicine.[37] Alberti assumed Stahl's chair in 1715, when the latter moved to Berlin. Juncker received his MD at Halle in 1717 at thirty-seven years of age, under Alberti. He continued to work at the *Anstalten* while he simultaneously served as *Doctor legens* (*Privatdozent*) for twelve years in the medical faculty. He received appointment as an ordinary professor only in 1729, after the death of the anatomist Coschwitz, and he never received an adequate salary.[38] Together, Alberti and Juncker contested the aging Hoffmann within the Halle medical faculty until the triumphal return of Wolff in 1740. Alberti in particular showed strident hostility to Hoffmann's theory, attacking it publicly in *Specimen medicinae theologicae* (1736), a fusion of Stahl and *Waisenhausmedizin*.[39]

Already from the beginning, as the architect of the curriculum for the medical faculty, Hoffmann had been committed to linking the pursuit of a medical degree with natural-philosophical training.[40] Most of his students took a strong interest in physics and chemistry and showed less concern with theology than the followers of Stahl. Hoffmann's connection with Christian Wolff grew closer after Stahl left Halle and Wolff assumed responsibility for the physics courses.[41] Wolff's forced departure in 1723 must have been a blow to this curricular commitment. After the expulsion of Wolff, Alberti took over both physics and botany, concentrating primarily on the teaching of physics, but he was not really up to that demanding role, especially in the era of Newtonian ascendancy.[42] Coschwitz remained responsible for anatomy and in fact showed more concern for botany than did Alberti, but he died in 1729. Thus, "in Halle through the forced exile of Wolff in the years between 1723 and 1740 there was temporarily a notable narrowing of the hitherto prevailing universality of natural-scientific educational opportunities [*Ausbildungsganges*]."[43]

In that measure, Alberti personified the "atrophy" (*Erstarrung*) of the university under the hegemony of Pietism.[44] He published his most aggressive disputation of Hoffmann in 1736 and followed it in 1740 with a polemic against all mechanistic medicine, *Medicinische Betrachtung von den Kräften der Seelen nach den Unterscheid des Leibes*, deriding mechanism as mere "puppet theater."[45] He insisted that there had to be a "principle of intelligence" in living things that transformed them from machines into organisms.[46]

Thus, Alberti reasserted in particularly polemical fashion the Stahlian/Pietist viewpoint. It was in these years that Hoffmann composed not only his final system of medical theory but also a formal discrimination of his position from that of the Stahlians (published only posthumously, in 1746).[47] Stahl and Hoffmann had been friendly rivals since 1694, and Stahl had been dead for some years, so why was it only so late that Hoffmann took this step? It was not Stahl personally that Hoffmann wished to dispute but rather his adamant and insolent epigones. When Alberti took over the field of physics with such limited efficacy, Hoffmann saw himself as the beleaguered defender of the natural sciences at Halle. Two of the three ordinary professors in the medical faculty after 1715 were adamantly Pietist Stahlians, and they had powerful allies in the wider university throughout the period in question. Hoffmann's commitment to extensive training in the natural sciences stood in stark contrast to the more spiritualist leanings of those, like Alberti and Juncker, committed to *Waisenhausmedizin*. Thus, the last years of Hoffmann's career at Halle can be characterized as a period of embattlement. In any event, he was hardly unhappy to see Wolff's return.

The royal reinstallation of Wolff in 1740 represented the thorough defeat of Pietism within the university. Combined with the death of Hoffmann shortly thereafter (1742), it also set the stage for a third generation in the development of the medical school.[48] These naturalists have come to be called the "rational physicians [*vernünftige Ärzte*]" in German historiography.[49] The most prominent figures were Johann Gottlob Krüger (1715–59) and his student Johann August Unzer (1727–99).[50] Primarily Hoffmann's students, they pursued a dual degree program: accreditation in natural philosophy (mathematics, physics, and chemistry) as well as in medicine.[51] The new generation of medical researchers at Halle were trained in an energized environment of rival medical and cultural theories, between Hoffmann's "mechanism" and Stahl's "animism," between Pietism and Wolffian rationalism, and between "eclectic" or "popular philosophy" and "school philosophy." After Hoffmann's death, they sought not only to reconcile these conflicting traditions but also to move beyond them into new terrain, especially in physiological psychology.[52] Krüger and Unzer saw themselves called upon to address the blatant preponderance of the "*Herrn Stahlianer*."[53] Krüger's key manifesto of the group, *Grundriss eines neuen Lehrgebäudes der Arzneygelarhtheit* (1745), needs to be seen as a response to just this controversy within the medical faculty.[54] Unquestionably, Krüger was the key figure in this Halle "constellation." As Carsten Zelle puts it, "Krüger mediated between systems-thinking and empiricism, mechanism and organism, philosophy and medicine, older and younger generations; his work stands between natural science (e.g., his *Naturlehre*) and literature (*Träume*)."[55] Unzer became his most distinguished student.[56] There was a crucial convergence between Hoffmann's protégés in medicine (Krüger, Unzer, et al.) and revisionist Wolffian philosophers at Halle (Alexander Baumgarten [1714–62] and especially Georg Friedrich Meier [1718–77])—namely, in "physiological psychology," "psychomedicine," and "philosophical pathology." At the core was a concern with what Christian Thomasius had already highlighted: *Affektenlehre*, the conceptualization of the feelings.[57] Feelings had both a philosophical and a physical register: they were a topic on which the disciplines could converge and, indeed, constitute a new "intermediate" science.[58]

Yet by 1750 the university had allowed this extraordinary pool of talent to disperse, and the medical faculty forfeited its leadership to other centers in Germany. Strikingly, the number of ordinary professors in the medical school stagnated at three from 1718 until late in the century. What happened? First and foremost was the underfunding of the university: virtually no increases in the original endowment for faculty salaries after the inauguration era. Few new ordinary (i.e., salaried) professorships were created—*none* in the medical faculty. That meant that existing salaries had to be cannibalized, faculty had to secure multiple appointments, and, when possible, seek outside income.[59] As Wilhelm Schrader reports, "the paltriness [*Geringfügigkeit*] of salaries for the majority of the professors, the call to others with no guaranteed salary at all, had to work adversely upon their status and industry, as well as upon the harmony of the whole faculty."[60] The upshot was that Halle faded from the forefront of the German Enlightenment and of German medicine, to be displaced by Frederick's Academy and Haller's Göttingen. The University of Halle experienced a dramatic decline in enrollments starting around midcentury. By 1775 the number of newly matriculating medical students had dwindled to thirty-seven.[61]

ORGANISM: THE PHILOSOPHICAL PHYSIOLOGY OF GEORG ERNST STAHL

From its moment of foundation, the Halle medical school enjoyed the presence of two of the greatest European medical eminences of the epoch: Friedrich Hoffmann and Georg Ernst Stahl. They were the only physicians in Germany who could be thought to approach the stature of Herman Boerhaave (1668–1738), who made the University of Leiden the mecca of medical education for all Europe. In Hoffmann, Halle had a scholar aligned with Boerhaave both in the ambition to integrate teaching and practice and in the avowed *mechanism* of his general theoretical orientation. His *Fundamenta Medica* antedated by several years Boerhaave's great manifesto of this position, the inaugural lecture of 1703. Like Boerhaave, Hoffmann wished "to have done with supposed 'forces,' innate powers, spiritual ephemera."[62] His iatromechanism aimed at the "establishment of a 'technomorphic model of the organism'"—that is, one that found in material movements sufficient causal patterns to account for organisms.[63] Even more notably, Stahl, Hoffmann's colleague in the Halle medical faculty, stood forth as the greatest opponent of the Boerhaavian orthodoxy in theory—an energetic and rigorous critic of mechanism.

Three vantages have dominated the appraisal of Stahl. One arose in connection with the rise of vitalism in the eighteenth-century French medical school at Montpellier, where the influence of Stahl was substantial.[64] A token of this, from 1806, was the hyperbolic estimation by the French medical theorist Pierre Cabanis (1757–1808) that Stahl was the greatest figure in medicine since Hippocrates![65] French scholars thereafter sought to grasp Stahl's position in order to clarify that of the Montpellier vitalists. Thus, a full translation of Stahl's major works into French had taken place already by the middle of the nineteenth century.[66] Joseph Tissot elaborated the relation between Stahl and the Montpellier vitalists in the editorial apparatus to that translation, and even more prominently, Albert Lemoine made it the theme of an important monograph.[67] Interest in Stahl's medical theory remained alive in France, as evidenced by the publication of a significant monograph in the 1930s.[68] This connection to Stahl remains important in recent scholarship on Montpellier vitalism by scholars like Elizabeth Haigh and Roselyn Rey.[69] In the 1940s, the German scholar Bernward Josef Gottlieb recognized this French appropriation and endeavored to revive interest in Stahl's work within German scholarship.[70] The great German historian of physiology Karl Rothschuh picked up this thread and carried it forward.[71]

A second vantage, focusing far more on Stahl's immediate context, has accentuated his connection with Pietism at Halle. Pietism shaped all three generations of the Halle medical faculty. That vantage has structured the account of the medical faculty just sketched above. The issue is whether this suffices to account for Stahl's role in the gestation of German life science. Concentrating on Pietism distorts Stahl's physiological contributions, making "animism" betoken a notion of organism that is not really adequate to his ideas. In what follows I will try to rescue organism—and even animism—from a too thoroughly Pietist reading.

Crucially for my account, a third vantage recognizes that Stahl's place in the history of German medicine came to be defined by the antipathetic appraisal of his work by the Göttingen physiologist Albrecht von Haller (1708–77).[72] From a position of professional dominance at the middle of the eighteenth century, Haller denounced Stahlian "animism" continually and harshly.[73] Haller detected Stahl's baleful influence elsewhere in European medicine: not only in contemporary Montpellier, whose key figures proved hostile to his own theories, but also in Edinburgh, where Robert Whytt (1714–66) elaborated ideas about animal physiology that directly contested those of Haller.[74] Whytt seemed just another Stahlian animist, in Haller's view.[75] Haller's polemic against Stahl shaped the historical account of German medicine thereafter, and largely to this day.[76]

This Stahl-Haller conflict shaped the consolidation of a new research program in German physiology over the course of the eighteenth century. Late in that century, as Haller's own stance came under criticism, aspects of Stahl's views received more favorable mention.[77] We find this evidenced in two of the most important German thinkers of the second half of the eighteenth century, Immanuel Kant (1724–1804) and Johann Friedrich Blumenbach (1752–1840), both of them avowed admirers of Haller. Kant lauded Stahl in a quirky text from the mid-1760s, *Dreams of a Spirit-Seer*, which has served as the epigraph for this chapter.[78] He took the same line in a brief rectoral address at Königsberg in 1786.[79] Blumenbach began as Haller's devoted disciple, and he proffered his positive assessment of Stahl only long after his master's death. In 1786, in his influential journal for contemporary medical theory, *Medizinisches Bibliothek* (Göttingen, 1783–88), Blumenbach wrote that Stahl was "without contest one of the greatest, deepest-thinking physicians the world has ever seen. His reputation deserves to be revived especially in a time in which seeds that he sowed so many long years ago are just now bearing riper fruit and in which his important principles, with some changes and limitations, have become virtually the ruling ones in the enlightened parts of Europe."[80]

Over the eighteenth century, within the German medical schools, the issue of organism would prompt a *diversion* from clinical training oriented to human health into experimental and theoretical investigations of general animal physiology. Stahl strongly opposed this diversion of interest, though ironically he became one of the key theoretical progenitors of its development, especially with regard to its central concept, organism.[81] With Stahl, if against his own intentions, began the crucial impetus toward autonomy of physiological inquiry within German medical schools. He epitomized the crucially generative exchange between physiology and philosophy, especially concerning what distinguished the living from the nonliving, on the one hand, and concerning how body and soul could constitute a unity in human life, on the other.[82] The massive synthesis of his views, *Theoria Medica Vera* (1708), became the point of departure for medical theory in Germany for the balance of the century, notwithstanding his readers' intense complaints about the stylistic impenetrability of its 1,500 or so dense Latin pages.[83]

Stahl was first and foremost concerned with the crisis of learned medicine by the close of the seventeenth century and its desultory consequences for clinical practice. Learned medicine had a venerable tradition, but Stahl urged that it needed to keep its focus on therapeutic efficacy, not theoretical speculation or bibliographical connoisseurship.[84] "Precisely here lies the foremost sin of this sort of medical study, that it has become the custom to rely more upon tradition, the opinions and achievements of others, than to pursue the truth for oneself."[85] All that really mattered, in his view, was how medical learning played out in the clinical situation.[86] Stahl worried especially about the influence on medical ideas from recent trends in metaphysics and in the physical sciences. Fascination with "theory" was proving the ruination of medicine as a profession, as Stahl saw it, and his publications aimed first and foremost to *insulate* clinical medicine from theoretical interventions (philosophical or natural-scientific) that would limit efficacious treatment.[87] He was confident that his own generalizations were soundly grounded in clinical evidence, whereas his rivals forced matters either on purely metaphysical grounds (e.g., Cartesian dualism) or in the haste to build a coherent system (thus, the iatromechanists after Descartes—Boerhaave and Hoffmann preeminent among them). Stahl became the most important German proponent of a contemporary uprising in clinical medicine led by the English physician Thomas Sydenham (1624–89). Like Sydenham (and, indeed, Boerhaave himself) Stahl evoked Hippocrates as the founding advocate of concrete clinical observations as against theoretical speculations.[88] But, quite characteristically, Stahl added: "it would be superfluous to revere Hippocrates as an authority since the matter lies before our eyes in countless examples."[89]

For Stahl, there was even the prospect of an *inverse relation* between theoretical sophistication and therapeutic efficacy.[90] He argued that young students of medicine ran the risk of getting so caught up in the theoretical pursuits of chemistry or anatomy, of physical natural science more generally, that they became inept at actual clinical practice.[91] Stahl invoked this hostile stereotype even as he elaborated 1,500 pages of *theory*! That paradoxical stance, as much as the specific theses he articulated, makes Stahl a crucial point of departure for the developments within the German medical faculty that would result ultimately in a separate science of biology. Quite simply, it was everything he wanted to avert, and yet he proved seminal for its elaboration.

The fundamental tone of Stahl's writing was disputatious, intended to shake conventional perspectives. Stahl understood himself as a rebel against orthodox medical thought and expected (and received) a hostile response starting from his earliest theoretical publications of the 1690s.[92] He observed acerbically: "Ten experts in the art, if called together for a consultation, will not only differ, condemning one another's manner of thinking and practice, but when it comes to treatment will recommend an entirely different, indeed essentially opposed course of action."[93] This cognizance of the crisis in learned medicine made Stahl radical both in substance and in tone, "as if I came from another world."[94] Thus, Haller's famous characterization of Stahl as "*homo acris et metaphysicus*," though unsympathetic, was not altogether unfounded.[95]

For all his emphasis on clinical outcomes, Stahl *was* a theorist. He set out from the ontological posit that matter was passive—*inert* in the strict, Cartesian sense.[96] This was a postulate shared widely in the mechanistic natural philosophy of the late seventeenth century.[97] From that vantage, matter had in itself no propensity to movement. Since motion arose from external impact, its ultimate source lay *beyond* the merely material.[98] Stahl showed no recognition of uniform rectilinear motion as equally inertial as a state of rest. Thus, he was less a critic of Newtonianism than *oblivious* to Newton's interventions in physics.[99] He showed no attentiveness to the idea of any interstitial void or empty space between bodies or, consequently, to the question of action at a distance, since for him matter was a plenum and motion was transmitted from one contiguous body to another. The minimal attunement to the *mathematical* in Stahl's discussion of physics is a further token of *absence*—rather than rejection—of Newtonianism.

Stahl, it must be remembered, was one of the most important chemical theorists of his time, generally credited with the invention of the phlogiston theory.[100] For Stahl, the material world presented itself in the form of chemical "mixtures" and aggregates of varying levels of complexity and integration.[101] The crucial point Stahl wished to derive from this conception of the material order of nature was that there was a decisive difference between inorganic compounds and living things.[102] This marked the insuperable divide between physics and medicine, as well as between artifice and nature. Physics could no more explain life through mixtures than artifice could concoct life from mixtures.[103] Stahl, like Boerhaave, was one of the pioneers of chemistry in the first half of the eighteenth century, but it is striking that *both* of them expressed grave reservations about the direct incorporation of chemical theories into medical practice.

This suspicion of physicochemical foundations for medicine drove Stahl's quarrel with mechanism. "The *economy of life* [*oeconomia vitae*] has its own laws. . . . [It is not a derivative enterprise] as if the sphere of medicine were so impotent that it had to look to the grace of others. . . . Only since the subtleties and inventions of Cartesianism were adopted into medicine have physics, mechanics, and related disciplines taken on such arrogance."[104] Thus, "it was a true maxim propounded by the older physicians that the physician begins where the physicist leaves off."[105] The "worst damage" to medicine arose with "the transplantation of a physical consideration of the body into the domain of medicine."[106] "The material theory of body and its mixtures offers no utility for the physician. . . . Knowledge of these material relations therefore belongs to physics, and if one expects from this an advantage in pathology, he is deluding himself . . . [and] confusing physical etiology with medical pathology."[107] Above all, Stahl insisted that the actual character of organic life—in humans as well as other living things—entailed an *immanent purposiveness* that could not be derived from merely physical-chemical processes. Even anatomy, because it dealt with dead things, could not access the complexity of life.[108] It was on this basis that Stahl so notoriously disputed the study of anatomy and chemistry—indeed, all physical sciences—for medicine.[109]

Life-forms were not only elaborately complex collections of mixtures and aggregates ("heterogeneous aggregates," in Stahl's terms) but also, by that very fact, far more vulnerable to degeneration.[110] And yet they *resisted* this ongoing decay.[111] This was the single most important empirical observation for life science, the key to a conception of life itself.[112] The capacity to resist an ongoing propensity to dissolution *defined* life. The crucial question for medical theory, then, was how life accomplished the resistance of a body to its immanent propensities to degenerate, to *die*.[113] Stahl made this the point of departure for his physiology: "Above all else, consequently, it comes down to this: to know, what is life?"[114] His physiological theory took as its point of departure developments associated with both iatromechanism and iatrochemistry—foremost, William Harvey's (1578–1657) discovery of the circulation of the blood and the work of Jan Baptist van Helmont

(1579–1644) and Thomas Willis (1621–75) on "fermentation."[115] This was already clear in his 1692 text, *De Motu Tonico Vitali*, where Stahl took up Harvey's circulation theory and suggested that it needed to be supplemented by an evaluation of the response of the tissues at the extremities served by the capillary system, whose differential porosity and pressure regulated the metabolic functions of the blood and consequently the general health of the organism.[116]

Living things ("animated" matter) could not be understood as passive; indeed, only an internal principle of self-organization and maintenance made sense of the empirically observable phenomena, Stahl insisted. This agency, *ex hypothesi*, could not be a property of matter, yet it also had to organize, control, and sustain the material composite that was the body of a living thing. It had to be causally effective. "Not the matter of the body—anatomy, chemistry, the 'mix' of fluids—but rather their interdependence" was essential.[117] Life was radically *more* than physics. It was this crucial concern for the agency of organic life that, Stahl insisted, no mechanical model could address.[118] Indeed, not only for Stahl but for an increasing segment of his contemporaries in the new century, mechanism simply could not account for all that.[119]

He chose to use the term *anima*, or "soul," for his own exposition, but he explicitly offered as synonymous the ancient Hippocratic notion of *physis* and numerous others.[120] For Stahl, *physis* and its subsequent elaboration in Aristotelian-Galenic discourse as *psyche* or *anima* quite clearly gestured to the same principle of life.[121] He was not fixed on the *term* but on the *function*.[122] It was the power to animate, to keep the living thing in process. *Movement* was its key feature: life or soul was the principle that moved matter within the living body.[123] It kept the processes of the living thing in constant motion, alimenting its tissues and removing their decay or waste (i.e., it was *motus tonicus vitalis*). Health was the routine success of this metabolic circulation, and even illness could be construed as a stressed accentuation of these movements, a more febrile effort to keep things moving and to free the organism of its degenerating elements—hence the interpretation of fever in Stahl's pathology. Illness was actually rare in a human life, he maintained; nature saw to it that humans stayed healthy for the most part. Indeed, this was the primary sense of *physis* in Hippocrates, that nature was the true guarantor of health, and the physician's role was secondary, to assist in this natural cure. Stahl characterized the first, natural force as *energy* and the second, therapeutic intervention as *synergy*, cooperation.[124] He enunciated this as his own essential program.[125]

Neither soul nor body could act without the other. Medically, it had never made sense to cut the body off metaphysically from the soul. This was a dead end, for without some way of accounting for their interaction, medicine was at a loss not only in its diagnostics but in its therapies. The interaction of soul and body had, accordingly, long been a preoccupation of medical theory, from the all-encompassing principle of *physis* in Hippocrates, through the *psyche/anima* of Aristotle, to the tripartite soul of Galen, with its vegetative, animal, and rational dimensions.[126] The modern iatrochemical tradition had elaborated on this construction with the conception of the *archaeus*; various theories of "animal spirits" thronged the medical literature of the later seventeenth century, seeking to bridge the metaphysical gap between material and mental substances so sharply rendered in the Cartesian system.[127] Neoplatonists like Roger Cudworth and Henry More offered similar notions of intervening principles between the divine and the natural order. Stahl disparaged proliferation of intermediate entities between soul and body; postulating them did not bring investigators one step closer to a theoretically viable account.[128] Thus, he insisted, all medical theory needed was a single notion of soul.[129] It had to be a *rational* soul, in his view, but Stahl distinguished between *logos* and *logistikos*, reason and ratiocination.[130] Rationality entailed purposeful order; it did not require explicit, intentional articulation, or what Stahl called ratiocination. The essential physiological operations of nutrition, circulation, secretion, and excretion rationally and purposefully organized the body's motions; hence, for Stahl the soul proved effective (i.e., rational), without requiring articulated forethought.[131] While, to be sure, the human soul operated in both registers, intentional articulation was not necessary for the rational efficacy of the soul in the body. This rational soul, as contrasted to the ratiocinating one, could be discerned as readily in the function of animals as in humans. In that sense, the souls of animals were every whit as "rational" as those in humans.[132]

Enormous controversy has beset Stahl's notion of "soul."[133] Traditionally, his position has been termed "animism," but "animism" is itself no unequivocal notion.[134] In Stahl's case it has opened out into two extremely problematic readings of his thought. First, there is the *religious* reading, which identifies the *physiological* theory Stahl articulated with the Christian notion of the soul—that is, a divinely oriented entity moving through the material world on its path to redemption.[135] Pietism certainly took Stahl's use of soul as extraordinarily congenial to its redemptive Christian mission, even in medical practice. But Stahl never invoked this specifically Christian sense of soul—sin and grace and conversion—in his discussions of life, health, or illness.[136] In his medical writings he had far less to say about the immortal soul than about the mortality of man. He did not dispute the Christian sense of soul; he was a devout Pietist in his personal and even his bureaucratic-political conduct. But it is not at all clear that this translated directly into his medical theory.[137]

A second, related but conceptually distinct reading of Stahl's doctrine of soul was its *metaphysical* construal, which insisted upon taking his notion as a *substance* theory and then finding fault with his ideas either because they were contradictory in postulating substance interaction or because tacitly they *materialized* the soul in making it an actual cause of physical phenomena. Leibniz, as we will see, made *both* arguments, charging Stahl with *faulty* metaphysics. Haller denounced *any* intrusion of metaphysics into empirical science.[138] Thus, Stahl came under fire from both flanks, yet he tried to distinguish his position from both of these metaphysical missteps.[139]

A decisive dualism did pervade Stahl's thought, but it differed from the Cartesian notion in an important and pathbreaking manner. Johanna Geyer-Kordesch formulates this aptly: "Stahl stringently separates not the soul from the body, but the organization (organism) of life from the inorganic."[140] Substantive dualism between soul and body struck Stahl as empirically and theoretically unproductive. In place of a discourse of *substances* and the impossibility of their interaction (a discourse of "opposites"), Stahl preferred a discourse of *reciprocity* between "agent" (soul or motive force) and "patient" (the material body), two different but mutually indispensable components of the living thing. For this ensemble, Stahl offered his crucial concept of *organism*. It betokened complex self-organization; moreover, this self-organization was *rational*: that is, it operated *purposefully* to preserve and enhance the organism. He emphasized that the "true sense of the term organic" was that it was "organized according to and because of specific purposes."[141]

Moreover, he insisted that his was above all an *empirical* analysis, even if it had some ontological posits (i.e., inertness of matter). The self-organizing activity of organisms could be observed only a posteriori, but it could be characterized methodically, and this was Stahl's theoretical ambition. That is, he argued that physicians had indisputable data for the interaction of soul with body. Any human could observe in personal bodily experience the utter interpenetration of physical and mental life, such that bodily functions and dysfunctions impacted mental dispositions (*Gemütsbewegungen*), and mental dispositions affected physical states.[142] Not only was there the commonplace actuality of involuntary perceptions and voluntary motions, but there were medically significant *psychosomatic* phenomena, that is, mental disruptions inducing physical ones, or physical ones inducing mental disorders. Thus, for Stahl, "the dynamics of illness and health were inseparable from mental and emotional states."[143] Indeed, human cogitation could as readily *disrupt* as order the bodily functions.[144] Stahl endorsed the widely shared prepossession that mental disorders in the mother during pregnancy would have physiological consequences for her fetus (so-called "maternal birthmarks").[145] Mental disorder occasioned many illnesses; moreover, psychic states accentuated many physical conditions through anxiety or misunderstanding and hence served as a peculiarly salient exacerbation of human disease.[146] Indeed, Stahl several times pointed out that animals were distinctly healthier than humans because the rational functions of the soul were not interfered with by the ratiocinating ones that characterized humans.[147] Stahl's concern was entirely with human health, but the consideration of animal life illuminated the issues, exposing in particular the mixed blessing of volitional and passionate human consciousness, which Stahl generally called "moral" influences in health.

NEGOTIUM OTIOSUM: THE DEBATE OVER ORGANISM BETWEEN LEIBNIZ AND STAHL

Stahl's theory of organism was the most important impulse to emerge from the Halle medical faculty in the first decades of the eighteenth century. The debate over organism between Stahl and Gottfried Leibniz in the second decade of the century brought this to decisive prominence, especially when Stahl published their exchange in 1720, four years after Leibniz's death.[148] Leibniz's own theory of organism would exercise considerable influence later in the century via a variety of channels.[149] Its articulation in the controversy with Stahl offers us a clear and determinate starting point for these two currents that would feed the mainstream of thought leading to the new science. As Huneman and Rey aptly state, "both Leibniz and Stahl took an interest in life-forms and medicine in response to and because of the dead end of the mechanist programs in physiology."[150] But they had drastically different senses of what these mechanist programs in physiology consisted in, and what a promising alternative could be. To say the least, "the respective arguments were founded on totally different analyses of the manifestations of nature."[151] Leibniz used his attack on Stahl to assert that his own system was the only adequate conception of divine providence.[152] Ironically, what he alleged to lie latent in Stahl's views—a vital materialism—*would* eventually get elaborated later in the eighteenth century, but its perpetrators (both French and German) would use Leibniz's ideas even more than Stahl's to conceptualize it.

Leibniz's Critique

Leibniz opened his commentary on Stahl with an affirmation of mechanistic physical science as alone consistent with the principle of sufficient reason. Immediately thereupon, however, Leibniz insisted that *final* causality had to have a place in a complete system of nature. The rejection of final cause was the great weakness of the "Epicureans," in his view. Final causality had a particular and a general form. The particular final causes were "machines of nature, or . . . organic bodies[,] . . . divine machines . . . [which can] preserve themselves and produce some copy of themselves."[153] This formulation of "divine machines" had been developed several years earlier in his correspondence with Lady Masham, and it is Leibniz's most important notion of *organism*.[154] But the whole formulation allowed Leibniz to advance to his most important *metaphysical* concern, that of a *general* final cause—namely, preestablished harmony as divine providence. A crucial corollary of this notion was that literally nothing in the system of nature was without divine purpose. Thus, "no one should suppose that the use of ends is proper only to the mechanism of living beings, while on the other hand they are of no use in unformed masses, and in general in inorganic bodies." Even in "unformed masses there are certainly machines of nature concealed."[155]

Leibniz wanted foremost to establish the implications of his notion of preestablished harmony for the crucial relation of soul to body. Against Stahl's physiology—a variant of physical influx, in Leibniz's schematization of hypotheses—Leibniz posed a "most perfect parallelism: on the one hand between the material and formal principles, or between body and soul, and on the other hand between the kingdom of efficient causes and the kingdom of final causes." That is, "the series of motions in the body perfectly correspond to the series of perceptions in the soul, and vice versa." Each followed rules that were restrictively authoritative in the respective "kingdoms." Thus, "the present state of the body arises from the preceding state according to the laws of efficient cause, and the present state of the soul arises from the preceding state according to the laws of final cause. In the former case the series of motions, in the latter case the series of appetites, take place." And so, Leibniz concluded: "it can . . . not be supposed that the soul, through its innate operations, to wit, perception and appetite, moves the body even in the least bit astray from its mechanical laws." Each order acted in perfect harmony but without causal interaction. "If this were not the case, there would be a continuous violation" of the principle of sufficient reason, he pronounced. On that basis, Leibniz offered thirty-one specific objections ("Animadversions") concerning Stahl's *Theoria Medica Vera*.[156]

The first "Animadversion" contended that the discrimination of chance objects from purposeful ones was misguided, for "in truth all things are directed towards an end." This was the upshot of Leibniz's principles of the system of nature and the purposiveness of divine providence. On that basis, "chance only plays a role in our ignorance."[157] It designated what had yet to be grasped by science. In his second set of criticisms, the so-called "Exceptions," Leibniz elaborated this first "Animadversion" in a starkly theological manner: "all things that happen arise from God. . . . Nor do I see in what way the contrary can be reconciled with divine providence, which embraces all things; or with theology, not only revealed but also natural."[158] To suggest there could be any chance occurrence in the world was to flirt with heresy, hence his nasty allegation that Stahl's views threatened not only revealed but even natural religion.

The upshot, as Leibniz put it plainly in "Animadversion II," was "all organism is in fact mechanism, but more exquisite, and so to speak, divine."[159] Stahl's whole project of a life science, which set out from an empirical distinction between living forms and inanimate ones, Leibniz dismissed as bad science. Leibniz elaborated in "Exception I" that "no mass is so unformed or so small that it does not contain in it some organic body or machine of nature."[160] He took this to be empirically confirmed by the microscopic discoveries of a panoply of life-forms in a drop of pond water, but Leibniz stated it as an a priori principle. At the same time, he added that there was "never any soul completely separated from some organic body."[161]

In the later "Animadversions," Leibniz elaborated his fundamental criticisms of Stahl's notion of soul-body interaction. First, he reasserted his principle of parallelism: "vital motions correspond exactly to the appetites of the soul," though these "are confused and remote from our attention."[162] To postulate actual causal interaction would open the door to absurdity, since the soul was so incommensurable with the body that there would be no proportionality of its causal impact in the response of the body: the soul "could command anything whatsoever."[163] To view the soul as a causal factor in material functions would be to have recourse to "something supernatural"; that is, "it would be some sort of miracle if the soul could bring about something in the body beyond its nature."[164] The only coherent notion of physical influx would be the *materialization* of the soul: "in this way the soul would be rendered corporeal and mortal."[165] That is, Stahl could be read as a reassertion of the materialism of the Epicureans and of Hobbes, whom Leibniz explicitly mentioned in this context.[166]

In "Animadversion III," he elaborated: "nothing happens in the body that is not based upon mechanical, which is to say intelligible reasons." This, he maintained, had become the established principle of modern science as articulated in "the very clear proclamations of recent thinkers." Organic motions were mechanical ones, though divinely subtle. They did not imply the intervention of the soul. Leibniz evoked the classic objection: a heart torn out from the organism still went on beating, and that could hardly signify that it carried its own "soul" with it.[167] He summed it all up in his "Exception XVI": "It must be known that a body cannot be actuated by a soul in such a way that the mechanical laws of bodies be the least violated. The soul gives no motion, nor any degree of motion, nor any direction to the body that does not follow from the preceding states and motions of matter. To allege the contrary is either to hold that the soul can be changed into the body, or to take recourse to inexplicable principles."[168] Thus, "the soul cannot violate the laws of corporeal nature, nor the body the laws of the soul," but it all worked out via preestablished harmony.[169]

Leibniz sharply criticized Stahl for seeking to restrict the importance of anatomy and chemistry for medical science. In "Animadversion X," he wrote: "I should . . . not wish that truths that are remote from present purposes [*usu*] . . . be perceived as useless, since greater usefulness [*usus*] may always be discovered."[170] As he elaborated in "Animadversion XI," if it was true that "until now medicine may not have sufficiently benefitted from the economy discerned in recent investigations," this seemed more a fault of medical training and government policy than of these basic scientific research endeavors.[171] He believed that both the new anatomy and the new chemistry showed great promise for medical application. In his prefatory remarks he made this clear: "there is hope that many things may be discovered in animal economy and in medical practice, by looking at the use of the parts and at the ends of nature."[172] In his "Exceptions X–XII" Leibniz elaborated these ideas. He claimed first that "the present state of medical science . . . is . . . in its infancy."[173] This was largely because "until now medicine has been exceedingly empirical."[174] Hence, "to the extent that physical reasoning is facilitated through mathematics or mechanics, and through microscopic experiments and chemistry, it is hoped that physics will grow little by little ... [and] it is hoped that pathology as well ... will make noteworthy advances."[175] Not only anatomy but chemistry held great promise: "I easily concede that up until now the utility of chemistry has not been great in explaining things that occur at an undetectable level in animals. But with the expansion of the science of chemistry, its application will likewise grow."[176]

Stahl's Defense

To grasp Stahl's defense, we must go back to the basic positions he felt called upon to justify: the distinction in the observable world between entities of chance and entities of purpose; the discrimination of motion from matter; the ascription of motion to an active agency; and, finally, the identification of this agency of motion, in organisms, with the soul.[177] As Stahl understood it, "mechanism is generically nothing else than mobility ... and specifically, ordered mobility, or aptitude for a determined sequence of motion."[178] Stahl has always been identified as antimechanist, but what he found off-putting in "the present sterile fight about mechanism" was the propensity to ascribe mobility to matter itself as an intrinsic property.[179] Instead, for Stahl, a *system of nature* (to use a crucial Leibnizian notion) would need to posit another reality besides matter, the principles or agencies that imparted motion. "Act arises from the agent, which similarly is not this matter itself, but rather something distinct from it, even at the very time as it acts immediately on the matter, or mediately through it."[180] For Stahl, everything hinged on the passivity of matter. He asked, rhetorically, "is body motion and motion body?"[181] He denied it. "Motion is different from matter," he insisted, "really something incorporeal, falling under no conception of dimension, extensive magnitude, or figure," which exhausted the actual properties of matter or body as such.[182] Because "motion is not matter itself," it was *immaterial* by definition. Moreover, the event of motion needed a causal agent; that is, "motion is not a naked act without an agent." For the eighteenth century, "force" would prove the decisive term for such considerations, though Stahl never made much of it.[183]

Stahl believed that any effort to ascribe motion to matter as an intrinsic property was the quintessence of atheistic materialism, a revival of Epicurean ideas, and the dead end of "mechanistic" physical theory. "The moderns ... say that all natural bodies are machines, endowed with inseparable, inherent motion of their own."[184] Thus, "those who call themselves 'mechanists,' or defenders of the mechanical philosophy, . . . [hold] that all of the least bodies of this universe already have their inherent movers . . . [and hence there is] no use to assign the motive energy to the soul. . . . For the body is able to do it spontaneously."[185]

According to Stahl, *organism* differed from mechanism in empirically discernible and theoretically decisive ways. The first point—empirical discernibility—was the key to Stahl's distinction of entities in the physical world that appeared to be random, chance configurations (i.e., aggregates or mixtures with no apparent purpose or function, like a heap of rocks) from other entities that appeared, again empirically, to be internally purposive or alive.[186] Purposiveness meant for Stahl the presence of a supervening order in the object, its principle of organization. The whole of life science, as he saw it, derived from this fundamental empirical discrimination of living things from inanimate congeries. This causal agency constituted the domain of inquiry.[187] If the difference was illusory, there could *be* no inquiry. In this crucial sense, Stahl was an empirical researcher. "I . . . do not appeal to conjectures or opinions, but simply to experience and observation."[188]

Stahl understood himself to be practicing and theorizing about medicine; that is, he believed his "type of business . . . is the physical science of the soul," which entailed "simply attending to accomplishments of acts of true conservation of the body for a vital end."[189] That entailed "the plain precise history of natural things, showing how they are and are produced."[190] As he put it in his "Summary of the Principal Points of Doubt," "I, along with the entire school of medical physiologists, attribute to the soul the energy for actively exciting the movements which are appropriate for both the conservation of its body and all the rest of its function . . . [, that is,] both vital and animal motions."[191] That is, "souls are the agent immanent to the organic body, exerting immanently such an act, which is the movement of the corporeal organs."[192] Stahl insisted that medicine was a *physics*, a natural science, based on observation and experience, not speculation. Hence, *soul* was an empirical concept: "not only the full nature of the soul, but also its entire destination, which we know, *in so much as we know and will ever know it*, only revolves around the affections of corporeal things. . . . *Such a science as we are capable of attaining* concerning the soul, *a physics supported by concepts and experience*, contains nothing other than those features that I stated" (my emphases).[193] Soul was an explanatory agency in physiology, not a spirit with a transcendent destiny.

The crucial *theoretical* point was that a soul was immediately associated with a particular body, that its primary function was the conservation and mechanical efficaciousness of that body; thus, its scope and efficacy were proportionally specific to the body with which it was in interaction. Its actions were purposeful/rational, even if not always consciously volitional. This last idea found expression in the distinction between *vital* and *animal* motions—between involuntary or spontaneous functions and voluntary actions. Unlike the "physicochemical" causal account that sufficed for inanimate mixtures and aggregates, the source of the purposeful movement that distinguished organisms appeared to be internal, not imposed from outside. It was this immanent—but, *ex hypothesi*, immaterial—ordering principle of movement that Stahl conceptualized by the term "soul." In short: "I recognize and declare the soul to be just such a principle energetically endowed with the faculty of moving."[194] Hence, "all motions . . . happen by the activity of the soul itself . . . voluntarily . . . and non-voluntarily, that is spontaneously."[195]

Leibniz disparaged Stahl's claim that the soul should be understood to be primarily, if not exclusively, the agent for the conservation of the physical body. He charged this claim with being dangerous to religion. He insisted that the (human) soul was in fact concerned not with conserving the body but with the knowledge of God and of its own ultimate destination.[196] Leibniz wished to stress what the later German Enlightenment would call the "destiny of man [*Bestimmung des Menschen*]." That is, "the mind is more tightly bound to God than to bodies . . . and destined . . . to knowing itself and through this to knowing the Author of things."[197] For Stahl, this had nothing to do with natural science: "I am certainly incapable of investigating whether by *physics* (or any science properly so called), the soul would be able to attain a knowledge of God himself as its author, a fortiori, whether it would have a closer and nearer access to him." This was "because in this terrestrial life the soul is very proximately and thus quasi-properly attached to corporeal affections . . . by her external and internal senses."[198] Moreover, "that the soul is destined to know herself, this occupation, as if physically relative to her essence, is not at all given or probable."[199] In this last sense, Stahl was writing not simply as a physician investigating the soul's physiological function but as a psychologist assessing the human soul both as *logos*, in which it was precisely *not* self-conscious, and as *logistikos*, in which it was not only quite finite and indeed flawed but, as he saw it as a clinician, far more embroiled in living than in knowing. For Leibniz, this appeared scandalously crass and inattentive to the highest human aspirations; but Stahl was defending a *medical* theory.

Against Leibniz's challenge to his rejection of experimental research in physics and chemistry in application to medicine, Stahl defended himself stoutly: "I advise that the more recent anatomy is most fecund in things that are alien to the scope of medicine."[200] Two points need immediate attention in this formulation. First, Stahl specified that his advice concerned *recent* anatomy, and second, he did not deny its fecundity in general but only as related to the scope of medicine—that is, clinical applicability. Stahl certainly recognized the pertinence of general anatomy for medicine and surgery, but he believed that "anatomical minutiae . . . form the large part of the more recent anatomy."[201] By this Stahl clearly meant work in microscopy as in the experimental research agenda of such figures as Jan Swammerdam (1637–80) and Marcello Malpighi (1628–94). Here is where Stahl strove to keep medical training focused on direct clinical outcomes. He disputed the scientific appeal of investigations in anatomy (or chemistry), not in themselves, but only as they failed to have immediate clinical application. Thus, he characterized recent anatomical research as "curiosity-inclined anatomy," contending that "minute anatomy in its overall capacity of favoring curiosity pertains to physical history"; "it does not also afford considerable use for medicine, if even any utility at all."[202] That is, "awareness of the deeper and more tender texture has absolutely no other use, neither medical nor surgical, if not perchance to eventually serve for prognosis." While he allowed for this eventuality, it was certainly not an actual factor in the present: "awareness of the texture correctly and strictly taken as such . . . brings no utility anywhere to the physician." In sum, "in the texture and inner and more subtle structure of the body, nothing subjacent is to be found that would provide anything for the scope of medicine."[203] For Stahl, this absence of clinical relevance resided in the very subtlety of these microprocesses: "all this work, absolutely and uniquely, is that of nature alone, or of a spontaneous energy in the living and active body . . . [to which] art has no special access." Thus, "art cannot bring any assistance to that act as such."[204] Roughly the same argument characterized Stahl's opposition to the application of chemistry to clinical medicine. As one of the leading chemists of his generation, Stahl did not dispute "the absolute necessity of a precise chemical science for treating the matter of mixtures."[205] Rather, he argued, "it is not clear that . . . if it were maximally known, it would bring considerable momentum to medical theory or practice."[206]

Thus, while all this research might be of interest in *physics*, it had no significance for medical *physiology or pathology*, in Stahl's view.[207] Leibniz found this, not unjustly, a rather-restrictive conception of the relation between basic and applied research. Deeply attuned to Malpighi's research program in "anatomical minutiae," Leibniz believed in its promise for medicine, but even more in its scientific and philosophical importance in its own right.[208] He wrote: "Care should be taken that there always be some eminent anatomists, botanists, chemists, who search into new things . . . not deterred or held in contempt under pretenses of uselessness. . . . We will never be overrun with a surfeit of Stenos or Malpighis."[209] In this, he was the visionary and Stahl a striking obstacle. Most significantly for the future, the impulse to explore "experimental physics" for its own sake, not for clinical application, would in fact characterize a set of investigators within the medical faculties of Germany whose physiological explorations would enrich the theory of organism (vegetative and animal, not merely human) and lay the foundations for the special science of biology. Haller would become the key German spokesman of this scientific agenda, and Stahl would serve as his preferred target of disdain.

CHAPTER TWO

Making Life Science Newtonian: Albrecht von Haller's Self-Fashioning as Natural Scientist

Haller's spiritual homeland was the early Enlightenment of the Netherlands and his esteemed mentor, Boerhaave, the man in whom the spirit of a Calvin, a Descartes, and a Spinoza met that of a Bacon and a Locke; Boerhaave, the man in whom the notions of iatrochemistry of a Sylvius linked up with anatomical thinking and the iatromechanics that originated from Newton; Boerhaave, the man who unified all this in one character [*Persönlichkeit*] that exerted enormous force on the young Haller.

TOELLNER, "Haller und Leibniz"[1]

Albrecht von Haller never attached great importance to working as a practicing physician: "Dieu me preserve de Pratique, je n'y saurois tenir [God keep me from practice: I wouldn't know how to keep going]."[2] "Haller, although he was the student of Hermann Boerhaave[,] . . . the primary founder of clinical medicine, was very rarely to be seen later in his life as the attending physician by a patient's bedside."[3] Similarly, though he was the foremost anatomist of Europe, he *never* performed surgery on live patients: he deemed it simply too risky.[4] "He was in the last analysis really not a physician. . . . Haller was the great natural-scientific scholar, who of course was involved in practice but who was not really all that interested in healing."[5] His interest and commitment fell altogether on the side of experimentation and research, not practical application. "Haller's whole project of experimental physiology consisted in the creation of a new science with its own research agenda. He thereby supported the separation of theoretical from practical medicine."[6] That is why he deserves a prominent role in the prehistory of biology, not just in the history of medicine. In this chapter I will trace the exemplary self-fashioning of Haller as a natural scientist (*Naturforscher*)—more concretely what *we* would call a research biologist (though, of course, he could never himself have used these terms). This self-fashioning gestated within the traditional medical faculty but established a new *Fachgebiet* quite distinct from it. I have a still more specific ambition within this larger one: namely, to assert that with Haller *experimental Newtonianism* became entrenched as the foundational method of this incipient new science. What experimental Newtonianism was in the eighteenth century and how Haller came to incorporate it into his identity as a naturalist, I propose, make his intellectual biography central to the genesis of biology.

THE RISE OF EXPERIMENTAL NEWTONIANISM AROUND 1700: HERMAN BOERHAAVE

In the history of science of the eighteenth century, enormous attention been given to the variety of "Newtonianisms."[7] I think it is essential to note the split, first articulated by I. B. Cohen in his classic *Franklin and Newton*, between the "mathematical" and the "experimental" Newtonian traditions, roughly identifiable with the two major texts *Principia* and *Opticks*.[8] More concretely, from the "Queries" to the *Opticks* arose what Thierry Hoquet has aptly termed "nonmathematical *physique*," and this proved the womb in which the nascent science of biology gestated.[9] While the tradition of "experimental philosophy" on a Newtonian basis emerged among the English in both chemistry and physiology before the publication of his *Opticks* (1704; Latin, 1706), the publication of this second of Newton's monumental works consolidated the experimental impetus mightily.[10] Published originally in English to reach a wider audience already fascinated by the "public science" of experimental demonstration, especially in London, the *Opticks* emphasized experimental, rather than mathematical, demonstration. The Newtonian cohort in and around the Royal Society and in the London public science sphere immediately took it up.[11] That experimental emphasis also shaped the crucial second edition of Newton's *Principia*, not least through Roger Cotes's preface.[12]

On the Continent, the Netherlands unquestionably played a decisive role in the reception of experimental Newtonianism, but this initially entailed *refashioning* Newton's person and work for local purposes. This aspect of the reception of experimental Newtonianism has become very important in recent scholarship. The main thrust of recent revisionism aims less at "experimental philosophy" and more at Newton's *religious* utility in the context of local Dutch controversies arising especially out of the association of "mechanical philosophy"—and even mathematics itself—with materialism and atheism, an association that emerged from the writings of Descartes and above all Spinoza.[13] That is, what initially interested the Dutch was Newton as "physicotheologian"—to take up the term John Ray invented for the "Newtonian ideology" in cultural politics that persisted over much of the eighteenth century.[14] Creating an "image of Newton . . . not just as a pious mathematician, but as a philosopher whose message was relevant to the whole of Christianity," they "fashioned" him into the safe and redemptive proponent of natural philosophy and mathematics and, indeed, more positively, into the architect of a compelling natural theology that rebutted the dangerous ideas of Descartes and Spinoza to smooth the troubled waters of early eighteenth-century Dutch cultural life.[15] Perhaps the central figure for this fashioning of "Newtonianism" in the Dutch Republic was Bernard Nieuwentijt (1654–1717).[16] His text *The Right Use of Natural Investigations* (1715) conceived of Newton as "the antidote to the poisonous Spinoza."[17] As Rienk Vermij puts it, "Newtonian philosophy . . . was seen as offering a decisive blow to the Spinozistic threat."[18] Thus, "the real triumph of Newton on the Continent started with the second edition of the *Principia*" (1713) and its much more explicit articulation of Newton's natural theology.[19] The ideological refashioning shifted into high gear with the publication of a pirated version of this second edition intended to propagate just these connections.[20] A key figure behind this pirate edition was Jean Le Clerc (1657–1736), who "really became a Newtonian after reading the second edition of the *Principia*."[21] While Le Clerc, in his review, praised Newton as "*the* greatest mathematician the world has ever seen," what was most important to him was that Newton "gave the *coup de grâce* to materialistic and atheistic speculations."[22]

Enthusiasm for "physicotheology" in the new natural philosophy remained a crucial feature of a distinctively Protestant, even Calvinist experimental Newtonianism, whose presence will be discerned across my entire narrative of eighteenth-century life science. In diametrical opposition, Spinozism and materialism will play an equally major role. Newton figured as a key resource in the tug-of-war between them.[23]

Some scholars have gone so far as to argue that "experimental philosophy" was already so well entrenched in the Netherlands that there was no need of a Newton to sponsor the notion.[24] To be sure, before Newton, there was a well-established tradition of experimental philosophy in the Netherlands, and at Leiden in particular, in the seventeenth century, drawing upon Descartes and Boyle. Burchard de Volder (1643–1709), professor of philosophy and mathematics at Leiden, turned to experimental physics under Cartesian auspices; he opened his experimental physics laboratory in

1675.[25] De Volder was not simply a Cartesian; he was stimulated by Boyle's air pump and the tradition of experimentation from Torricelli through Pascal to Boyle.[26] Another major source of Dutch experimental philosophy was the invention of the microscope and its use in the Low Countries in anatomical and botanical research by such figures as Jan Swammerdam (1637–80) and Antonie van Leeuwenhoek (1632–1723).[27] Harold Cook urges that it is even misleading to distinguish natural philosophy from natural history in the Dutch context: it was all one "big science" of laborious, detailed natural investigation.[28] Moreover, the revisionists argue that, even in the Netherlands, Newton was initially taken as a mathematician, not an experimental philosopher: "before 1715, in academic circles, Newton was admired as a mathematician, but not as a physicist."[29] That, of course, concerned the *Principia*, above all, and the revisionists contend that Newton's *Opticks* had a "somewhat neglected reception in the Netherlands," even in Latin translation.[30]

Against the revisionists, I propose to take Edward Ruestow's claim, *without amendment*, as my point of departure: "an outspoken admirer of the British school of natural philosophy, Boerhaave prepared the way for the emergence of the University of Leiden as a leading center of Newtonian and experimental science on the continent."[31] In Holland, notwithstanding overstated revisionism, Herman Boerhaave proved a key exponent of the *experimental* Newtonianism articulated in the "Queries" to the *Opticks*.[32] "Through the influence of Newton, Boerhaave soon became an enthusiastic adherent of the method of experiments in natural science, that is, of the so-called 'experimental philosophy.'"[33] For Boerhaave, Newton meant science with a rigorously empirical-experimental agenda, needing no metaphysical first-philosophy along either Aristotelian or Cartesian lines: "The drive to the ultimate metaphysical and the original physical causes is neither necessary nor useful nor possible for the physician," he proclaimed.[34]

In a famous oration in 1715 Boerhaave was the first Dutch academic to publicly enunciate his adulation of Newton.[35] He had become a Newtonian far earlier, however.[36] How did he get there? One key is de Volder. Boerhaave was his student, composing his doctoral dissertation in philosophy under de Volder. Around 1689, Boerhaave took a course from de Volder "on 'the geometric synthesis of the ancients' and 'the analysis of the moderns'—terms with striking resonance in the context of Newton's *Principia*."[37] As Ruestow observes, "the precise character of the mathematical physics taught by de Volder . . . remains disappointingly obscure."[38] According to Sassen, "in the years in which Boerhaave attended his lectures de Volder had not yet read or had only very recently seen Isaac Newton's *Philosophiae naturalis principia mathematica* (1687)."[39] But Ruestow notes that Jean Le Clerc, who knew de Volder quite well, "wrote that he [de Volder] applied himself to the new techniques of differential and integral calculus and had undertaken a close reading of Newton's *Principia mathematica* soon after it was published."[40] Indeed, Le Clerc claimed that the great mathematical physicist Christiaan Huygens (1629–95) came to de Volder seeking clarification of Newton's mathematics.[41] That gains credibility when we learn that Huygens willed his mathematical papers to de Volder and further that in de Volder's own papers later were "found calculations of some of Newton's proofs."[42] If we recall that John Locke turned to Huygens for clarification of Newton's mathematics in the *Principia*, perhaps to help with his own crucial review for a Dutch Francophone journal, then de Volder's standing as a mathematician becomes quite impressive.[43] Boerhaave was good at math, earning a living in the early 1690s by tutoring in that subject.[44] Whether his math skills were adequate actually to grasp the technical proofs of the *Principia*, one might still doubt. Yet Boerhaave could read and understand the arguments without being put off by the math, and that proved decisive.

While the connection with de Volder is central, there is another connection that may well be equally important for Boerhaave's Newtonianism: Archibald Pitcairne's brief stint at Leiden. Pitcairne (1652–1713), a Tory, High-Church Scot, matriculated at the University of Edinburgh in theology in 1668 but took his time obtaining a degree, wandering back and forth to the Continent, dabbling in law and medicine, before obtaining an MD at Rheims in 1680. Around 1675, probably in James Gregory's Edinburgh mathematics course, he became a close friend of David Gregory (1659–1708), who was to become one of Newton's most sycophantic disciples.[45] His uncle, James Gregory (1635–75), was among Newton's earliest mathematical admirers, and when he died in 1675, David inherited all his papers, including the Newton correspondence. The younger Gregory saw to it that he ingratiated himself with the great man at Cambridge. In 1683 Gregory succeeded his uncle in the chair of mathematics at Edinburgh. When he received Newton's *Principia* in the summer of 1687, he set about systematically digesting it. Pitcairne was rooming with him, and they worked through the mathematics together.[46]

In 1691 Pitcairne received a call to the chair in the practice of medicine at Leiden, though he had published little and "the Leiden authorities must have chosen [him] largely on the basis of word of mouth and manuscripts."[47] En route to Leiden, Pitcairne visited Newton in Cambridge, in the spring of 1692, and Newton entrusted to him his new manuscript, *De Natura Acidorum*, "on alchemy, chemistry, and the theory of matter."[48] In this work, Newton "unequivocally set forth the case for short-range attractive forces analogous to gravity."[49] While this became widely known only with the publication of *Opticks* in 1704, and especially with the insertion of "Query 23" (31 in later, standard editions) in the Latin version of 1706, "Newton had early recognized that short-range forces had obvious implications for physiological explanation."[50] Pitcairne immediately saw the relevance of Newton's theory of matter for the elaboration of a more sophisticated iatromechanism in the manner of Lorenzo Bellini (1643–1704). "Beginning with Pitcairne, early Newtonian physiologists relied on the physics of attracting particles in their theories of animal function."[51] Indeed, Pitcairne's inaugural lecture at Leiden in 1692 and his subsequently published Leiden "dissertations" demonstrate the key "relationship between Pitcairne's ideas on medicine and physiology and Newton's concept of the microcosm *before* the publication of the *Opticks* in 1704" (my emphasis).[52] To be sure, Pitcairne's new iatromechanics synthesizing Bellini and Newton was "a slightly uneasy union," as Anita Guerrini puts it.[53] Thus, she queries: "Was Pitcairne a Newtonian? In his close attention to Newton's method, ideas of causality, and theory of matter, Pitcairne showed himself more closely acquainted with Newton's thought than many, or most, of his contemporaries." For our purposes more importantly, "he introduced these ideas, in a medical context, to an entire generation of physicians, among them Herman Boerhaave."[54] Pitcairne's inaugural lecture at Leiden proved both "the first rigorous statement of English Newtonian medicine" and "the starting point of Dutch mechanistic, but non-Cartesian physiology."[55]

Boerhaave attended Pitcairne's inaugural lecture, which was generally received with enthusiasm, and it is altogether likely that he read Pitcairne's "dissertations," especially the methodologically crucial first one (1693), concerning the circulation of the blood, with its heavy invocation of Newtonian methodology and theory of matter.[56] Boerhaave's Newtonianism clearly demonstrated strong parallels to the "flourishing school [inspired by Pitcairne in Britain] which developed theories of function based on Newton's ideas of atomistic matter and short-range force." However, this British school "effectively dissolved by the time of Pitcairne's death" (1713), whereas Boerhaave carried his own tradition of "Newtonian Physiology" forward another several decades.[57]

"Newtonian Chemistry" emerged in England in the years after 1700. John Freind (1675–1728) published *Prelectiones Chimicae* in 1709, which was based on lectures delivered in 1704.[58] Gerrit Lindeboom believes that even before Freind, "Boerhaave was . . . the first to apply the Newtonian principles to the advancement of chemistry."[59] He was at least contemporary with Freind on many issues, and he clearly exerted a far more long-lasting influence.[60] He started teaching private courses in chemistry in 1702 at the request of his British students, notwithstanding the presence of a chaired professor of chemistry at Leiden.[61] He eventually succeeded to the chair in chemistry in 1718, upon that scholar's death, and held it until 1729. Although he became a great pioneer in chemistry, he strove to keep it at some distance from medicine.[62] In his 1718 inaugural lecture, "On Chemistry Correcting Its Own Errors," Boerhaave expressed the high esteem he had long felt for English pioneers in scientific chemistry, especially Boyle and Newton, but he "considered chemistry as an independent science, not as an *ancilla* either of pharmacy or medicine."[63] Peter Shaw's pirate textbook, *A New Method of Chemistry* (1727), which was based on Boerhaave's lectures, enjoyed enormous success all over Europe.[64] Reacting to this very familiar piratical provocation, Boerhaave issued an authorized edition, *Elements of Chemistry*, in 1732, which became the dominant text in the discipline for much of the rest of the century.[65]

Boerhaave *remained* a Newtonian throughout his long and illustrious career. Rina Knoeff has been taken to have established that Boerhaave "never introduced Newton's ideas into his courses, nor paid much attention to the Englishman."[66] In fact, her contribution can be taken in quite a different sense. Knoeff's earlier monograph on Boerhaave stresses two crucial dimensions of his career—his Calvinism and his chemistry—but she also notes there, and makes the starting point of her later essay on Boerhaave, that "we hardly find references to Newton in Boerhaave's works."[67] Moreover, his most noted disciples—with one exception, whom Knoeff fails to cite—seemed not to have discerned any Newtonianism in his work: "it seems as if either Boerhaave's alleged Newtonianism was not recognized by his pupils, or Boerhaave's medical teaching was not very Newtonian at all."[68] When he did take up Newton in his own works, "Boerhaave mainly referred to the *Opticks*, first published in 1704, and he was particularly impressed with the thirty-one speculative queries at the end of the book."[69] For Knoeff, this diminishes Boerhaave's Newtonianism; for me, however, it makes the essential point. When Knoeff alleges—erroneously, I think—that "Boerhaave's alleged 'Newtonianism' eventually led to a decline of Newtonian medicine across Europe," that privileges a highly problematic conception of "Newtonian medicine" and renders as "decline" what I will endeavor to show as highly creative. Instead, I wish to salvage Knoeff's notion that Boerhaave "taught a particular kind of Newtonianism," with the *Opticks* and especially the "speculative queries" as its source, and this proved decisive for the development of the life sciences in eighteenth-century Europe.

Knoeff argues that Boerhaave "changed his mind" over the course of his career, becoming increasingly skeptical of uncovering laws of nature in scientific inquiry. However, cautious empiricism was a centerpiece especially in those domains of inquiry where mathematicization appeared most unlikely, as was the case in both medicine and chemistry, the fields Boerhaave developed most extensively over his academic career. Knoeff reasons that "Boerhaave's change of mind almost seems to reflect Newton's changing focus from the mathematical approach of the *Principia* towards the experimental method of the *Opticks*.[70] Now, we should not so simply equate the *sequence* of Newton's publications with a change in his *focus*. After all, key parts of the *Opticks* were developed long before the *Principia*. But the difference in thrust between the two great works did prove central to the emergence, in the eighteenth century, of a decidedly *nonmathematical* research program under Newtonian auspices, which is what I mean by "experimental Newtonianism." Boerhaave's trajectory, in my view, paved the way for that development. Knoeff continues: "Boerhaave was so impressed by the *Opticks* that he stated: 'I never saw a book where [there] were stronger arguments drawn from experiments: it is the best pattern in the world and deserves the highest honour.'"[71] That would be the inspiration of researchers in chemistry and physiology over the balance of the century. "Boerhaave was particularly pleased with Newton's promotion of chemical methods in order to uncover the workings of the powers of nature," though "Boerhaave's chemistry was essentially different from the chemistry advanced by Newton in the *Opticks*."[72] But this can—and *should*—be read as development in a field that Newton himself recognized as only beginning to get its bearings. Boerhaave's chemistry *advanced* the field, relative to its seventeenth-century roots in "chymistry"; it did not misdirect it.[73]

More immediately and bluntly, with Boerhaave—Theodore Brown notwithstanding—there was no decline of "Newtonian medicine."[74] One need not accept a highly speculative and failed project—that of Pitcairne and his circle, which Knoeff calls "Newtonian mechanical physiology"—as the only authentic form of medicine that could be called "Newtonian."[75] This school reached its peak in Britain with the publication of James Keill's *Account of Animal Secretion* in 1708, which argued that "the whole Animal Oeconomy does . . . depend upon this Attractive Power[,] . . . the only Principle from which there can be a satisfactory Solution given of the *Phaenomena* produced by the *Minima Naturae*."[76] As Knoeff notes, "Boerhaave was particularly critical of the Newtonian physiology of the British Newtonian physiologist James Keill (1673–1719)."[77] That, it seems to me, was a sound judgment, not a denigration of Newtonianism. As Boerhaave himself and many since have argued, not all Newtonians promoted what was most fruitful in Newton.[78] Against Brown and Knoeff (and behind them, Robert Schofield), the fact that Henry Pemberton was a Boerhaave student does *not* betoken that Pemberton disordered the Newtonian endeavor.[79] Newton entrusted him with work on the third edition of the *Principia* with good reason, and Pemberton's subsequent characterization of Newton's program was in fact a decisive and positive force in the development of natural inquiry—and the life sciences in particular—over the second half of the century, particularly for Albrecht von Haller.[80]

In the Dutch academic context, the two figures most famous for their advocacy of experimental Newtonianism were Willem 'sGravesande (1688–1742) and Peter van Musschenbroek (1692–1761). Boerhaave personally transmitted his enthusiasm for Newton to 'sGravesande and Van Musschenbroek during their student years, and they carried it with them on their extended visits to England, to their respective encounters with Isaac Newton in person, and to their lifelong advocacy of experimental Newtonianism. It is with Willem 'sGravesande that the entrenchment of Dutch Newtonianism is generally associated.[81] It "changed the climate of Leiden university in many respects."[82]

Notably, as Vermij suggests, 'sGravesande paid less attention to natural-theological argument than to experimental demonstrations in his teaching.[83] More emphatically, Ad Maas writes: "In fact, 'sGravesande . . . managed to dissociate Newton's natural philosophy from the metaphysical and theological concerns of Newton's Dutch followers."[84] Originally, he had been much closer to these impulses, especially in connection with Nieuwentijt. While 'sGravesande was studying for his law degree in Leiden, he also pursued a lively interest in mathematics. In 1707, law degree in hand, he moved to The Hague, where he was central in the development of the key Francophone *Journal littéraire de la Haye*. This connection made 'sGravesande prominent enough to be included in the Dutch party that went to England in 1715 to celebrate the coronation of George I and the Hanoverian succession. He extended his stay for two years, meeting and working with Jean Théophil Desaguliers (1683–1744), the leading figure of British experimental Newtonianism, then meeting Isaac Newton himself and getting elected to the Royal Society. Newton was so impressed with 'sGravesande that he praised him to the leaders of the Dutch delegation, and they persuaded the curators back home to appoint 'sGravesande to the chair in mathematics in Leiden in 1717, notwithstanding his lack of standard credentials.[85] While the physics lab that had been created by de Volder remained in the charge of the aged natural philosopher Sengueridius until the latter's death in 1724, immediately thereafter 'sGravesande took it over and worked with the talented instrument maker Jan van Musschenbroek (1687–1748), Peter's older brother, to design a remarkable series of experimental demonstrations, primarily oriented to Newtonian principles in physics.[86]

In 1720 'sGravesande produced an important book, *Physices Elementa Mathematica, Experimentis Confirmata, sive Introductio ad Philosophiam Newtonianam*, the earliest and one of the most important Newtonian explications on the continent of Europe. A second, revised and expanded version of this work appeared in 1725.[87] The British Newtonians loved the book. John Keill immediately rendered an unauthorized translation.[88] This irked 'sGravesande, and Desaguliers worked with him to produce an authorized one.[89] There was a clear sense of alliance between these British Newtonians committed to "experimental philosophy" and 'sGravesande in Holland. As Ruestow puts it, among the British "the Dutch Newtonians enjoyed an influence surpassing that of any other continental scientists in the early eighteenth century."[90] The French reception was markedly cooler: "the first edition of the *Physices elementa mathematica* had been ignored by the prestigious *Journal des sçavans* and was reviewed unsympathetically elsewhere."[91] But by the end of the 1720s, Frenchmen interested in Newtonian ideas recognized 'sGravesande as an important authority. Voltaire contacted him early in the 1730s for advice in comprehending Newton. Maupertuis, the other key proponent of Newtonianism in France in the 1730s, explicitly formulated his conception of Newton in juxtaposition to that of 'sGravesande in order to win over the Paris Academy of Sciences, but later he visited with 'sGravesande and Desaguliers and appears to have moved substantially toward their orientation.[92]

As Vermij puts it, "'sGravesande and Van Musschenbroek aimed to completely restructure natural philosophy on the foundation of Newton's theories."[93] Peter van Musschenbroek was Boerhaave's student when the latter gave his famous address of 1715 celebrating Newton, and we have good reason to believe that Van Musschenbroek's Newtonianism owed not a little to that introduction. "All his life Van Musschenbroek was a faithful follower of Boerhaave."[94] He did his dissertation in medicine with Boerhaave, then went on to Britain in 1717 to learn more about Newtonian science. When he returned, he took a second degree in (natural) philosophy. After a brief stint at Duisburg in Germany, he returned to a position first at Utrecht, where Haller met him, then together with 'sGravesande, whom he admired highly, at Leiden. Together, they were renowned throughout the Continent as the foremost advocates of experimental Newtonianism.[95]

While even in 'sGravesande we can detect over the course of his career a movement away from a mathematical to a more instrumental and experimental approach to natural inquiry, this was far more prominent in Van Musschenbroek from the outset. Kees de Pater draws some important contrasts: "Van Musschenbroek's textbooks are less mathematical in approach than the various editions of 'sGravesande's textbook. . . . [A] comparison of the third edition of 'sGravesande's textbook (1742) with the second edition of Van Musschenbroek's *Elementa Physicae* (1741), which appeared more or less simultaneously, is illuminating. . . . Van Musschenbroek . . . pays attention to magnetism, electricity, heat, meteorology and the strength of materials, topics that are largely ignored by Newton and 'sGravesande. . . . Van Musschenbroek pays attention to chemistry which is ignored by 'sGravesande."[96] De Pater adds: "In Van Musschenbroek's work forces play an important role, perhaps even more than in the work of other Newtonians in and before his time."[97] In short, Van Musschenbroek was attuned to the nonmathematical *physique* of Herman Boerhaave, which took inspiration primarily from the speculative "Queries" of Newton's *Opticks*. That became the main concern of late eighteenth-century natural inquiry. A key figure in that impulse was Boerhaave's greatest medical student, Albrecht von Haller.

ALBRECHT VON HALLER BEFORE GÖTTINGEN: THE MAKING OF AN EXPERIMENTAL NEWTONIAN

Albrecht Haller was born October 18, 1708, in Bern, Switzerland, the youngest of several children; his mother died when he was very young. He grew up speaking French and (Swiss) German, and he swiftly mastered the classical languages.[98] Haller's father was a lawyer who wanted Albrecht to study for the clergy, but after he died in 1721, the fourteen-year-old decided for himself on medicine.[99] In 1722 he moved in with his step-uncle, Johann Rudolf Neuhaus (1652–1724), a doctor in Biel, where he had his first encounter with Cartesian philosophy.[100] Neuhaus was a convinced Cartesian, but the young Haller would question him at every turn: "[Descartes's] physics was too arbitrary and speculative for the critical young man, and he pestered his master with questions as to how one could know 'that the particles of the second element were round, that the specks [*Stäubchen*] that composed the matter of magnets were helical.'"[101] Much later Haller wrote of his first reading of Descartes that "every page revolted me."[102]

After this brief "apprenticeship" Haller betook himself to university—initially at Tübingen (1724–25). Haller's key instructor was Johann Georg Duvernoy (1691–1759), an anatomy professor, who first inspired Haller's lifelong and defining interest in anatomy. But Duvernoy had little to work with in Tübingen: "Keine Leichen, keine Bücher [no cadavers, no books]," as Gerhard Rudolph succinctly puts it.[103] He did get Haller interested in a dispute he had initiated with an anatomist from the eminent University of Halle, Georg Daniel Coschwitz, who claimed to have discovered a new duct of the salivary glands. Duvernoy believed he had disproved this experimentally, and he encouraged Haller to present a formal disputation on that controversy, based entirely on Duvernoy's own research.[104]

Tübingen proved intellectually inadequate not only for the fledgling Haller but for his teachers as well. When the Saint Petersburg Academy of Sciences opened in 1725 and sent out invitations to scholars in the West, preeminently in the Germanies, Duvernoy took up the invitation in the spring of that year. So did the other important instructor for Haller in Tübingen, the philosophy professor Georg Bernhard Bilfinger (1693–1750).[105] Since Bilfinger was a Wolffian and Haller definitely took a course from him, there has been some effort to link Haller with the German school-philosophical tradition.[106] That seems highly dubious. Haller had other ideas. In his medical lecture course, Duvernoy had used Boerhaave's *Institutes* as his textbook, so Haller knew exactly where to go for the cutting edge in medicine.[107] After sixteen months at Tübingen, he moved on to Leiden in April 1725.

When he arrived, the Leiden University medical school was "in flourishing condition," Haller noted in his *Tagebücher*: "in our field of science . . . we have everything we could wish for. . . . Nowhere are the endeavors in natural, anatomical, and the other arts carried to a higher level."[108] He pointed first and foremost to Herman Boerhaave and the latter's younger colleague, Bernhard Siegfried Albinus (1697–1770), and continued: "as concerns the [anatomical] theater, the library, and the [botanical] garden, we were provided with everything."[109] Leiden University, established in 1575, had raised the medical school to prominence in these areas already by the middle of the seventeenth century, rivaling or surpassing the medical school of the University of Padua, which it sought to emulate.[110] By the time Boerhaave began his career, "Leiden's medical faculty was considered the best in

Europe, attracting many students from all over the Continent."[111] It is important to register the scale and cosmopolitanism of Leiden medical school by 1725. In his first class, Haller discovered he had more than 120 classmates, over half from beyond Holland: 40 British, 20 German, a smattering from all over the rest of Europe.[112] Conspicuous as an exception, there was virtually no one from France.[113] Such a medical school class size was substantial. Haller's own classes at Göttingen drew only 50–60 students per year at most, according to Hubert Steinke.[114] Among German universities, only Halle had enrollments anywhere near the size of Leiden, and most had very few students matriculating in medicine. At the moment Haller arrived, Boerhaave towered over the whole enterprise. There were technically five chairs in the Leiden medical faculty, but there were only three professors, because Boerhaave held down three of the five chairs! "He taught the preliminary subjects, botany and chemistry, then he taught physiology and pathology, then special pathology and therapeutics, and finally clinical medicine."[115]

Haller made many lifelong friends while studying at Leiden, and anyone who had studied at Leiden before or after him took him up as one of the fellowship. No fewer than fifty Boerhaave students corresponded with Haller between 1725 and 1777, and in the years before his Göttingen appointment they formed the lion's share (twenty of twenty-two correspondents) of his network.[116] Perhaps most important among them for Haller personally was Johannes Geßner (1709–90) from Zurich. Haller met him in Leiden and they studied together not only there but in Paris and then in Basel as well. Together they undertook a mountain tour in the summer of 1728 that would lead not only to their lifelong collaboration in Swiss botany but to Haller's famous poem *Die Alpen*.

When we ask after Haller's encounter with Dutch Newtonianism at its source, we can draw upon evidence from his Holland diaries.[117] As Knoeff notes, there is no mention of *any* association of Boerhaave with Newton in that diary. Interestingly, by contrast, Haller's great rival, Julien Offray de La Mettrie, was emphatic about his experience with Boerhaave in 1733: "Boerhaave étoit Newtonien, convainçu et convainquant." For La Mettrie, Boerhaave was an active proponent of the Newtonian philosophy.[118] But we get no sense of that from Haller. In the diaries, Haller did mention 'sGravesande, in what appears to be his revised version of the original diaries, as "a diligent, creative, but not very academic mathematician."[119] More extensively, later in his diaries, Haller wrote: "in the philosophy faculty there is . . . 'sGravesande. A lawyer from The Hague who, upon becoming acquainted with Newton, had understood his teachings so well that now he has constructed a complete system of it. What is best about him is the experiments, which because of his beautiful instruments, he can perform often and precisely. It is all presented in his new edition of the *Physics* text."[120] From this we can infer that Haller probably attended 'sGravesande's lectures in 1725. He met Jan van Musschenbroek, the instrument maker who worked with 'sGravesande on his experiments.[121] He even met Jan's brother, the famous experimental physicist Peter van Musschenbroek, in Utrecht.[122]

While Haller, thus, encountered all the Dutch Newtonians during his student years, there is no evidence of any explicit identification with *Newtonianism* on his own part. Rather, he seemed altogether focused on his *medical* training, specifically on *anatomy*, the domain of Boerhaave's young and brilliant colleague Bernhard Siegfried Albinus. Notably, while we have little correspondence between Haller and Boerhaave, Haller corresponded with Albinus from the moment he departed Leiden. Albinus, Haller's diaries avow, "should become the greatest figure in anatomy that Europe has ever seen. Already he dissects with such precision that he does not disturb a single hair."[123] But Albinus was more than a technical master in dissection. He was also proving a bold theorist in physiology, willing, indeed, to part company with and criticize his grand colleague.[124] His private physiology lectures "on human nature" were especially important in this regard, and Haller attended these.[125] They offered a version of "mechanism" in medicine that already encompassed a great deal of what would later be termed "vitalism," and in this we can see Albinus as a very important predecessor of Haller's own transformative impact on the discipline. Later in their careers, Albinus would complain bitterly and publicly that Haller had stolen many ideas from him. The prefaces to many of the sequential volumes of Haller's monumental *Elementa Physiologiae Corporis Humani* throng with his defenses against these charges from Albinus.[126] In addition to Albinus, Haller was fascinated with the anatomical innovations of Frederick Ruysch (1638–1731) of Amsterdam. Already in his eighties when Haller met him, Ruysch had developed a secret technique for wax infusions in blood vessels that had led to substantial advances in circulatory anatomy.[127]

We have few details of Haller's course work at Leiden, and it appears that he advanced very swiftly toward a degree, using the work done in Tübingen to his advantage. His dissertation was a reworking of his Tübingen thesis.[128] During a summer 1726 visit to Halle, which he reckoned "the premier university in Germany," he had sought out Coschwitz to discuss the salivary duct issue central to both thesis and dissertation, and he also observed the anatomical work of the scholar.[129] This only confirmed his negative opinion.[130] When he took his degree in May 1727, Haller was only nineteen years old. It was typical, in those days, for still-green medical students (those who could afford it, that is) to round out their training with a study tour, a *peregrinatio medica*, as it was called.[131] Leaving Leiden, Haller did this, seeking "encounters with the most important doctors and researchers of his time."[132]

In Leiden, Haller had befriended John Pringle (1707–82), who would complete his Leiden MD in 1730. Pringle provided Haller with some contacts for his visit to England.[133] Haller had no English and this imposed significant limitations on the balance between tourism and training during his time in England, July 25–August, 28, 1727.[134] To be sure, he encountered James Douglas (1675–1742), royal physician.[135] He met with William Cheselden (1688–1752) and observed his operations at the hospital where Cheselden served as chief surgeon, and he visited the extensive natural-history cabinet of Hans Sloane (1660–1753), which Haller esteemed the grandest in the world.[136] It would seem that his main contact and guide in London medical-scientific circles was Sloane's librarian/assistant, Johann Caspar Scheuchzer (1702–29), scion of an eminent Zurich scientific family.[137] Scheuchzer took Haller all over London and arranged his visits with a number of key figures, including Cheselden.[138] From Bern some years later, in his newly acquired English, Haller wrote a letter of thanks to Sloane for his and Scheuchzer's hospitality during his London stay.[139]

We have two versions of Haller's travel diaries for his stay in London. The one contains considerable revisions that he entered in the period after he had returned to Switzerland (primarily in Basel, 1728–29) and, indeed, after he had acquired some command of the English language (and literature).[140] The other, apparently unrevised version seems to contain all the evidence of his medical training.[141] The most interesting discrepancy between the two *Tagebücher* is that what seems to be the untouched diary has virtually nothing on Isaac Newton, whereas the edited and revised diary (completed in Basel, 1728–29) is full of Newton and develops a very important assessment of English scientific ascendancy. Haller's characterization is worth citing fully:

> In the sciences it would appear no country takes precedence over England, except perhaps in the matter of law [since they follow common, rather than Roman, law]. However, in the study of nature, pointed experiments and all that which is encompassed by geometry and the nature of things, they top all previous ages and all contemporary lands. The causes are (1) the wealth of the land and the good government, which are the preconditions for the apprecia-

> tion and compensation of scholars; (2) the reflective and creative nature of the people, which sets about appraising everything with which it is presented, good or bad, in every detail; (3) the admiration of scholarship. How much science is valued at court is shown in the controversy between Newton, Clarke, and Leibniz, when the queen [actually, the princess] presided over the exchange. Newton's monumental funeral procession and his grave in Westminster Church [*sic*], and in particular the extraordinary admiration of the entire people for this great mind, prove that here one admires distinguished scholarship as much as in other lands one admires aristocrats or generals. . . . These are the reasons why Wallis, Newton, Hawksbee, Keil, Boyle, and so many others among our scholars have achieved such great things, and Desaguliers, Raphes, Pemberton, Clarke, and others will bring further fame to their island with new accomplishments.[142]

A bit later, Haller returned to the matter of Newton's fine marble grave and the enormous honor the English people paid to Newton, "making a demigod [*Abgott*] out of him."[143] As Richard Toellner has observed, the marble grave was completed only in 1731, so Haller could not have seen it but only read of it later in the press.[144] Indeed, much of what Haller wrote in Swiss retrospect about his London tour might well, as he himself noted in another context, be found in the *Guide de Londres*—in short, cribbed from published materials he had to hand in Basel, including a good bit of material from England that, having learned English, he could now read.[145] On the question of England more generally, it is very interesting to ask when Haller actually read *Lettres sur les Anglois et les François* (1725), by the Swiss author Beat Ludwig von Muralt (1665–1749).[146] In celebrating English civilization over that of France, Muralt stole a march on Voltaire's famous *Letters on England*.[147] Haller mentioned Muralt in his revised *Tagebücher*, but it seems more likely that this was an insertion from the Basel revisions than that Haller brought that perspective with him when he landed on English shores.[148]

It seems that Newton's prominence in the English environment (he had died only a few months before Haller arrived in London) was palpable to the young visitor even in the absence of sufficient linguistic competence. That *impression* was an authentic part of the initial experience. But its *significance* developed for Haller later. It is noteworthy that in the meticulous inventory of his possessions and purchases that Haller composed for his return to Switzerland, there are no books relating to Newton. I believe Haller's new conception of science in the England he had visited so briefly in 1727 only later proved decisive for his intellectual identity. In his revisions, Haller twice mentioned the Leibniz-Clarke correspondence.[149] That deserves further consideration. Already in the unrevised diary, Haller noted his visit with Pierre Desmaizeaux (1666–1743) in London.[150] Desmaizeaux was the translator and compiler of the French version of the Leibniz-Clarke correspondence.[151] In the original diary, Haller touched on Desmaizeaux's literary endeavors but did not mention the Leibniz-Clarke correspondence; however, he inserted it seamlessly into the same sentence in his revisions of 1728–29.[152] But was Haller already aware of Desmaizeaux's crucial role in the propagation of Newtonianism in 1727? Did he see a copy of Desmaizeaux's two-volume edition in the latter's rooms? Everything argues, rather, that it was his vastly more informed vantage in Basel that made this connection significant. Quite simply, retrospective fascination with Newton is the most significant element in Haller's self-formation as a natural scientist in these years. Ironically, I believe it was his time in Paris that crystallized Haller's understanding of Newton's epochal centrality.

Paris, September 1, 1727, to February 21, 1728

An enormous amount of research has appraised Haller's medical training in Paris, but I suggest there is another aspect of his stay that needs to be considered: namely, the impact of Paris on Haller's sense of the state and future of natural science more generally and how his own pursuit might fit into that larger impetus. No one seems to have seriously considered *why* Haller, already a credentialed MD from the foremost medical school in Europe, and clearly advancing in his "journeyman" acquisition of additional technique in anatomical dissection, would suddenly opt for a term of study in *advanced mathematics* with Johann (I) Bernoulli (1667–1748) in Basel in 1728. That had little to do with Haller's pursuit of *medicine* and everything to do with Haller's pursuit of *science*—in a word, with his adoption of "Newtonianism."

The Medical Aspect

Toby Gelfand has observed: "From a technical standpoint, French surgery assumed a position of European leadership in the late seventeenth century and the first half of the eighteenth. French, or to be more precise, Paris surgeons built this reputation on major operations, new instruments, and anatomical work[,] for which cadavers were in plentiful supply. Their publications dominated the literature, and Paris became the outstanding center for learning anatomy and surgery."[153] In a more focused paper, Gelfand has examined what was called "the Paris manner of dissection" in the early eighteenth century.[154] That is, Paris was noted as the place where students could get hands-on experience in dissection. While medical students everywhere could *observe* dissections in lecture demonstrations and surgical theaters, what they wanted—and Haller most adamantly—was to *do* them. This drew a lot of British medical people to Paris over the first half of the eighteenth century.[155] Notably, Boerhaave gave explicit advice to two of his young students, the Geßner brothers, to supplement the theoretical training in medicine he had given them with the practical experience that would be put at their disposal by the Parisians.[156] They followed his advice, arriving in Paris two weeks before their good friend Haller. Gelfand has established that the bulk of the hands-on training in anatomy in Paris was in fact made available, particularly to foreign students, only in private courses (for a fee).[157] Jacques-Bénigne Winslow (1669–1760) was at that time "the most highly regarded anatomist of France."[158] What the Geßners learned, upon arriving in Paris, was that Winslow was not at that moment willing to offer "private courses" in dissection for any sum of money.[159] Accordingly, they turned to the distinguished chief of surgery at the Paris Charité hospital, Henri François LeDran (1685–1770). Haller followed suit when he arrived in Paris on September 1, 1727. They were all nonplussed to learn of a Paris regulation that delayed official anatomical dissections and demonstrations until after October, and they sought, successfully, to find ways around it, primarily via autopsies at the Charité, presided over by LeDran. When even this proved insufficient, Haller made other arrangements, and in emulation the Geßners negotiated with LeDran to get secret instruction before the ban on dissection was lifted.[160]

Unquestionably, the Paris stay was of "extraordinary importance" in Haller's technical advancement.[161] Without it, Erich Hintzsche has aptly observed, Haller would never have been able to take over the anatomy course at the University of Basel in 1728/29.[162] But as he also and famously pointed out, a lot of misinformation has been propagated by the standard sources on Haller about what he did in, and learned from, his stay in Paris. Controversy revolves especially around Haller's relation with Winslow. Gerhard Rudolph, for example, claims: "Particularly fruitful was Haller's activity with the initially not very highly esteemed Winslow in Paris, with whom he learned to study organs in their relative topographical situation."[163] Winslow also impressed Haller with his structural-functional thinking, according to Rudolph. But Hintzsche has established that Haller never worked with Winslow, though he may have attended one or two of his anatomical demonstrations. Instead, he made a systematic collection of student notebooks from Winslow's lectures, which he collated carefully.[164] Hintzsche has published an inventory of the newfound notebook of Haller's time in Paris, the *Manuscripta Winslowiana*, to document this more critical assessment of the Winslow connection.[165] While Haller's disciple and authorized biographer, Johann Georg Zimmermann (1728–95), reported that Haller seemed excessively complimentary of Winslow in his conversations at Göttingen, and there are remarks of high praise in Haller's *Methodus Studii Medici* and in the *Bibliotheca Anatomica*, there are also strikingly negative judgments about Winslow in Haller's final commentaries, his articles for the *Encyclopédie d'Yverdon*.[166] The upshot is that Winslow was important for Haller, but more as a transmitter of certain techniques than as a direct teacher or model for anatomical research.

By contrast, Haller attended and carefully documented a number of autopsies and surgical procedures by LeDran. Strikingly, most of these surgeries proved unsuccessful. Urs Boschung calculates, on the basis of the diaries, that of nineteen patients whom Haller observed LeDran treating surgically, sixteen died; in Haller's later account in the *Bibliotheca Chirurgica*, out of seventeen surgeries there were fifteen fatalities.[167] Many Haller scholars, including Boschung, surmise that this may well have established his determination never to practice surgery himself, though he taught the field for seventeen years.[168] Haller also noted the relative lack of sanitation in Paris hospitals as contrasted with English ones and the correspondingly higher death rates there.[169]

In early December 1728, Haller had a falling out with LeDran.[170] Immediately he set about working independently by arranging with a "Duverney" (son of the aged and famous anatomist?) to purchase a cadaver and other anatomical specimens on which to work.[171] As Boschung notes, Haller meant to make the most of his opportunities in Paris, "self-consciously choosing his own way, even without the defining influence of a teacher."[172] As Hintzsche puts it, "Haller's extensive anatomical studies of the Paris period are largely independently undertaken labors."[173] In any event, there is a striking gap in documentation concerning Haller's anatomical work from December 19, 1727, to February 21, 1728, the date he left Paris.[174] Haller was meticulous in documenting what he did in Paris, and that has raised the question whether this gap in the record betokens a suspension of research.[175] That gets linked to Haller's much later claim that he had to leave Paris prematurely because he had acquired a cadaver from a grave robber and been denounced for it by a neighbor.[176] Hintzsche notes that the time span between the last date of his accounts of anatomical labors in the Paris notebooks and his date of departure from Paris is too extended to make this story very plausible.[177] But the last phase of Haller's stay in Paris, in any event, remains extremely obscure.[178]

So is the motivation for abandoning his ostensible plan to continue the *peregrinatio medica* to Italy—a major locus of medical research and practice, with which, later in his career, Haller would prove substantially concerned.[179] Instead, suddenly, he resolved to go to Basel to study mathematics with Bernoulli.[180] That has simply been taken, supplemented by the notion that Haller was somehow "homesick" for Switzerland, as a self-evident course of action.[181] I submit that won't do. No one has really explored the fact that Johannes Geßner was already planning to go to Basel to study mathematics with Bernoulli. Geßner would get his MD from Basel in 1730, and upon his return to Zurich he would teach primarily mathematics and physics, which suggests that this training with Bernoulli was a centerpiece of Geßner's academic preparation. Haller may have decided to go to Basel simply in order to stay with his friend.[182] But even that is not enough. We need to think more deeply about this intellectual interlude.

Why mathematics and Bernoulli *then*? Rudolph assures us simply: "He wanted to fill a hole in his education."[183] Hintzsche offers us something roughly similar: "he felt his limited knowledge of the higher mathematics was a hole in his medical education."[184] But that is hardly an adequate explanation. Why did *that* "hole" matter? Why then? Why did Haller respond to it in that way? R. R. Beer observes: "Clearly it was under the influence of the English that Haller drew near mathematics, the stern queen of the age. Newton was of course the great name, which had dawned upon him. In order to penetrate fully into his world, he had to supplement the natural sciences with differential and integral calculus."[185] This is the closest I have seen anyone come to a useful conjecture about the decision. Let us see if it can be developed more compellingly.

The Science Aspect

An important aspect of Haller's Paris stay needs to be explored: the happenings in the Paris Academy of Sciences and the discourse of natural philosophy within the academy. Bernard de Fontenelle's famous *Éloge* of Newton was read at the open session of the Academy of Sciences on November 12, 1727.[186] Haller was in attendance: that much he tells us in his diaries.[187] What else did Haller know about the discussions swirling in Parisian scientific circles at that moment? A recent major study has dramatically enhanced our grasp of these questions. J. B. Shank's *The Newton Wars and the Begin-*

ning of the French Enlightenment (2008) thoroughly updates the classic work from Pierre Brunet, *L'introduction des théories de Newton en France au XVIIIe siècle avant 1738* (1931).[188] While Shank has other agendas than those to be pursued here, he offers us a sound footing for the pursuit of our question about Haller's experience of the discourse of natural philosophy in Paris in 1727–28.

The most important point Shank establishes is that the French assimilated Newton's *Principia* immediately upon its publication, but into a very specific and established framework. From the initial review of the *Principia* in the *Journal des sçavans* in 1688 through Fontenelle's famous *Éloge* of 1727, the French discriminated between Newton the mathematician and Newton the physicist.[189] While they welcomed and celebrated Newton for his formidable *mécanique*, they were unequivocal in dismissing his *physique* as, in Shank's stark summation, "an epistemological category error."[190] Physics (natural philosophy), in this line of thought, was about *physical causes*, whereas mathematics was about "relations between quantitative and spatial magnitudes."[191] What made Newton's physics so problematic was his claim, in book 3 of the *Principia*, that attraction was an actual physical force, not merely a mathematical model. Nor were the French alone in dismissing "action at a distance" as a preposterous recursion to "occult qualities."[192] That proved the response of the foremost Dutch physicist, Christiaan Huygens, and, of course, of Leibniz and his followers as well.[193] The French did not even think that aspect of "Newtonianism" needed to be taken seriously—until the 1720s.[194] It was in the period surrounding his *Éloge* that Fontenelle "explicitly identified the Newtonian enemy as the principle of gravitational attraction across empty space."[195]

Shank is fascinated with the tradition of mathematical Newtonianism, especially in France. The result, in my view, is an underestimation of experimental Newtonianism elsewhere.[196] Shank is certainly correct that for most advanced practitioners of "analytical mechanics" in the aftermath of the *Principia*, the important game was to restate Newtonian mechanics in the new and more efficient language of (Leibnizian) calculus, since Newton himself had chosen to formulate it primarily in the "classical" form of geometric synthesis. Shank goes on, however, as follows: "That so many Englishmen after 1690 were eager to see a new, 'analytical' *Principia* developed using the newest infinitesimalist mathematics reveals the extent to which the 'Continental' understanding of Newtonian science was not so exclusively 'Continental' after all."[197] What gets obscured in this formulation is whether and when English Newtonians adopted Newton's *physical* theory, regardless of which mathematics was used to formulate it. Shank contends: "few if any defenders of Newtonian gravitational attraction existed anywhere in Europe before 1710." That is, the French discrimination between *mécanique* and *physique* was "the most common interpretation for everyone . . . [and was] true even in England. . . ."[198]

The dating, here, is crucial, and Shank equivocates even in continuing this very sentence: ". . . even if Newtonian physics found its first advocates here." Just *when* did that happen? Shank himself gives clear evidence that it was well *before* 1710, notwithstanding his own dating. We are now eminently aware of a "Newtonian physiology" that set in during the 1690s (Pitcairne et al.) and of a "Newtonian chemistry" that set in around 1700 (Freind et al.), and we know that in his important *Astronomiae et Geometricae Elementa* of 1704, David Gregory—in Shank's own words—"was representative of the attractionist approach to Newton's *Principia* then growing in importance in Britain and Holland."[199] Most significantly, public experimental demonstrations of Newtonian physics had become a major feature of the London scene before 1700.[200] In a word, the "experimental philosophy" associated with Isaac Newton did not have to wait until the appearance of his *Opticks* in 1704. Rather, Newton sought to capitalize upon and advance this already-thriving tendency in Britain by guiding it with the "Queries" of the *Opticks*.[201] While the mathematical and the experimental elements in Newtonianism may have been in some tension among the British Newton enthusiasts, what Shank's story underestimates is the entrenchment and power of the "experimental" orientation among these figures. Perhaps this is because in France it was not taken that seriously—until the late 1720s.

Shank teaches us that there was an important shift in the orientation of Parisian science and especially of the Academy of Sciences in the years after 1715. While analytical mechanics had been a mainstay of the academy in the period 1690–1715, thereafter its importance waned dramatically. As the leaders of the early generation died off, they were not replaced by figures of comparable stature.[202] Instead, the Academy of Sciences came to be dominated by two new figures, closely allied in orientation: René-Antoine Ferchault de Réaumur (1683–1757) and Jean-Jacques Dortous de Mairan (1678–1771), who shifted the attention of the institution and of Parisian science generally away from analytical mechanics to natural history and pragmatic technologies.[203] The perpetual secretary of the academy, Fontenelle, found it expedient to go along. The result was that by the 1720s, the status of mathematics in the academy had eroded profoundly. The only significant representative of the higher mathematics was François Nicole (1683–1758), but he had only a peripheral influence.[204] The mathematicians at the Paris Academy of Sciences over the last decade of the seventeenth and the first decade of the eighteenth centuries had been deeply influenced by Johann (I) Bernoulli. But now Bernoulli had become substantially marginalized, much to his annoyance; his submissions for prizes at the Paris Academy were consistently rejected, leading him to suspect a "cabal" there actively intent on thwarting him. He complained, as John Greenberg reports, that "mathematicians in Paris in the 1720s handled the infinitesimal calculus clumsily, awkwardly, and cumbersomely and perhaps sometimes even misapplied it when using it as a tool to solve mathematical problems."[205]

Things began to change, however, with the appearance of the young Pierre-Louis Moreau de Maupertuis (1698–1759), who attached himself to Nicole, the one mathematician in Paris to earn Bernoulli's praise. Maupertuis made his breakthrough in December 1723, when he became "an academician and a member of the class of mathematicians."[206] In 1726 he delivered his first original mathematical paper to the academy.[207] This reinvigorated Fontenelle's longstanding loyalties to the analytical-mechanical tradition.[208] For decades he had been working on his own synthesis of the calculus, but he had not found it politic to publish it in the environment dominated by Réaumur and Dortous de Mairan. In 1724–25, however, he contacted Johann Bernoulli asking for assessment of his revisions of the book manuscript. In August 1726 Fontenelle submitted it to the Academy of Sciences for accreditation, and it was approved in February 1727. In November, at the same public session where his *Éloge* of Newton was read, Fontenelle himself read the preface to his newly appearing work.[209] By the late 1720s Bernoulli seemed to have regained prominence in the Paris scene, as "the pace of research in the infinitesimal calculus began to quicken and its quality began to improve there, largely as a result of the interest, zeal, and personal ambitions of Maupertuis."[210] What is most important to derive from all this is that the infinitesimal calculus was a *hot* topic in Paris in 1727–28, and that everyone acknowledged the preeminence of Johann Bernoulli in that field. Indeed, the most important evidence of this is that *Maupertuis himself* decided that he had to go to Basel and study with Bernoulli to advance his own mathematical understanding. This was quite a sacrifice in status for Maupertuis, who was not only an aristocrat of means but also already a noted mathematician and member of the Paris Academy of Sciences, and this betokens the importance he attached to studying directly with Bernoulli. As the foremost mathematician of the day in Paris, this was a tremendous acknowledgment of the gap between Basel and Paris.

We have, in short, uncovered the context into which the young Albrecht Haller stepped in Paris in 1727–28, inducing his interest in calculus and Bernoulli. The question that remains is: what has this to do with "Newtonianism" for Haller? In what measure was the revival of calculus in Paris in the late 1720s associated with the question of Newton? Maupertuis, again, serves as the key. In May 1728 he wrote to Bernoulli, asking to study with him in Basel. Then he set off to *London* for a three-month stay, "meeting everyone from Desaguliers and Samuel Clarke to the journalist Pierre Desmaizeaux."[211] He encountered Henry Pemberton, who had assisted Newton with the final edition of his *Principia* and then, after Newton's death and relying on his close consultations with the master, had published *A View of Sir Isaac Newton's Philosophy* (1728), one of the most highly regarded formulations of Newton's ideas from the British-Dutch vantage point.[212] Maupertuis also became acquainted with the best British mathematicians—Brook Taylor, James Stirling, and Abraham de Moivre—through whose support he was admitted into the Royal Society. In short, Maupertuis was motivated in just this moment not simply to go work with Bernoulli but to steep himself in British "Newtonianism." Taylor, Stirling, and de Moivre, it bears pointing out, were committed to Newtonian *physics*, not just Newtonian *mathematics*.

My suggestion is that in Paris in the 1720s the question of formal mathematics could no longer be kept apart from the question of Newtonian physics (and metaphysics), and that as Maupertuis advanced in his sophistication in mathematics, he was also, and decisively, advancing to his pathbreaking advocacy of Newtonian physics (though *not* metaphysics). After 1732 he would emerge as the first and most important scientific advocate of Newtonian physics in France. Yet Maupertuis was the "direct heir to the analytical mechanics program in France," and he engineered the acceptance of Newton's long-despised physics in that context.[213] Shank demonstrates aptly that in the famous *Discours sur les différentes figures des astres* (1732), Maupertuis drew upon Malebranche for a ubiquitous philosophical skepticism that rendered equally problematic Cartesian vortical theory of motion and Newtonian attraction, thus mitigating the epistemological horror with which the latter had so long been viewed in Paris. At the same time, however, Maupertuis chose to "avoid altogether 'sGravesande['s] application of experimental philosophy to the principle of Newtonian attraction." He "privilege[d] mathematics as an epistemological guarantor of empiricism in a manner very different from English and Dutch 'experimental philosophy' . . . [revealing his] pedigree in the Malebranchian world of French analytical mechanics."[214] Haller did not follow *that* line at all. He was never a disciple of Malebranchian analytical mechanics. Studying with Bernoulli, instead, would enable him to understand and embrace the English and Dutch tradition of "experimental philosophy."

Basel

Haller left Paris for Basel well before Johannes Geßner, perhaps because the latter came down with a serious illness in late January. Reaching Basel March 15, Haller began his studies with Bernoulli on April 1.[215] He studied under Bernoulli together with Johannes Geßner both in the spring of 1728 and again in the 1728/29 academic year.[216] While he studied with Bernoulli, Haller also worked with Johann Rudolf Mieg (1694–1733) in anatomy and with Benedikt Stähelin (1695–1750), a physics professor who was also avid for botany. Stähelin proved crucial for another aspect of Haller's education in Basel: learning the English language. Stähelin was a major source of Haller's intense new Anglophilia, giving Haller English poetry to read and helping Haller commit himself to "philosophical poetry"—that is, didactic poetry along the lines of the English Augustan, Alexander Pope.[217] His "resolution to write à l'Anglais dates from 1728."[218] He set about "to do for German poetry what had been done in English verse—namely, to write philosophical poems."[219] The result was *Versuch schweizerischer Gedichte* with its four great didactic poems: *Die Alpen* (1729); *Gedanken über Vernunft, Aberglauben und Unglauben* (1729); *Falschheit menschlicher Tugenden* (1730); and, in the second edition, *Über den Ursprung des Übels* (1734).[220]

Die Alpen brought Haller immediate international fame because his celebration of the wild beauty of the Alps evoked one of the most important aesthetic innovations of the age, attunement to the "sublime" in nature.[221] The Alps after Haller would enjoy a distinct prominence in European literature; the Bern highlands became a pilgrimage place for the age of sensibility.[222] But his poetry also contained a substantial element of cultural criticism, connected with Muralt's condemnation of decadence.[223] This stressed the superiority of England over France, to be sure, but even more, it glorified the rustic simplicity of Swiss mountain villagers.[224] On that score, there is much to compare with the Genevan Jean-Jacques Rousseau, writing some decades later. *Die Alpen* is the direct fruit of Haller's first Alps tour, in the summer of 1728.[225] He set out July 7 with his friend Geßner, after they had completed their first Bernoulli course. One of the central purposes of the expedition was to study alpine botany, and they planned to write up their findings jointly.[226] Graciously, Geßner, who would remain a very active botanist over his entire career and an enthusiastic Linnaean, unlike Haller, eventually allowed Haller to publish their collective discoveries.[227]

In August 1728, from Bern, Haller wrote to Mieg, offering to substitute for him in the University of Basel anatomy course because of the latter's illness, and Mieg accepted.[228] Accordingly, while continuing his studies of mathematics with Bernoulli, Haller (twenty years old!) assumed all duties at Basel in anatomy—both dissections and lectures—for the academic year September 1728 to May 1729. This experience served to fix Haller's scientific "calling."[229] Still, he had no regular appointment, and he was forced to return home to Bern after the end of the term. Only a few months after Haller left for home, Maupertuis arrived in Basel to enroll in the university to get retrained in the higher mathematics by Bernoulli. He began his course work in September 1729, enjoying as well the privilege of living with the distinguished mathematician and his sons, who would become lifelong friends. Maupertuis subsequently became Bernoulli's most important ambassador in the Paris Academy of Sciences, a locus of consummate importance to the prickly mathematical master.[230]

Mathematics was at the heart of much of Maupertuis's early career. But what of Haller? What did studying with Bernoulli mean for him? Certainly, he developed enormous respect for the genius of the man. In the dedication of his Boerhaave edition, 1740, Haller wrote of Bernoulli: "one day as an old man, if indeed God shall grant [me] the years, I shall be roused by the love of my teacher to explain to my students, not without a blameless exaltation of mind, that I heard . . . Johan Bernoulli [lecture]."[231] He made a more extended comment in a letter, cited first by Zimmermann, then by H. M. Jones: "Leibniz was a Columbus, who caught a glimpse of a few islands of his new world, but a Newton, a Bernoulli were born to be the conquerors of [that new world]."[232]

Still, what actual place did the higher mathematics have in Haller's intellectual identity and work? According to Rudolph, "the new science carried him in its current for many years."[233] More insight arises, in my view, from the correspondence of Haller with Nils Rosén von Rosenstein (1706–93), eventually "the leading Swedish physician of his day."[234] In 1729–30 Rosén was residing in Geneva and he opened a correspondence with Haller, whom he regarded as a rising star in Swiss medicine, with a query concerning the place of mathematics in medicine. Though Haller's response has not survived, the ensuing letters from Rosén indicate that Haller stressed the inadequacy of any "iatromathematical" approach. The Swede agreed "that the flawed state of medical knowledge was hardly adequate yet for reliable mathematical calculations."[235] In short, Haller *never* believed—not even at the very time he was immersing himself in its study—that (higher) mathematics had any immediate application in medicine or in his own anatomical/physiological research program.[236] What *was* the point? In my view it was Haller's effort to establish for himself what kind of "Newtonianism" he would embrace.

Haller's Experimental Newtonianism

Richard Toellner has it exactly right: "Haller recognized Newton's true importance for science only in 1728, in Basel."[237] There is good reason to believe that the crucial moment for the establishment of Haller's intellectual identity came in Basel in 1728–29, while he was revising his travel diaries. The burst of poetry associated with just these years, I believe, is best construed as the hammering out of this intellectual identity. Beer, notably, has a good biographical sense of this moment of reflection and integration: "Now, with distance, he could bring into order what he had read, what he had heard, what he had experienced, and thus develop himself into the figure who would be able to take on an important part in the progress of his century. . . . Only now in many ways did he achieve the appropriate proportions, not least in his judgment of contemporary English literature. . . . England was the midwife of this new becoming."[238]

Haller was the first poet in German—perhaps the first major poet in Europe (even before Pope)—to celebrate Newton.[239] "The name of Newton rings through all of Haller's early work."[240]

> A Newton exceeds the limits of created minds,
> Finds nature at work and appears as master of the universe;
> He weighs the inner force, that is active in bodies,
> That makes one fall and moves another in a circle,
> And he breaks open the tables of the eternal laws,
> Once made by God and never broken.[241]

In the poem *Falschheit menschlicher Tugenden*, Haller proclaimed that Newton "fills the world with clarity" and "is a continual source of unrecognized truth."[242]

Toellner makes clear that it was not just poetic enthusiasm for Newton that Haller achieved in Basel but a critical grasp of his technical physics, thanks to Bernoulli.[243] Now, Bernoulli as Leibniz's partisan might seem a poor source for a correct understanding of Newton, but that would be misguided. Bernoulli was one of Newton's most assiduous readers. He had worked through every mathematical proof and corrected several of them. Though he was a Cartesian vorticist by disposition, he was not doctrinaire.[244] He would pander to the true believers in Paris, to be sure, but as he wrote in a confidential letter, he did not really believe so strongly in this view that he could not make good sense of Newton's alternative physics. He was a mathematician, but he was not hostile to experimentalism. Indeed, for a stint, he served as a professor of natural philosophy at the University of Groningen in the Netherlands, where he gave courses in experimental philosophy. That would be why Maupertuis would find in him the ideal mentor after 1729. Haller found him this already in 1728. So Toellner is right when he writes: "With the help of the mathematician Bernoulli Haller really understands the 'Principia' and makes sure that Newton teaches the right idea of the construction of the world, the right physics and, above all, the right theory of the forces, and that he follows the right method."[245]

Toellner makes yet another point about Haller and Newton in 1728. Haller was a very young but extraordinarily ambitious man. He had already suffered "deeply from the contempt towards science in his home town" of Bern, and would again.[246] "Newton had not only shown the way to the right cognition of the world as being the way of scientific research; he had also proved that by this new science there was fame to gain. . . . In the field of science one could harvest what had until then been the privilege of birth and property, of the powerful only. . . . Scientific achievement was ennobling."[247] That shaped Haller's identity permanently. From Newton he learned that "to achieve fame through scientific achievements is possible and real."[248]

The upshot of all this is that there were two eminent idols of his early intellectual biography: Boerhaave and Newton. As François Duchesneau puts it, "Haller is a follower of Boerhaave, and Boerhaave himself felt that empiricist requirements of Newtonian methodology had to be extended to physiology."[249] In the review he published of his own book on Swiss plants in the *Bibliothèque raisonnée* in 1742, Haller wrote: "After the Newtons and the Boerhaaves it is permitted to say, I do not know, I was wrong, I have not completely arrived at the truth."[250] Boerhaave and Newton offered Haller science *and* piety. "Where a Hobbes doubted, a Newton believed; where an Offray [de La Mettrie] scoffed, a Boerhaave worshiped."[251]

What were Haller's aspirations by 1732? He was just about finished writing serious poetry; in any event, he never saw that as his calling, indeed terming it a "sickness" that would overtake him from time to time. He was practicing medicine, but that was never fulfilling for him. Research was what fascinated him. He was an experimenter, not a healer. What obsessed him was technical anatomy and physiology. Haller wanted to pursue physiology as *theoretical comparative zoology*, not as a propaedeutic for human pathology. In short, Haller resolved to make a decisive step from practical medicine to a still-not-institutionalized *Fachgebiet* of experimental physiology. Steinke summarizes the situation aptly: "Haller was himself part of the experimental tradition. As Richard Toellner has stressed, his intellectual home was the early Enlightenment of the Netherlands with Boerhaave as its master. Iatromechanism without rigidity and the encouragement of the empirical approach were the two main elements Haller adopted as his own fundaments of physiology. The detailed study of Newton's and Boyle's work after his medical training reinforced this position and shaped the outline of Haller's world of science before he had performed any considerable research himself."[252]

By 1732 Haller had become a committed *experimental* Newtonian. In my view the best discussion of this is the section aptly entitled "Haller's Newtonianism" in Shirley Roe's masterful essay "*Anatomia Animata*," of 1984. At Leiden, Roe writes: "Haller was inculcated with a markedly empirical approach to anatomical and medical studies, with a negative view of Cartesian rationalism and its nonobservational method, and with a strong emphasis on the usefulness of mechanical reasoning in physiological explanation."[253] There are good grounds for recognizing the strong empirical-observational approach to medicine that Boerhaave stressed, and the practical anatomical work with Albinus was unquestionably experimental. There is further reason to reckon Boerhaave an explicit critic of Descartes, and 'sGravesande clearly emphasized this point in his "introduction to the Newtonian philosophy." The key novelty Roe introduces here is that this was consistent with adherence to a generically "mechanist" approach to physiology. But this was a "Newtonian rather than Cartesian mechanism," in that what Boerhaave and 'sGravesande taught was a "force mechanics," and this was what Haller would develop and make his own. "Haller thus received through Boerhaave and 'sGravesande, if not his introduction to Newton, a strongly positive grounding in Newtonian physics and philosophy," Roe writes.[254]

What I take this to mean is that Haller was given a *basis* for embracing Newtonianism at Leiden. He would actually achieve that identification only later, in Basel after 1728. Since Haller worked through 'sGravesande's course in 1725, he was of course familiar with his *Physics*, because it was the textbook. But one can study a textbook and still not penetrate to the spirit of the science. Instead, it was Pemberton's book—and Haller's preparation to read it receptively—that proved the turning point. Haller ordered Pemberton's book in 1730 from Bern, which means he had learned of its importance in Basel, probably from Stähelin and perhaps from Bernoulli himself. Haller's personal bibliographical annotation concerning the book dates perhaps as late as 1733–34.[255] Notably, Haller characterized the work in his newfound English: "written principally for those curious that not possed [i.e., did not possess] Mathematiks enough to peruse [Newton's] profond [*sic*] Demonstrations by themselves."[256] Pemberton gave Haller the key tenets of Newtonian "experimental philosophy" in a form he could digest: "Philosophers are to be content, if they attain one of the intermediat [*sic*] marshes [*sic*] of the scale of cause's [*sic*], and noway oblig'd to come to the first, hidden in an impenetrable obscurity."[257] Haller specifically deemed it an "injustice," which Pemberton properly exposed and corrected, to accuse Newton of proposing occult qualities; that is, Haller accepted Newton's *physics*, not just his mathematics.

Haller could then appreciate and deepen his grasp of 'sGravesande and experimental Newtonianism.[258] From this point forward, Roe is correct that "Haller consciously sought to emulate the Newtonian program in his scientific work."[259] He took up an "experimentalist empiricism" and a "nonreductionist mechanism." Steinke opts for "non-reductive mechanism" over "Newtonian physiology."[260] While he sees much that would make a Newtonian association justified in terms of Haller's methodology—"its experimental approach, its search for natural laws instead of hidden qualities, and its refusal of unfounded hypotheses"—Steinke suggests that Haller had a different sense for the ontology of force from Newton. As Steinke understands the argument from P. M. Heimann and J. E. McGuire, Newton's "forces were mathematical powers with an ontological nature of their own. They had no material existence and were more closely linked to the space in which they operate."[261] Steinke claims that Haller "transformed Newton's concept." He believed that forces were "properties of a substance" and that "matter was not passive but active."[262] Thus, forces were part of the physical world, not some immaterial intervention, though they were ultimately unknowable in any causal-ontological sense, because they were not directly accessible via experiment. Humans could discover only empirical regularities of the "outward appearances of nature; the 'inner nature,' the ultimate laws, structures and purposes were known only to God."[263] Still, even Steinke recognizes: "The way of thought of the age was ruled by Newtonianism. This had its impact, too, in Haller's treatise."[264]

Haller never developed a "comprehensive medical system with irritability and sensibility as its cornerstones," Steinke observes.[265] He "advocated the *esprit systematique*, the building of theories based on experience, but he refused the *esprit de système*, which was solely nourished by conjectures."[266] The distinction of *esprit systematique* from *esprit de système*, first articulated by Étienne Bonnot de Condillac (1714–80), was made famous by Jean Le Rond d'Alembert (1717–83) in his *Preliminary Discourse to the Encyclopédie* (1751). As Lester King puts it, "Haller acted as if he were putting into practice the doctrines that Condillac described and advocated."[267] "Haller represented an enlightened empiricism."[268] His "critique of the metaphysical foundation of science is radical," Simone De Angelis argues compellingly.[269] He was therefore obliged to work out a philosophical and methodological alternative. Its core was experimental Newtonianism, "according to which concepts are inferred on the basis of the sensible perception of physical phenomena and constructed in the understanding."[270] This conception "makes [man's] capacities for knowledge altogether dependent upon the anatomical structure of the brain and the psychophysiology of sense perception."[271]

Of course, this raised questions of the adequacy of knowledge: its "warrant [*Evidenz*]" or "certainty." According to De Angelis, it was 'sGravesande who provided Haller with the proper response: 'sGravesande "in his 1736 logic of the empirical sciences formulated a probability and hypothesis theory based on analytical methods derived from Newton that Haller adopted and applied as the methodological vehicle in his physiology."[272] De Angelis, drawing on the prior work of Giambattista Gori, further claims that 'sGravesande's own "constitution of a concept of warrant [*Evidenzbegriffs*] for the domain of the empirical-experimental sciences developed in the frame of a juridical and natural-law context" drawn from Pufendorf and Cumberland, who had themselves evoked the seventeenth-century neurophysiology of figures like Lower and Willis to explain knowledge in terms of human physiology.[273] De Angelis believes there is good evidence of Haller's grasp of this natural-law literature.[274] Moreover, she contends that Haller kept pace with an "anthropological" transition around the middle of the eighteenth century, which moved beyond the sensualist epistemology of Locke, Hume, and Condillac to a much more aggressive focus on "the examination of the material substrate of sense-physiological processes of knowledge" as worked up "at the intersection of medicine, psychology, and theory of knowledge."[275]

The core feature of a "middle position" between dogmatic rationalism and radical skepticism, for Haller, was commitment to physicotheology: embracing both the evidence of nature in confirming the presence of a providential divinity and the converse "guarantee" that such a providential divinity provided for human inquiry into the order of nature.[276] This, De Angelis argues, was already the case with 'sGravesande.[277] "It is decisive that 'sGravesande supplements [Hume] with a deistic underpinning."[278] This confidence in God's purpose in the world was the ground for "moral certainty," the central epistemological posit of 'sGravesande's scientific method, which lent force and promise to the pursuit of knowledge even in the context of human limitations.[279] So, too, "the certainty of knowledge required for Haller a supplemental theological foundation."[280]

HALLER AND THE GERMAN *GELEHRTENSTAND*

More than an experimental physiologist, Haller played a momentous role in the wider German academic world, the *Gelehrtenstand*, as a "universal scholar" whom everyone had to recognize.[281] He set the standard for academic achievement and general learning—hence Johann Gottfried Herder's striking phrase: Haller carried "an alp of learning [*Alpenlast der Gelehrsamkeit*]" on his shoulders.[282] After Haller's death, Johann Friedrich Blumenbach commented: "I think it says a lot—but, as I see it, not too much—when I maintain that Haller was the greatest among all recently deceased scholars who have been working in Europe since Leibniz's death. He was the greatest scholar as concerns variety as well as quantity and depth of his knowledge."[283]

Stephen D'Irsay has argued that Haller should be understood as the paradigm of the pious German Leibnizian metaphysician, in contrast with Voltaire as the paradigm of secular French mechanism.[284] Haller *was* pious, and he worked in Germany (for a time); he was indeed very important for German Enlightenment self-conception. Yet D'Irsay is wrong. To be sure, Haller came to see Voltaire as his essential foe, but not because Haller saw himself as a German or a fortiori as a Leibnizian. It was not the *mechanism* but the *irreligion* of the French *esprits forts* that occasioned Haller's lifelong antagonism. Voltaire epitomized for Haller French *Freigeisterei*. The essential chord here is religious and even, I would like to suggest, more Swiss than German.[285] We must insist with Richard Toellner that "Haller was remarkably circumspect about Leibniz," and it is clear that "Leibniz was and remained alien to Haller. Never did Leibniz gain any influence over Haller's thought."[286] Concretely, Toellner makes clear that "Haller did not come to grips with German rationalism until Göttingen at mid-century, after he had already established his world-view." In fact, "on the basis of experimental science he rejected rationalism in the form of the Leibniz-Wolff metaphysics."[287]

Haller's early poem on the origins of evil, composed in 1734, drew upon the only book Leibniz published in his lifetime, the famous *Theodicy* (1710), but, as Karl Guthke has shown, Haller's annotations upon reading the *Theodicy* were hardly uncritical.[288] Leibniz was certainly not unfamiliar to Haller, and as a student of Bernoulli, Leibniz's key ally, he would have clearly learned of his importance.[289] Yet equally clearly, Haller did not appreciate Leibniz as of the same stature with Newton—or even Bernoulli![290] The issue was crucial to Haller's sense of scientific identity: "Nature did not intend us to be universal. Our life is too short, our mental powers are too limited, our memory is too unreliable."[291] Leibniz seemed *too* universal for Haller, never plunging deeply enough into any one field.[292] Whether this is a fair assessment one can certainly contest, but that it was an important verdict for Haller is nonetheless unequivocal.

Roe agrees with Toellner: "Haller's intellectual home was in fact not Germany but the early Enlightenment of the Netherlands."[293] The best evidence we have of Haller's considered judgment of German school philosophy came in an extended review he published in the *Bibliothèque raisonnée* (1746) of the work of his closest friend on the Göttingen faculty, the philosopher Samuel Christian Hollmann (1696–1787).[294] Hollmann was an outspoken critic of Wolffian rationalism, and Haller supported his view with vigor.[295] In his review, Haller wrote of a "Germany . . . where, led by Christian Wolff, philosophers have taken a step backward."[296] Wolff, like Descartes a century before, believed in the primacy of pure reason. For Haller, this was to indulge in philosophical romances.[297] In the preface to the first volume of the German translation of Buffon (1750), Haller resorted to brutal sarcasm in characterizing the undertakings of such rationalism:

> Almost a hundred years have elapsed in Europe since the explanation of natural events, and the construction of arbitrary systems, became the esteemed priorities of great scholars. Accordingly, once René Descartes put forth the organization and construction of the world by means of mechanism, and had taken the liberty to give whatever figure and motions to the smallest parts of matter he needed for his explanations, all of Europe regarded this constructive power as an inalienable privilege of natural philosophy. Worlds were constructed, elements, vortices and screw-like particles were created, and it was believed that this generally was the best thing to do, when the actual events

> in Nature allowed themselves to be explained only in part through the alleged constructs which had appeared.[298]

In his address to the Göttingen Royal Academy of Sciences in 1751 as well, Haller made clear his allegiance to Francis Bacon and empirical inquiry as against Descartes and the rationalist project.[299] The contempt for Descartes extended as well to Leibniz and to Wolff.[300] For Haller, as for Leonhard Euler, whose views he knew and admired, "Leibniz and Wolff resolve the problems of motion not within the frame of a natural-scientific inquiry" but metaphysically.[301] Haller found their idealism (in the technical sense) irreconcilable with empirical inquiry.[302]

All this is crucial for what shaped the course of emergent life science in Germany. We are now in a position to assess Haller's impact on emergent *German* life science: precisely because by training and disposition Haller was *not* steeped in what some historians of science insist upon viewing as injurious German "metaphysical" impulses. His role in German eighteenth-century thought can help illuminate and delimit the actual character of life science in Germany (and rescue it from its significantly erroneous later reputation). If Haller proved seminal for that emergence, we must think of it in terms other than *Schulphilosophie*.

Yet for Haller personally this raises another crucial question: if it was not German metaphysics that animated his vitriolic attacks on freethinking, whence did that furor arise? The answer is to be found rather in the distinctive Protestantism that Haller carried with him lifelong from his origins in Bern, but which found its most aggressive expression after he returned home from a stint at Göttingen that proved monumental for the German Enlightenment and for German life science.

CHAPTER THREE

Albrecht von Haller as Arbiter of German Medicine: Göttingen and Bern (1736–1777)

Whatever his theoretical self-conception by 1730, Haller had yet to begin building his own scientific career. His Basel teaching and research in anatomy and physiology were based on Newtonian science, and he knew that he would pursue that path. It remained only for him to establish where and how he was going to do it. In the eighteenth century there was no place for a researcher interested in comparative animal physiology except in a medical faculty. That was why in 1736 he decided actively to pursue a position at the new university in Göttingen (Georgia Augusta University), but only after and because he had failed to establish a position for himself in his hometown of Bern.

After Haller returned to Bern from Basel in the summer of 1729, he experienced continual difficulties establishing himself professionally over the better part of seven years (1729–36).[1] As Lester King puts it tersely, in these Bern years Haller "did not receive much recognition."[2] He did give anatomy demonstrations and sought to open an anatomy theater, but he received little support. During 1731, Haller conducted numerous dissections on dogs, cats, and rabbits, especially delving into the physiology of the diaphragm.[3] He also did some microscopic work in botany, though not in zoology. His first outside recognition came in a letter (December 9, 1730) from a Leiden classmate, Georg Nikolaus Stock (1701–53), a contributing writer for the Nuremberg *Commercium Litterarium ad Rei Medicae et Scientiae Naturalis Incrementum Institutae* (1730–45), inviting him to submit material.[4] Starting in 1731, Haller there began publishing short notices of his botanical discoveries in the Swiss highlands.[5] In November 1733 Haller wrote to the managing editor of the *Commercium*, Christoph Jakob Trew (1695–1769), seeking regular publication of his botanical research in the journal, and Trew accepted a series of submissions (sixteen of Haller's nineteen publications before he went to Göttingen appeared in this journal).[6] These proved very strategic for the advancement of Haller's career.

During the autumn and winter of 1732, Haller wrote up his anatomical findings, and in 1733 he published his study *De Musculis Diaphragmatis* . . . (Bern, 1733), "the first independent anatomical work by Haller."[7] Ostensibly on this basis, in December 1733 Haller was named to the Academy of Sciences at Uppsala. Why would the academy in Uppsala take any notice of a minor anatomical essay published in Bern, Switzerland? The probable answer is that Johannes Geßner was already corresponding with Linnaeus and had informed him of his collaborative work with Haller on Swiss botany, a few indications of which were appearing in the *Commercium*. Linnaeus was always keen on creating a network promoting botany.[8] That may be why Uppsala recognized Haller before any other place in Europe did, including his own Bern. In 1734, perhaps because of the publication and Uppsala's recognition, he finally received authorization from the Bern authorities for an anatomical theater, but it became operational only in 1735.[9]

THE ESTABLISHMENT OF THE UNIVERSITY OF GÖTTINGEN

Göttingen proved the perfect locus for Haller to realize his ambitions. To understand that, we need to recall the goals behind founding the new university in 1734. Its key architect and curator for the balance of his life was the eminent privy councillor in Hanover, Gerlach Adolf von Münchhausen (1688–1770). His vision of the new university was that it should "lead neither to atheism nor to naturalism . . . nor introduce enthusiasm or any evangelical papacy."[10] Münchhausen and his colleagues had the example of the recently established University of Halle constantly before them. With its founding in 1694, it had revolutionized university life in the Germanies, but it had also fallen into difficulties that were obvious by 1730. After Christian Wolff's expulsion from Halle in 1723 because of Pietist allegations of his atheism, Halle became notorious for nasty squabbles and dogmatic entrenchment. It was this, above all, that the founders at Göttingen wanted to avoid. Münchhausen saw to it that the university faculty had no right to make appointments; the state administration took control, thus protecting Göttingen from the outset against the "scholarly guild mentality and academic nepotism [*gelehrtem Zunftdenken und akademischen Nepotismus*]" characteristic of other German universities.[11] He further arranged that the theological faculty would have no oversight of the other university faculties. Because "the theological faculty from the very beginning had no recognized right of censure," Rudolf Vierhaus has noted, the founding of the University of Göttingen "brought to an end the epoch of confessional control over universities."[12]

While he was suspicious of the sectarian fanaticism of the Halle Pietists, Münchhausen was equally concerned about Wolff. The latter had emerged as an obvious candidate for a position at the new university in Göttingen. The question was thoroughly discussed, including all the allegations against Wolff—that he "'might be a *naturaliste* or even a Socinian,' ... [perhaps] an atheist of the Spinozist sort."[13] After hearing all sides, Münchhausen decided not to recruit Wolff, though he insisted that his ideas should be taught among others at Göttingen.[14] At the same time, Münchhausen sought "an orthodox-Lutheran but highly irenic theological faculty and one that played a relatively quiet, unassuming role within the university."[15] He had a strong preference for religious moderates, disapproving of anyone "of a sharply etched intellectual profile."[16] The emphasis should be on hard work, teaching excellence, and the avoidance of disputes. Münchhausen was guided in this by the eminent theologian at nearby Helmstedt, Johann Lorenz von Mosheim (1694–1755).[17] In persuading Münchhausen to restrict the theology faculty's authority in the university, Mosheim was seconded by another key adviser, the Hanover royal physician Paul Gottlieb Werlhof (1699–1767).[18]

Of even more direct relevance, Werlhof set the program for the formation of the Göttingen medical school.[19] He argued that a good medical school had to offer training in five areas: anatomy, botany, chemistry, *theoria medica*, and practice.[20] He advised Münchhausen to recruit medical faculty embracing the Boerhaave tradition: "Without all sectarianism, with sufficient physical and mathematical science, [students] might be sent forth into the world by their professors with a more plausible *nexus ratiociniorum de rebus medicis* [rational organization of things medical]. In that manner Mr. Boerhaave in Leiden has generated quite a demand, as have Mr. Hoffmann and his opponent Mr. Stahl at Halle, in a lesser measure. If *antiqua eruditio medica* [knowledge of the classics of medicine] is also present, as in the case of Mr. Boerhaave, so much the better. That is rare, however."[21] Another key adviser regarding the medical school, a Dr. Textor from Karlsruhe, stressed the importance of observation and experience; Halle had made only a modest beginning in that direction, and Göttingen could do more.[22]

Haller's appointment at Göttingen has gained an element of legend in retrospect that the actuality simply will not confirm.[23] Though he turned out to be epoch-making both in his field and for the university, he was hardly an established eminence in medicine when the new university began to construct its faculty. In fact, Haller was most famous as a poet, not a credential that would be conducive to a medical school appointment. He was no obvious candidate in 1736. Hintzsche has this just right: "Haller's actively solicited appointment as professor of anatomy, surgery, and botany presumed some achievements, but these were also hardly so grand as to make this call something automatic."[24] The first call to the chair of anatomy, surgery, and botany at Göttingen went to Johann Wilhelm Albrecht (1703–36),[25] but he died suddenly on January 17, 1736, after only a year at the medical school. Meanwhile, other appointments were made to the medical faculty. Georg Gottlob Richter (1694–1778), a student of Boerhaave's, arrived in 1735 to teach courses in practical medicine, and he was established as the official leader of the medical faculty.[26] A major effort was undertaken to recruit the eminent iatromechanist/iatromathematician Georg Erhard Hamberger (1697–1755) from Jena.[27] He eventually declined but offered in his stead his student, Johann Andreas Segner (1704–77), who came to Göttingen in 1735, taking up the teaching primarily of mathematics, physics, and chemistry.[28] With Albrecht's sudden death, Segner took over most of his ongoing courses, but the need for a replacement was dire.[29] An offer was promptly made to the Tübingen anatomist Burchard David Mauchart (1696–1751), but he declined.

That finally opened the path for Albrecht Haller. He applied for the chair in correspondence with August Johann von Hugo (1686–1760), royal physician in Hanover, who served as adviser to Münchhausen concerning the medical faculty at Göttingen. Earlier, in February 1732, based on reading Haller's first botanical articles in the *Commercium*, Hugo had made contact and invited Haller to exchange botanical resources, and this they did over 1732–33. In January 1736 Hugo let Haller know that the position in medicine at Göttingen had again come open, and Haller then asked for his support in securing the appointment.[30] Hugo actively oversaw the negotiations.[31] Like his close associate Werlhof, he was committed to recruiting Boerhaavians for Göttingen, and Haller fit the bill.

The position, as with its first holder, encompassed the three fields of anatomy, surgery, and botany. It is striking that, eager as Haller was for the position, he nonetheless posed firm demands for the infrastructure of the new medical faculty. Clearly having in mind the resources that had made his own training in Leiden so effective, he demanded the creation of a botanical garden and an anatomical theater, and he succeeded in getting commitments for both of these.[32] He made no similar demands regarding clinical-training facilities; it is clear his concerns were for research resources at Göttingen.

It did not take Haller long to get what he demanded for the medical school, because Münchhausen's faith in Haller quickly became entrenched. Haller "vastly exceeded the expectations that Münchhausen, the father and *spiritus rector* of the new university, had placed in the young Boerhaave-student."[33] Haller fit Münchhausen's ambitions for Göttingen perfectly, and the curator made every effort to accommodate him. "[Haller's] personal satisfaction in Göttingen was a high priority for Münchhausen. He was always highlighting [Haller's] accomplishments on behalf of the university, and he attempted, by expressions of concern and particular accommodations, to fend off the repeatedly threatened departure of [Haller] (1747–53)."[34] When, nonetheless, Haller eventually departed, Münchhausen termed him "irreplaceable."[35]

Haller was never happy at Göttingen, however monumentally productive he proved there.[36] In fact, over the years Haller quarreled with *most* of the Göttingen faculty, including *all* his medical school colleagues.[37] His student Johann Georg Zimmermann (1728–95) made the following notation in his private journal: "The enemies of Mr. Haller have everywhere been of a very great number."[38] Indeed, his authorized biography (1755) made no secret of Haller's many conflicts, and this displeased his mentor.[39] Haller suffered not only professional rivalries but continual personal tribulations, beginning with the death of his first wife as the result of an accident on the day of their arrival in Göttingen. Remarriage brought its own tragedies. Haller's diaries reflected his personal *Angst*, full of psychological and religious storm and stress, of which there will be more to say later.[40]

Within a few years of his appointment, "Göttingen, thanks to Haller, became the foremost center for medical education in Germany," eclipsing not only Halle but even Leiden in its reputation and recruitment.[41] Indeed, "the medical faculty at Göttingen without question deserves to be considered the premier and most outstanding such faculty in Germany in the eighteenth century," Julius Pagel argues, since "at no other university were the natural-scientific disciplines that bore on medicine taught with greater completeness and thoroughness[,] . . . not merely as supportive disciplines . . . but . . . as independent sciences."[42] Ulrich Tröhler makes the same point: "for the first time at a university medicine and practical-experimental natural science were institutionally united."[43] The crucial relation between what in the German eighteenth century was called, generically, *Naturlehre* and the more specifically medical fields like pathology and clinical medicine achieved great clarity in the Göttingen medical faculty, thanks to the specialization of faculty. Segner, for example, concentrated primarily on math and physics.[44] Richter taught practical medicine.[45] Haller worked to create rigorous discipline in the intermediate fields—botany, anatomy, and physiology. They all affirmed the appropriateness of animal study for human medicine.[46] That would be central to Haller's new experimental physiology. Botany, to be sure, was also important: how could it not be in the age of Linnaeus?[47] Notably, Haller's inaugural lecture at Göttingen dealt with how to study botany (without academic supervision!).[48] He was also adamant about creating a substantial botanical garden for the new Göttingen medical school. And his first major scientific publication at Göttingen was in botany, not anatomy.[49] But as he explained to Linnaeus himself in correspondence from the mid-1740s, he elected to turn away from botany to anatomy and physiology.[50] This self-conscious specialization (of one of the "last universal scholars of the West") is decisive.[51]

That Haller was an awesome scholarly presence is unquestionable. But how was he as a teacher? "Haller's fame attracted many students, but his teaching seems not to have been suited for beginners. . . . He rather neglected the younger and more idle students, but took all the more care of the advanced and promising among his pupils."[52] Richard Toellner reports that "Haller's lectures were dry and went completely over the heads of beginners, as is revealed in the reports of his students [in] *Von und über Albrecht von Haller*."[53] In his Ceremonial Lecture for the Göttingen Academy, "the way Haller talked about teaching duties makes evident that he had grown sour about this, on account of how much time he had to spend on it; also, in terms of oral presentations, he wasn't an accomplished or inspiring teacher."[54] In fact, Haller conformed much more effectively to the model of the graduate laboratory director of today. He described his own practice quite clearly:

> As I saw many diligent young people pass through the new academy on their way to achieving the highest honors at the Georgia Augusta, I came to advise the more ambitious among them, each in accord with his talent, to dedicate themselves to a difficult topic in anatomy or a physiological issue, and to work on that project over two winters [academic years]. There were enough specimens; I provided support as much as I could for the efforts of these young people and conducted most of the certainly unbelievably large number of experiments on various animals myself. Thus, since simultaneously many students exerted intense efforts upon single parts of the noble craft, more could be undertaken than I could have set in motion by myself with the most strenuous endeavor.[55]

Haller produced approximately thirty-eight dissertation students.[56] They concentrated primarily on anatomical research, and eight specialized in animal vivisection.[57]

The three most famous were Johann Friedrich Meckel (1724–74), Johann Gottfried Zinn (1727–59), and Zimmermann. Meckel earned his MD in 1748, then joined the Berlin Collegium Medico-Chirugicum around 1750, becoming Haller's key informant on happenings at the Berlin Academy and the court of Frederick II, especially concerning the French in Berlin: La Mettrie, Maupertuis, and Voltaire.[58] Zinn earned his MD in 1749, specializing in the physiology of the sense organs, especially the eye. He went to Berlin upon graduation but returned to Göttingen after 1753, to replace Haller. It was an unhappy arrangement, for he felt oppressed by others on the medical faculty, who may have taken out their resentment of Haller on his prize student.[59] Zimmermann earned his MD in 1751 with a dissertation, "On Irritability" (1751), widely taken to express Haller's own views though there were in fact significant differences. He went on to a successful career as a practicing physician, *Stadtarzt*, and eventually royal physician in Hanover, replacing Werlhof.[60] He was also a well-known author, among whose achievements was the authorized biography of Haller in 1755.

Beyond his own immediate teaching, Haller was inextricably connected with two institutional innovations that consolidated the prominence of the Georgia Augusta University at Göttingen. The first was a new and influential scholarly journal, the *Göttingische Zeitung der gelehrten Sachen*, better known under its revised title *Göttingische Gelehrten Anzeigen* (*GGA*). Established in the mid-1740s, it took off when Haller became its managing editor in 1747. Not only did he direct the journal (1747–53), but he also wrote a great deal of its material. The estimate is that over the years 1747–77, Haller published no fewer that nine thousand reviews in the *GGA*![61] He wrote on the natural sciences, of course, but he also commented on literature, politics, religion, and history—in short, he set his stamp on the whole intellectual universe of Enlightenment Germany. Since a primary goal of the journal was to introduce to Germany the most important achievements of the broader European Enlightenment, Haller's journal and his own reviews represented one of the most authoritative and complete accounts of the intellectual world of the mid- to late eighteenth century.[62] This made Göttingen one of the key centers of the German Enlightenment and Haller one of its decisive arbiters. Above all, the *GGA* gave the world Haller's take on everything.

Perhaps even more important for Göttingen was the second institutional innovation: the Royal Society of Sciences at Göttingen. Already as he was creating the new university, Münchhausen had the idea of establishing a research academy to bolster the reputation of the scholars at Göttingen, but it did not immediately come to fruition. In the late 1740s, as the university's success became clear, the idea was revived. Münchhausen sought Haller's views, and Haller was blunt: he would participate only in an academy that he designed and directed. Münchhausen thereupon turned the whole thing over to Haller. The latter knew exactly what he wanted, and he set the terms for the new academy. First, he wanted the academy to be royally sponsored, on the model of the Paris Academy. Second, the sitting members of the academy should be drawn from the faculty of the Georgia Augusta, and each would be expected every year to contribute a short paper reporting on new research. Haller wanted these contributions to the academy to be separately rewarded. Academy honoraria should supplement established faculty salaries. Finally, he wanted a direct relation between research in the academy and teaching at the university. This was a distinct departure from the relatively clear division of labor established in eighteenth-century Germany between academies, which were research institutes, and universities, in which teaching was a sufficient end in itself.[63]

The combination of these two features, making research integral to university faculty endeavors, was the most important innovation in German higher education, canonized in Wilhelm von Humboldt's University of Berlin in 1810. Göttingen, where Humboldt himself studied, provided a model for that new idea of university education. Haller, in that measure, seems a crucial predecessor of Humboldt. On the other hand, Humboldt's university vision was preeminently humanist, and Haller's academy was clearly not. In his memo to Münchhausen, Haller famously observed: "The taste of the world is not at all disposed toward languages, toward philology, but solely toward mathematics and physics."[64] He made no place for the humanities in his academy, because he did not believe they were in a position to achieve *new* knowledge. That was the distinct precinct of the natural sciences. It was also his view that specialization was absolutely essential for the advancement of scientific knowledge, and therefore, he wanted a program for the academy that would allow—indeed, *force*—scholars to work up manageably narrow research problems for presentation at the academy. The key was that *new* knowledge be presented. Haller recognized the importance of mastering what was already known, but organizing and recounting that was the role he assigned to teaching. Advancement of learning came with new—for Haller primarily with *experimental*—research. "For it is readily apparent how far removed a paper read before the Paris Academy of Sciences is from the lecture in a professor's classroom."[65] As perpetual president of the new Göttingen Academy, starting in 1751, Haller autocratically enforced his vision.[66]

The academy "was expressly intended by its founders to enhance the quality and reputation of the university."[67] And it did become a presence, attesting to the power of Haller's achievement and the execution of his vision. It helped make Göttingen the most important university in Germany and a foremost center for the advancement of the Enlightenment. Already by 1764 the prominent Hanoverian statesman Georg Brandes could observe that the academy "had provided the university the greatest acclaim and given it an advantage over all its sister universities."[68] That was a testimony to Haller's monumental impact. The academy became his vehicle for influencing the German and wider European scholarly world. Even after he departed Göttingen in 1753, he exerted control in absentia via the academy secretary, Johann David Michaelis (1717–91). But the academy would languish without him. His own *Dissertation on the Sensible and Irritable Parts of Animals*, published in 1753, was the high point of its scientific contribution for several decades, and only in the 1770s would a different agenda breathe new life into the academy.

Not surprisingly, as Haller became the dominant figure at the university, he began to receive solicitations from others in the field, seeking positions at Göttingen itself or elsewhere through Haller's patronage. Johann Gottlob Krüger wrote to him from Halle starting in 1747 seeking support for a call to Göttingen or Helmstedt, as it became clear to him that he could not advance at Halle itself.[69] He eventually moved to Helmstedt in 1751, and in his later works he declared himself, not least out of personal gratitude, an enthusiastic supporter of Haller's theories.[70] Even after he left Göttingen, Haller received such requests—for example, from Johann Heinrich Lambert (1725–77), who wrote to Haller in 1758 seeking an appointment in the philosophy department at Göttingen.[71] Haller supported his case from Switzerland, but to no avail. Lambert then continued his career first at the Bavarian Academy of Science (1760–62) and then at the Royal Prussian Academy of Sciences (1765–77), becoming the philosopher Immanuel Kant admired most in all Germany.[72]

HALLER'S RISE TO EUROPEAN PROMINENCE IN PHYSIOLOGY

Haller's first great physiological intervention was his critical edition of Boerhaave's lectures: *Praelectiones Academicae in Proprias Institutiones Rei Me-*

dicae (7 vols., 1739–44): "Haller's edition includes two sets of notes, one explaining Boerhaave's doctrines and one containing Haller's own amplifying and critical remarks."[73] According to Erna Lesky, Haller used this edition as a path to his own mastery.[74] Hubert Steinke makes the same point: "the detailed assessment of Boerhaave's statements was a means of distancing himself from his teacher."[75] Haller's explicit criticisms drew the outraged protest of a Dutch scholar named Noortwyck. Provoked, Haller replied to Noortwyck that Boerhaave's physiological theories frequently lacked anatomical foundation, and he revealed "no less than twenty-one anatomical errors in the first third of Boerhaave's *Institutiones Medicae* (1734) alone."[76] Progress in science, he clearly believed, overrode the constraints of misconceived politeness or uncritical deference to the authority of Boerhaave. Haller believed he could venerate Boerhaave without losing his own independence. In this he was, I suspect, in step with many of Boerhaave's Dutch disciples, and in a sense he was carrying forward publicly an agenda that he had witnessed in Albinus's "private course" in Leiden already in 1725.[77] Indeed, the question arises how much of his entire project vis-à-vis Boerhaave was in emulation of Albinus and grounded, as with the latter, in a better physiological sense derived from richer anatomical research.

Haller was an irascible man, and controversy spanned his entire career.[78] "My fate has determined that—although I always mean the best by other scholars and never seek to demean their names, no matter what they think of me—nevertheless, I must . . . virtually year by year witness the appearance of bitter diatribes against me."[79] Actually, as Lesky has aptly observed: "Haller was not only convinced of the correctness and utility of his agonistic principles, but he saw it rather as a necessary desideratum of a progressive, fruitful and fructifying [program of] research."[80] There is no question he relished "the honor of being in the right."[81] As he observed, "controversy teaches us to select a portion of the field which we will cultivate more assiduously, and if it is disputed, we fortify it rigorously."[82] Each of Haller's many controversies teaches us something very important about the prehistory of biology.

Haller's criticisms of Boerhaave landed him in a controversy with Gerard van Swieten (1700–1772), fellow student of Boerhaave's and author of a commentary on Boerhaave's other great medical work, *Aphorisms* (1742–72).[83] Van Swieten published in the Netherlands, where there was a legal prohibition against reprinting Boerhaave's lectures, based on the latter's protests about pirate reproductions all through Europe. Therefore, Van Swieten had to publish a *commentary*, stressing his own supplements to Boerhaave's spare remarks. Still, Van Swieten, in the early volumes especially, saw fit only to explicate the master's text, never to offer his own ideas. He took great offense at a comparison with Haller's work, made by an anonymous reviewer (Haller himself), that asserted that Van Swieten wrote as a Catholic (with deference) while Haller wrote as a Protestant (with independence).[84] The controversy, which lasted until 1753, began with an angry letter from Haller, dated May 26, 1748, accusing Van Swieten of fomenting his brother-in-law Noortwyck's nasty review in a Dutch periodical.[85] Van Swieten denied that and tried to avoid a public squabble, but it ensued nonetheless.

An even more formidable dispute over Boerhaave's legacy arose with Julien Offray de La Mettrie (1709–51). Worlds different from Haller, but probably Boerhaave's other most famous student, La Mettrie studied with Boerhaave in 1733 and became his foremost advocate in the Francophone community. Initially, he translated Boerhaave's *Institutes* into colloquial French in two volumes (1740). Then, in six volumes (1743–48), he adapted into French Haller's critically annotated edition of Boerhaave—"incorporat[ing] much of the content of Haller's notes" but without crediting Haller. Thus, Haller certainly had grounds for his charges of plagiarism. However, Kathleen Wellman suggests that there was a real difference "in both style and substance. La Mettrie did not simply replicate a disjointed series of Latin footnotes. Instead he provided a connected commentary written in French in a conversational style."[86] In any event, the *formal* differences pale beside the *philosophical* departures. "Boerhaave, in La Mettrie's hands, must be acknowledged as a crucial figure in defining objections to Cartesian physiology and as a primary source for the interpretation of Locke as a materialist and a medical figure."[87] That is, La Mettrie turned Boerhaave into a materialist. Haller was outraged and he charged La Mettrie with a perversion of Boerhaave's ideas. He reviewed La Mettrie's translations savagely in *GGA*.[88] He also reviewed La Mettrie's *Histoire naturelle de l'âme* (1745) in the same vein.[89]

La Mettrie retaliated. "The dedication of *L'homme machine* to Haller was La Mettrie's revenge for this second review."[90] He turned the tables on the plagiarism charge by publicly acclaiming Haller as his source of inspiration, but the dedication was also "an exercise in mockingly exaggerated praise with erotic innuendoes that parody a youthful and well-known poem of Haller's."[91] This mockery sparked Haller's outrage anew. His immediate review of *L'homme machine* (December 1747) in the *GGA* solemnly made known the extreme distaste he felt at finding himself linked in any way to the doctrines expressed in the anonymous volume.[92] "La Mettrie had known perfectly well that his action would undermine the vulnerable position of a man who, in the eyes of all Europe, was one of the most respected personifications of the fusion of scientific eminence and orthodox piety."[93] Two years later, still seething, Haller wrote an open letter to the editors of the *Journal des savants*, March 12, 1749, declaring: "I disavow this book as entirely opposed to my views. I regard its dedication as an affront. . . . I have never had any connection, acquaintance, correspondence, nor friendship with the author of *L'Homme machine*."[94] But overall, "Haller's serious-minded response to La Mettrie's jests caused him to be regarded by an amused public, despite his being the morally injured party, as in some measure deserving of the ridicule he met with."[95]

In terms of the ideas at issue, "Haller . . . was content to describe *L'homme machine* as an 'epitome of the poem of Lucretius . . . merely augmented by a few observations and discoveries of the modern period.'"[96] He insisted that "everything *L'homme machine* had said about irritability had been derived either from his own publications or from certain experiments performed by Bernhard Siegfried Albinus."[97] Even La Mettrie's death did not placate Haller's rage. His Academy Address of 1752 included denunciations of La Mettrie, and it compromised fatally the efforts of Maupertuis on behalf of Frederick II to recruit Haller to the Berlin Academy. In all these contexts, it is notable that Haller's *religious* concerns decisively shaped his academic or scientific ones.

THE CONTROVERSY WITH GEORG ERHARD HAMBERGER

In 1747 Haller published his pathbreaking *First Lines of Physiology*. That work was largely the fruit of Haller's controversy with Georg Erhard Hamberger of Jena. A celebrated "iatrophysicist," Hamberger believed that medicine not only could but should be derived logically from mathematical-physical principles. Thus, he "applied his principles of physics to medicine."[98] Hamberger had a theory about the role of the superior and inferior intracostal muscles in respiration based on the angles of motion of the rib cage. Haller criticized this in a comment in his Boerhaave edition, setting off a decade-long controversy over the relative merits of mathematical derivation and empirical observation in physiology.[99] Somewhat later, Haller noted that, in order to confirm his position, he "was compelled to carry out new experiments and to repeat them often."[100] That is, the controversy triggered his substantial experimental program in the 1740s, especially involving animal vivisection. He had conducted animal experiments starting in the 1730s, but the controversy with Hamberger in the 1740s led him to systematic experimentation.

Based on massive experimental evidence, Haller won the argument with Hamberger, but even more, he redefined the field of physiology.[101] *First Lines* elaborated the "superiority of experimental physiology over mechanical hypotheses," as Haller proclaimed in an article summarizing the controversy in *Nouvelle bibliothèque germanique* in 1748.[102] Moreover, "the controversy on respiration shows Haller's shift between anatomical and physiological experimentation."[103] *First Lines* articulated the two key elements of Haller's *anatomia animata*: functionalism and the essential role of animal experimentation.[104] Anatomy as the doctrine of the structure of the body and physiology as the doctrine of its activities represented for him an indivisible unity.

THE CLIMAX OF HALLER'S GÖTTINGEN YEARS: *DE PARTIBUS CORPORIS HUMANI SENSILIBUS ET IRRITABILIUS*

The most important achievement of Haller's Göttingen years came at the very close, his monumental lecture to the Göttingen Academy, *De Partibus Corporis Humani Sensilibus et Irritabilius*, delivered in April 1752 and published in 1753.[105] It triggered a seismic shock in European physiology and became the point of departure for a new sense of the *Fachgebiet*. The problem that had preoccupied the mechanistic tradition in medicine from the seventeenth century forward was involuntary animal motion, preeminently in the heart.[106] Already in the seventeenth century, experiment had indicated that the heart would keep beating when cut out of some organisms, and that it could be started up beating again by a stimulation even in that condition. This militated against theories of any "spiritual" faculty operating throughout the body. More local explanations seemed called for. That, and the fact that muscle contraction vastly exceeded in its manifest force the physical stimulus that provoked it, suggested that there was something intrinsic at work in these tissues. Thus, Haller's experiments aimed to establish "another motive faculty to be distinguished from elasticity and the *vis nervosa*," that is, from motion associated with strictly physical interaction (elasticity) and motion consciously induced by nervous stimulation (*vis nervosa*) or by "sensibility" in its voluntary, not receptive, capacity.[107] In his *Elementa Physiologiae Corporis Humani* he would call it *vis insita*, but he was careful to avoid any claim that it was an essential property.

Haller's methodological starting point clearly reiterated his established position: "I am persuaded that the greatest cause of error [in physiology] has been that most physicians have made use of few experiments, or none at all, but have substituted analogy instead of experiments."[108] On the one hand, Haller insisted on a drastic austerity concerning his claim that "I only relate facts which I have actually seen."[109] This claim implied virtual muteness regarding theory: "I have never accepted hypothetical speculation and I often wonder why 'Haller's system' is cited, for I said simply that those parts feel or move, which I have seen feeling or moving. . . . It could be that the experiments contain errors, but there is, however, not even the shadow of a hypothesis within the system."[110] On the other hand, the academy lecture did clearly propose what he himself termed a "system," that is, "a new division of the parts of the human body" based on the distinction of the properties of irritability and sensibility.[111] He defined these two properties extremely narrowly: "I call that part of the human body irritable, which become shorter upon being touched," and "I call that a sensible part of the human body, which upon being touched transmits the impression of it to the soul; and in brutes, in whom the existence of a soul is not so clear, I call those parts sensible, the Irritation of which occasions evident signs of pain or disquiet in the animal."[112] His paper to the academy, as he rhetorically organized it, aimed simply to report which body parts exhibited which property under experimental investigation. "From all these experiments collected together it appears, that there is nothing irritable in the animal body but the muscular fibre."[113] Similarly, "we have seen that the sensible parts of the body are the *nerves* themselves, and those to which they are distributed in the greatest abundance."[114] Moreover, those parts of the body that were sensible were not irritable, and conversely.

Beyond this observational claim, Haller was adamant that he meant nothing more: "But the theory, why some parts of the human body are endowed with these properties, while others are not, I shall not at all meddle with. For I am persuaded that the source of both lies concealed beyond the reach of the knife and the microscope, beyond which I do not chuse [*sic*] to hazard many conjectures."[115] Irritability was "a physical cause, hidden in the intimate fabric, and discovered through experiments, which are evidence enough for demonstrating its existence."[116] Haller was a skeptic about ultimate causes. He permitted himself to draw an analogy between irritability and gravitational attraction in Newtonian physics, and he made the argument that an inference from the macroscopic observation to the microscopic level of causation was allowable as analogical conjecture. That is, "experiments have taught us the existence of this property [i.e., irritability], and doubtless it is owing to a physical cause which depends upon the arrangement of the ultimate particles, though the experiments that we can make are too gross to investigate them."[117] The exact ontology did not concern him; Haller "never presented a precise analysis of the relation between force and matter. What was important to him was the disclosure that a force—call it corporeal or physical—was acting *in* the fibre and that there was no need to suppose another force—physical or immaterial—acting from outside upon the fibre."[118] Thus, Haller was willing to conjecture that there might be some ultimate connection between irritability and "gluten," but he would not press that further.[119] Still, this conjecture indicated that Haller favored the "hypothesis" that irritability might be associated with the fundamental chemical form of organic matter, which was called "gluten" in the eighteenth century.[120] However, gluten was a component of *all* organic fibers, and Haller proposed to identify irritability strictly with *muscle* fibers. His recourse, in modern parlance, was to consider irritability an emergent property of a higher level of tissue complexity.

As Steinke aptly phrases it: "the essential mark of his physiology consisted in identifying specific structures and their specific functions."[121] Haller insisted that "only experiments can enable us to define what parts . . . are sensible or irritable," and that his results could be refuted only by contradictory experimental results. He was entirely certain "no such experiments can be produced; nature is too constant."[122] Haller announced at the very outset that he had performed experiments on 190 animals in the last year alone, constantly replicating his experiments to discount any eccentricity of a particular case.[123] One distortion in the experimental procedure that Haller explicitly acknowledged and discounted was the action of powerful chemical reagents: "this corrosive force has nothing in common with life."[124] On the other hand, in his address, Haller did not provide a wealth of detail about the experimental regimen of his work. To be sure, protocols of many of the experiments were maintained, yet even these proved indeterminate at critical junctures, especially for those not actually present at the experiment. Considering some of his technical discussions, elaborate preparation was sometimes required in order to enable the definitive "observation" for which an experiment was intended.[125] Subsequent controversy would reflect this indeterminacy of experimental regimen, as problems of replication and the resulting variability of experimental outcomes plagued the debate.

The timing of Haller's definitive determination of the irritability thesis is important. "In 1750, Haller was not one step closer to his later ideas than he was in 1740."[126] But in November 1750 he embarked on "systematic experimentation" with one goal: to isolate irritability within specific tissues of the organism. Haller kept at it relentlessly, with the collaboration of his graduate students, particularly Zimmermann, until the end of 1751.[127] One possible stimulus for this new precision and renewed intensity of application may have been correspondence with Paul Werlhof, who expressed concern about the scope and intent of the research program.[128] Another potential source of the redoubled effort may well have been competitive publications and reports about progress on "irritability" by colleagues in Leiden in the post-Boerhaave era: such figures as Johannes de Gorter (1689–1762), Hieronymus David Gaub (1705–80), Frederik Winter (1712–60), and Abraham Kaau Boerhaave (1715–58).[129]

While it was irritability Haller wished to grasp, his experiments began with something he conceptualized more clearly: sensibility. "Haller had always considered sensibility as the conscious reception of impressions transmitted through the nerves."[130] Because that seemed so self-evident, "nerves and sensibility did not occupy any considerable position in his research agenda before 1750."[131] Haller's assumption regarding sensibility may have misled his graduate student Zimmermann, whose dissertation tended to blur the differences between sensibility and irritability, leading to inferences strikingly out of keeping with the view Haller himself articulated a year after Zimmermann's dissertation appeared.[132] Zimmermann took his dissertation to Paris and to the Netherlands to present his findings to leading scholars in both communities and make a name for himself, but as was the wont of such texts in that day, many assumed that the dissertation had been written by Haller himself. Meanwhile, a review of the dissertation appeared in *GGA* (by Haller) in 1752, alerting the German community to the key ideas. Thus, all three research communities learned of Haller's investigations of irritability *before* he even gave his Göttingen Academy Address, much less circulated his published text. That only made the discrepancies between Zimmermann and Haller a constant sore point in the ensuing controversy. In his address, Haller went to some lengths to distance himself from his own student's views—not only on experimental details but much more importantly on their theoretical implications. Zimmermann, for example, did not require consciousness in sensible response, as Haller would.[133] More generally, "while Zimmermann envisioned in irritability a newly discovered principle of life, Haller restricted his sphere of activity to muscle fibers capable of contraction."[134] In the ensuing controversy, "Haller [would] find himself in a dilemma: how might he affirm [strict] empiricism to his disciples and even to himself as a convinced Baconian, and yet present himself as the founder of a new system of ideas?"[135]

Haller knew his results were controversial. This motivated his extensively repeated experiments. One result, that tendons were neither irritable nor sensible, flew in the face of the entire medical tradition of the West, and to confirm it Haller expressly stated that he had conducted more than a hundred experiments since 1746. He was utterly confident that "the tendons have no organ neither of motion nor sensation."[136] Moreover, he drew a practical corollary that had enormous significance to surgeons and physicians dealing with wounds: "we need no longer be afraid of wounds of the tendons, of whatever kind they are."[137] This practical inference provoked furor in the surgical community, led by the eminent practitioner Anton de Haen (1704–76) in Vienna.[138] Another controversial outcome was that Haller deemed insensible the dura mater encasing the brain.[139] Further, with overt disdain, Haller dismissed the assertion of Théophile Bordeu (1722–76) that the glands were served with nerves.[140]

His strongest rhetoric targeted the various "animists" and "materialists" who diverged from his own view, generally by reducing the two distinct properties of sensibility and irritability to one fundamental life-force. Among the animists, he singled out Robert Whytt (1714–66), insisting against the latter that sensibility could not be the source of all movement, since irritability was manifest in muscle tissue totally removed from an organism.[141] Haller rejected the idea that there could be unconscious nerve action, the heart of Whytt's alternative physiology.[142] Sensibility simply *meant* conscious awareness, for Haller. "The soul is a being which is conscious of itself."[143] If there were motions in organic tissue that did not arise from conscious volition, as there obviously were, then sensibility could not account for them. "Irritability therefore is independent of the soul and the will."[144]

This, of course, seemed to the partisans of "animism" little short of an endorsement of "materialism." Already, even before Haller published his Academy Lecture, Claude-Nicolas Le Cat (1700–1768) and Heinrich Friedrich von Delius (1720–91) had made such allegations, on the basis of their reading of Zimmermann's dissertation, which they took to be a statement of Haller's own view. Le Cat began correspondence in 1752 with the accusation that Haller's experimental method led to "materialism" or "Spinozism." That deeply offended Haller.[145] Similarly, Delius published *Animadversiones in Doctrinam de Irritabilitate, Tono, Sensatione et Moto Corporis Humani*

(Erlangen, 1752), before the publication of Haller's Academy Address. He took the published dissertation of Zimmermann as Haller's own view and found it suspiciously materialistic. Haller was, of course, most sensitive on this count because of La Mettrie, and he insisted that the latter's "impious system" had been "totally refuted" by the experiments reported in the *Dissertatio*.[146] Sensibility—consciousness—could no more be reduced to irritability as a mechanical phenomenon than irritability could be accounted for by the soul. The *experimental* dualism of Haller's results confirmed a *metaphysical* dualism between the body and the soul.[147] Dualism was at the heart of Haller's lifelong harangue against Georg Ernst Stahl and anyone else who took an "animist" view.

Haller had become the most famous life scientist in Europe by 1753.[148] He would be the most often-cited living physician in the Paris *Encyclopédie*, primarily because of the work of his Göttingen years.[149] In a sense, all roads to biology in Germany over the course of the eighteenth century went through Haller, warranting the historian's particular attention in reconstructing its gestation. The path was through him, but others would sometimes have to *run over him* to get where they were going.[150] Emergent biology needed to go where Haller refused to go.

Toellner has suggested the fruitfulness of applying Thomas Kuhn's notion of a "paradigm shift" to what took place in medicine (specifically, physiology) around the middle of the eighteenth century, via Haller's Göttingen Academy lecture on irritability and sensibility.[151] Toellner argues that Haller, though an adamant advocate of the *mechanist* approach to medicine, triggered what would be a hundred-year period of *vitalist* ascendancy.[152] Of course, Haller's impact needs to be conceived as part of a larger shift "vitalizing nature in the Enlightenment." This shift had its origins as much in the reception of the Leibniz-Newton controversy over "force" in the metaphysics of science, and in French materialist ("Spinozist") thinking about organic life around midcentury (Maupertuis, Buffon, La Mettrie, Diderot), as in Haller alone. Nonetheless, Toellner's stress on Haller is apt.

Haller's ideas about irritability set off a ferocious Europe-wide debate. All the epistemological and methodological presuppositions of the disciplines of anatomy and physiology came under acute reexamination. It was a moment of enormous disciplinary exertion toward self-clarification. Over 140 scholars, writing from some sixty locations, made significant contributions to the debate.[153] It dominated the seven years after Haller's publication, triggering an unprecedented barrage of experiments. Two massive collections documented the debate: *Sulla Insensitività ed Irritabilità Hal-*

leriana: Opuscoli di Vari Autori, edited by Giacinto Fabri (3 vols.; Bologna, 1757–59), and *Mémoires sur la nature sensible et irritable des parties du corps animal*, edited by Albrecht von Haller (4 vols.; Lausanne, 1756–60). Europe had never before seen such a mobilization of study and debate on a topic in life science.

Haller, to be sure, was disconsolate that "irritability" became generalized into a system of explanation. "Irritability is becoming a sect," he wrote his intimate friend Samuel Auguste Tissot (1728–97). "That is not my fault."[154] Perhaps not, but Tissot himself was a major propagator of just that systematic generalization that Haller found too bold. He came very close to proclaiming Haller "the Newton of physiology."[155] Throughout his introduction to the French translation of Haller's *Dissertation on the Sensible and Irritable Parts of Animals*, Tissot stressed the analogy between irritability and gravity.[156] But he also and more importantly elevated Haller to a prominence for European life science that would be confirmed both in the furor of the debate over irritability and in the retrospect of physiology as a *Fachgebiet* within German thought. Tissot's Haller was far more in keeping with his significance for the times than Haller's own estimation of his contribution.

The debate over irritability was remarkably indecisive; both sides produced a body of experimental evidence and theoretical argument, and both sides clung to their respective convictions and found the evidence supporting them sufficient.[157] No unity emerged: there was agreement neither about method nor about result. But we must also attend to the kind of question that physiology by the mid-eighteenth century sought to address and that Haller's theoretical abstemiousness seemed to neglect. Deliberately, he "did not answer the needs for a comprehensive explanation of all physiological processes and the fundamentals of life."[158] Not only was this resistance the outcome of the longstanding division between the "theoretical" orientation of physiology and the "mechanical" orientation of anatomy—the contrast of the "pen" and the "sword" that Andrew Cunningham has so tellingly explored.[159] But Steinke argues that the controversy also "demonstrated the aporia of this new science."[160] "In the mid-eighteenth century, it was still the exception to approach a physiological problem by experimental investigation," Steinke concludes.[161] Thus, "no Haller school of experimental physiology" survived his return to Switzerland in 1753.[162] Indeed, "the rejection of Haller's specific notions . . . has hitherto been underestimated."[163] That is, "Haller's *themes* dominated German physiology of the second half of the eighteenth century, but not his concepts."[164]

One of the key issues posed by the controversy turned out to be whether "*animal* experimentation [was] a valid means of investigating *human* physiology."[165] Haller's critic Anton de Haen in Vienna asserted that "man and animal differed in their whole nature (*tota natura*)."[166] But Haller was very strongly committed to the continuity of animals and humans in physiology. He "even expected that a comparison of the brains of animals with their behavior might be of use for the investigation of the human mental faculties."[167] Above all, Haller insisted that animal vivisection was the essence of physiology as a research science.[168] "The overall amount of animal experiments and their repetition all over Europe was a new phenomenon. No previous debate—not even Harvey's discovery—had provoked such widespread experimental investigation."[169] There was, in fact, no continuous tradition of animal experimentation in the late eighteenth century—except, perhaps, in Italy.[170] In describing the experimental fervor that gripped the country, one commentator wrote of "a kind of general irritation of the whole of Italy [*communis quaedam totius Italiae irritatio*]."[171] Indeed, uniquely in Europe, Italy for the rest of the century continued to pursue this experimental trajectory in the work of its greatest physiologists—Leopoldo Caldani (1725–1818), Felice Fontana (1730–1805), and Lazzaro Spallanzani (1729–99)—to reach a new peak in the famous controversies over "animal electricity" at the turn of the century.[172] Though Haller was convinced that vivisection was the method proper to physiology, the next generation of German physiologists—Johann Friedrich Blumenbach, Carl Friedrich Kielmeyer, and Johann Christian Reil—were not particularly dedicated to animal vivisection.[173]

Notwithstanding all Steinke's pertinent reservations, the practice of physiology was set on a new footing thanks to the engagement with Haller's challenge. If the experimental focus in physiology was not continuous in the aftermath of Haller, this did not betoken the rejection of its findings. Rather, to take up a discrimination introduced by Roselyn Rey, a concern with the larger "animal economy" based on "observation" displaced Haller's exclusive focus on the "experimental" discrimination of types of tissue and their responsiveness to stimuli.[174] This could appear a loss of determinacy, as when Steinke writes: "The laboriously established division of irritability from sensibility via countless experiments got softened (Gaub, Unzer, Bordeu, Barthez), and irritability received new associations of meaning (Reil, Cullen, Brown) or it disappeared into numerous 'newly discovered' life-forces (Wolff, Hunter, Medicus, Blumenbach)."[175] But the powerful objections entered by the French vitalists, notably the insistence of Henri Fouquet on a greater holism in the assessment of animal function, and the shift over the eighteenth century from the question of *movement* to the question of *sensibility* as the defining issue for a grasp of *life* as a scientific problem, should not be seen simply as a regression.[176]

HALLER'S SWISS YEARS (1753–1777)

Haller abandoned Göttingen in 1753. His restiveness had clearly begun much earlier, however. He received many invitations to take up a chair at other universities, but none tempted him. Six calls came from Halle alone (for some of which Leonhard Euler, no less, served as key negotiator).[177] Haller received two calls from Utrecht, one from Oxford (1747), and one from the Berlin Academy. Frederick II wrote to Maupertuis in 1749: "je vous donne carte blanche pour Haller."[178] Negotiations concerning the Berlin offer extended through the early 1750s and are documented in Haller's correspondence with Maupertuis and others.[179] Eventually, Maupertuis tired of Haller's equivocations and forced Haller to make a decision. Haller opted not to accept, offering two reasons: first, his family's future required a Bern connection, and second, he had promised Münchhausen and the English king that he would stay at Göttingen five more years after his ennoblement, which occurred in 1749.[180] Of course, these reasons should have kept Haller from nibbling at the offer in the first place.

In any event, it seems the ennoblement of 1749 did serve sufficiently as "golden handcuffs" to keep him at Göttingen a few more years. There was also something to the claim that Bern was important. He seems to have had a desperate need to be welcomed into the Bern elite. "What seething unrest must have driven Haller. Apparently the one thing he wanted more than anything in the world was recognition in his own community."[181] In 1745 Haller had been elected in absentia to the "Great Council" of Bern: perhaps a token gesture in light of local political unrest about elite domination.[182] In any event, he was thrilled, and his Bern correspondents urged him to return and take up a political career. Haller set his hat for a position on the "Small Council," the actual ruling body of Bern. He never made it. "Haller strove—largely in vain—to become 'accepted' by the best Bern society and to join the true ruling class."[183] The best he got was an administrative post at Roche, 1758–64, managing the saltworks.

His return to Switzerland imposed significant limitations on his research resources: he had no dedicated anatomical theater and no large supply of specimens.[184] Nonetheless, he did complete a monumental work in experimental embryology, *Development of the Chicken Heart* (1758).[185] This became the pivotal text for the reassertion of preformation theory in embryology, propagated by his associates Charles Bonnet and Lazzaro Spallanzani.[186] The composition of *Elementa Physiologiae Corporis Humani* (8 vols., 1757–66), his masterpiece, entailed exercise of the "pen," not the "sword," to apply Cunningham's distinction.[187] Later, in the 1770s, Haller composed *massive* bibliographical guides: for example, *Bibliotheca Anatomica* (Zurich, 1774–77), two volumes, 1,700 pages, discussing 7,000 authors from all times and places having to do with anatomy.[188] A central feature of the last phase of Haller's career was massive correspondence with his followers and friends, his "network of networks."[189] Notably, Haller corresponded with a substantial group of Italians. In 1745 Haller opened correspondence with Giovanni Battista Morgagni (1682–1771; professor at the University of Bologna until 1707 and then at the University of Padua from 1711 to 1771), and Morgagni became a key intermediary to Italians.[190] Haller viewed Morgagni as the founder of pathological physiology; they never quarreled because the Italian suppressed any differences.[191] Ignazio Somis (1718–93; professor at Turin) proved another major intermediary for Haller in Italy and was a staunch defender of Haller's irritability there (187 letters between 1754 and 1777). Carlo Allioni (1728–1804; professor at Turin) became a key middleman especially in botany; their correspondence spanned from 1757 to 1776.[192]

Renato Mazzolini has made the interesting conjecture that the Italians, because they were not caught up in the metaphysical issues that preoccupied the Germans, proved particularly receptive to Haller's experimental redirection of physiology.[193] The three most important experimental physiologists in Italy in the late eighteenth century were prominent in Haller's network. Correspondence with Caldani (professor at the University of Padua after 1765) began in 1756, with 125 letters to Haller and 99 from him.[194] Fontana (professor at the University of Pisa and the University of Florence) began corresponding with Haller in 1759.[195] Spallanzani (professor at the University of Reggio Emilia, 1757–63, then Pavia, 1763–99) began corresponding with Haller in 1765; they were staunch allies on preformation but conflicted over blood circulation, especially in 1776.[196]

The most controversial question concerning Haller's return to Switzerland has to do with the relation of his piety to his science. After his return, Haller took up a prominent role in what might be termed a *Protestant* Enlightenment: "a circle of scientists [*Naturforscher*], theologians, literati . . . primarily ensconced in Calvinist cultures . . . in Switzerland, England, the Netherlands, Göttingen, and Berlin," who saw themselves called to campaign against "materialists and atheists."[197] The Genevan Charles Bonnet orchestrated his association with Haller from the outset as an anti-

philosophe crusade, targeting especially Buffon as "Epicurean."[198] The program is captured in the title of an important study by Mazzolini and Roe: *Science against the Unbelievers*.[199]

In the 1930s Margarete Hochdorfer composed an important monograph on the tension between Haller's religiosity and his science.[200] In response, Richard Toellner mounted a very impressive case for the coherence of Haller's life and work.[201] Karl Guthke was not convinced, however.[202] Scholarship has proliferated since, yet the problem of Haller and religion has not been resolved.[203] Perhaps it is a matter of asking the right historical question. Toellner suggested that twentieth-century scholarship had blown out of proper historical proportion Haller's intimate religious anxieties, reconstructing them into a conflict of principle between his science and his religious faith.[204] From the adamantly secular vantage of twentieth-century reception, it seemed important to infer that Haller, the scientist, could never accept the dogmas of positive religion, attesting to the incompatibility between science and religion.[205] Toellner insisted this was bad history regarding Haller personally and even a misunderstanding of the relation between science and religion in principle.[206]

The secularizing interpretation pounced upon a notorious scandal that arose at Haller's deathbed, when he announced to the circle of eminent and religious figures attending him that he had never been able to believe in the dogmas of his faith.[207] That "confession" created a culture shock in the German intellectual community. Some simply denied that the reported confession took place. Others took grim satisfaction in the contradiction that it ostensibly revealed in Haller's character, rendering him a thoroughgoing hypocrite because he had long been alleged to be hyperorthodox.[208] Clearly, it seemed a deep blow to the hope that scientific practice and religious piety could be embraced simultaneously: Haller had been in many ways the symbol of this fusion for his times, and he had explicitly praised his great predecessors Newton and Boerhaave for preserving just this unity.[209] To rescue Haller's piety in the wake of the deathbed confession, a Bern scholar, Johann Georg Heinzmann, had immediately undertaken to publish materials from Haller's personal diaries that attested to his spiritual devotion. The whole work was entitled *Tagebücher seiner Beobachtungen über Schriftsteller und über sich selbst*, but the most important material was gathered in a section in the second volume of the publication, "Fragmente religiöser Empfindungen." The editor selected material from manuscripts that were in the family's possession at Haller's death and that have not survived, so that scholarship cannot confirm the principles of selection or retrieve omissions that might bear upon the meaning of the whole. Even as it stood, the publication left a substantial gap in the record, from 1747 to 1772, with the editor's excuse that material in this interval had been written in English![210] Such philological blemishes acknowledged, it remains clear from these documents and others that Haller wrestled with spiritual issues his entire life, and in similar terms.

Accordingly, it makes no sense to conceive of any fundamental shift in Haller's inner spiritual life or to think that his scientific work induced an estrangement from his religious origins.[211] In that measure, Toellner has successfully debunked the secularist projection of twentieth-century scholarship. Especially in his more recent essays, Toeller richly elaborates the authentically Calvinist salvation anxiety that marked Haller's spiritual struggles. The doubts and self-accusations that throng in Haller's diary entries, always intended for his private use only, marked vividly the sort of spiritual fear and trembling that, in the Calvinist tradition, a Christian in a condition of sin had to feel in clinging to the hope of God's redeeming grace.[212] Precisely what such Christians should never presume was their worthiness of salvation, even as it was their most desperate hope. Faith was precisely the struggle to overcome the intransigence of sin and the "hardness of heart" that came of entanglement in the world. Haller knew himself for a sinner and a man of relentless ambition; the world was very much with him, and he felt his waywardness from God and the peril for his soul. He constantly bewailed the recalcitrance of his "heart" to both his reason and his faith.[213] Toellner argues correctly that this was hardly a *lack* of Christian piety but rather its essential character in the Calvinist tradition. In a word, we have no grounds to doubt Haller as a Christian believer. Toellner is correct about Haller's anguished personal piety.

Yet, important as that is, it is not the right historical question. Where Toellner falls short, I contend, is in the historical interpretation of Haller's *public* religiosity, which was marked by a polemicism that can only be construed as dogmatic and even reactionary. As his friend and Swiss compatriot Johann Georg Sulzer privately observed, there was something "childish" about Haller's public religious professions.[214] Others viewed them still less sympathetically. His former student Zimmermann had come to loathe what he saw as his teacher's turn to "hyperorthodoxy" and even went so far as to blame it on opium addiction.[215] The young Goethe, as a spokesman of the new generation of the Sturm und Drang in their year of dominion over the book review section of the *Frankfurter Allgemeine Zeitung* in 1772, made it quite clear how out of pace Haller's public religiosity was with the more advanced sentiments of the day, and not just in France.[216] Haller liked to profess that Protestantism was a religious orientation of tolerance, and he celebrated the Dutch Republic generally and the stance of his great mentor, Boerhaave, more particularly as the embodiment of this spirit.[217] Yet he reveled in a complacent juxtaposition of two extremes—the backward Catholic Church, sunk in superstition (*Aberglauben*), versus modern freethinking (*Unglauben*), verging on atheism—each, in opposite ways, threatening the stability of state and society.[218] Only a Protestant culture could guarantee good order. He saw himself and his coreligionists as the golden mean between and, above all, *against* these two extremes.

That *did* have an impact—both on his personal science and on the epoch. However authentic Haller's religiosity, it became a force that impeded important impulses toward modernity in the science and the culture of his time. In that context, I suggest Haller's return to Switzerland in 1753 *did* exacerbate just this public and negative dimension of Haller's religiosity. It has long been contended that his outrage over La Mettrie and materialism triggered a change in the tone of his writings both on science and on culture by 1750.[219] Clearly, he accentuated these issues in an important preface he composed in 1751 for the translation of a French text already targeting freethinking as socially dangerous.[220] I believe his new alliance with Charles Bonnet, which only commenced in the Swiss years, exerted particular influence.[221] Another solidification of ideological agenda can be associated with the arrival in Bern of Fortunato Bartolomeo De Felice (1723–89), who was instrumental both in propagating Haller's views on science and religion and in creating the institutional culmination of the whole impulse, the *Encyclopédie d'Yverdon*. In 1757 De Felice published a major statement of the experimental Newtonian tradition as he understood it from 'sGravesande and Haller: *De Newtoniana Attractione Unica Cohaerentia Naturalis Causa Dissertatio*.[222] He went on to become the founding editor of the *Encyclopédie d'Yverdon*. As Marita Hübner notes, "The *Encyclopédie d'Yverdon* was the most important Protestant successor [*Folgewerk*] of the *Encyclopédie* of Diderot and d'Alembert."[223] More than a successor, it was unquestionably conceived and received as a *rejoinder* to the Paris *Encyclopédie*, a rearguard campaign against French vital materialism, offering a piously Protestant version of Enlightenment, with reference in particular to natural philosophy.[224] Haller was a central figure in the Yverdon project with his many contributions to the *Encyclopédie*.[225]

Not even Bonnet would prove orthodox enough for Haller, almost ruining their crucial alliance, which they salvaged only by desisting from discussion of their religious differences for the sake of their common cause.[226]

This polemical turn in Haller is manifest in the enormous amount of published work that he devoted to the attack on his religious adversaries in the last decade of his life: the three volumes defending revelation and attacking freethinking.[227] It was already manifest in his increasingly explicit concerns over Buffon's latent Epicureanism in the 1750s, eagerly fomented by Bonnet; in his celebratory review of Reimarus's work on natural religion as a German riposte to French materialism; and in the religious twist to his debate with Caspar Friedrich Wolff over preformation and epigenesis—all of which will be the object of our consideration in a later chapter. Shirley Roe concludes that Haller self-consciously *restricted* science so that it could not conflict with religion: "Where there is a danger of scientific theories forming a basis for materialism and atheism they must be rejected." Roe calls this, aptly, "science within the limits of religion."[228] The Kantian resonances are particularly apposite. That was *exactly* what Haller (and, I would argue, Kant as well) endeavored to uphold. But the whole position proved untenable, both intrinsically and historically.

CHAPTER FOUR

French Vital Materialism

Might one not willingly believe that there has never been but one original animal, prototype of all the animals in which nature has done nothing more than elongate, shorten, transform, multiply, obliterate certain organs?

DIDEROT, 1753[1]

The notion of "French materialism" in the eighteenth century is hardly new.[2] But perhaps the modifier "vital" might make more salient the decisive feature of this materialism as it informed the development of life science in France and thus help us to understand the contours of the German reception of this critical phenomenon.[3] I wish to argue that a quite specific French challenge at midcentury decisively inflected the course of the German life sciences thereafter. This chapter will articulate this French challenge, and the next will take up the German figures who were provoked or inspired by this bracing new way of thinking.

The most sustained historical analysis of this paradigm shift is the work of Peter Hanns Reill. In an extensive series of essays, followed by a synthesizing monograph, Reill has offered a persuasive account of the new scientific model of the mid-eighteenth century.[4] In an essay suggestively titled "Between Mechanism and Hermeticism," Reill demonstrates the complex strategy of the new approach to avoid both dead ends of reduction to strict mechanism and of appeal to the "occult qualities" of the Neoplatonic/Hermetic tradition and to endeavor instead "to create a middle realm with its own inherent structure." Thus, "the principles of mechanical natural philosophy which had supported Cartesian, Leibnizian, Wolffian and early Newtonian science were attacked in the name of a reanimated nature filled with matter imbued with active, vital forces." Reill illuminates how centrally a metaphysical commitment—mind-body dualism—had figured in the mechanistic model, and how the new science needed to challenge its premises. "The new mid-eighteenth century alternative to mechanism denied both of the principle [*sic*] characteristics of matter ascribed to it by the mechanists, namely its inert nature and its aggregative composition." Since "none of the postulated active forces could be seen directly, nor could they be [directly] measured," Reill notes, "a new form of reasoning was demanded," a kind of semiotics: "comparative, functional analysis and analogical reasoning." The scientists of the second half of the eighteenth century had "a strong commitment to close observation," and their real contributions in physics came not in theory but in experimentation: they "designed elaborate experiments and instruments to register what was not immediately perceptible."[5]

In a very short span of years around 1750, French naturalists—Pierre-Louis Moreau de Maupertuis, Georges-Louis Leclerc de Buffon, Denis Diderot, and Julien Offray de La Mettrie—opened the way for a new "vital materialism" in the life sciences. This theoretical mutation began in the early 1740s, led by Maupertuis and Buffon, who were in close contact at the time. Shortly before departing France for his new post as president of the Berlin Academy, Maupertuis published (anonymously) various versions of his provocative *Vénus physique* (1746).[6] Around the same time, Buffon composed the essay that would eventually become the "Preliminary Discourse" to his *Histoire naturelle, générale et particulière* in 1749.[7] Together they challenged the adequacy of the mechanist paradigm for the life sciences. La Mettrie had been even more provocative, with his *Histoire naturelle de l'âme* (1745) earning him exile from France to Holland, then with *L'homme machine* (1747), which earned him exile from Holland to Frederick II's Berlin.[8] Diderot made the most dramatic synthesis of these impulses in his *De l'interprétation de la nature* (1753). But the decisive breakthrough came with the publication of the first three volumes of Buffon's monumental *Histoire naturelle* in 1749.[9]

Meanwhile, parallel developments were taking place more specifically in physiology at the medical school in Montpellier, led by Paul Joseph Barthez (1734–1806) and Théophile Bordeu (1722–76).[10] Provoked by Albrecht von Haller's 1752 address to the Göttingen Academy on irritability and sensibility, "Bordeu and the other doctors from the medical school at Montpellier criticized the distinction that Haller made between irritability and sensibility. Bordeu claimed that all living matter was sensible and that irritability was only a special case of sensibility."[11] Barthez and Bordeu brought about a decisive shift in medical thought at Montpellier from the "iatromechanical" to the "vitalist" orientation.[12] After Bordeu moved to Paris in 1751 and joined the circle of the *Encyclopédie*, the two French currents merged powerfully. Diderot gave theoretical expression to the whole program in his *Pen-*

sées sur l'interprétation de la nature (1753) and brilliant literary evocation a decade later in his *Rêve de d'Alembert* (composed in the summer of 1769).[13]

Central to this revisionism was a reflection on the animal-human boundary, especially the ancient issue of the "animal soul."[14] One salient moment was the controversy between Étienne Bonnot de Condillac and Buffon over animal nature, but this was a transnational impulse of the European Enlightenment, not unique to France.[15] In the Scottish Enlightenment, Robert Whytt (1714–66) published a major study on the character of animal functions and their neural connection in 1751.[16] It was followed by *A Comparative View of the State and Faculties of Man with Those of the Animal World* (1765), by the Scottish philosophical physician John Gregory (1724–73).[17] The work of the very important Hunter brothers, William (1718–83) and John (1728–93), belongs squarely in this domain.[18] We will follow the specifically German inflection of this matter in detail in later chapters.

PIERRE-LOUIS MOREAU DE MAUPERTUIS AND THE *VÉNUS PHYSIQUE*

The first major intervention of the new approach to life science came in an anonymous text by Pierre-Louis Moreau de Maupertuis, *Dissertation physique sur l'occasion du Nègre blanc* (1744). Its ostensible stimulus was the appearance that year in the Paris salons of an albino black child. The actual stimulus, scholars have established, was a longstanding debate within the Paris Academy of Sciences over the origin of "monsters"—that is, congenital deformity.[19] And behind that debate stood the whole quandary of preexistence in embryology. By the mid-eighteenth century, the theory of generation had become the flash point both for natural science and for physicotheology. All the methodological issues of a naturalistic epistemology for experimental inquiry, on the one hand, and all the metaphysical issues of a credible natural religion, on the other, assumed maximum acuity just there. When Maupertuis transformed his first endeavor into a longer book a year later, he published it under the title *Vénus physique*—clearly a rhetorical sally addressing his argument not to the Academy of Sciences but to the salon and the public sphere.[20] In his alienation from the academy he was seeking a new role as *esprit fort*, writing for the public sphere in the manner of Voltaire.[21] But that did not mean his work was not intended as a serious scientific challenge. It was, and it succeeded in drawing considerable attention. The original edition of *Vénus physique* went through three printings by 1750, and with significant shifts in content.[22]

The thrust of the *Vénus physique* was, as Michael Hoffheimer has contended, primarily critical, not constructive.[23] It claimed that the available empirical evidence could not support *any* version of preformation theory: "neither ovism nor animalculism could reasonably be accepted."[24] Evidence from heredity and from embryological observation compelled a return to the ancient view that the fetus arose from the mixture of the seminal fluids of both sexes. The idea of *emboîtement*—the complete preexistence of germs from the moment of divine creation of the world—had to be replaced by the idea that Harvey had enunciated, *epigenesis*: that is, that "organs . . . are progressively formed from, or emerge from, an originally undifferentiated, homogeneous [material]," in the terms of C. U. M. Smith.[25] The two claims that observation and experience warranted—biparental contribution to and epigenetic development of the embryo—left a great explanatory void: *by what mechanism* did these phenomena take place? "Could the ordinary laws of motion suffice, or should we call upon new forces for help?"[26] Maupertuis noted that already in chemistry Étienne-François Geoffroy (1672–1731) had argued that chemical reactions demanded a theory that surpassed the impact-transmission of force. Geoffroy proposed *affinity* (*rapport*) as the essential principle. Maupertuis accepted this.[27] Indeed, already in physics a departure from strict mechanism had become necessary: the recognition of *attraction* ("action at a distance") as a fundamental force.[28] "Why should not a cohesive force, if it exists in Nature, have a role in the formation of animal bodies? . . . [I]f these particular particles had a special attraction for those which are to be their immediate neighbors in the animal body, this would lead to the formation of the fetus."[29] Maupertuis speculated that in the parental seminal fluids there were elementary particles that were keyed to particular parts of the parental organisms. In the mixture of parental seminal fluids at conception, these particles were "attracted" to one another in some naturally lawful manner to form the same parts in the offspring.

Maupertuis supposed that in the plethora of elementary particles floating about in the seminal fluids, most derived from the immediate parent, but there were others of more remote heritage or altered by circumstance. "Chance or a shortage of family traits will at times cause variant combinations," he wrote.[30] Moreover, "I do not exclude the possible influence of climate and food. . . . I simply do not know how far this kind of influence of climate and food may go after many centuries."[31] Yet drastic variants, left to nature, tended to degenerate and vanish and Nature returned to her original patterns: "after a few generations, or even in the next generation, the original species will regain its strength."[32] Such a theory, Maupertuis argued, could handle all the anomalies that had accumulated so fatally in the theory of preformation: evidence of inheritance of traits from both parents, hybrid animals, the emergence of "monsters," and so on.[33] Maupertuis turned from the level of individual generation to the question of heredity across generations, and therewith to the question of variation in species. It was clear from human breeding of animals that "Nature holds the source of all these varieties, but chance and art sets them going."[34] Breeders produced plants and animals that "did not exist in nature. At first they were individual freaks, but art and repeated generations turned them into new species."[35]

Such were the conjectures Maupertuis felt prepared to offer in 1745. Buffon took them up and celebrated them in the first volumes of his *Histoire naturelle* (1749). His comment is worth citation:

> The general difficulties common to both systems [ovism and animalculism] have been seen by a man of spirit who seemed to me to have reasoned better than all those who have written before him on this matter; I mean the author of the *Vénus physique*, printed in 1745; this treatise, although very brief, gathers together more philosophic ideas than there are in many large volumes on generation; as this book is in everyone's hands [!], I will not analyze it; it is not susceptible of analysis; the precision with which it is written does not permit an extract; all that I may say is that one will find there the general views that do not differ greatly from the ideas which I have given and that this author is the first who has begun to approach the truth from which we were farther than ever since eggs were imagined and spermatic animals were discovered.[36]

Maupertuis welcomed the alliance in his 1752 "Lettre sur l'organisation des animaux," writing that Buffon "believes that each part of the body of members of both sexes furnished organic molecules, of which the seminal fluids are reservoirs, and that these molecules arrange themselves after the mixture of fluids and join together by attractions in the internal mold."[37] Thus, Maupertuis acknowledged Buffon as the leader of the new direction in life science.

DENIS DIDEROT, *MÉDECIN-PHILOSOPHE* AND *SPINOSISTE*

"We are on the threshold of a great revolution in the sciences," Denis Diderot wrote in 1753.[38] While his anticipation exceeded the actuality, a significant paradigm shift *was* at hand, and his philosophical vision did give it distinguished expression. Diderot's crucial text, *Pensées sur l'interprétation de la nature* (1753; rev. ed., 1754), was a manifesto for the paradigm shift to vital materialism in the mid-eighteenth century.[39] Microscopy had revealed a thriving world of life at the most elementary level, which Buffon and the Francophone eighteenth century termed the "molecule."[40] Moreover, experimental observation had generated stunning anomalies that made established theories of generation (preformation) seem problematic: Abraham Trembley's polyp (1741), Charles Bonnet's parthenogenesis (1741).[41] John Turberville Needham (1713–81) even persuaded many that he had experimental evidence reviving ostensibly discredited "spontaneous generation."[42]

Diderot's *Pensées sur l'interprétation de la nature* attacked the preponderance of the mathematical (and mechanist) paradigm in natural science, in favor of a *physique expérimentale* that was emerging in the new fields of "chemistry, physiology, and biology."[43] We can most effectively contextualize Diderot's impact by taking up a curious mistake in ascription by the eminent German theologian from Halle Sigmund Baumgarten in 1745, when he identified the author of the anonymously published *Natural History of the Soul* as "the physician [Denis] Diderot."[44] Baumgarten was wrong, of course, on two counts: the book was not by Diderot (but by La Mettrie), and Diderot was no physician. Yet the mistake was thoroughly comprehensible because the book *sounded like* Diderot, and Diderot *sounded like* a physician.[45] He was shortly to translate into French one of the most imposing tomes of English medicine with the aplomb that only a medical doctor should have possessed.[46] More decisively, for my purposes, Diderot believed that medicine offered a distinctly privileged entrée into the key philosophical issues of his day. "It is very hard to think cogently about metaphysics or ethics without being an anatomist, a naturalist, a physiologist, and a physician," he asserted.[47] That is, philosophy could best conduct its affairs under the rubric—or in the (dis)guise—of medicine. He adopted, in a word, the persona of a *médecin-philosophe*.[48] La Mettrie made the persona of the *médecin-philosophe* an unequivocal stance for "radical Enlightenment."[49] His bold provocations made urgent the methodological and the metaphysical issues of the paradigm shift to a "vital materialism." Thus, Baumgarten had very sound instincts in identifying Diderot with La Mettrie.

From his *Pensées philosophiques* (1744) to his *Lettres sur les aveugles* (1749), Diderot famously made the passage "from deist to atheist"—and landed in the prison of Vincennes for his pains.[50] In prison, Diderot set himself to the systematic study of the just-published first three volumes of Buffon's *Histoire naturelle*.[51] His detailed prison notes on the text were confiscated and subsequently lost, but a reprise of his understanding of Buffon is to be found in the article "Animal" in the *Encyclopédie*.[52] Perhaps the most important and irreversible turning point for Diderot in the *Lettres sur les aveugles* was the determination that natural science should be the ultimate test of the validity of thought.[53] Diderot firmly believed philosophy was undertaking a most important task for the liberation of science—"to find an order in nature that does not come from God"—hence, "science ought to respond to questions of metaphysics."[54] While not a practicing naturalist, Diderot read naturalist and especially medical publications carefully, especially after 1749. He was, in fact, a *philosopher* of natural science, and what he did publish, especially the *Pensées sur l'interprétation de la nature* and his articles in the *Encyclopédie*, suffice to situate him at the cutting edge of the *philosophy* of natural inquiry of his day.[55]

The transition from prison at Vincennes back to the ambitious editorial venture of the *Encyclopédie* entailed very close collaboration with the leading mathematical physicist of the day, Jean Le Rond d'Alembert (1717–83).[56] The latter's *Preliminary Discourse to the Encyclopédie* (1751) constituted a major statement of Enlightenment philosophy and natural inquiry.[57] Yet it stood in stark contrast to the one Diderot discerned in the writings of Buffon. A crucial passage from the introduction to d'Alembert's *Traité de dynamique* (1758) asserted a definite hierarchy in cognitive worth, setting rational mechanics, with its a priori mathematical formalism, above mere experimental *physique*.[58] D'Alembert was quite convinced of the intellectual superiority of mathematical physics, but he faced significant opposition, especially as Buffon outlined the alternative path of "experimental physics" and brilliantly embodied it in his own work in natural history. Diderot clearly preferred Buffon's program. "A new domain seemed to have been won for science which required a fresh, direct contact with things, new methods for their investigation, and which promised the discovery of the concrete 'individual qualities of things.'"[59]

In criticizing the "mathematical" method, Diderot faulted its perfect abstraction, which bore little resemblance to the actualities of observational science when they were "brought back to earth." There was about the mathematical natural sciences a penchant for reason "to consult itself instead of nature."[60] Diderot proposed instead to extend the idea of "interpretation"—a philological or hermeneutic approach—to the natural sciences. To that end, Diderot invoked the precedent of Bacon, taking his title from Bacon's *Interpretatio Naturae*.[61] At the same time, Diderot evoked *génie*, suggesting that gifted experimentalists simply had a knack for knowing which trail to follow, "an inexplicable intuitive awareness of the workings of nature."[62] Diderot made a place for hypothesis in empirical science that was neither illicitly metaphysical (as Newton claimed) nor transcendentally a priori (as Kant would claim) but simply imaginative.[63] One had to find a place in the methodology of science for "imagination, analogy, and every individual creative and inventive faculty."[64] That was why Diderot used the term "interpretation." He envisioned an empirical inquiry that not only fit the endeavors of "the newly rising branches of natural and medical science" on the basis of a "concept of experimentation which made the older one seem less 'concrete'" but assimilated those endeavors to the "humanistic" ones involved in the "science of man."[65]

This is what Sergio Moravia has termed the "epistemological liberalization" of the Enlightenment.[66] Diderot's *Pensées sur l'interprétation de la nature* offers one of the clearest statements of the methodological program of this "liberalized" new empiricism of the eighteenth century. Notably, he distinguished between the mere *observateur* (the alleged Baconian "empiric") and the *interprète*, for the latter sought general principles behind the phenomena. Dieckmann pulls this together in a summation: "If one now links the conception of the *interprète*, who by his conjectures transcends the endless dependence of one phenomenon upon the other and arrives at a determining cause, to the conception of the *genius of experimental science*, who alone is capable of creative conjectures, we seem to have in the *interpreter of nature* the scientist Diderot expected for the new investigation of things outlined in the *Interprétation*."[67] This reconstruction tallies closely with the representation of the scientific culture regarding objectivity in that era in Lorraine Daston and Peter Galison's study.[68]

Diderot's fundamental theoretical ambition was coherence: a unified view of nature. Continuity was indispensable, theoretically: "the absolute independence of one single fact is incompatible with the idea of the whole; and without the idea of the whole, there can be no [natural] philosophy."[69] Diderot's inspiration came largely from Buffon and Maupertuis and perhaps, less congenially, from La Mettrie.[70] He also made explicit reference to the experimental achievements of Trembley and Bonnet.[71] From Buffon and Maupertuis Diderot explicitly drew his grand hypothesis: "It appears that nature is disposed to vary the same mechanism in an infinity of different ways. It never abandons a type of product until it has proliferated its instances into all possible forms."[72] In particular, this led Diderot to conjectures about organic forms. "Might one not willingly believe that there has never been but one original animal, prototype of all the animals in which nature has done nothing more than elongate, shorten, transform, multiply, obliterate certain organs?"[73] Thus, continuity prevailed: "one sees the successive metamorphoses of the accoutrements [*envelope*] of the prototype, as it once had been, bring one domain into proximity to another by insensible degrees."[74] Indeed, Diderot advanced to the idea that not only in the kingdom of the living but throughout all nature "there has never been but one prototypical being for all beings."[75] "For Diderot there was no difference between the organic and the inorganic except in the degree of organization. His whole world was dynamic. The universe was a great animal, and it was also one enormous elastic body conserving *vis viva*."[76] Diderot wrote to Charles Pinot Duclos (1704–72) in 1765: "Sensibility is a universal property of matter, a property that lies inert in inanimate objects [but one] that becomes active in the same objects by their assimilation into living animal substance."[77]

Diderot noted that Maupertuis, in his *Système de la nature*, virtually endorsed this idea, while Buffon seemed to find it too bold. His own disposition was to drive Maupertuis's tentative conjecture to its radical conclusion, despite the hesitations of both his predecessors. His vision for the revolutionary new science was that it should establish one unifying force encompassing all the new forces that "experimental physics" had discovered.[78] So he pressed, under the disguise of religious distress, to unearth the ultimate significance of Maupertuis's argument.

> It is here that we are surprised that the author . . . has not seen the terrible consequences of his hypothesis. . . . I would ask him accordingly whether the universe, or the general collectivity of all the sensible and thinking molecules, forms a whole or not. . . . If he agrees that it is a whole . . . as a result of this universal copulation, [then] the world, like a great animal, has a soul; . . . this world soul—I don't say *is* but *might be*—an infinite system of perceptions and . . . the world might be God.[79]

That was blatant "Spinozism" as the age understood it.[80] Diderot was imputing it to Maupertuis left-handedly, and equally underhandedly embracing it himself, a tactic not at all lost on Maupertuis, as his rejoinder subsequently reflected, nor lost on the age or on our reception.[81] As the editor of Diderot's text observes, *both* Maupertuis and Diderot were part of "that neoSpinozist movement which propelled the [entire] century."[82]

This raises a second, striking aspect of Sigmund Baumgarten's mistaken identification of Diderot with La Mettrie. He employed (this time, aptly) a momentous epithet: "Spinozism."[83] Indeed, notwithstanding Baumgarten's anxieties, Spinoza's metaphysics had become a theoretical resource for the articulation of a more subtle and dynamic materialism. Diderot and La Mettrie did not simply replicate Spinoza's metaphysics; indeed, one can query the depth of their reading of the seventeenth-century metaphysician by 1750.[84] La Mettrie wrestled with the allegation of "Spinozism" in the introduction to his collected works (1751), noting that he had never seriously studied Spinoza and that if they had come to the same conclusions, it was by entirely different routes.[85] Rather, contemporary developments in life science and medicine had a decisive bearing on what he (and Diderot) chose to make of "Spinozism."

In short, "Spinozism" was a pejorative hurled at, but then—much more interestingly—defiantly *embraced* by, the "materialists" of the new life science.[86] If they were "Spinozists," they were of a quite different stripe: "not abstract speculators; they [we]re scientists; taking their point of departure from precise experiments in embryogenesis and animal physiology, they profess[ed] to have found in matter itself the laws which preside over the origin and development of life."[87] The thrust of this new Spinozism was to "refashion a monism more in accordance with the findings of science."[88] The new Spinozism found its most succinct and decisive formulation in Diderot's remarkable little article "Spinosiste" in the *Encyclopédie* (vol. 5; 1765). The entirety of that text deserves citation:

> One should not confuse the old *Spinozists* with the modern *Spinozists*. The general principle of the latter is that matter is sensible, which they demonstrate by the development of the egg, an inert body that, by the sole instrumentality of gradual warming, passes into the state of being sensitive and alive, and by the growth of every animal, which at the outset is nothing but a point and which, by the nutritive assimilation of plants—in a word, of all substances that provide nutrition—emerges as a large, sensitive, and living body occupying a large space. From this they conclude that there is nothing but matter and that it suffices to explain everything. For the rest, they follow the older Spinozism in all its consequences.[89]

Diderot was responding, in short, to a new breakthrough in science. The novelty in this Spinozism was twofold: a vitalism entailed in the proposition that all matter was sensible and a stress on organic growth. Far from Spinoza's "geometric" mechanism, these thinkers emphasized an immanent creativity in nature that mechanism simply could not account for. It was Leibniz, not Spinoza, among the traditional metaphysicians, who offered them the most conceptual scope.[90] As Paul Vernière classically summarizes it, through Leibniz, "nature as a whole appeared like a vast living organism." Thus, "to the geometric monism of Spinoza succeeds a vitalist monism dominated by the idea of nature."[91]

These impulses achieved their most brilliant articulation in 1769 in Diderot's *Rêve de d'Alembert* (D'Alembert's dream).[92] The arrival of Théophile Bordeu in Paris and their association in the *Encyclopédie* project brought Diderot into intimacy with the "most modern philosophical physicians" and the breakthroughs of the school of Montpellier.[93] He drew the best minds of French medicine—most from Montpellier—to compose articles on life science for the *Encyclopédie*. "Without the physicians of Montpellier—Bordeu, Fouquet, Ménuret de Chambaud—Diderot would surely not have written the *Rêve de d'Alembert*," Jacques Roger comments.[94]

In *D'Alembert's Dream*, Diderot fictionalized Bordeu to expound the new ideas of vital materialism.[95] Bordeu's role was to "demonstrate the connection between the 'systematic philosophy' of the dreamer and the most solid observations of science."[96] As the character "Diderot" asserts in the initial conversation, "We have no more idea of what [animals] have been in the past than we have of what they will become. The imperceptible worm wriggling in the mire is probably on its way to becoming a large animal; the huge beast whose size terrifies us is perhaps on its way to becoming a worm. Perhaps they are each a momentary production of this planet and peculiar to it."[97] "Who knows whether everything is not tending to degenerate into the same great, inert, motionless sediment? And who knows how long this inertia will last? Who knows what species may once again evolve from such a huge mass of sensitive and living particles? . . . [E]ach possibility exists equally, they merely depend upon the motion and various properties of matter. . . . You have two great phenomena: the transition from a state of inertia to one of sensitivity, and spontaneous generation; let them satisfy you."[98]

The thrust of *D'Alembert's Dream* was materialism. But at the same time it was a work of vitalism. This vitalism, however, did not entail any specifically progressive developmental element.[99] Instead, Diderot affirmed a perennial spontaneous generation. As Roger puts it, "the universe of Diderot is one gigantic game of dice, where everything is determined, where nothing is known, where the dice themselves change form in the course of the game."[100] Yet Diderot remained a *vitalist*, sharply differentiating himself from reductive materialism, whether in the crude form of Helvétius or even in the formulation of his close friend, Baron d'Holbach.[101] Such vital materialism found its most important articulation not in the latter's *Système de la nature* but in Buffon's monumental *Histoire naturelle, générale et particulière*.

THE "BUFFONIAN REVOLUTION"

The first three volumes of Buffon's *Histoire naturelle, générale et particulière*, the most important publication in natural history of the eighteenth century, appeared in 1749. Buffon's "Preliminary Discourse" to the *Histoire na-*

turelle, entitled "On the Manner of Studying and Expounding Natural History," probably composed as early as 1744, has frequently been compared with Descartes's famous *Discours de la méthode*, in the sense that it was self-consciously offered as foundational for the scientific practice of the times. It contained the most thoughtful and important formulation of the vison of natural inquiry at the foundation of the paradigm shift associated with vital materialism. Indeed, Phillip Sloan has argued that it is appropriate to think of a "Buffonian revolution."[102] That idea will guide this discussion.

The first volumes of *Histoire naturelle* appeared in the aftermath of the ferocious "wars" of the 1730s and 1740s within the Paris Academy of Sciences between Newtonianism and Cartesianism, in which Buffon unequivocally aligned himself with the Newtonian faction.[103] He began his career as a mathematician, and his approach to natural history was grounded in his earlier methodological and philosophical studies in mathematics, probability theory, and epistemology.[104] One of his earliest achievements came as the translator and commentator of Newton's *Fluxions* (1740). As a correspondent of the Swiss mathematician Gabriel Cramer, the editor of the papers of Bernoulli and Leibniz, he was exposed to the most advanced mathematics of the age.

Buffon's particular inflection of "Newtonianism" was sharply at odds with the predominant, mathematical reception of Newton in France. As was evident in his first major publication, the translation of Stephen Hales's *Vegetable Statics* (1727; Fr. trans., 1735), one of the most adamant texts of English experimental Newtonianism, Buffon adopted this Anglo-Dutch approach far more wholeheartedly than other French interpreters of Newton.[105] As he put it in the preface to his translation of Hales, the work was excellent precisely because it was "only experiment and observation," and "experiments must be sought and systems feared." To be sure, the goal of natural science was systematic knowledge, but "the system of nature perhaps is dependent on several principles . . . unknown to us, and their combination is [also] unknown."[106] That was not what the Paris Academy wanted to hear—not even French Newtonians. As J. B. Shank puts it aptly, "the other French Newtonians inside the Royal Academy[,] Maupertuis and Clairaut, for example, were anything but English-style experimental philosophers."[107]

In this context it is useful to consider Thierry Hoquet's rather reckless essay "History without Time: Buffon's Natural History as a Nonmathematical Physique."[108] Let me specify why I have termed Hoquet's essay "reckless." First and foremost is his insistence that Buffon was not a Newtonian. In the abstract of his essay Hoquet asserts: "Buffon never claimed to be a Newtonian and should not be considered as such."[109] In the body of the essay he elaborates: "if Buffon himself is expected to be a Newtonian, his *physique* should be mathematical."[110] Hence, for Hoquet, *physique* without math signified Buffon without Newton.[111] This is a shockingly crude appreciation of the meaning of "Newtonianism" in the eighteenth century, on whose *diversity* there is by now a vast and compelling literature.[112] It ignores grossly the tradition that began with Newton's *Opticks*, the prime vein for a nonmathematical *physique*, as many, starting with Cohen's *Franklin and Newton*, have amply demonstrated.[113] Above all, Hoquet's view excludes peremptorily the very idea of an *experimental Newtonianism*, which, particularly as Shirley Roe developed it with specific reference to Albrecht von Haller, can be discriminated as perhaps the most important development in the natural sciences in the eighteenth century.[114] I take this *European* impulse as salient and, most pertinently, Buffon's project as an exemplary instance, fostering a veritable paradigm shift in the second half of the eighteenth century. Experimental Newtonianism in continental European thought is just what Hoquet terms "nonmathematical *physique*." Hoquet himself registers an ensemble of experimental Newtonians—'sGravesande, Cramer, Buffon, Musschenbroek—and this discursive community is central to my reconstruction.[115] He makes another important point in suggesting that the "Epicurean" tradition provided significant resources for this nonmathematical *physique*. At the same time, as he makes clear, this brought with it a very dangerous ideological connotation of "materialism" and "irreligion."[116]

So, what was nonmathematical *physique*? Hoquet gets this exactly right: nonmathematical *physique*, in the French eighteenth-century context, was essentially the concern to identify and explicate "the number of forces operating in nature."[117] Above all, nonmathematical *physique* was intended to be "real *physique*," that is, an *explanation* of nature, not merely a catalog of its objects.[118] "The avowed goal of Buffon's *Histoire naturelle* was to present a real natural philosophy[,] . . . 'to raise [natural history] to something greater.'"[119] Buffon's preface to the *Histoire naturelle* called for an *interpretation of nature*, as Diderot properly understood and embraced.[120] By elevating the place of *interpretation* or *theory construction* in natural history, Buffon wanted both to *redefine* natural history and to *diversify* natural philosophy. He insisted that natural history not be content to remain merely a "science of describing" but rather take on the aspect of *interpretation*, to become *physique* (i.e., explanatory science), by aggressively pursuing aspects of induction and analogy that Bacon had adumbrated in the *Novum Organum*. Natural history should become part of natural philosophy, or more precisely a natural science *of its own*: "we must try to raise ourselves to something greater and still more worthy of our efforts, namely: the combination of observations, the generalization of facts, linking them together by the power of analogies, and the effort to arrive at a high degree of knowledge."[121]

Buffon did not propose *reducing* natural history or experimental philosophy to mathematical physics. On the contrary, "the true goal of experimental physics is . . . to experiment with all things which we are not able to measure by mathematics, all the effects of which we do not yet know the causes, and all the properties whose circumstances we do not know. That alone can lead us to new discoveries, whereas the demonstration of mathematical effects will never show us anything except what we already know."[122] By arguing that mathematical truths were "only truths of definition," Buffon drew a drastic line between mental models and physical actuality. "Mathematical truth is thus reduced to the identity of ideas, and has nothing of the real about it."[123] While it was elegant and self-evident, mathematical reasoning remained abstract, and "there is no more in that science than what we have put into it."[124] This was to break radically from a dominant vein of natural philosophy emanating out of the seventeenth century, whether Cartesian or Newtonian, which embraced the mathematical approach.[125]

One of Buffon's crucial interventions was to claim that the sciences were utterly plural; there could be no one all-encompassing *physique*. Everything then hinged on how natural philosophers parceled out the domains of inquiry and theorized their delimited claims. The division between mathematical and nonmathematical *physique* cannot simply be mapped onto that between natural philosophy and natural history. There were crucial overlaps. That is why I find more promise in the notion of *experimental Newtonianism*. Buffon's *scope* for his version of natural history was quite ample vis-à-vis traditional natural history. He concerned himself with the formation of the planets, the development of the earth, and the history of life-forms, opening a (quite wide) space for nonmathematical *physique*. But he aimed, to say it again, to make of this a distinct field of *natural philosophy*.

Hoquet notes that the title of Buffon's work defied expectations; readers were expecting a *catalog* of the Jardin du Roi; instead they got *natural philosophy*.[126] Yet Buffon called his text "natural history," so Hoquet accuses Buffon of "cheating" by using that title.[127] (Reckless, again!) Hoquet asks why Buffon "disguised" his project: why didn't he simply call the work "natural philosophy," since that was what he intended?[128] The question is in a measure *mal posée*. Instead, we should note the immediate and persistent allegation that Buffon was *not entitled* to be a natural historian.[129] He was disparaged as either (or simultaneously) a speculative (and/)or a popularizing "stylist."[130] This effort to write Buffon off as an entertainer, a stylist, but not a researcher of substance, began with the first reviews of the *Histoire naturelle*. These alleged that Buffon wrote for the salons, for the ladies: that valets found him more interesting than scholars did. This was simply not true, in terms of either the reception or the intention of the work, and its monumental success gave the contention the lie.[131]

The adoption of experimental philosophy in explicit opposition to mathematical physics was the decisive theoretical basis upon which the autonomy of the life sciences had to be founded. Buffon was central in promoting a "fundamental break of this group of sciences, and of philosophy more generally, away from the physicists' paradigm of seventeenth-century science." That is, "quality, process, historicity and concreteness are elevated in [Buffon's view] above mathematical abstraction, quantification, mechanism and rigorous, deductive analysis."[132]

For Buffon this methodological approach brought up crucial issues in the conception of natural inquiry. "The manner of properly conducting one's mind in the sciences is yet to be found," he argued, and that raised "the most delicate and the most important point in the study of the sciences: to know how to distinguish what is real in a subject from what we arbitrarily put there in considering it."[133] What natural science sought was "physical truths," and these always "depend only on facts," that is, on observation and experience. "One goes from definition to definition in the abstract sciences, but one proceeds from observation to observation in the real sciences."[134] To be sure, this science of the concrete or real was limited. While one might presume "there is a kind of order and uniformity throughout nature," Buffon observed, "all that is given to us is to perceive certain particular effects, to compare these with each other, to combine them, and, finally, to recognize therein more of an order appropriate to our own nature"—that is, in conformity with our own finite rationality.[135] We are "obliged to admit that causes are and always will be unknown to us, because our senses, themselves being the effect of causes of which we have no knowledge, can give us ideas only of effects and never of causes."[136] At best we can attain "only a probability." However, "frequent repetition and an uninterrupted succession of the same occurrences constitute the essence of [physical] truth," a "probability so great that it is equivalent to certitude."[137]

Buffon insisted that it would be "impossible to establish one general system, one perfect method, not only for the whole of natural history, but even for one of its branches."[138] This very ambition was a "metaphysical error," because "nature proceeds by unknown gradations, and, consequently, it is impossible to describe her with full accuracy by such divisions, since she passes from one species to another, and often from one genus to another, by inseparable nuances."[139] The artificial regularities of human logics, and the determinacy of the categories of any human ordering schematic, cannot map accurately the prolific dispersion realized by nature. Working from one or even several indicative parts of any natural entity will always fail to capture all the nuances compounded in such a natural product or exhaust its relations with all others. Thus, Buffon concluded, classification schemes "should be used only as systems of artificial signs[,] . . . only arbitrary connections," serviceable for finite human discourse.[140]

Articulating this shifting conception at the point of maximal perplexity between a conceptual and a causal continuity, Buffon challenged the abstract classificatory scheme of Linnaeus, demanding a methodologically more defensible empirical connection to actual particulars. He disparaged Linnaeus unremittingly for the formalism of his classification systems, and especially for building his system on one specific feature of plant life, the system of reproduction.[141] Then he went on: "the systems which have been devised for animals are even more defective than the systems of botany."[142] By contrast, he asserted, a "true method" would "involve the complete description and exact history of each particular thing."[143]

Buffon proposed that an empirical approximation of his ideal would begin with the detailed description of the particulars of a given organism—not only its external features but also its internal functions, though he added that "it would be foreign to the purposes of natural history to enter into very detailed anatomical examination," which should be reserved to a separate treatment in "comparative anatomy."[144] After such a detailed description, Buffon continued, a proper natural history must offer a *history*—not only of the individual organism but of its species. This practice, as executed in the *Histoire naturelle,* would be "general and particular" as the title had it: *systematic* (i.e., aimed at comprehensive coverage) but not *system building* (i.e., explaining everything from a single principle or set of principles).[145] In this systematic approach, he indicated that there were some obvious and essential procedural principles. First, he argued, there must be a discrimination between organic and inorganic matter. Second, there must be a discrimination of plants from animals. This engendered the traditional tripartite schema of mineral, vegetable, animal, which natural histories had employed since antiquity. Buffon then suggested that *environment* constituted the next obvious ordering principle: earth, air, and water established habitats whereby organisms were differentiated. But Buffon was not interested in any abstract schema; he insisted that the perspective of the inquiry was always human. Man was the measure insofar as proximity—both spatial and morphological—and usefulness to the human drove the interest of inquiry. For Buffon, natural history was always from the vantage of man and for the purposes of man. More, he came increasingly to believe that man could and needed to intervene in nature to impose order beyond what it provided in itself.[146]

Buffon maintained that "the question [of biological species] is ontological and not simply criterial," in the words of Phillip Sloan.[147] Buffon "required a literal material continuity of forms in relation of true generation of like by like in historical time."[148] That is, Buffon's entire enterprise was to "achieve some kind of immanent, connected understanding of [actual] phenomena,"and "[t]he recurrence of the empirical particular organism, perpetuating itself by the 'eternal round' of generation, could satisfy, at least qualitatively, the necessary conditions for this calculus of physical truth."[149] He wrote: "it is possible to descend by almost imperceptible degrees from the most perfect creature to the most formless matter; from the most perfectly organized animal to the most inert [*brut*] matter."[150] Essentially, Sloan infers, Buffon represented an "opening to true historicity in the concept of species."[151] Accordingly, "whereas preformationism had rendered the relations of organisms purely occasional, Buffon's theory, relying on the immanent continuity of the *moules intérieures* and the *molécules organiques*, required a literal material continuity of forms in relation of true generation of like by like in historical time."[152] Thus, "natural history" shifted from a descriptive-classificatory pursuit to a genealogical-causal one. The "law of continuity" in its Leibnizian derivation, and the even older notion of the "great chain of being," came to be reinterpreted in terms of a dynamism that was altogether new. Arthur Lovejoy calls it the "temporalization of the great chain of being," the shift from a mere classificatory schema to a postulation of relation and development.[153]

We best do justice to the "Buffonian revolution" if we link it not only with experimental Newtonianism and nonmathematical *physique* but with vital materialism, or adamant *naturalism* in inquiry into nature. Particular formulations in Buffon expressed his own personal style and cultural preferences, but he was articulating views congenial to a new range of naturalist practices that would emerge as special sciences with methodologies and theoretical concerns of their own. Lyon and Sloan offer a rich conspectus of such emerging fields: "Studies of plant and animal distribution patterns in relation to geographical change, historical cosmology and geology, the renewed interest in the descriptive study of the stages of embryological development, the conceptualization of comparative anatomy, studies on plant and animal hybridization, all begin to take shape as discrete disciplines, narrowing their focus, sharpening their inquiries."[154] Taken together, these led to the gestation of biology over the balance of the eighteenth century, especially as German thinkers elaborated them.

THE "WAR" OVER NATURAL HISTORY: BUFFON VERSUS RÉAUMUR

Buffon clearly aimed his title and his text against the prevailing form of natural history in France: the practices of René-Antoine Ferchault de Réaumur and his school.[155] He despised the thorough imbrication of traditional natural history with physicotheology, as in the work of Abbé Noël-Antoine Pluche (1688–1761).[156] He meant to extricate natural history—and, all the more, natural philosophy—from this preoccupation with "insect theology," that is, the marvels of the invertebrate world as token of divine providence.[157] His audience got that clearly. Réaumur's key disciple, Joseph Lelarge de Lignac, made that the centerpiece of his denunciation, *Lettres à un Amériquain* (Letters to an American; 1751ff.).[158] It was also a central provocation for Charles Bonnet and Abraham Trembley, two of Réaumur's true believers. They all saw Buffon as "Epicurean," in the sense of materialistic-atheistic, but also in the sense of the despised "spirit of system."[159] By contrast, just what they decried Diderot and the party of the *Encyclopédie* celebrated.[160] Diderot adopted the phrase "interpretation of nature" from Bacon to describe and advocate the new nonmathematical *physique* he identified with Buffon.

The tension between Réaumur and Buffon is well known, yet it bears revisiting, if only to set the stage for what followed.[161] They represented competing visions for the future of natural history.[162] Their rivalry generated a remarkable theoretical disputation between their two "parties" concerning its character and purpose, and those of natural philosophy more generally.[163] In the public sphere, this was complemented somewhat farcically by the scandalous pronouncements of La Mettrie and the outraged protestations of Haller.[164] Mediated significantly by Maupertuis, on the one side, and the Swiss pair of Haller and Bonnet, on the other, the factional war would have a decisive influence on the self-conception of German life science for the second half of the eighteenth century.

What neither faction disputed was the revolutionary significance of the experimental discoveries of Trembley and Bonnet for life science.[165] While the experiments of Buffon and John Needham stirred intense interest as well, for a time their infelicities put a significant damper on the epigenetic movement with which they were identified.[166] But no one could dispute the polyp and its staggering implications.[167] The extravagant language in which the Paris Academy couched its announcement of Trembley's discovery, replete with grandiose classical allusions and a sense of radical rupture, gives a sense of the amazement that was felt, extending well beyond expert scientific circles.[168] The widely debated question of the "animal-plant" originally stimulated Trembley's investigation of the unobtrusive little green hydra in his windowsill specimen jar.[169] He had presumed, especially because of their color, that these were plants.[170] But observation of their movement around the jar toward the light began to draw him to the opposite conclusion. Trembley concluded that the "polyp" *was* simply an animal, and perhaps the simplest of animals.[171] But he then established experimentally that these animals had capacities for regeneration, reproduction, and hybridity that surpassed any notion of animality currently under consideration. What made Trembley's experimental observations epochal was their *theoretical* implications.[172] Georges Cuvier later claimed that Trembley "changed, so to speak, all the ideas that had been entertained about the physiological anatomy of animals."[173]

What might have been a gracious generational transfer took on the edge of contestation because Buffon did not hesitate to employ the resources of the new public sphere of the *philosophes*, including ridicule, to achieve the displacement of what Réaumur represented.[174] Buffon wished to promote a very different agenda for natural history. He pressed for an unrelenting *naturalism*—that is, the exclusion of "the moral, the theological and the metaphysical" from natural-scientific practices. Above all, one had to "get beyond 'the hand of God' as much as possible" in conducting scientific inquiry.[175] At the same time, Buffon pressed for the elevation of natural history into a natural science—an explanatory and generalizing set of knowledge claims. In both regards, he took Réaumur to stand for everything he wished to dispel.

Still under the sway of an exclusively *descriptive* conception of its practice and presented with unheard-of novelties crying out for such description by the revelation of the microscopic world of life-forms, natural history from 1650 to 1750 maintained as its consummate ideal simply to recognize and characterize with austere precision the features and the behaviors of these new organisms. Moreover, a Baconian fervor to avoid leaping to premature generalizations—epitomized in the Newtonian conceit of "feigning no hypotheses"—disparaged the failings of *l'esprit de système* (Cartesianism) in many facets of natural science. Hoquet has given us a rich backdrop for this conflict by juxtaposing the approaches of Claude Perrault (1613–88) and Pierre-Sylvain Régis (1632–1707) at the close of the seventeenth century.[176] While Perrault represented the epistemological modesty of the Baconian program and thus "historical" knowledge, the patient accumulation of empirical data, Régis represented the "spirit of system," the ambition for a comprehensive theoretical structure of knowledge, a unified system of physics, inspired by Descartes. As Hoquet observes, "The dichotomy between the historical and the philosophical translated into the two competing literary forms: the memoir (or essay) and the system."[177]

When, with the first volumes of *Histoire naturelle* in 1749, Buffon made to seize the leadership of natural history in France, Réaumur mobilized his party to attack Buffon with all the vehemence and scientific authority he could muster. Réaumur identified entirely with the stance of abstemious empiricism, and he and his followers construed any departure from that stance as a fall into the pernicious "spirit of system." They propounded an almost blind denunciation of any generalization or hypothetical synthesis as "system"—that is, as excessive speculation or imaginative projection.[178] Pervasive within this adamance was a strong commitment to a *religious* function for science: *physicotheology*, or the discovery of God's hand everywhere and as the only possible explanation for observed data. This was a centerpiece of Réaumur's approach, and it animated the denunciations of Buffon and the *philosophes* as "Epicureans" by his spokesman, Lelarge de Lignac. Réaumur, Bonnet, and their camp indiscriminately lumped Buffon with the Encyclopedists and all of them with "materialism," indeed atheism.[179] As Hoquet aptly observes, de Lignac "addresses neither the set of texts of Buffon, nor those of the historical Epicurus, but rather lumps together a cast of figures in which Buffon is detached from all his orthodox professions and where Epicurus becomes the representative of a physics without God, governed by no Providence."[180] In their fierce antipathy, the disciples of Réaumur construed Buffon and Diderot as "Epicureans," just as they construed the *philosophes* as a coterie of *esprits forts* deeply threatening to the traditional fabric of French culture. These partisans welcomed the successful parody by Charles Palissot, *Les philosophes* of 1760, and they embraced the slur *cacouac* to characterize these *philosophes*.[181] This partisan rhetoric proved highly congenial to the Swiss alliance of Haller and Bonnet as well, with important implications for the German reception of the dispute.

Rhetorical disparagement of all theoretical induction became the hobbyhorse of the Réaumur school of natural history, and the key *target* was Buffon's approach. In these circles the "art of observation" in fact came to be invoked *against* Buffon. In a famous essay submitted originally to a prize competition sponsored by the Haarlem Society of the Sciences, the Genevan Jean Senebier (1742–1809) celebrated Charles Bonnet as his model for the "art of observation," construing this as the preeminent style of inquiry for the naturalist of the eighteenth century.[182] Bonnet himself had proposed the competition topic ("How the art of observation might contribute to the perfection of the mind") upon his election to the Haarlem society.[183] Senebier had been inspired by Bonnet's writings on the art of observation starting with the publication of the latter's *Essai analytique* in 1760, and he regarded Bonnet's *Considérations sur les corps organisés* (1762) as the epitome of such an approach.[184] In 1782 Senebier wrote to Bonnet of his admiration: "It was in studying you that I learned how to read from nature, and to paint her, but in studying you I despaired forever of resembling you; you depict [literally, "use a brush"] as only a Buffon, and you share alone with Réaumur and Spallanzani the art of seeing well."[185] For Bonnet and for Senebier, "a scholar like Réaumur understood how to bridle premature enthusiasms and to uphold the primacy of experience and the patience of observation against the coryphaeus of natural philosophy, of *biophilosophy*."[186] The observer, Senebier wrote, "regards Nature as a book, in which it is necessary to attempt to read the characters with rigor without presuming to imagine what signification they ought ultimately to have."[187]

In the key preface to his first major intervention in natural history, the *Traité d'insectologie* (1745), Bonnet appealed to the famous quarrel of the ancients and the moderns, siding emphatically with the moderns and their "renewal of philosophy."[188] In his *Considérations sur les corps organisés* he proclaimed: "How many of the marvels unknown to the sage [Augustine] and to the ancients have not our instruments and our methods revealed to us!"[189] Bonnet waxed rhapsodic about the advancement of science in his day: "how many forces, properties, modifications of matter have revealed themselves to our senses, to our understanding!"[190] In his major works, Bonnet expounded the experimental identity he encoded in the phrase "the art of observation." In "Réflexions sur les progrès de l'histoire naturelle," chapter 8 of part 9 of his *Contemplation de la nature* (1764), Bonnet reconstructed a modern tradition of experimental science. "Consider the rapid progress of natural history in the last thirty years," he proclaimed. The discipline had languished for centuries before being "awakened by the voice of *Redi*, animated by those of *Malpighi* and *Swammerdam*, sustained, encouraged, excited by that of *Vallisnieri*, of *Réaumur*, [and] it has dispelled the night of chaos, crushing the ignorance, error, prejudice that like so many monsters guarded the approaches to nature."[191] Above all, Réaumur, "the ornament of France and of his century," stood as the epitome of the "art of observation," Bonnet proclaimed.[192] "I had until [reading Réaumur] observed only by instinct; M. de Réaumur saw to it that reflection took the place of instinct: he taught me to see and made me an observer."[193]

Bonnet claimed that this progress resulted from breaking the grip of premature generalization through the rigorous appeal to observation and experience. "The philosophical naturalist ought, above all, to insist upon the exceptions to the rules that are taken to be general."[194] Naturalists had been too tempted "to judge the unknown by the known and to restrict nature within the narrow limits of actual knowledge."[195] But the known was not a good guide to the unknown.[196] "*Analogy*, which is one of the torches of natural science, cannot dissipate all the shadows . . . [and] one is left to grope along with the fingertips of experience."[197] Thus, Bonnet pronounced his key methodological commitment: "natural history is the better logic."[198] "The path of observation ought always to be preferred as more reliable."[199] "The course of a Réaumur, of a Trembley, tells us more than [philosophers of] the likes of [Pierre] Nicole or [Christian] Wolff."[200] Thus, "the truths of natural science, the fruits of observation and experiment, multiply and perfect themselves without end."[201] By contrast, "the spirit of system gives birth sometimes to theories that one then forces experiments to confirm. Our century has furnished us with celebrated examples."[202] Bonnet enunciated the fundamental conviction of the party of Réaumur in a simple disparagement of "those daring geniuses . . . who invent theories before they have made observations."[203]

Writing after Réaumur's death in 1757, Bonnet celebrated "that great observer whom I will always hold so fondly in my memory."[204] To Haller, in that same moment, Bonnet observed of Réaumur: "there was never a man who carried to a higher degree the spirit of observation."[205] Of Haller himself, especially after his elaborate experiments with the chicken embryo, he penned equally glowing comments: "M. de Haller excels in the art of observation."[206] To Haller, Bonnet wrote: "I thank you for the observations you have done me the honor of communicating to me. You were born to enlighten the human race; the philosophic system builders mislead it. While the latter arrange nature according to their fantasy, you observe what is there; and I take as of greater import one of your observations than all the ideas of a Buffon or a Diderot."[207]

If Réaumur, and then Haller, stood for Bonnet as exemplars of proper method, of the "art of observation," Buffon epitomized everything he opposed.[208] He considered Buffon above all a system builder, and he contrasted Buffon with "another naturalist [*physicien*], one who imagines nothing before he sees it, and who sees nothing but what is there" (i.e., Réaumur).[209] To those of his own party who suggested that he gave Buffon's system of "dreams" too much attention, Bonnet replied, "I have believed it appropriate to devote something to the celebrity of this dreamer and to the singularity of his dreams."[210] Bonnet disavowed pleasure in criticizing Buffon: "I respect this great writer, but I respect still more the truth."[211] "It is with regret that I bring up again that author whose genius and whose talents I admire; but I should be allowed to forewarn my readers against the impression, all too common, of his great celebrity. He himself admits in part (*Hist. nat.*, II, 168) that his theory preceded his experiments, and one knows how much the way one sees depends on the way one thinks."[212] Bonnet went to great lengths to make his pervasive critique of Buffon palatable to his readers, praising "the eloquent author of the *Histoire naturelle*, whom I have criticized with regret, and whose rare talents and sublime genius I sincerely admire."[213] "If nature did not make him an observer, instead she endowed him with the most brilliant of gifts and made him the most eloquent man of his century."[214]

In any event, Bonnet took very seriously the *experimental* work that Buffon reported in his texts and that Needham had already articulated from his own point of view in 1747.[215] In the seventh chapter of his *Considérations*, composed in 1749, Bonnet gave careful summaries of all the experiments Buffon reported that he had conducted with Needham and of Buffon's inferences from these experiments.[216] He expressed serious reservations even in 1749, but the definitive refutations came later: based first on Haller's experiments on the chicken embryo and ultimately on the experimental replications, directly confuting Needham, undertaken with compelling rigor by Lazzaro Spallanzani (1729–99). In a note from 1778, Bonnet observed that Spallanzani had been able to "demonstrate the falsity of the opinions of the two most celebrated epigenesists of our century."[217] Another note from 1778 called the experimental reports and theoretical inferences Buffon developed merely an "ingenious novel."[218] A third note from 1778 asserted that all the experiments Buffon reported "were badly done" and that Buffon was caught up by "deceiving appearances." The upshot, Bonnet reported with clear satisfaction, was that "there remains nothing, absolutely nothing for M. de Buffon of all the principal facts upon which he based his system."[219]

Bonnet could then quite complacently conclude: "I leave it to informed naturalists to decide between the eloquent writer and those observers whom I have cited."[220] He recommended that Buffon subject what "his fecund genius was pleased to invent" to the judgment of a more "severe reason" and abandon it all.[221] The celebrated author would have understood more properly if he had "consulted nature more than his imagination. It is beautiful and rich, but nature is worth still more."[222] "In general, M. de Buffon does not appear to possess the spirit of analysis, or if he does, his imagination has not let him use it in a happy manner."[223]

What Bonnet could not endure in Buffon and Needham (or Maupertuis, the still anonymous author of *Vénus physique*) was the idea that matter should of itself generate organic life.[224] This was to hark back to the refuted ideas of spontaneous generation or occult properties:

> What should astonish us is to see naturalists [*Physiciens*] who in a century as enlightened as our own take up once more these errors and deploy all the force of their genius to persuade us that an animal forms itself like a crystal and that a mass of flour will transform itself into weevils [*Anguilles*]. They revive the *occult* qualities that good philosophy banished from natural philosophy [*la Physique*]. They take recourse to *instincts*, to *forces of attraction* [*forces de rapports*], to *chemical affinities*, to *organic molecules* that are neither vegetable nor animal and that form by their combination [*réunion*] the vegetable and the animal.[225]

Needham—but also Buffon—seemed to be "reviving spontaneous generation [*générations équivoques*], the falsity of which is so well proven."[226]

Haller's confirmations of Bonnet's commitments were particularly welcome, and Bonnet cited a characteristically harsh judgment Haller sent him in a letter dated November 8, 1767: "Count on it that M. de Buffon is wrong. . . . These philosophers believe none of what we believe. They believe instead everything that their imagination furnishes them in support of their cause."[227] In an earlier letter, Haller set up the contrast in even balder terms: there were, he opined, "two classes of scientists [*savans*]: there are those who observe, often without writing; and there are those who write without observing. . . . A third class is even worse: it is that of those who observe badly."[228]

Virginia Dawson contends that in the eighteenth century France did not in fact determine the scientific horizon of the rest of Europe, nor Réaumur that of natural history more specifically. To complement Réaumur, Dawson points us to an essentially *non-French* line of filiation: "the education that Trembley and Bonnet received at the Academy of Calvin in Geneva and . . . the influence of Dutch Newtonian science through 'sGravesande and Boerhaave, and Leibnizian ideas, possibly derived from contact with other Swiss scientists."[229] Crucial here is the role of Gabriel Cramer, the teacher of Trembley and Bonnet at Geneva and a major figure in European science and mathematics in the first half of the century.[230] It was Cramer who helped the young Bonnet define his pursuit in natural history, and it may have been Cramer as well who guided Trembley in this direction, though the evidence is sparser.[231] Of great significance, Cramer was steeped in and devoted to the Anglo-Dutch experimental Newtonianism that, as we have already seen, proved crucial in the intellectual formation of Albrecht von Haller.[232] Cramer had visited the Netherlands and worked closely with 'sGravesande and Boerhaave; he was also a close associate of the Bernoullis in Basel, from whom he received the commission to edit the key correspondence between Leibniz and the elder Johann Bernoulli. In Cramer we get further confirmation of this powerful current of experimental Newtonianism, especially its Dutch transmission and Swiss reception.[233] But it is striking that in these same years one of Cramer's most assiduous correspondents and admirers was Buffon.[234] The circle of experimental Newtonianism, of which Buffon was the preeminent French exponent, seems to have been very tight, and Cramer seems to have been well connected on all tangents. This was a tradition with which Réaumur had nothing to do.

CHAPTER FIVE

Taking Up the French Challenge: The German Response

French philosophers have confused everything so much in their preoccupation with a few apparent peculiarities in animal and human nature, and German philosophers order most concepts of this sort more for their own system and according to their own perspective than with a view to avoiding confusions.

HERDER, 1772[1]

The stimulus of *vital* materialism from France achieved maximal intensity around midcentury, as it penetrated to the very heart of German cultural life—to Berlin and its academy—in the person of Pierre-Louis Moreau de Maupertuis. He served as president of the Berlin Academy from 1746 until his death in 1759 (though he was absent from 1756 onward).[2] In addition, closely affiliated with Maupertuis in Berlin, Julien Offray de La Mettrie fomented continuous outrage from 1748 to even beyond his death in 1751 (thanks to the posthumous publication of his collected works).[3] Finally, at just this moment came the translation and reception of Buffon's monumental *Histoire naturelle*, a phenomenon of capital importance for German science.[4] The confrontation of the allied views of Buffon and Maupertuis with those of Haller, as expressed in his various commentaries on the *Histoire naturelle*, took place under the glare of concern over Epicurean naturalism and French vital materialism.[5]

My claim is that this confrontation set the research agenda for German life science from the early 1750s forward. One response, intensely religious, came from the Hamburg scholar Hermann Samuel Reimarus (1674–1768). Another, far more experimental and radical, came from a Berliner, Caspar Friedrich Wolff (1735–94). Over the 1750s Reimarus set himself to refute French materialism—and even the French natural history of Buffon and Condillac—by delving deeply into the problem of animal instinct. Haller found this highly congenial.[6] Concurrently, in 1759, Wolff published his dissertation, *Theoria Generationis*, a pioneering achievement in German experimental life science, then plunged into controversy with Haller over epigenesis, especially with his *Theorie von der Generation* (1764).[7] In the wake of all this, a former member of the Halle circle of *vernünftigen Ärzte*, Johann August Unzer (1727–99), offered a major synthesis: *Erste Gründe einer Physiologie der eigentlichen thierischen Natur thierischer Körper* (1771).[8] Equally significant were two philosophical uptakes in the 1770s, by Johann Gottfried Herder (1744–1803) and Johann Nicolaus Tetens (1736–1807). That is our formidable itinerary in this chapter.

PIERRE-LOUIS MOREAU DE MAUPERTUIS (AND JULIEN OFFRAY DE LA METTRIE) IN BERLIN

The Berlin Academy had already been reorganized before Maupertuis arrived as its new president in 1746. He persuaded the king to allow him to reorganize it yet again. He wanted to raise the academy to the standard of the great academies of science he had known already in Paris and London. The only other really European eminence in the Berlin Academy was Leonhard Euler (1707–83), who had moved from his Saint Petersburg base to Berlin before Maupertuis arrived, but only because he mistakenly believed that Maupertuis had already agreed to come. The other key protagonist in the academy was the Berlin-Huguenot preacher and author Samuel Formey (1711–97), who had assumed the important role of secretary for the academy by the time Maupertuis arrived. Around Euler and Formey respectively a polarization of the academy (between "Newtonians" and "Wolffians") had already entrenched itself that would prove fateful for Maupertuis's tenure as president.

It has long been presumed that Maupertuis was the Newtonian archenemy of Wolff, bent on undermining the Leibniz-Wolff philosophy.[9] That role really belongs to Euler.[10] To be sure, Maupertuis sympathized with Newtonianism, but he tried to play a role above the fray befitting the president of the academy.[11] That was not simply an *institutional* commitment, however. There was a good deal of Leibniz in Maupertuis himself, perhaps more than he openly or even subjectively acknowledged.[12] The metaphysics that became the central preoccupation of Maupertuis by the end of the 1740s, articulated in his *Essai de cosmologie*, and the life science that he developed over that same period, culminating in his *Système de la nature*, have an unmistakably Leibnizian cast.[13]

The first great controversy Maupertuis had to negotiate was already under way when he arrived: the famous "monads" prize competition of

1747. Maupertuis could neither contain nor redirect its course.[14] The impetus had come, ultimately, from Frederick himself. It seemed a direct affront to Christian Wolff, aimed to undercut the prominence of the Leibniz-Wolff philosophical system, and it triggered emergent cultural-national sensitivities. The newly established scholarly review journals and the wider press in the German lands took up the topic and made it a matter for independent assessment by the public (*Öffentlichkeit*). The issue spilled beyond the frame of the Berlin Academy—indeed, by the actions of its own leading members. The main enemy of *Schulphilosophie* within the academy, Euler, circumvented the prize proceedings by publishing—anonymously for a time—a work of his own (printed, not so subtly, by the publisher for the king and the academy in Berlin) that addressed the prize topic in a resoundingly hostile attack on Leibniz's ideas.[15] By the time his opponent in the academy, Formey, published a response to Euler in December 1746, it was part of a wider wave of German media commentary on the controversy.[16] When the contest winner, J. H. G. Justi, subverted the procedures of the academy still further by publishing his essay in his own journal, instead of awaiting official publication, and, in addition, reasserted all the anti-Leibnizian arguments that Euler had propounded, a new furor swept the German media. The academy had become an object of public judgment, its authority compromised and its assessment of intellectual substance second-guessed. The outcry over the prize competition of 1747, mobilizing the new German-language public sphere, looks in many ways like a dress rehearsal for the dismal events of the "Koenig affair" a few years later. The story of Maupertuis in the 1750s has been consumed with Samuel König's charges of plagiarism from Leibniz, Maupertuis's excessively authoritarian response backed by his king, and especially the ensuing scurrilities of Voltaire.[17] That affair permanently damaged Maupertuis's reputation in Germany and across Europe (thanks to Voltaire). Maupertuis was hardly faultless. He chose to deal with König's charges in a peremptory manner and thus outraged the German public. But his achievements with the Berlin Academy have been overshadowed by this incident, leaving us with a false impression of the balance.

There are three institutional dimensions to Maupertuis's tenure that deserve note, alongside his own intellectual achievements.[18] The first was his effort to beef up the section for speculative philosophy of the academy, an institutional feature without parallel in the academies of science in the West. It was beset with mediocrity and dogmatic provincialism, so Maupertuis systematically sought to improve its tone, not only by actively recruiting talent from throughout Europe to bolster its ranks, with a modest success in the persons of Jean Baptiste Mérian (1723–1807) and Johann Georg Sulzer (1720–79), but also, and more importantly, by working extensively in this area himself.[19] The second institutional initiative was his aggressive reorganization and disciplining of the Collegium Medico-Chirugicum, which was attached to the academy and drained away substantial funding for very little intellectual value-added. Maupertuis recruited to Berlin one of Haller's star students, the anatomist Johann Friedrich Meckel the Elder (1724–74), in order to revitalize the program.[20] He also brought other life scientists to Berlin to bolster that aspect of the academy, including a new director for the Berlin botanical gardens. Finally, he tried to develop standards for academic productivity and for scientific research by discerning and articulating the frontiers of inquiry of the day and defining the conduct befitting a productive research scholar.[21]

One of the most interesting interventions of Maupertuis in Berlin was his solicitation of Frederick II's rescue of La Mettrie from the outraged authorities in Holland upon the publication of *L'homme machine* in 1747. Although later Maupertuis tried to lead Haller to believe that this was the king's own act, without any intervention on Maupertuis's part, the evidence points rather to his active role in La Mettrie's relocation.[22] Maupertuis was, sometimes uncomfortably, the longstanding patron of his countryman from Saint-Malo, first in Paris and then in Berlin. In Paris La Mettrie had paid Maupertuis the "compliment" of dedicating to him the scandalous *Natural History of the Soul* (1745). While there must have been times when Maupertuis rued this relation with his rowdy countryman, it is clear that he also supported him at crucial junctures and that he took seriously some of the ideas that La Mettrie dared to thrust into the public arena.

Maupertuis had to be circumspect in his recruitment efforts with Haller. That does not, however, gainsay a substantial and creative convergence between La Mettrie and Maupertuis concerning ideas on life science. La Mettrie's presence in Berlin helped stimulate this most important intellectual development in Maupertuis's years there. Certainly, they had numerous conversations, and organic form could hardly have failed to emerge as a point of important agreement. The very fact that La Mettrie went so far in print made it perhaps possible for Maupertuis to venture farther on his own than he might have otherwise. Indeed, David Beeson, one of the best scholars on Maupertuis, can discern no ultimate difference between them on this score, concluding that "both La Mettrie and Maupertuis seem to suggest that materialism is the inevitable conclusion of empiricism."[23] *Both*

La Mettrie and Maupertuis appeared in the German context to be promoting Epicurean materialism. At least from the vantage of its commentators and especially of its enemies, the "Epicurean" or "Spinozist" coterie clearly included the three names La Mettrie, Diderot, and Maupertuis, becoming extended, perhaps to his displeasure, to Buffon.[24]

In 1751 or 1752 Maupertuis published what Mary Terrall describes as "a little Latin book, obscurely titled *Dissertatio Inauguralis Metaphysica de Universale Naturae systemate*, purported to be by one Dr. Baumann of the University of Erlangen."[25] Diderot read this Latin version and discussed it with Maupertuis in Paris in 1753, then "accused" it of Spinozism in his *Pensées sur l'interprétation de la nature*, published at the end of that year.[26] Indeed, Maupertuis had been accused of all this already upon the publication of *Vénus physique*, in the review of that work in the *Bibliothèque raisonée*, 1745.[27] The connection with Spinozism in 1753 really came as no surprise to Maupertuis, since his friend La Condamine had raised it in their correspondence, and Maupertuis himself speculated, before his contact with Diderot, that it was this Spinozist tinge that may have put off Buffon from responding to the *Système*, though he had sent him a copy personally.[28] He made all the public gestures of disavowal of Spinozism, but those in the know were quite sure he did not mean it.[29]

Terrall believes that Diderot's discussion in his *Pensées sur l'interprétation de la nature* prompted Maupertuis to allow a French version (probably the original composition, but purportedly a "translation" of the Latin) to be published under a new title more befitting its content: *Essai sur la formation des corps organisés* (Berlin and Paris, 1754). In 1756 the text appeared in the new edition of Maupertuis's *Oeuvres*, under a more inclusive title *Système de la nature: Essai sur la formation des corps organisées*, together with his somewhat disingenuous *Réponse aux objections de M. Diderot*.

Even as Maupertuis was publishing the little "Baumann" venture, the first volumes of Buffon's *Histoire naturelle* began appearing in German translation. The alliance of Buffon and Maupertuis on issues in life science proved central for the German reception. In his work, Buffon praised the *Vénus physique*, which by 1752 was clearly known to be the work of his friend Maupertuis, as a pioneering study in the theory of generation.[30] To make their common ground more apparent (and acceptable), Maupertuis published the brief essay "Lettre sur la génération des animaux" (Letter on the generation of animals) in his key text of 1752, the *Lettres*.[31] This short essay, plus the *Vénus physique* and the various versions of his *Système de la nature*, made Maupertuis the most visible spokesman in Germany for the new impulse in "natural history" that was being launched by the French vital materialists.

MAUPERTUIS AND LIFE SCIENCE

Michael Hoffheimer's assessment of Maupertuis and life science concludes that his work, for all its empirical problems, "nonetheless marks an important turning point in the history of biology."[32] The *Système de la nature* sought, in the words of David Beeson, "to bring the whole natural world within a single system," articulating a *general* theory of active matter and reconciling that with the theological and philosophical concerns of the day.[33] As Mary Terrall maintains, in the most productive years of his life "Maupertuis was working on . . . questions of the generation of organisms, the properties of matter, and the possibility of a science of life."[34] A single, blind force of attraction in the universe could not suffice to explain the distinctive character of organized life-forms. "One will never explain the formation of a single organized body strictly by the physical properties of matter."[35] On the other hand, the irreducibility of living forms to mechanism made the question of integrating them within a coherent theoretical framework all the more critical for his thought.[36] How, in short, could he draw the realm of organic forms into a *system* of nature?

The semantic field of scientific thought in the eighteenth century was peculiarly divided over the notion of system. First, there was the great enterprise of Linnaeus to develop a unified and coherent *classification scheme* for all of the natural order. His notion of *System of Nature* represented one of the crucial poles of scientific language with regard to the natural order and its unity, namely, system as classification. But, over against this famous idea of a "natural system" articulated by Linnaeus, Maupertuis introduced "system of nature" to signify a *unified theory* of natural-scientific explanation. Harking back to the seventeenth-century masters, he construed system as an integrated *explanation* of nature. The models, here, were Newton's elegant laws of celestial and terrestrial mechanics or the earlier systematization of Descartes. These, Maupertuis averred, needed to be supplemented both to account for the generalizations of the new chemistry and, above all, to take into account the enormous domain of life-forms that the mechanical sciences simply could not yet explicate. These two divergent notions of "system of nature" would operate side by side over the second half of the eighteenth century to shape the conception of life science. Maupertuis shared the view of his close friend and theoretical ally, Buffon, that what was needed was nonmathematical *physique*. Experimental Newtonianism was the essence of their approach to natural inquiry. Its crucial features would be extensive empirical inquiry ("observation and experiment") *combined with* a demand for synthetic (theoretical) interpretation. This *philosophy* belonged to the generic domain of "natural philosophy" in the traditional sense that it was concerned with explanatory regularities, but it went off in a quite specific direction—namely, toward "vital materialism."

As Terrall argues, Maupertuis was engaged in "synthesizing evidence from anatomy, natural history, animal breeding, and travel literature" to propose a "theoretical model of active matter," ascribing to it "properties responsible for organization and heredity" and even affirming the "possibility of change in organic forms."[37] Thus, "his challenge to preexistence theories went beyond anatomy and physiology to a theory of organization (on a submicroscopic scale) *and* a theory of heredity (on the macroscopic scale)."[38] It was a matter of empirical observation that preformation simply did not save the phenomena. To propose an alternative, on the other hand, one need not offer a definitive account, Maupertuis claimed, but only one with fewer difficulties than its rival. The most sensible theory was that "the formation of the first individuals having been miraculous, those that succeeded them are no more than the effects of those properties . . . [with which the Creator endowed] even the smallest parts of matter."[39] Extending "intelligence" to matter was *necessary* for a scientific explanation of lifeforms. "If one wishes to offer a conception about this, though it works only by analogy, it would seem necessary to have recourse to some principle of intelligence, to something similar to what we call *desire, aversion, memory*."[40] Only in this way could one answer the question: *by what specific mechanism* did certain parts become eyes, and others ears?[41] Thus, in embryology, each particle in the seminal fluid of father and mother retained "a kind of memory of its original placement [in the adult organism,] and it will seek to take it up again as often as it can to form the same part in the fetus."[42] This language of desire, aversion, and memory was intentionally analogical and metaphorical. Maupertuis offered, as an alternative, a language of "instinct."[43] He was even prepared to surrender all the metaphors, so long as the phenomena he had identified received the scientific attention they merited, which meant that a mechanistic physics could not suffice.[44] Maupertuis argued that his new hypothesis escaped all the difficulties that had beset preformation: "resemblance to [both] parents, the production of monsters, the birth of hybrids, all this is simply explained."[45]

The *Essai de cosmologie* contained two passages of extraordinary significance for his thoughts about organic forms. In the first passage, Maupertuis offered a remarkable conjecture:

> Yet could one not say that in the chance combination of natural products, given all that present themselves possess a certain aptness [*convenance*] enabling them to survive [*subsister*], it is no marvel that such aptness is to be found in all the species that actually exist? Chance, one might say, had produced a numberless multitude of individuals; [only] a small number found themselves constructed in a manner that the parts of the animal could satisfy its needs; in another, infinitely larger number, there was neither aptness nor order: all these last have perished; animals without mouths cannot live; others which lacked reproductive organs could not perpetuate themselves: the only ones that remain are those where order and aptness happened to arise, and these species which we see around us today are but the smallest part of that which a blind destiny produced.[46]

The second passage, from late in the third part of the *Essai*, reflected on speculation that collision with a comet might have been responsible for the Deluge. Maupertuis stressed the epistemological frailty of man, which would be particularly exacerbated by such a catastrophe:

> Before, all the species would have formed a sequence of beings which would, in a manner of speaking, be little more than the contiguous parts of one single whole: each one linked to its neighboring species from which it would differ only by insensible nuances, thus forming a mutual communication which would extend from the first to the last. But once this chain is broken, the species we could know only by the mediation of those which had been destroyed would have become incomprehensible to us: we live, perhaps, amid an infinity of such beings of which we can discover neither the nature nor even the existence.[47]

Even among the species we did observe, we would be deprived of any pattern of order to comprehend them. "Each species in isolation cannot complement nor help understand the others: the majority of beings would appear to us like monsters and we would find nothing but obscurity in our judgments."[48] The evidence from fossil geology, he suggested, made this dread much more vivid, for it demonstrated drastic change in the history of the earth.[49]

Reflecting again on catastrophe theories in geology in his *Système de la nature,* Maupertuis speculated that with any such catastrophe, "new unions of elements, new animals, new plants, or things utterly unprecedented" might well ensue.[50] Indeed, Maupertuis picked up the thread of the argument from chance for the variety of species and situated it in this general theory of dynamism in nature:

> Might one not explain in this manner how from a mere pair of individuals the multiplication of the most diverse species might have ensued? They will not have owed their initial formation to anything more than random productions, in which the elementary particles did not maintain the order that prevailed in the male and female parents; each degree of error will have created a new species, and as a result of repeated deviations there will have arisen the infinite diversity of animals which we see today, which may grow further with the passing of time, but to which the passage of centuries may [just as well] supply only the most imperceptible augmentations.[51]

What was Maupertuis's ultimate position? Maupertuis conceived "a dynamic nature full of changing forms."[52] The challenge was to establish the laws governing the process of change. Thus, Maupertuis "made activity fundamental to matter" even as he "historicized the problem of organization."[53] Generation should be construed as part of a universal dynamism of nature, advancing to higher and higher degrees of organization from inanimate matter to living organisms to man.[54] He found evidence of this self-formative capacity in mineral crystallization, as in the "tree of Diana."[55] He conjectured that in the early phases of the formation of the earth, the globe had a fluid surface in which all elements floated freely. The least active elements eventually gave form to metal and stone, while the most active emerged as animals and man.[56] This plasticity persisted. He concluded that plants and even crystals exhibited some elements of the self-organization that he discerned definitively in animal sexual reproduction, and accordingly he offered the prospect of a unified theory of natural process.[57] Maupertuis wove chance and lawfulness together into a theory of process and emergence; his was a distinctly *naturalist* and *historicist* philosophy of nature.[58] He propounded boldly a theory of *active* matter, of "vital materialism." It was *materialism,* in that it imputed to nature the power of self-formation.[59] But it was *vital* in that it recognized that the mechanical principles of physics did not suffice to account for the phenomenon.[60] In a word, he set out to establish a plausible theory of *hylozoism.*[61] We can scarcely conceive a more revolutionary agenda in life science at the middle of the eighteenth century.[62] It made the dramatic paradigm shift shared with contemporary writings of La Mettrie, Buffon, Diderot, and the school of Montpellier all the more salient for a German academic audience concerned with the questions of life science. That German audience was caught up at the very moment with the translation of Buffon's monumental *Histoire naturelle* into German, especially under the influence of Albrecht von Haller's reception of that text.

ALBRECHT VON HALLER AND THE TRANSLATION OF BUFFON'S *HISTOIRE NATURELLE*

Haller played a prominent role in the publication of the German Buffon translation by providing introductions to the first two volumes, though he did not himself translate any part of the work.[63] The project was conceived and executed by scholars primarily at the University of Leipzig but was published in Hamburg.[64] Haller was in contact both with the publisher and likely translator, Berthold Joachim Zinck (1718–55), and with the supervising editor and commentator, Abraham Gotthelf Kästner (1719–1800).[65] He clearly believed in the venture and promoted it in glowing terms via his review in the *Göttingische Gelehrten Anzeigen* (*GGA*).[66] It was Zinck who persuaded Haller to provide the introductory matter. Haller composed an original work as the preface for the opening volume of the translation. The preface to the second volume was, in fact, a work that he had published in the interval in French but that Zinck was able to induce him to reprint in German in slightly altered form.[67] That Haller had already commented extensively in French showed how closely he was following the debates around the issue of generation in France.[68]

The Buffon translation venture proved a great success in general, and Haller clearly benefited from the publicity surrounding his two prefaces for the project. Although he played no further direct role in the enterprise, he provided a frame in which Germans would read the *Histoire naturelle*.[69] He had established himself as the leading life scientist in Germany, not only with his edition of Boerhaave but with his own *First Lines of Physiology* (1747). More, he had become one of Germany's decisive intellectual arbiters, especially once he took control of *GGA* in that same year. That made his commentaries on the Buffon translation salient.

That Haller took so active a role in the Buffon project signals several issues of personal concern. First, Haller used the prominent venue of the Buffon translation to articulate before the German naturalist community his clearest statement of what I have been calling "experimental Newtonianism." Hence, the first preface was frequently reprinted under a general title, "On Hypothesis," as a contribution to the general methodology of natural philosophy, not as something tied specifically to Buffon.[70] It discussed Buffon directly only toward the end, and really only as an illustration of the general argument about research method that Haller had been developing at length, which signals that Haller's personal agenda in this preface went beyond introducing Buffon.[71] To be sure, he noted that Buffon had produced a "great and magnificent work," but he added quite pointedly that "the author always goes somewhat further than his information, experiments, and insight."[72]

That brings us to a second issue: Haller's intense reevaluation of his own orientation to the problem of generation. He could no longer subscribe to his mentor Boerhaave's spermist version of preformation. Clearly, Haller knew all about the Trembley experiments with freshwater hydra and their disruptive implications.[73] Moreover, he was attuned to all the criticisms emanating from the French scene (e.g., Maupertuis's anonymous *Vénus physique*) that put preformation, ovist as well as spermist, in question. He was, for a time, drawn to epigenesis.[74] The Needham-Buffon experiments of the late 1740s represented the most important European endeavor to justify epigenesis, and Haller—like his future ally Bonnet—devoted very close attention to the results and their various interpretations.[75] "Whatever kind of hypothesis Mr. de Buffon may make, the experiments maintain their value."[76] It seems Haller had hoped for something more definitive on the question than he believed Buffon's proposal actually provided. Buffon proposed to derive from these experiments "nothing less than a total overthrow of the generally accepted opinion" favoring preformation. In Buffon's version, "nature herself enjoys the right of forming herself, of organizing herself, and of passing freely from the inanimate state to that of a plant, or that of an animal."[77] Haller disagreed; "matter may tend toward certain configurations without having thereby the power to do so of itself." That is, while it was driven by forces, it could not itself account for them: "these qualities are not part of the essence of matter. They are alien to it. . . . A first cause has thus dispensed to the various classes of matter powers and forces calculated according to a general plan."[78] Divine providence was essential, and that led Haller back to preformation. His conclusion was: "After having granted everything to Mr. de Buffon, do we still not need a directive intelligence, in order to place the organic particles appropriately . . . ?"[79] That is, "Mr. de Buffon needs a force which has foresight, which can make a choice, which has a goal, which, against all the laws of blind combination, always and unfailingly brings about the same end."[80]

A third important issue for Haller was Buffon's disparagement of the taxonomical approach of Linnaeus. That Linnaeus had been one of his earliest supporters was not lost on Haller. He was still corresponding with Linnaeus and actively publishing in botany. He informed Linnaeus of his decision to concentrate foremost on anatomy only in 1747, as we have seen.[81] Moreover, he respected the Linnaean endeavor, considering the possibility of any botanical research utterly dependent upon a classificatory system to navigate the bewildering profusion of plant life.[82] The acerbity of Buffon's treatment of Linnaeus struck Haller (and the rest of the German research community) as gratuitous.[83]

Finally, Haller took careful note of the Sorbonne's alarm over the religious propriety of the *Histoire naturelle*. Interestingly, in the commentaries of 1750–51 Haller *defended* Buffon, not only on general methodological grounds ("the truth does not know how to be impious") but also in terms of Buffon's personal intentions: "I do not believe that the theory of Mr. de Buffon tends toward such evil ends."[84] "Is this system dangerous? Is it in the interest of religion that it should turn out not to be true? Ought a book be suppressed which proposes such systems?"[85] Certainly, Haller knew of such fear for religion, "but is this fear founded, and would faith be lost if from experience we were to grant productive forces [*bauenden Kräfte*] to Nature? . . . I am without fear in this matter."[86] There was one notable exception, however: the first segment of the *Bibliothèque raisonée* review expressed very strong objections concerning Buffon's theory of the earth.[87] Haller upheld a strictly biblical sense of time, strikingly out of step with the state of research in emergent geology.[88] At least these hypotheses introduced by Buffon aroused his deep religious commitments.

While, already in 1750, Haller was obviously hostile to "freethinkers" of the stripe of La Mettrie, he did not wish to extend this suspicion to Buffon (or even, at least overtly, to Maupertuis). In these very years Haller was involved in extensive negotiations with Maupertuis about joining the Berlin Academy. His loathing for La Mettrie made him reluctant to move to Berlin, even after the death of his flippant foe, yet there is no evidence in these negotiations that Haller took Maupertuis for a dangerous freethinker. It appears that, for Haller, Buffon did not merit such suspicion either, though de Lignac made that the essence of his 1751 polemic against Buffon, *Letters to an American*, with which we can be sure that Haller was quite familiar.[89]

Haller's attitude changed dramatically, however, by the mid-1750s. One source of this shift may well have been his move back to Switzerland (1753) and his new association with Charles Bonnet, especially upon receipt (1754) of the latter's defense of preformation, which appealed to Haller both scientifically and religiously. But another source was the publication of a key work by Hermann Samuel Reimarus (1754) that clearly implicated Buffon in the threat of Epicurean naturalism. The *GGA* published two very extensive review essays on Reimarus's work (in all likelihood by Haller), which celebrated it precisely for sounding the alarm over the threat of Epicurean naturalism—not just from the scandalous La Mettrie but from Maupertuis and Buffon as well.[90]

HERMANN REIMARUS AND "EPICUREAN" LIFE SCIENCE

In 1754 Reimarus published *Die vornehmsten Wahrheiten der natürlichen Religion*.[91] Kant regarded it as the foremost effort regarding natural religion to appear in Germany.[92] It would be followed by two other major works before the close of the decade: *Vernunftlehre* (1756) and *Allgemeine Betrachtungen über die Triebe der Thiere, hauptsächlich über ihre Kunsttriebe* (1760). The three works showed two crucial aspects of German life science of the day: first, a sharp alarm over the propagation of a "materialism" of distinctly French origin; and, second, a fascination with the question of animal instinct, pivoting on the question of the animal/human boundary in terms of the longstanding debate over the "animal soul."[93]

French materialism and *Freidenkerei* elicited concerns all over Europe, but particularly in Germany, about the relation of the human mind and soul to animal life. Animal behavior was a central issue in the mid-eighteenth century, pivoting around the unpleasant contention that if animals were machines, then humans had to be as well: a position very few, if any, thinkers of the epoch could countenance.[94] Hence, the contrapositive had to be explored: if humans were not machines, then neither were animals. But then, what were they? And how could they be explained? The questions divided the intellectual community between those who believed that natural science *could* provide this explanation and those who believed that only a *supernatural* recourse was possible. Certainly, Reimarus embraced the latter view. By contrast, it was the thrust of the French vital materialists to uphold the former position.

In the preface to the *Vornehmsten Wahrheiten*, Reimarus articulated a key provocation for his work: "not without distaste have I observed that in the last few years a very unusual group of small writings, for the most part in the French language, have been spread about the world, in which not just Christianity but far more all natural religion and morality are mocked and attacked." For Reimarus, the calling of philosophy was to debunk such "contemporary freethinking [*jetzigen Freydenkerey*]."[95] In a two-part article published in 1974, Julian Jaynes and William Woodward construe *Vornehmsten Wahrheiten* as a rebuttal to the "Epicureans," or materialists. Revealingly, having worked with the English translation (1768), they did not realize that the subtitle attached to the English version was the invention of the translator, not Reimarus himself.[96] That subtitle was extraordinary. It read: "Wherein the objections of Lucretius, Buffon, Maupertuis, Rousseau, La Mettrie, and other ancient and modern followers of Epicurus are considered, and their doctrines refuted."[97] Reimarus was appalled at this liberty, alleging, in a letter to the London *Monthy Review*, that the translator had no call to include Buffon, Maupertuis, and Rousseau among the followers of Epicurus.[98] Though he disagreed with Buffon and Maupertuis (the case of Rousseau is less relevant here), Reimarus denied that he presented them as Epicureans.[99] The translator, Richard Wynne, replied: "I frankly confess, I added the words in question to the title; but . . . I had good reason for inserting them." According to Wynne, Reimarus *did* treat Buffon and Maupertuis as Epicureans, whether he wished to admit it or not.[100] He turned the query back on Reimarus: "How came you to tax the gentlemen above mentioned (I own not unjustly) with maintaining the principal opinions of *Epicurus* and *Lucretius*, and rank them with *La Mettrie*?"[101]

Freethinking Epicureans were indeed the obsession of Reimarus: "These people maintain, all of them, in their main propositions, that there is a physical world active outside ourselves, . . . an autonomous, necessary, and eternal world, and [that] this [is] a certain truth upon which further ideas can be developed."[102] Reimarus approached the whole question from the vantage of the principle of sufficient reason. For him, the physical world had no *intrinsic* reason to be so and not otherwise; its lawfulness was, in that sense, imposed from without, not self-generated, and therefore, it was an imperfect being, dependent upon something that *did* have the power to create from its own resources and that could *give purpose* to the inert material world in the form of living beings.[103] Reimarus insisted not only upon a transcendent creation but also upon the teleological structure of this creation: the physical world "is not for itself but for the sake of the living, of man."[104] More explicitly: "the usefulness for living things constitutes the essential reason according to which all reality and essential properties of the material [of the world] get determined."[105]

Once he had disposed of the case that the world or nature could be eternal or self-sufficient, Reimarus turned to animals and humans. They, too, had to have an *origin*, but the notion that this would be the physical world or natural process *alone,* Reimarus contended, was absurd. He scoffed at the notion that the sun's warmth, working on the "slime" of the earth, could have awakened the impulse to life that, through many missteps, eventually generated viable organisms of many distinct species, culminating in man.[106] Such was the lore of the ancient materialists, Epicurus and Lucretius, but it appalled Reimarus that in his enlightened times figures like La Mettrie should try to revive "this collapsed body of doctrine [*dieses verfallenen Lehrgebäude*]."[107] He lapsed into outrage in place of argument: "I cannot understand how nowadays people can permit themselves to uphold before the court of their own reason such arbitrary concoctions [*willkürlichen Erdichtungen*], which are formed from raw ignorance of nature and from numerous contradictions and errors."[108] The ancients were faulty enough, but the moderns had no ground to stand on, for spontaneous generation had been definitively refuted in modern times by "the best naturalists [*Naturkündiger*]."[109]

Above all, Reimarus could not countenance the notion that life-forms could have arisen from the material world by chance.[110] Authors like La Mettrie, as he saw it, preferred chaos and chance to the intervention of an intelligent designer: they would rather ascribe "all, instead, to unfeeling, dumb forces of a dead material, and to give themselves over to blind fate or random chance."[111] Nor had he any sympathy for the "new pantheists" who wished to impute life to nature itself—"a penetrating force that animated the entire body of the world and ruled with understanding," a thinking matter or world soul.[112] For Reimarus, undertaking to remove final cause from natural philosophy was misguided,[113] and it was not surprising that those who denied teleology in nature denied religion.[114] He accused Buffon of reviving Epicureanism and aiding and abetting atheism by rejecting teleology.[115] Nor would he concede anything to Buffon's argument that natural philosophy should avoid metaphysical hypotheses: "if one were to take away all of Mr. B.'s arbitrary propositions and hypotheses, what would be left of his whole system?"[116]

All this confirms Wynne's contention about Reimarus linking Buffon and Maupertuis with Epicureanism, notwithstanding his later embarrassment.[117] For Reimarus, there was no point to knowledge of the physical world if nature had no purpose we could discern.[118] The world made no sense without God. Thus, he upheld Swammerdam, Réaumur, and others who found a divine hand in the marvels of invertebrates and infusoria, against Buffon and Maupertuis, who scoffed at all that.[119] Reimarus went so far as to offer a detailed defense of the English author who found God's providence in the thick skin of the rhinoceros, the example Maupertuis used as the reductio ad absurdum of physicotheology.[120] Nothing was absurd for Reimarus in the natural order if it gave sign of God's Providence. Anyone who did not see it that way was an "Epicurean"—that is, tantamount to an unbeliever.

In that light, Reimarus offered a detailed critique of the experiments of Buffon and Needham.[121] He avowed that his intent was only to establish the uncertainty of their observations and the impropriety of founding a theory of generation upon them.[122] He noted that Buffon had set the whole endeavor in motion in an effort to demonstrate that Leeuwenhoek had earlier misconstrued microscopic observations into a theory of generation that was too hasty and ill-founded.[123] Reimarus deftly rejoined: "perhaps many of the arguments that Mr. Buffon used against Leeuwenhoek can be brought against himself."[124] Three arguments put in doubt the findings of Buffon and Needham. First, observations generated by microscopic instruments were problematic and unreliable because their objects were too infinitesimal and the instruments themselves too crude.[125] Second, Buffon and Needham drew drastically different theoretical systems from the *same* observations.[126] Finally, replication experiments by Peter van Musschenbroek and by the anonymous author of *Letters to an American* (de Lignac) had radically different results.[127] In the third edition (1766), Reimarus drew the essential conclusion: the issue was not really about *observations* but rather about the *inferences* [*Schlussfolgerungen*] drawn from the observations.[128] Buffon had committed the same error that he had criticized in Leeuwenhoek: premature theoretical generalization based on equivocal observations.[129]

Only concrete living things gave the world purpose, and these only because they were ensouled.[130] In the third edition, Reimarus made this claim even more emphatic: "the bodies of living things are perfect only in the measure that they serve as appropriate instruments which correspond to the nature or the endeavor of each soul after its own way of life."[131] He believed that "this way of seeing things fills the great gap which we still have in philosophy and links . . . knowledge of the natural world [*Naturlehre*] with metaphysics."[132] The soul—animal as well as human—was the lynchpin of this metaphysical connection and of its theological concomitant. Reimarus's line of thinking consigned all plants to the lifeless, material world.[133] They had no "inner perfection"; they had no intrinsic purpose but existed merely for the sake of animals.[134] To exclude plants from living forms was a drastic consequence of his metaphysical orientation, and it begged essential questions about organic form and intrinsic organization. But this was consistent with his express ambitions: "my project here is to demonstrate the divine wisdom and intention not [merely] in physical organization but at the same time immediately in the forces and capacities of the soul."[135]

THE CONDILLAC-BUFFON DEBATE ON ANIMALS AND REIMARUS'S *TRIEBE DER THIERE*

Reimarus's greatest work, the *Triebe der Thiere* (1760), was his effort to refute "empiricists" who, along the lines of Condillac's *Traité des animaux* (1755), sought to explain animal behavior on the pattern of human learning, as acquired from sense experience.[136] Against such empiricists, Reimarus insisted on the *innate* character of animal behavior, "instinct" or "drive [*Trieb*]."

The abbé de Condillac published his *Traité des animaux* under the impression that Buffon was spreading rumors that he had plagiarized from Buffon's earlier volumes for his own *Traité des sensations*.[137] In 1754 Buffon published the fourth volume of his *Histoire naturelle*, which included crucial essays on the nature of animals and on human nature.[138] These essays seemed to equivocate between a mechanistic and an animist notion of animal life. That equivocation also characterized the summary of Buffon's views in a very famous article by Diderot for the *Encyclopédie* ("animal") that had appeared in the interval between this fourth volume and the three original volumes of the *Histoire naturelle* from 1749.[139] That made Buffon's theory of animal nature and its relation to human nature highly charged for the political culture, as reflected in de Lignac's fierce polemic, *Letters to an American* (1751). In the *Traité des animaux* Condillac alleged that Buffon's position was bewildering (i.e., incoherent) because he both ascribed sensibility to animals and insisted on treating them as Cartesian machines.[140]

To clarify the matter, he proposed to work *by analogy* from a consideration of human experience to that of animals.[141] Condillac asserted that *sensibility* not only distinguished animals from the rest of nature but assimilated them, at least analogically, to humans. Thus, "animals compare, judge, . . . have ideas and memory." He cited passages where Buffon, despite himself, recognized all these features in animal behavior.[142] Animals *learned* how to sense and they *judged* among sensations and objects on the basis of pleasure or pain. At the outset they had only an indeterminate, internal sense of pleasure and pain, which needed to be localized and associated with external stimuli.[143] "Thus, animals owe to experience the habits that one takes to be theirs by nature."[144] They learned, but without the intention of learning. Associations of pleasure or pain became nodes around which memory organized experiences. "Everything in this depends on the same principle, need [*besoin*]; everything is accomplished by the same means, the association of ideas."[145] Animals did not learn by imitating their conspecifics but rather from their individual experience, yet since the needs that drove these experiences were the same in each distinct animal form, the resulting behavior appeared uniform.

Provocatively for Reimarus, Condillac dismissed the term "instinct" as an empty word without theoretical significance.[146] "Instinct is nothing, or it is the beginning of knowledge."[147] It was wrong, he claimed, to take instinct as anything other than a matter of degree of knowledge, which could be explained in terms of learning. Only the simplicity of the needs and habits constitutive of instinct made it so inerrant. "One sees that instinct is only sure if it is restricted."[148] Thus, Condillac elaborated a continuity between instinct and reason consistent with his sensationalist theory of human learning as acquired in sense experience, accrued in memory, and sorted by reflection.[149] Distinguishing between a "self of habit" and a "self of reflection," Condillac consigned animal instinct to mere self of habit. "That measure of reflection that we have above and beyond our habits is what constitutes reason."[150]

All these issues preoccupied Reimarus in the *Triebe der Thiere*. Already in his *Vornehmsten Wahrheiten*, Reimarus offered a clear conceptualization of the essential features of the concept *Trieb* ("instinct" or "endeavor" [*Bemühung*]—with echoes of *conatus*): it was the capacity to pursue a self-beneficial goal "without any individual reflection, experience, and practice, without any training, example, or model, from birth onward, with an artfulness ready from birth that was masterful in achieving its end."[151] He promised his readers that he would elaborate his ideas on instinct and reason in a subsequent work.[152] He fulfilled that promise in his magnum opus of 1760.[153]

Using extensive documentation of animal instinctual behavior, Reimarus worked up a classification of its distinctive features. His massive synthesis of the data gave his conception of *Trieb* a warrant that made it indisputable. He distinguished globally among "mechanical," "representational," and "elective [*willkürlichen*]" drives, articulating ten classes and fifty-seven subclasses.[154] By *elective* Reimarus meant "an inclination or aversion of the will toward present, though indistinct, representations—that is, toward the feeling of sensual pleasure or pain—whence the elective actions [*willkürliche Handlungen*] arise that are appropriate to the inclination or aversion."[155] Among the elective drives, he distinguished between the general form—self-preservation—and more particular manifestations, which he labeled *Kunsttriebe* (skillful drives).[156] These were his primary concern.[157] Indeed, the notion of *Kunsttrieb* proved the most theoretically fruitful concept he articulated.[158] He began by characterizing *Triebe* in general as "all natural efforts toward particular actions [*alles natürlichen Bemühen zu gewissen Handlungen*]."[159] He characterized as *Kunst* (artifice; skillful activity) all "regulated [i.e., rule-governed] capacities for certain actions."[160] Thus, the decisive features of the *Kunsttrieb* were that it was natural and yet it was agential; that is, it entailed a measure of *choice*. In short, Reimarus arrived at a notion of animal behavior as immanent purposiveness. It was *purposive* precisely in that it aimed "at the maintenance and well-being of the animal and its species [*auf die Erhaltung und Wohlfahrt des Thieres und seines Geschlechtes*]."[161] It was, however, altogether *innate*: this was his overriding theoretical insistence. By virtue of this innateness, all *Triebe* were "endowed by the Creator [*von dem Schöpfer eingepflanzt*]."[162]

Triebe der Thiere elaborated this interpretation explicitly against the work of Condillac and Buffon. In §112, Reimarus followed Condillac in accusing Buffon of simultaneously adopting Descartes's animal-machine hypothesis and nonetheless ascribing sensibility to animals, and hence of falling into contradiction.[163] However, his main opponent was Condillac, for the thrust of his argument was to deny Condillac's central thesis that animals learned by experience on analogy with human development.[164] Instead, Reimarus built a powerful case for the *innateness* of instinct, different *in kind* from human learning. In tacit response to Condillac's claim that these were unanalyzed terms, Reimarus undertook to articulate *reason* and *instinct* into clear and cogent concepts via juxtaposition.[165] "I demonstrate from the powers of animals and from the properties of the creative drives that they do not consist in an effectiveness that the animals achieved for themselves through experience and reason, nor through a degree of reason, but that these are inherited capacities that arise out of the determinate natural powers of animals [*angeborne Fertigkeiten sind, welche aus den determinierten Naturkräften der Thiere entstehen*]."[166]

Reimarus saw himself as no mere philosophical speculator since he engaged richly with the natural-historical literature.[167] Thus, he allowed himself to disparage Condillac's view as "one of those hypotheses that can appear valid only in the scholar's study" and that would have been overturned had Condillac paid more attention to observing nature.[168] At the end of §103, Reimarus reiterated this methodological scruple: "without sufficient knowledge of actual nature, all hypotheses turn into idle figments of the mind arising in the scholar's study."[169] While the interaction between soul and body was ultimately a "mystery, and how it actually works will always be impossible for us to know," nevertheless, experience and observation made it clear that these things happened, both in animals and in humans.[170] "We thereby must presume the actual connection between the soul and the body, merely according to experience, even if we cannot explain the nature of this mutual interaction."[171] This was the fundamental methodological maxim of empirical research science, Reimarus concluded: "For the original essential forces are, in accordance with their regular determination, the first ground of all the actualities of nature: and all philosophers must acknowledge that it is impossible to determine a priori any further philosophical or mathematical proof concerning the fundamental forces of things and their determinate laws, but they must simply assume them in accordance with experience in order to carry out inquiry."[172] At the close of his chapter on the *Kunsttriebe*, Reimarus made this clear: "by the term 'skillful drives' I designate the matter itself, which is before the eyes of everyone to observe, but not the cause or the manner of its possibility. . . . One must first come to be acquainted with the matter itself according to its actual properties before one can ask how all that has really been observed in animals can be possible."[173] Thus, Reimarus formulated *methodological-empirical* responses to crucial epistemological and ontological questions about the nature of forces and drives that bore upon human nature, not just animal nature.

His empirical project of classifying animal instinctual behavior came to be integrated into a clearly school-philosophical (Wolffian) project of "faculty psychology," as Reimarus rigorously distinguished the conscious content and operative faculties of the human mind from those of animals.[174] It was this philosophical proximity that, as Condillac so clearly noted, drove the entire eighteenth-century inquiry into the animal soul: "It would seem strange to inquire what animals are if it were not a means to understand better what we ourselves are."[175] For Reimarus, animal behavior was neither acquired by experience nor derived from reason, and thus there was a systematic difference in kind between animals and humans. Yet the "rule according to which the sensual representation of animals proceeds appears to be completely the same as the rule of our lower faculties of soul [*niederen Seelenkräfte*]."[176] Animals shared with humans certain features of sensibility—external sense organs, imagination, memory, and an internal sense for pleasure and pain and inclinations derived from these. Still, all these had a different significance in the whole structure of awareness, in that animals were never capable of conceptual relation or inference. "It is from this, then, that we can grasp how the animals know things and discriminate among them, as well as how they are aware of what they are conscious of. Everything is merely indistinct and confused, and yet very lively."[177] These are the key categories of German school-philosophical faculty psychology. While humans classified conceptually, "the animal discrimination of species and genera has an entirely different basis and must be essentially different from our own."[178] Humans operated discursively: it was from a general concept that they were enabled to discriminate instances of commonality.[179] But in animals, "sensibility in the vast majority of instances suffices to recognize and to distinguish individual things as well as types."[180] Reimarus then worked through the levels of human reasoning—from concept formation through judgment to inference—as developed in German school-philosophical teachings and demonstrated that animals could do none of these. Instead, they acted from conditioned reflex and habit. "If, then, all thought consists in concepts, judgments, and inferences, then we cannot, in the literal meaning of the term, say that animals think."[181] Humans had the capacity for reflection, which was the true source of reason. And reason was what made them truly human.[182] That was reflected in their unique capacity for language.[183] Reimarus insisted that no animals, not even apes, could ever achieve language.[184]

Yet there were compensations in animals for this lack of reflection: they were by instinct far more effective in achieving the end of self-preservation. This efficacy was immediate, inveterate, and sufficient for its ends. Enhanced senses, exquisitely intricate and appropriate behaviors, and restricted but functional needs all worked to make animal drives efficacious far beyond anything that the considered and labored achievements of individual humans could attain.[185] It took the generational accrual of implements and skills through the trial and error of rational reflection to bring humans to a superior estate in the natural order. "The same behaviors appear among these animals, from the start of their lives, a capacity that, without slow and awkward experimentation, without preliminary errancy and confusion, from the very first produces masterpieces."[186] It would be odd indeed should animals have accomplished this through reason or reflection, or even through learned habit, for then they would clearly exceed humans in all these capacities. That was absurd, and so the notion that there was anything learned or rational about this had to be false. Instead, Reimarus concluded, it had to be completely *innate*.

Only after his exhaustive classification of the various types of *Triebe* did Reimarus turn to the fundamental issue of explanation. In chapter 8, he surveyed "the opinions of the ancients," and in chapter 9, he examined "the hypotheses of the moderns" on the question.[187] After his learned conspectus of other views, he brought his work to its theoretical conclusion in chapter 10: "The Probable Characteristics of Animal Skillful Drives [*Kunsttriebe*]." The opening line of that chapter reiterated his key claim: "we can offer no grounds to ascribe reason or any grade of the same to animals, as contrasted with us humans, nor even to ascribe to their brains any natural images or innate figures that would be of use for this faculty [*einige diesem Vermögen behülfliche Naturbilder und angebore Figuren in dem Gehirne beylegen*], or even to take into account any extraordinary intervention by God."[188] What, then, *could* explain the phenomenon?

> How is it then conceivable that the animals, with such lowly physical and spiritual powers, in part without any external experience, without upbringing, guidance, examples, or verbal instruction, but above all without any reflection and actual thinking, without concepts, judgments, inferences, and the discoveries that flow from these, without themselves knowing any purpose or the ability to recognize the relationship of means to that purpose, without experiments and long practice, nonetheless find themselves at all times capable of constructing completely and masterfully the most ubiquitously useful and clever artificial actions [*Kunsthandlungen*] for the many needs of every aspect of life and for the preservation of their species, and notably oftentimes from the very moment they come into the world?[189]

Reimarus argued that there were only four possible sources: (physical) mechanism, external sense and responsive imagination, inner experience, or "implanted blind inclination [*eingepflanzte blinde Neigung*]"—or, most likely, all these in interaction.[190]

In a series of contributions to the key Berlin Enlightenment journal *Briefe, die neueste Litteratur betreffend*, Moses Mendelssohn subjected Reimarus's work to careful scrutiny. He concentrated his criticism upon the four sources of instinctual behavior that Reimarus articulated in chapter 10 of *Triebe der Thiere*, deeply dissatisfied with the way in which, having found the first three sources—bodily mechanism, sense and imagination, and internal sense or disposition—inadequate, Reimarus resorted to "implanted blind inclination."[191] For Mendelssohn, this was no more of an explanation than those intermediate forces of nature (e.g., the "hylarchic principle") that Reimarus had appropriately written off as "empty noises" in his review of the philosophical tradition.[192] Mendelssohn did not dispute that Reimarus had *described* something essential about instinctual behavior—most prominently, artificial constructions (*Kunsttriebe*) like beehives, spiderwebs, and beaver dams—but he reiterated the distinction Reimarus himself articulated between description and explanation and insisted that Reimarus had failed in his *explanatory* ambitions.[193] He argued that the notion of innate inclinations elucidated nothing.[194]

HERDER BETWEEN THE FRENCH AND HALLER: PHYSIOLOGICAL PSYCHOLOGY

In his *Abhandlung über den Ursprung der Sprache* (Treatise on the origin of language; 1772) Herder made the famous quip that Condillac had tried to make animals human, while Rousseau had tried to make humans animals, hence the task was to establish just what humans really were by discriminating them from animals.[195] His point of departure was the work of Reimarus in comparative ethology: "*That the human being is far inferior to the animals in strength and sureness of instinct, indeed that he quite lacks what in the case of so many animal species we call innate abilities for and drives to art* [*Kunstfertigkeiten und Kunsttriebe*], is certain."[196] The challenge was to be true to Reimarus's careful description of the features of instinct distinguishing animal from human behavior and yet to offer a more convincing explanation for why these features arose. Herder argued that the laborious human development of capacities, relative to the ostensible "perfection" of animal behavior grounded in drives or instinct, could be explained in terms of a fundamental difference in *environmental situatedness* (what he called the "sphere" of the animal's life) and hence the intensity of the stimuli attending it.[197] Other animals had drastically narrower spheres of life than humans. Their ecological niche—to use modern terminology—was circumscribed both in space and in survival constraints, so that *within* that narrow field, Herder contended, animal experience was sensually more intense and focused, hence more immediately effective and complete. He enunciated his theory in a proposition: "*The sensitivity, abilities, and drives to art* [*Kunsttriebe*] *of the animals increase in strength and intensity in inverse proportion to the size and diversity of their circle of efficacy*."[198] The narrowness and intensity of focus of sensual experience in animals made language redundant.[199] While they uttered sounds in response to their feelings, they had no need and hence no capacity for the elaboration of formal language; thus, "if we wish to call these immediate sounds of sensation 'language,' then . . . their origin . . . is clearly animal: *the natural law of a sensitive machine*."[200]

In contrast with the specificity of every animal niche, humans had literally a *global* environmental orientation.[201] Human capacities had, accordingly, to be far more undifferentiated and open to adjustment for difference in context and challenge. This made humans develop more tentatively but also enabled them to become more complex and reflexive in their assessment and action. "The *instinctless, miserable* creature which came from nature's hands so abandoned was also from the first moment on the *freely active, rational* creature which was destined to help itself, and inevitably had the ability to do so. All his shortcomings and needs as an animal were pressing reasons to prove himself with all his forces as a human being."[202] Uniquely, "the human has no single work, . . . but he has free space to practice in many things, and hence to improve himself constantly."[203] Herder wanted to use this to establish a species differentiation: "in the hollow of that great bereftness of drives to art, the *germ of a substitute* . . . would be a genetic proof that '*the true orientation of humanity*' lies here, . . . that the human species does not stand above the animals in *levels* of more or less, but in *kind*."[204] Herder offered his own term for this holistic, integral capacity in humans: *Besonnenheit*, "*the total determination of his thinking forces in relation to his sensuality and drives*."[205] Or rather, it is "*the single positive force of thought*, which, bound up with a certain *organization of the body*, is called *reason* in the case of human beings, just as it becomes *ability for art* in the case of animals, which is called *freedom* in the case of the human being, and in the case of animals becomes instinct."[206] Herder offered this theory as an instance of a general principle of compensation that he termed a "great relationship, which runs through the chain of living beings," whereby the absence of one endowment or capacity occasioned the elaboration of another.[207] This idea of a "general economy of animal life" would prove seminal for the development of life science, especially as taken up and elaborated by Carl Friedrich Kielmeyer.

Across his entire oeuvre Herder aimed to distinguish humankind "on naturalistic and physiological grounds alone, with reference to the structure and organisation of the human body."[208] "Reason is no compartmentalized, separately effective force but an orientation of all forces that is distinctive to his species."[209] Accordingly, "no psychology is possible which is not step for step determinate physiology."[210] That is, a "science of the soul must become entirely natural science"; it must seek out empirically the evidence of the action of forces and establish as coherently as possible how they interact to produce the complex ultimate phenomenon of human experience.[211] What he called "physiological psychology" represented his most encompassing contribution to life science. He endeavored to understand concretely and in detail how the specific senses served as organs of emergent lucidity. "It is a very difficult matter to trace every science in all its concepts and every language in all its words back to the senses in which and for which they arose, and yet that is essential for every science and every language."[212] Here was the most original and powerful aspect of his approach. He worked up a theory of developmental psychology grounded in the specificity of each of the senses, which he found confirmed in the characteristics of the particular forms of fine art that appealed to them.[213] He first articulated this theory in 1769, in *Viertes kritischen Wäldchen* and "Zum Sinn des Gefühls," and he developed it fully over the 1770s in *Vom Erkennen und Empfinden der Menschlichen Seele* and in *Plastik*.

Herder drew his inspiration significantly from explorations of sense capacities by Frenchmen, especially Condillac and Diderot.[214] He had begun to study Condillac in 1764 and followed their writings avidly.[215] He sought to link the insights of the French psychologists with his own theory of the development of the senses as reflected in the character and emergence of the forms of art. In *Traité des sensations*, Condillac attempted to derive everything in consciousness from sensory perception; even reflection should be explained strictly from the juxtaposition of sensations. He expressed this in a famous analogy of a marble statue coming to life, as in the ancient myth of Pygmalion and Galatea.[216] Herder was especially fascinated with this analogy and with Condillac's stress on the sense of touch. "Touch is at once the first, certain, and faithful [*treue*] sense that emerges: it is already in its first stages of development in the embryo, and only gradually over time do the other senses distinguish themselves from it."[217] "It is exactly the same in the history of art among the peoples as with the history of human nature. Formation for the sense of touch had long been in place before representation for the sense of sight could emerge."[218]

Far from rejecting the impulses of his century emanating out of France, Herder embraced them.[219] In addition to the work on the psychology of the senses, he was drawn to the natural history based on comparative anatomy developed by Louis-Jean-Marie Daubenton and theorized by Buffon, Maupertuis, Diderot, and La Mettrie. The ancient materialism of Epicurus and Lucretius proved prominent here, as did the modern version of Spinoza.[220] There was a *choice* in aligning himself with Lucretius, with Spinoza, with Condillac and Diderot, with Buffon and Daubenton. Such drastic naturalism seemed to betoken that "mankind would have to be regarded as abandoned by universal providence to its own resources."[221] For Bonnet and Haller and Reimarus—even for Buffon and Kant—divine providence was *not* dispensable. Herder of course believed thoroughly in divine providence, but his method of exposition consistently eschewed it.[222]

The wholeness of man as revealed in the developmental psychology of the human faculties became the theme of Herder's decisive work of the 1770s, *Vom Erkennen und Empfinden der menschlichen Seele*. For Herder, the task was to get beyond the arid formalities of Wolffian school philosophy.[223] He hoped to replace metaphysical dualism with a sensual self immersed in the real world via space, time, and force.[224] Above all, he insisted, "the human being is feeling through and through."[225] "What are all the senses but mere modes of representation of a single positive force of the soul?"[226] "We are a single thinking *sensorium commune*, only touched from various sides."[227] The essential point was to establish *continuity* between the physical and the mental. Herder turned to the most important physiologist of his day, Albrecht von Haller, to exploit his theory of *Reiz*, or irritability, to establish this thoroughgoing unity.[228] Yet he needed to reconstruct Haller as well: *physiological* dualism of irritability versus sensibility created yet another obstacle to the holistic understanding of man.

The crucial move of Herder's text was to focus on "the broad region of sensations, drives, effects, of action" at "the heart of our being."[229] *Heart* proved the rhetorically decisive term. It operated on the two registers Herder essentially sought to fuse: the heart is a *muscle*, and hence it fell legitimately within the most literal sphere of Haller's physiology of *Reiz*, and yet, of course, metaphorically, the heart is the center of feeling, emotion, passion, spirit. It has long stood as the symbol of the embodied vitality of human experience. Thus, it became the figural "seat of the soul" in the rhetoric of Herder's exposition. It was always "around" the heart that Herder pitched his claim that "the soul with all its forces feels itself *living*. . . . [I]t is only *present* in the universe through action and reaction on this body full of sensations."[230] "The entire inner man is one. All passions ring round the heart."[231] "Our entire internal, excitable Self, from that inexhaustible fount of excitation, the heart, down to the smallest fiber animated by *Reiz*, follows these simple laws."[232]

In his text of 1775, Herder wrote, "Perhaps dead matter has wound through all the stages and steps of mechanism and raised itself to the little spark of life that is only the beginning of organization, and yet how powerfully it still surges in the feelings of a human soul!"[233] *Reiz* was Herder's key to this crucial metamorphosis. "Quite generally, nothing in nature is separated, everything flows onto and into everything else through imperceptible transitions."[234] "All life expands and leaps to higher stages."[235] "In the abyss of irritability lies the seed of all sensibility, passion and action."[236] "The nerve proves more subtly what has been said concerning the fibers of irritation generally."[237] "Sensibility is just the aggregate of all the dark irritations, just as thought is the bright aggregate of sensibility."[238] Thus, Herder blatantly and deliberately overrode the primary distinction of his source, Haller.[239]

Herder's treatment did not arise from lack of knowledge of Haller's ideas. He had studied Haller thoroughly by 1771.[240] As Simon Richter has argued, Herder meant from the outset to "deliberately misinterpret" Haller.[241] If Haller "refused to recognize irritability as a form of life," Herder, by contrast, used Haller as evidence for the "transformation of dead matter into moving life."[242] Via the polysemantic valences of the term *Reiz,* Herder was deploying a crucial *metaphor* binding medicine with aesthetics. In medicine, *Reiz* signified simply "that which causes a physical sensation of pleasure or pain." But in aesthetics, *reizend* had a dramatically different register. It was an adjective used preponderantly to refer to "allure," the "gentle violence" of "feminine beauty," as in the phrase *reizendes Mädchen.*[243] That is, *reizend* betokened that which aroused (erotic) desire.[244] This was emotional, aesthetic—a matter, assuredly, of a *sentient* agent. Thus, "[Herder] grasped at the analogy [between the medical and the aesthetic senses of *Reiz*] as a means to bridge the gulf between irritability and sensibility."[245] That was a methodological as much as a metaphorical undertaking.[246] It used the tropic resources of language to make physiological psychology plausible.

Herder's whole invocation of *Reiz* aimed to use and abuse Haller, to misread him strongly against his own intentions. The best evidence of this lies in the passages from both 1775 and 1778 where Herder evoked Haller specifically by name. In 1775 he wrote: "Haller's work is Pygmalion's statue grown warm in the hands of a lover of humanity."[247] In 1778 he wrote: "Haller's physiological work raised to psychology and enlivened with mind like Pygmalion's statue—*then* we can say something about thinking and sensation."[248] Two things are crucial to grasp here. First, Herder claimed in each instance that Haller required a decisive *supplement* to be adequate for physiological psychology. And, second, that supplement could be found in the metaphor of Pygmalion's statue, a metaphor we could never find in Haller's work. It was, however, the decisive metaphor of Condillac and the French psychological school. Herder thus announced his subversion of Haller in the cause of vital materialism.

After midcentury the divide separating progressive from regressive "research programs" (in the Lakatosian sense) came to be willingness or unwillingness to *explore and explain* vital materialism.[249] By that test, the greatest German scientist and the greatest German philosopher of the eighteenth century (Albrecht von Haller and Immanuel Kant, respectively) appear strikingly conservative. Some intellectual rebellion was in order, and Herder sought to provide it. His rebellion against Kant will concern us later.[250] It is his subversion of Haller that is the topic here. I suggest that in his essays of the 1770s Herder did to Haller what La Mettrie had done to him in 1747: deliberately to read him in a vital-materialist manner deeply offensive to Haller *and yet* consistent with the most important impulses emergent in his own research.[251]

CASPAR FRIEDRICH WOLFF: THE EMPIRICAL ASSERTION OF EPIGENESIS

There is a great deal that remains to be done in accurately placing Caspar Friedrich Wolff in the history of life science in the eighteenth century. The two guiding lights for what I will attempt here are the works of Ilse Jahn and Shirley Roe. Jahn has opened up important new questions about the origins of Wolff's thought.[252] Roe has made the most complete effort to reconstruct and assess its character and impact.[253] While I will draw on others as well, these two scholars have framed the investigation.[254] Jahn has raised a curious anomaly to strategic significance for the reconstruction of the life science of the era of Wolff: why did a supervising professor's name not appear on Wolff's 1759 dissertation, in striking noncomformity with academic protocols of the era? While that absence had been noted and explored before, particularly by A. E. Gaissinovich, who used newly discovered archival documents to begin shedding light on the matter, Jahn has been able to raise some crucial considerations that go beyond what anyone hitherto surmised.[255] Roe has proposed that Wolff must be understood as a philosophical rationalist, deeply immersed in the way of thought of Christian Wolff and *Schulphilosophie*. I agree that we must understand Wolff's work from the vantage of his philosophy of inquiry, not simply his experimental results or their theoretical interpretation.[256] But I will try to nuance Roe's reconstruction by suggesting that the methodology embraced by Wolff was in fact more eclectic than Roe presumes and that this, too, sheds important light on the life science of his moment. Indeed, I will suggest that Wolff might have been far closer to the experimental Newtonianism that Roe has been so decisive in imputing to his great adversary, Albrecht von Haller.[257]

We know very little about Caspar Friedrich Wolff's childhood and adolescence beyond the facts that he was born and raised in Berlin, the son of a tailor.[258] He seems to have had an excellent education for someone from such a background (reflected in his competence in Latin and his familiarity with ancient authors), enabling him to enroll in studies at the Collegium Medico-Chirugicum of Berlin in 1753.[259] There he studied, above all, with the anatomist Johann Friedrich Meckel the Elder and with the botanist Johann Gottlieb Gleditsch (1714–86), the curator of the royal gardens in Berlin. In 1754 he advanced to the normal course of studies for medical students at the University of Halle, where, after nine semesters, he defended his famous dissertation, *Theoria Generationis*, in November 1759.

Wolff matriculated in the University of Halle medical faculty in the very last year of Christian Wolff's life and rectorship of that university, and well after the great luminaries of the third generation of the Halle medical faculty—Krüger and Unzer—had departed. The question is: who at Halle was left to guide and inform the young Wolff's work? How did he come to the topic and the approach of his dissertation? No one in the medical faculty of the late 1750s evidenced the boldness and vision that could have inspired the young Wolff to such a daring dissertation.[260] In particular, we have had occasion to note the mediocrity of anatomical and botanical resources at Halle, yet these were the overweening preoccupations reflected in Wolff's experimentally driven dissertation. The only Halle faculty member we know to have been present at the defense of Wolff's dissertation was the dean, Andreas Elias Büchner. Yet Wolff did not mention him in the archival document concerning his dissertation defense.[261] The most likely director of studies for Wolff, as far as Jahn can establish, was Philipp Adolf Böhmer (1717–89). "Böhmer . . . announced for summer 1756 a public lecture [series] '*de homine generatione*' [on human generation], in winter 1757/58 led a private course on anatomical preparations and demonstrations, and in summer 1758 gave a lecture series on the *Institutiones physiologicas* of [Christian Gottlieb] Ludwig [1709–73], whom [in his dissertation] Wolff explicitly named as a representative of epigenesis."[262] But there is little reason to think that Böhmer, any more than Büchner (the figure Gaissinovich believed to be Wolff's most likely mentor at Halle), would have been a theoretical supporter of the line of inquiry that Wolff undertook.

Jahn's work suggests that the inspiration may not have come from Halle at all but rather from his earlier years in Berlin, in the person of the president of the Berlin Academy, Pierre-Louis Moreau de Maupertuis. Jahn dwells on the quite specific "issues and . . . formulations of research programs concerning 'generation'" that appeared in Letter 7 ("On the Generation of Animals") and in Letter 19 ("On the Progress of the Sciences") of Maupertuis's work of 1752, which presented a "bounty of forward-looking research programs, questions and tasks to be undertaken," that can be quite specifically linked to what Wolff tackled in his dissertation and thereafter.[263] Jahn writes:

> When one considers the daring and conviction with which Wolff went about his microscopic observations after the example of Buffon and Needham, and the originality with which he interpreted theoretically his results concerning the epigenetic development of plant germs and chicken embryos, whereby he introduced a new "theory of generation," one can certainly surmise that he carried out the commission of Maupertuis and simultaneously redeemed the latter's legacy to the Berlin academicians. This impression grows yet stronger when one also takes into consideration the later works, [especially] the work on the formation of the intestine in the fertilized chick (Wolff 1766–1767), which was composed while he was still in Berlin, [and] which Karl Ernst von Baer characterized as "the greatest masterpiece that we know in the field of observational science."[264]

Particularly revealing is that Maupertuis celebrated the microscopic experiments of Buffon and Needham as setting the terms of investigation for a whole "new nature."[265] As we have seen, Reimarus and Bonnet were arguing for the utter inadequacy of these same experiments and their theoretical interpretation. Far more pertinently, similar objections had been raised by Albrecht von Haller.[266] That Wolff nonetheless took up this specific experimental tradition suggests a positive inspiration that could not have come from any plausible source in Germany besides Maupertuis.

For evidence that Maupertuis was actively eliciting an endeavor along such lines in Germany, Jahn points to a remarkable prize competition announced around 1755 for the section of experimental natural philosophy: "Whether all living things, as much in the animal kingdom as in the vegetal kingdom, arise from an egg fertilized by a germ, or from a prolific matter, analogous to the germ [*une matière polifique, analogue au germe*]."[267] Jahn notes that *no* answers were submitted, even when the competition was extended to 1759. A renewed call in 1761–63 elicited four submissions but no prize was awarded.[268] Quite simply, no one in the radius of the Berlin Academy wanted anything to do with this question—no one, that is, but Caspar Friedrich Wolff!

Thus, crucially, Wolff emerges as a German naturalist directly and positively inspired by the innovations of French vital materialism. His radical departure from the preponderant way of thinking that characterized German natural philosophy, culminating in his ruinous controversy with Albrecht von Haller, may also help to account for the fierce hostility that he met when he sought to teach in his alma mater, the Collegium Medico-

Chirugicum in Berlin, after his service in the Seven Years' War as a Prussian military surgeon.[269] His former instructor, Meckel, had not only his professorial privileges to uphold but also his personal allegiances to the figure whom Wolff's work most directly challenged, his own mentor Albrecht von Haller.[270] The bitterness of the conflict over Wolff's private course offerings in Berlin explains not only his failure to obtain a position at the collegium, though two came open in the mid-1760s for which he applied, but also his decision to accept a call to the Saint Petersburg Academy of Sciences in 1766–67.[271] That call had been instigated, notably, through his recommendation to the Russian authorities as early as 1760 by Leonhard Euler, who clearly knew of and praised highly his dissertation work.[272] Euler had assumed direction of the Berlin Academy upon Maupertuis's departure in 1756. Jahn notes the strong criticism and intervention that Maupertuis brought to bear upon the Collegium Medico-Chirugicum, starting in the late 1740s, for laxity in scholarship and research.[273] Maupertuis appointed both Meckel and Gleditsch to the collegium in order to raise its intellectual tone.[274] But that does not imply that they would show any loyalty to Maupertuis or to his agenda for research, especially after 1751—though, to be sure, they would not overtly resist Frederick II, who upheld Maupertuis's institutional agenda at least. Euler stands in clear contrast to the members of the Berlin Academy associated with the collegium, whose conservatism, I am suggesting, was not only institutional but intellectual.

What exactly do I mean by such intellectual conservatism? To get at that, we must turn to the miserable efforts of Caspar Friedrich Wolff to secure the approval of the dominant force in German life science at the time, Albrecht von Haller. To even suggest intellectual conservatism in regard to Haller seems outrageous: he was one of the most creative scholars of his time. And yet, he was a very problematic figure, especially after his return to Switzerland. Even as he continued his pioneering work—here most centrally on embryology—he was becoming more and more punctilious about his personal authority over life science in Europe (especially after the enormous controversy spawned by his Göttingen Academy Address on irritability and sensibility of 1752) and adamant in his opposition to the "flood of freethinking" he saw closely associated with the rival camp in life science headed by Buffon and Maupertuis.[275] Here, his new alliance with Charles Bonnet played a decisive role, as revealed in the specifics of his treatment of Wolff.

After a seemingly generous opening line in his 1760 review of Wolff's dissertation, Haller proceeded over the balance of his review to browbeat his young rival in no uncertain terms.[276] He was part of the same academic culture in the medical-scientific world of midcentury Germany that spurned Wolff as an insouciant beginner. Ironically, Wolff hoped for better from Haller. I believe that Wolff really expected that Haller would find his work not only experimentally but theoretically congenial.[277] When Wolff began his investigations, he was not entirely aware of Haller's fateful conversion to preformation.[278] The major articulation of that conversion appeared little more than a year before Wolff defended his dissertation and probably before Wolff could adapt his experiments or their interpretation to incorporate Haller's views.[279] To be sure, Wolff would never have countenanced preformation, even under the auspices of Haller; his experiments and their epigenetic interpretation remained incontrovertible, as he saw it, and as he would establish beyond question in his later work on the development of the intestine.[280] Alas, Haller summarily dismissed even that work which would later set the standard for experimental embryology.[281]

The whole protracted controversy shows how long Wolff held out the hope that Haller would see the light in the experimental findings and return to an epigenetic orientation. It was only upon the publication of volume 8 of Haller's great compendium of physiology, where the issues of generation were treated in a bluntly preformationist manner, that Wolff saw how intransigent Haller had become. "There Is Nothing of Epigenesis," the title of Haller's chapter read.[282] The theological origin of all generation was stated here in a balder form than anywhere else in his works, together with a ridiculous calculation of all the humans that must have been encapsulated at the original creation of Eve. For Wolff, this must have appeared strikingly reactionary.[283] He gave up eliciting Haller's approval and sought, instead, the common academic truce of agreeing to disagree.[284] Haller ignored this gesture from his young opponent; it hardly mattered after the latter withdrew to Saint Petersburg.

Wolff's response to Haller's irritated warning in 1766 that epigenesis was dangerous to religion suggests that by that point he had come to recognize an ideological dimension that went beyond the experimental uncertainties and theoretical divergences that had driven their controversy earlier.[285] Wolff conjectured bitterly that Bonnet had an influence on the hardening of Haller's position—a suspicion that we can now verify from the Bonnet-Haller correspondence and from Bonnet's extensive invocations of Haller in his works of the 1760s and even more in their revised editions published after Haller's death.[286] The last thing Wolff needed, in his professional difficulties, was to be lumped with materialists and atheists. He answered the Haller-Bonnet religious scruple in a manner that was strikingly in line, nonetheless, with Maupertuis. A rational-scientific explanation of the world, in his view (as in that of Maupertuis), in no way gainsaid the need for an intelligent creator at its origin.[287] While he admitted that preformation was peculiarly suited to religious exploitation, he denied that epigenesis threatened religion. As he wrote in his letter to Haller, "against the existence of divine Power, nothing is demonstrated if bodies are produced by natural forces and natural causes; for these forces and causes themselves, and indeed nature itself, require an author in the same way as organic bodies."[288] The essential point, here, is that Wolff set out from the view that natural science need not seek or serve religious agendas but rather that its point was to be *natural*—that is, empirical and rational.

We have come to the domain of Shirley Roe. What was Wolff's philosophical orientation to natural inquiry and how did he get there? "One of the most striking aspects of Wolff's writings, particularly the *Theoria generationis*," she notes, "is their scholastic, deductive style of presentation. Wolff's dissertation is a model of the 'mathematical method' championed by Christian Wolff as the universal language to be used by philosophers and scientists alike."[289] Thus, she judges, "Christian Wolff, and the 'Wolffian philosophy,' was of enormous influence on Caspar Friedrich Wolff."[290] What the latter endeavored to create with his dissertation was "the first rationalist embryology," in the form of "a deductive scheme based on principles, definitions, scholia, and syllogistic reasoning."[291]

Anyone who takes up the *Theoria Generationis* will find this immediately persuasive. But, I wish to suggest, that may well be too easy. By the time of Christian Wolff's death in 1756 a considerable amount of external criticism and internal mutation had befallen the "Wolffian philosophy."[292] Above all, the suitability of the "mathematical method" for "philosophers and scientists alike" had come into question.[293] We need to ask whether Caspar Friedrich Wolff's rationalism was so simply "Wolffian" as Roe maintains. That he *was* a "rationalist" is not in question; but whether he was a rationalist in a sense that rendered building inferences drawn from observation and experience superfluous, or that construed rational deduction a priori as the ultimate and only "science"—these are questions that need to be examined a bit more carefully. Indeed, one might wish to proceed with caution even in ascribing such views to Christian Wolff himself!

In both the *Theoria Generationis* and the *Theorie von der Generation*, which elaborated and defended it in 1764, one finds much that fits Roe's characterization, but also some things that don't. Certainly, Caspar Friedrich

Wolff embraced Leibniz's core principle of sufficient reason. Its role, as Roe writes, "cannot be overestimated."[294] What that principle prescribed, in her reconstruction, is that "explanation should consist not simply in describing a phenomenon but in showing the reason *why* it exists as it does."[295] The problem, here, is that one need hardly be Wolffian or Leibnizian to embrace this characterization of explanation. Everything hinges on what the phrase "showing the reason *why* it exists as it does" means. It can mean, for example, *any* account that makes a claim to determinate knowledge, or it can mean the high metaphysical claim that the account is exhaustively and immutably certain. As Roe understands German school philosophy, "the key to explanation is logical demonstration" that "something 'must be so and cannot be otherwise, it must necessarily behave thus.'"[296] The phrase "logical demonstration" betokens the deductive model of proof in the "geometric method."[297] The key element Roe picks up from Wolff's own words is the claim to *necessity*.

Now, what was happening in scientific argumentation in the mid-eighteenth century was precisely the mutation of this kind of claim to metaphysical necessity into a claim of intersubjective warrant, grounded in consilience of evidence: an a posteriori but no less emphatic claim to valid knowledge. This, I suggest, was the epistemological core of "experimental Newtonianism." If Roe is right, Caspar Friedrich Wolff could not be associated with such a development. But I think Wolff was in fact more eclectic than the explicit pronouncements thus far considered might suggest. There was more going on in his method. In this, I think Gaissinovich has a better insight in linking Wolff to experimental Newtonianism.[298] To go still further, Olaf Breidbach may not be too far afield in seeing elements of Spinozistic immanence in Wolff.[299] I don't want to suggest that Roe was wrong in finding metaphysical deductivism in the dissertation (and in the 1764 defense), but only that this was not everything that was going on in Wolff's natural philosophy. This should not imply, either, that Wolff was confused. I think one can be eclectic and coherent, and I think Wolff was. The burden of proof, of course, falls on me.

Let me start with what Roe herself offers: Wolff's embryology "also had its roots in the mechanism-vitalism controversy," which she believes lingered in the atmosphere of the University of Halle from the days of Hoffmann and Stahl.[300] Certainly, the final argument of Wolff's dissertation, with its striking claim that his work had demonstrated that "mechanical medicine" was an illusion, suggests not only that this was prominent in his concern but that he took a very definitive stance on the question.[301] But we need to be careful not to take this really to mean complete rejection of mechanism. Roe maintains—correctly, in my view—that "Wolff attempted . . . to steer a middle course between the reductionism of mechanism and the inexplicability of vitalism."[302] That, certainly, is where the leading minds of the third generation of Halle medicine—Krüger and Unzer—came out.[303] It would not be surprising if less than a decade after their departure (and less than that since the publication of some of their key works), such ideas *did* still float about in the milieu of the Halle medical faculty. The key patron of the third generation, Büchner, was now dean of the medical faculty and rector of the university. Certainly, he is conventionally regarded as a partisan of Hoffmann and mechanism at Halle, but as I have argued in chapter 1, this was a matter of restoring balance against an excess of Stahlianism—and of a Stahlianism infused with *Waisenhaus-Medizin*—something we have no reason to think Wolff would find appealing. Roe herself points out a crucial error in the German translation of Wolff's dissertation that misunderstands Wolff's repudiation of Stahlianism as in fact his affiliation with it:[304]

> All of those functions of the body that I have denied to be mechanical, I have not explained in any way, inquiring in fact into the connection that exists between the machine and life, but by no means searching further for the causes of this where it has no dealings with the machine. If therefore you should wish to interpret my mind on this, benevolent reader, you could easily err in this. And certainly indeed and especially I would suffer [*paterer*], if you should impute to me the opinion of Stahl, or that received from him and slightly altered, which Whytt and others have recently proposed, in which, namely, the functions that occur in our body are attributed to the power of an immaterial soul, whether acting directly and freely, or coerced by the inconvenience inflicted upon it.[305]

Roe's Russian colleague Gaissinovich repeatedly insisted that Wolff was no vitalist.[306] Of course, that all hinges on exactly what "vitalism" means. There is a case to be made that even Stahl was not positing a totally separate metaphysical entity intervening in bodily processes.[307] And the question of immanent teleology in the organism is one that even today philosophy of biology has not resolved.[308]

The question of mechanism-vitalism is a theoretical issue *within* empirical inquiry into life-forms, not an epistemological issue about the *warrant* of the inquiry itself, and so we have still not resolved the question posed by Roe. To get at that, I think, we need to get deeper into Wolff's texts, not only to take up his explicitly philosophical pronouncements but to reconstruct the actual character of his argumentation. I want to start with the second, because this is where Wolff's actual approach to inquiry complicated his ostensibly school-philosophical, "mathematical method." Wolff explicitly claimed that, of the three parts of the dissertation, he had composed the first, which dealt with botany (longer than the two following parts combined), so that he might lay the groundwork for the more complex and essential topic of generation in animals. In other words, we must see the long treatment of plant generation as a *model* for the later exposition of animal generation. Crucial, here, is Wolff's insistence that plant generation and animal generation followed the same causal principle: that there was only *one* theory of generation that explained all organic life.[309]

He began by citing Stephen Hales on plant physiology.[310] The key point he derived from Hales was the ubiquitous transfer of fluids from the environment surrounding the roots to every part within the plant. This fluid *movement* required, Wolff adjudged, some explanatory *force*.[311] Attraction was not sufficient for the purpose.[312] Since fluid transfer was essential to the metabolism of the plant, he termed the requisite cause simply an "essential force [*vis essentialis*]." The effect was ubiquitous and central to plant physiology, so the cause had to be equally ubiquitous and central—in a word, essential. There is, here, nothing particularly metaphysical. It is a simple inference that for a given effect one should presume a given cause and, parsimoniously, the *same* cause for the same effect, as Newton prescribed.[313] Wolff then proceeded to an examination—macroscopic and microscopic—of this fluid propagation in the plant and of the substance of the fluid itself. He noted that especially in new plant tissue at the microscopic level this fluid substance was unstructured. He could isolate drops of this fluid, clear and free-flowing, and he could agglomerate these drops to one another or separate them into droplets arbitrarily, since they had no inherent determinacy.[314] But as this fluid moved or settled, it congealed into more viscous and eventually solid structures in the plant, accounting, in that process, for a growth within or an extension outward of the existing plant structure.

The key point I wish to stress, here, is that there is no derivation a priori in this account. It is entirely a matter of experimental observation and concrete description. The first twenty propositions of part 1, then, are *entirely* empirical, not derived from axioms in some logico-deductive manner. In §21, Wolff did draw an inference that he claimed was necessary ("thus it follows with necessity"), namely, that the assimilation of fluids was the source of the growth of plant tissue.[315] I can take this only as an *inductive* inference, not a deduction, and the "necessity" Wolff claimed can only be a matter of what we would call inference to the best explanation, not demonstrative logic.[316] Given that inference, Wolff felt entitled to ask after the *cause* of this established regularity. It was, of course, the "essential force" he inferred from Hales at the outset, "through which the compressed fluids penetrate and expand the plant substance."[317] The *cause* was recognized in and through its effects; its "essence" could not at all be specified, nor the mechanism of its effectiveness.

In §27, Wolff noted that the fluid substance naturally became viscous and eventually formed more solid structures; thus, he imputed to it a *propensity* to solidification: *solidescibilitas* (German, *Erstarrungsfähigkeit*). Therefore, this fluid was not a simple liquid (water) but rather a compound of elements that nourished and helped plant tissue to grow—a *Nahrungssaft*, or nutritive fluid. Wolff insisted that this same kind of basic nutritive liquid was present in animals, citing Boerhaave and Haller's *First Lines of Physiology*.[318] Nutrition and growth, then, were processes of continuous transformation of liquids into solids. This whole approach in terms of the relation of the fluid to the solid was central to the medical physiology taught by Boerhaave—which Wolff knew, as his references indicate, from the Haller compilation and translation. Wolff drew his crucial conclusion: "the parts develop in the manner that their substance first arises as a simple mixture without any internal organic structure, and then in this substance in the manner explained above (§29) vessels [*Gefässe*] and nodes [*Bläschen*] get formed."[319] This was the core of Wolff's epigenetic theory. In §34, he explained that in science when two phenomena were in constant conjunction, the effort turned to finding what the causal relation between them had to be: which was the source of the other. In the case of the fluid substance and the tissue nodes, Wolff argued, clearly the former had causal precedence.[320] This was "logical," to be sure, but it hardly betokens some grand metaphysical system of deduction. There is a "principle of sufficient reason" at play, but we could call it something less grandiose, such as sound inference. My point is not that Wolff did not invoke the principle of sufficient reason. It is, rather, that he didn't have to. In invoking it, Wolff was putting on a mantle of "*philosophical* knowledge" that dressed him in Leibnizian purple, but what he was doing was simply experimental scientific inference.

The question then arises: why did he feel he needed this authorization? That has to do with what he understood as "theory" and as "science." We get intimations of that in the opening propositions of the *Theoria Generationis*, where he was confident that his readership already agreed with him on philosophical knowledge, that this was conventional wisdom, as indeed school philosophy *was* in academic Germany at that moment.[321] Wolff's full defense of his way of thinking came only in the *Theorie von der Generation*, when he recognized he needed to defend himself from strong criticism of his procedure, his results, and their warrant. The overt argument about the method of inquiry in the latter work is thus more revealing than the "scholastic" presentation of the dissertation, which I am suggesting offered a modern empirical inquiry in traditional costume.[322] One of the provocations that Wolff explicitly articulated in the preface to his *Theorie von der Generation* was Charles Bonnet's somewhat-cavalier use of the technical term "demonstration" throughout his *Considérations sur les corps organisés* (1762).[323] As a scholar trained in German logic, Wolff took Bonnet to task for not understanding the true meaning of this term. That made him eager to show his own, more precise understanding of, and proper implementation of, such logical forms in his own work.

The point of departure of his argument was that notwithstanding all that had been written before on the problem of generation, no true *theory* had ever been offered.[324] Instead, what had proliferated were descriptions—"*histories*," as Wolff termed them, in keeping with the distinction prominent in his time.[325] He professed wonderment that scholars should believe that their *histories* could count as *explanations*.[326] Wolff specified what he took to be the insufficiency of "historical" (i.e., descriptive) accounts: "not to be concerned with why things have these properties and not others, and why matters transpire in this way rather than another."[327] For Wolff, effective explanation required that one grasp something in terms of its reasons and causes (*Gründen und Ursachen*).[328] This was "philosophical knowledge," or "science" (*Wissenschaft*).[329] For Wolff, a *theory of generation* specifically should "construe everything . . . from the quality [*Beschaffenheit*] of the forces through which the organic bodies are formed, and from the properties [*Eigenschaften*] of the substance from which they are to be formed."[330] That is to say, an explanation would derive the ultimate character of organic forms from a *causal account* that necessarily produced the forms from these two originating elements of force and (primordial) material.

This was where Wolff invoked the Leibnizian term "principle of sufficient reason" as the character of such an adequate causal account deriving "an organic body from its causes."[331] He termed anatomy merely descriptive ("historical") knowledge but deemed a theory of generation to be *philosophical* (scientific) knowledge. Physiology, in Wolff's view, depended upon the structure of the organic body described in anatomy and therefore could not be the science that would explain how that anatomy itself took form.[332] Thus, "physiology relates to anatomy precisely [*accurat*—one of Wolff's favorite words] as a corollary to its theorems, from which it derives; my theory relates to anatomy as a demonstration of these theorems to the theorems themselves."[333] In this formulation, Wolff was invoking philosophical logic *analogically* to explain the relation among scientific inquiries in which his own specific pursuit should be situated. He drew an explicit analogy to the Wolffian system: "The doctrine of generation is related to anatomy as rational psychology to empirical."[334] This is what grounds Roe's interpretation, but from our vantage, this is not as transparent as the young Wolff believed, for we do not find the relation between empirical and rational psychology in Christian Wolff quite so unequivocal.[335]

Wolff's experimental observations concentrated on the fertilized chicken embryo, following upon the work of Malpighi (and, without realizing it, paralleling Haller's concurrent research, which appeared in 1758). The details of his observations, the contrast with Haller's, and the ensuing disputes about the technicalities of the embryological observations have been discussed elegantly and thoroughly by Roe and others, notably Maria Monti, and I need not go into these.[336] In §166, the opening section of part 2, on the development of animals, Wolff asserted the parallel of fluidity at the origin of organic form in animals that he had developed in his account of plant growth: "a mass that . . . in general consists merely of a few loosely conglomerated little globes [*Kügelchen*] and simply amassed one upon another, transparent, movable, and almost fluid."[337] Wolff insisted that the cause of organic formation was, just as with plants, the "essential force," which induced movement of the nutritive fluid to all the parts of the organism—notably, at the outset, without the need of a vessel system or a heart to pump it.[338] He drew comparison to the work on the movement of lymphatic fluids by Meckel and Alexander Monro.[339] In §242, Wolff enunciated his core thesis succinctly: "Thus, the essential force, together with the solidification propensity of the nutritive fluid, constitutes a sufficient principle of all development both in plants and in animals."[340] The crucial difference between animal organization and plant organization was that the solidification propensity in animal tissue worked far more slowly than in plants.[341]

Roe asks, "what *is* the essential force, and how does it operate?"[342] She observes, "Wolff does not explain the nature and operation of the essential force in any greater detail. He simply claims that it exists and then proceeds to show how, together with secretion and solidification, the essential force produces different structures."[343] It is true Wolff did not explain that, and he was quite aware that he did not: "It is enough for us to know that it [the essential force] is there and to recognize it from its effects."[344] That is, Wolff adopted the methodological posture of eighteenth-century experimental Newtonianism. Forces were in "essence" or origin "obscure" and beyond the reach of science, yet they could be postulated on the basis of their palpable effects. Of all his methodological principles, this is probably the most important, and it has nothing to do with Wolffianism and everything to do with experimental Newtonianism.[345]

In §161, Wolff posed an argument against the claim common among preformationists that the entire structure of the organism was already present at conception but simply still invisible.[346] While this was a view commonly ascribed to Malpighi, Wolff noted correctly in his later *Theorie von der Generation* that this was not in fact the Italian researcher's view but rather one that Haller developed on the basis of Malpighi's observations.[347] He argued that even with a medium-power microscope the basic particles (*Kügelchen*) were clearly discernible, and thus, he found it incongruous to claim that the parts that they should compose could remain invisible, concluding vigorously: "Thus, that parts are hidden because of their infinite smallness and only gradually appear is a fable."[348] Before he spoke so boldly, Wolff had offered a disclaimer: "in general one cannot say that what is not accessible to our senses for that very reason is not present." But he argued that this protest was more "clever [*spitzfindig*] than true" when applied to the microscopic observations in question.[349] Thus, he opened himself up to what became Haller's most insistent criticism: that Wolff's whole methodology was grounded upon the claim that what one didn't see could not be there.[350] Wolff struggled to establish that this was an aside—a *scholium*, in school-philosophical language—not a principle on which his argument was grounded.[351] But the criticism remained, and Bonnet would trumpet it over and over again in his works of the 1760s and thereafter.[352]

The final section of Wolff's dissertation elaborated his general theoretical insights and, tangentially, related them to the work of others. In §231, the opening section of this part, Wolff explained that he deliberately set about articulating his own experimental observations and conclusions without mention of parallels in the work of others so that his results should be judged by the reader on their own merits.[353] In the next section, he acknowledged C. G. Ludwig and J. T. Needham as predecessors associated with epigenesis, linking them all the way back to Aristotle and, more recently, Harvey.[354] He claimed in the following section that all these figures tacitly worked with something like his own "essential force." He noted that Needham wrote of "an expansive and vegetative force": "By expansive force Needham understands the force of the growing matter through which a specific point of the latter attempts to remove itself to the outermost [*ins Unendliche*] from those nearby."[355] Wolff argued that his own theory of essential force was more concrete in its characterization of the material and the process of fluid movement and tissue formation. Wolff's close attention to Needham's theory, but not his distinctions from it, allowed Haller, in his review, to claim that Wolff's work was just an experimental elaboration of Needham's ideas.[356] That led Wolff to articulate a far more extended and critical assessment of Needham in his *Theorie von der Generation*. There, he argued that Needham was not really after a theory of propagation of organisms but rather a theory of spontaneous generation, of the formation of the organic directly out of inorganic materials.[357] By contrast, Wolff claimed always to work within an organism's self-formative processes.[358] For him, all generation was nutritive growth.[359] Hence, his research project was, by and large, indifferent to sexuality in reproduction or even to the moment or role of fertilization.

Theorie von der Generation not only inveighed against the preformationist notion, by which, it alleged, all organic bodies became blunt miracles, but offered a clear sense of the *theoretical* alternative offered by *epigenesis*: "a nature that destroyed itself and that created itself again anew, in order to produce endless changes, and to appear again and again from a new side," "a living nature, which through its own forces produced endless changes."[360] Wolff's *Theorie von der Generation* aimed not only to persuade the learned public of his epigenetic approach but also to (re)convert Haller. He wrote: "I am confident that [Haller], when I have only had the opportunity to present my reasons, [which are based on a knowledge of animal physiology] with which he is quite familiar, will soon completely agree with me and allow the argument for continuity [epigenesis] to go forward. *From Mr. Bonnet* I cannot hope the same. He seems to me, like many who take themselves for physiologists, to be quite far from such a knowledge of the nature of animals."[361] Wolff had little respect for Bonnet: "everything that Mr. Bonnet has written is for the most part derived from Mr. von Haller. Here and there he has mixed in a little bit from other authors. Of his own there is, on the other hand, nothing."[362] This withering judgment occasioned Haller to chastise Wolff for impertinence toward his elders![363] Wolff was harsh, to be sure, but Bonnet had it coming. Bonnet was a preformationist "true believer," and by 1760 he did not really *care* what Wolff had found, since the whole matter had been settled in his view by the incomparable Haller's work on the chicken embryo. During his last years in Berlin Wolff worked tirelessly on the formation of the intestine in the chicken embryo to prove he was right. He only published these materials in Saint Petersburg. Haller at least bothered to read them, then dismissed them. Karl Ernst von Baer, certainly an experimental biologist of a class with Haller, had the last word, but it was too late for Wolff ever to hear it.

JOHANN AUGUST UNZER AND THE BALANCE OF GERMAN PHYSIOLOGY BY 1770

In 1771 the former Halle "philosophical physician" Johann August Unzer published the culminating work of his life, and a milestone in the development of German physiology: *Erste Gründe einer Physiologie der eigentlichen thierischen Natur thierischer Körper* (The first principles of a physiology of the authentically animal nature of animal bodies).[364] In the interval since leaving Halle, Unzer had lived in the Hamburg area, participating in the key science journal *Hamburgisches Magazin*. Then, from 1759 to 1764, he produced his own journal, *Der Arzt*, an important vehicle for "securing a proper estimate of physicians and of the art of healing, and of extending sound medical knowledge" to the wider German public.[365] That is, Unzer's journal served the primary professional ambition of the medical community in Germany, as Thomas Broman has reconstructed it for this period.[366] But Unzer was also deeply committed to theoretical physiology, carrying forward the research program developed at Halle in the 1740s under the leadership of his mentor, Johann Gottlob Krüger, and concentrating on neurophysiology. This theoretical interest moved Unzer first to publish a series of popular articles in *Der Arzt*, then to articulate his views in a more rigorously scholarly form. In 1768 he published *Grundriss eines Lehrgebäudes von der Sinnlichkeit der thierischen Körper* (Outline of a theory of the sensibility of animal bodies), but he found it necessary, over the next three years, to revise and extend the ideas developed there, culminating in his decisive publication of 1771.[367]

Unzer belongs in a distinctive line within German professional medicine that moved away from concerns with practice toward experimental physiology as a comparative zoology. Unzer made this shift in orientation clear in the final sentence of his preface to *Erste Gründe einer Physiologie*: "Man is by no means the only subject of this work[;] . . . it contains rather the principles of a Zoology, or natural history of every species of animal . . . according to their peculiar animal forces."[368] He put theoretical physiology before therapeutic medicine: how could physicians develop therapies, he demanded, when they had no conception of the physiology they were attempting to affect?[369] Indeed, his shift away from considerations of clinical practice was even more blatant, since Unzer explicitly declined to elaborate pathological diagnostics or therapeutic applications of his theoretical neurology.[370]

Unzer's monograph represented the first major departure in German physiology after Haller had consolidated his eminence with his grand compendium *Elementa Physiologiae Corporis Humani* (Physiological elements of the human body; 8 vols., 1757–66). Unzer acknowledged Haller as "the greatest anatomist and physiologist of the day" and used copious citations from and references to Haller throughout his study.[371] Haller had characterized "the mechanisms of all parts of the animal body," especially in the greater *Physiology*, "in a manner almost impossible to be surpassed."[372] Yet at the same time Unzer offered a searching and thorough critique of Haller's "new doctrine in physiology" concerning irritability (*Reizbarkeit*; *vis insita*).[373] Haller's theory held that muscle action was "neither animal nor sentient." It was mechanical—indeed, strictly speaking, it was *inanimate* (and, as such, it had nothing to do with life). The result was a starkly reductive dualism of animating "soul" and (not quite inert) matter. "Haller seems to be of the opinion, that no movements except the voluntary are produced by the soul. . . . He recognizes, nevertheless, the action of the imagination, sensations, instincts, and emotions. . . . It necessarily follows that the sensational conceptions, desires, instincts, etc., are not mental but corporeal. . . . But no sound metaphysician can grant such a confusion of ideas."[374] The "empirical psychology" of Wolffian school metaphysics was more plausible than Haller on this score, Unzer believed.[375] The "lower faculties" of mind (*Gemüth*) were really mental, not mechanical, though they were stimulated by and reacted upon physiological phenomena, and this was as true for the "higher" animals as for humans.

Unzer offered a radically divergent *reconstruction* of Haller's experimental results. He disputed, not what Haller *observed* in muscular contractions, but rather how Haller *interpreted* these observations: the issue was not (ambiguous) *observations* but *inferences* from them. Haller's experimental evidence needed another interpretation. Unzer offered a coherent alternative: "that which Haller terms *vis insita* [*angeborne Kraft*] . . . is nothing else than the *vis nervosa* [*Nervenkraft*] of external impressions exciting direct nerve-actions."[376] Thus, Unzer could both recognize the power of the experimental evidence Haller had amassed concerning irritability and yet redirect that evidence to support a neurophysiological alternative to Haller's theory. While Haller was obviously paramount in his theoretical environment, Unzer was situating his work within the larger area of the relation between consciousness and physiology. The controversy had become more pointed in the exchange between Haller and Robert Whytt.[377] Unzer denied the diffusion of the soul throughout the body, Whytt's major contention against Haller.[378] Simultaneously, he rejected the materialist thesis that all physiological phenomena could be reduced to physicomechanical causes.[379] But he was interested in theorizing how sensation and especially volition could work *physiologically*, and this meant developing a theory of the brain acting in consonance with conscious states (and that led him, even more importantly, to questions of the nervous system acting even without consciousness).

Unzer's vantage was expressed pointedly in his book title: what was it that made animals *distinctive*? For him, the answer lay in neurophysiology. The study of the brain and nerves represented the key domain of Unzer's investigation.[380] He regarded these as the distinctive "animal machine" in the complex "animal economy."[381] Brain and nerve functions "must be considered as the most fundamental and most general *principles* of the whole animal machine."[382] Unzer aimed to establish neurophysiology as "a branch of science altogether distinct," thus "separating authentically *animal* physiology from the general physiology of the entire animal economy."[383] While neurophysiology was part of the *general* physiology of which Haller was the grand master, it delved more deeply than Haller had done into the essentially directive role of the nervous system in animals.[384] The proper object of neurophysiology was to reconstruct, on the basis of experimental and observational evidence, how brain concomitants of sentience ("material ideas") communicated, via the nerves, with the rest of the bodily "machine." There had been significant advances in the field since Haller's great Academy Address on irritability and sensibility in 1752, in particular in the work of Alexander Monro II in Scotland and Felice Fontana in Italy.[385] On the other hand, Unzer had only disdain for the physiological psychology that Charles Bonnet had developed in the same period.[386] Unzer proposed to develop his own synthesis around the central conception of *Nervenkräfte*, or what Laycock rendered for his English translation by the Latin phrase *vis nervosa*.[387] Unzer's goal was twofold: first, to displace Haller's *vis insita* theory of muscle irritability with a theory of nerve-action (*vis nervosa*) and, second, to show how nerve-action worked, even without the "soul" (consciousness or sentience), to drive animal functions. His fundamental conclusion asserted: "all the processes required for the life and preservation of an animal organism, can be effected by the *vis nervosa* only."[388]

Unzer agreed with Haller that some animal functions did not require the active role of consciousness or the "soul." But while Haller wished to infer that this made these responses merely mechanical, in accordance with his dichotomy between irritability (*Reizbarkeit*) and sensibility (*Empflindlichkeit*), Unzer concluded—and this was his key theoretical innovation—that the nerve-force could operate autonomously of the directive intervention of consciousness or the "soul." Unzer postulated independent nervous function both within sentient animals (*beseelten Thieren*)—either in collaboration with voluntary nervous functions or in the absence or interruption of these—and within "un-souled" animals. Hence, he was pioneering what has become the conception of the "autonomic nervous system," but only as part of an even wider comparative physiology of nerve function that could account not only for bodily reactions in higher animals despite the interruption of voluntary nerve impulses but also for the animal functions of organisms without brains. His key thesis was that animal life manifested an independent, nonconscious, nonvolitional nerve action that could not be explained by physical-mechanical laws but functioned sui generis.[389] The strategy Unzer proposed was to work from experimental observations of nerve function both with and without consciousness.[390]

Unzer repeatedly emphasized the limits of empirical knowledge concerning neurophysiology.[391] Yet he also insisted that there was observational evidence sufficient to infer from discerned effects to causal forces active in the animal organism.[392] Thus, he had no doubt that animals endowed with brains were capable both of feeling external impressions and of exerting conscious interventions in the animal economy.[393] Humans construed their experience as agential in precisely this way, and they projected it onto the wider animal kingdom.[394] This was the crucial purview of the notion of the animating "soul," going back all the way to Aristotle. Unzer was not so much interested in rejecting this idea as in localizing it, and in two senses. First, he insisted that the "soul" could act upon the body only via the brain.[395] Second, he theorized that this interaction (two-way, since the brain/body also affected the "soul") required, however mysteriously, a one-to-one correlation between mental content and brain event (a "material idea," as it was termed at the time).

For Unzer, the "nervous fluid" or force was "neither aqueous, nor glutinous, nor elastic, nor ethereal, nor electrical." He conceded, "it is not known how and when it contributes to the animal actions."[396] And yet it needed to be postulated to account for the indisputable—and extremely rapid—transmission of nerve impulses from the external organs of sense or the internal organs of the body to the brain and from the brain to all of these in turn.[397] Indeed, the directionality of these nerve transmissions was a cardinal empirical datum upon which to build neurophysiology. Unzer suspected, as the most plausible explanation of directionality of transmission, that different nerve fibers served motor and sensory functions.[398] Afferent and efferent nervous impulses, as we now call them, could be empirically discriminated not only at the extremities of these impulses—in sense organs or the brain, for instance—but they could be isolated and instigated experimentally in the nerve fibers themselves, so that their directional impulse could be demonstrated as active apart from these origins.

Even in sentient animals, Unzer noted, the heart and circulatory system as well as the digestive system appeared to operate without conscious direction.[399] These are today known as domains of the autonomic nervous system. But the really interesting question for Unzer's neurophysiology and theory of animal nature came with the lower animal organisms, which did not manifest a brain coordinating a central nervous system. "Animals destitute of brain as sea-anemones, tape-worms, etc, and . . . microscopic animals, polypes, etc . . . although without thought or sensation, appear to act as designedly, spontaneously, thoughtfully and volitionally, as animals really endowed with mind."[400] Thus, "is not the whole life of an oyster, a sea-worm, a polype, a snail, a spider, a flea, an ant, a bee, etc.—is not the whole of their acts, or part of them, solely an operation of the *vis nervosa*?"[401] "Polypes may be enabled to perform all their animal movements, solely by means of external impressions on their nerves, without having feeling or thought, and without either brain or soul. . . . These animals do not act as mere machines, as Descartes supposed, but according to purely animal laws, which cannot be deduced from either mechanical or physical principles or explained by them. . . . As Haller observes on the last page of his introduction to the translation of Buffon's 'History of Nature,' they are animals whose life consists simply in irritability."[402] To conceive of simpler animal organisms as *in*animate was simply misguided; nor could they be taken as identical with plants, which were largely "mechanical" organisms. Instead, they belonged to the animal kingdom.[403] To think that these organisms had "soul" distributed through their entirety (as did Whytt) was absurd and resulted in the further absurdity that such souls could be divided to engender subsequent souls in their fragments (as in Trembley's polyps).[404]

"But do *true insentient animals exist*?"[405] Was the "soul" essential to all animals? Was it really necessary to explain animal functions even in animals *with* a "soul" (consciousness)? Unzer realized that this challenged one of the most entrenched ideas about animals—that the very notion of animal required the "animation" of a soul.[406] Thus, he recognized the central resistance in his intended audience. Unzer considered all the objections and replied: "all the processes required for the life and preservation of an animal organism, can be effected by the *vis nervosa* only."[407]

What could be fully conscious/volitional in sentient animals could also arise as unconscious nerve-action; hence, the involuntary nervous system could often substitute for—or, in lower organisms, physiologically replace—volitional nerve-actions. Unzer traced the same continuity for *instinct*.[408] Strikingly, he located this whole phenomenon of instinct *within* consciousness, a *sensed* impulsion.[409] Animals came to feel the force of these instincts as something incontestable acting upon them (from within); knowing neither their proper object nor their purpose, yet they had to comply, and nature had seen to it that the environmental circumstances that triggered instinctual impulses also accorded with their fulfillment.[410] The initial thrust of instinct did not arise *in*, though it was felt *by*, consciousness, but it could become increasingly volitional. What sentient animals experienced initially as a blind drive they could develop into a volitional pursuit; that is, they could do *willingly* what they were initially compelled to do willy-nilly.[411] In that sense, "instinctual passions" arose.[412] More generally, Unzer recognized the physiological possibility that voluntary control could take over initially involuntary processes.[413] Respiration was the most interesting case.[414] It led Unzer to conceive of ontogenetic development within the higher animal organisms that exhibited a shift from involuntary to voluntary.[415] This created the possibility, which he did not pursue, that this could be viewed as a development *across organisms*, from lower to higher: an idea with enormous potential for the balance of German life science over the later eighteenth and beginning of the nineteenth century.

In 1771 Haller was Unzer's principal interlocutor, much as he had been for Caspar Friedrich Wolff in the 1760s. Unsurprisingly, in his review (more accurately, his dismissal) in the *GGA*, Haller responded to Unzer's revisionism with the same magisterial condescension he had shown Wolff.[416] Unzer responded in detail to the review in his *Physiologische Untersuchungen: Auf Veranlassung der Göttingischen, Frankfurter, Leipziger und Hallischen Recensionen seiner Physiologie der thierischen Natur* (1773), though without identifying the reviewer explicitly as Haller.[417] Phrase by phrase, Unzer defended himself against each point in the review, often cleverly enough faulting the reviewer for misunderstanding Haller (i.e., himself)! He did not back off one iota from his disputation of Haller's theory of irritability or from his advocacy of the neurophysiological alternative he had propounded.

JOHANN TETENS: THE PHILOSOPHICAL UPTAKE OF ISSUES FROM LIFE SCIENCE

A very important effort in German philosophy to come to terms with issues from life science came with the two-volume compendium by Johann Nicolaus Tetens, *Philosophische Versuche über die menschliche Natur und ihre Entwickelung* (1777).[418] The work had immediate and widespread impact in the German philosophical community.[419] It is certainly no coincidence that J. G. Hamann could report in a letter to his friend Herder (May 17, 1779) that in the years leading to the completion of his *Kritik der reinen Vernunft* (1781) Kant had this work by Tetens constantly open on his writing desk. Tetens showed a remarkable attunement to the essential issues of his time across a wide variety of fields, including much we have considered in this chapter. He discussed in discriminating detail the ideas of Süßmilch and Herder on the origin of language, of Reimarus and Unzer on instinct and physiology, of Bonnet and Wolff on preformation and epigenesis. He related all of this to the confrontation of Wolffian school philosophy with British empirical psychology. And, as well, Tetens was cognizant of the challenge of (French) "materialism," especially as this was propagated, somewhat ironically, by the *psychological* speculations of Charles Bonnet.[420] From Tetens, then, we can discern, more clearly than from any other single source, what sophisticated German philosophy considered the essentially contested issues of life science by the close of the 1770s.

Tetens proved the most discriminating critic of Bonnet within the German psychological community.[421] Did the new neurophysiology leave any room for an immaterial soul, and if so, how could it be conceived to interact with the body? These were problems at the metaphysical core of the Leibnizian tradition, but neither Tetens nor Unzer (nor even Reimarus) believed they could be resolved by a metaphysical doctrine, not even by Leibniz. Instead, Tetens affirmed with Unzer that this had to be a question of empirical inquiry, following the methods of observation and experiment as far as they could reach.[422] Unzer represented the most advanced level of physiological theory for Tetens, even beyond the work of the "greatest physiologist" of the epoch, Haller.[423] The question that Unzer made salient helped Tetens to advance to a more complex perspective, as the title of his thirteenth essay ("On the Essence of the Soul in Humans") betokened. The fundamental issue was the relation between the soul and the brain, the neurophysiological domain in which Unzer had made his most distinguished contribution (and in which Bonnet's mechanistic psychology also situated itself prominently). Pointedly, Tetens urged that Bonnet's brain-fiber account of human consciousness represented not empirical science but pure speculation—indeed, it was *metaphysics* without acknowledgment.[424] In this, Tetens affirmed positions that Unzer had explicitly taken against Bonnet.[425] Unzer provided, by contrast, the most plausible empirical evidence and interpretation. Unzer took the view that the lower animals literally had no soul. For Tetens, this went too far in one direction, just as Bonnet's commitment to soul—or, indeed, *many* souls—in a polyp went too far in the other.[426]

In *Philosophische Versuche* Tetens took up the issue of preformation versus epigenesis in what many consider the most discriminating form in his day. He identified the extreme positions with Charles Bonnet (preformation, or, as Tetens and his age termed it, *Evolution*) and Caspar Friedrich Wolff (epigenesis). But in fact he found Wolff's experimental results unconvincing for the most part, and he turned to the earlier ideas of Buffon as a more convincing representation of the merits of epigenesis.[427] By contrast, he endorsed much of Bonnet's position, only faulting it for going too far.[428] Thus, Tetens believed that Bonnet and Haller had made a compelling case for the necessity of a preexisting *form* for the reproduction of organisms; otherwise, the species continuity and the organic coherence of each new specimen seemed to him too inexplicable.[429] On the other hand, he believed that, via the epigenesis proposed by Buffon and Wolff, organisms developed, incorporating into themselves alien, even inorganic materials and creating new forms, even developing through stages, especially in the embryo, which would be displaced in the ultimate, mature organism.[430] The question of "new forms" was, he asserted, the essentially contested issue between *evolution* and *epigenesis*.[431] An adequate theory would have to incorporate both the need for a guiding design and the recognition of emergent properties. What he liked about Buffon's version of epigenesis was precisely the latter's insistence on an "interior mold," a preexisting, necessary form for the organization of the "organic molecules" of each new organism.[432] As always, Tetens sought an intermediate position. Thus, he proposed "evolution via epigenesis": "an evolution from within that can occur through epigenesis."[433] "[T]his epigenesis through evolution appears to be the general form of emergence of organized beings."[434] On the basis of a fixed original form, the organic matter of a new organism formed itself and grew into the mature form via the mechanical processes characterized by epigenesis.

Epigenesis would become the central point of contention between the philosophies of life science of Herder and Kant in the 1780s. With Johann Friedrich Blumenbach's elaboration of the notion of a *Bildungstrieb* from

1781 forward, epigenesis became dominant.[435] These ideas fused with questions of *Lebenskraft* and animal instinct—*Trieb*, out of Reimarus—to consolidate the theoretical frame of emergent life science, culminating in Carl Friedrich Kielmeyer's pathbreaking address of 1793 delivered at the Karlsschule in Stuttgart.[436] The balance of this inquiry will trace this trajectory.

CHAPTER SIX

From Natural History to History of Nature: From Buffon to Kant and Herder (and Blumenbach)

Natural history has been, until now, really the *description of Nature,* as Kant has very correctly remarked. He himself uses the name "natural history" for a particular branch of natural science, namely, the knowledge of the gradual alterations which the various organisms of the Earth have suffered through the influence of external nature, through migrations from one climate to another, and so forth.

SCHELLING, 1799[1]

"The parallel between the history of the earth and the history of the animal world"—to invoke the title of an essay by Reijer Hooykaas—proves decisive for the gestation of biology in the eighteenth century.[2] Historicizing geology would have a direct bearing on the historicizing of life and hence on the emergence of a science of biology. Both histories were *new,* Hooykaas notes, but "the new discipline of the 'history' of the earth influenced the still younger discipline of the 'history of life.'"[3] In the course of the eighteenth century, "natural history" in its classical sense of natural description (*Naturbeschreibung*), which had found brilliant systematization in the work of Linnaeus, underwent a mutation to "history of nature" (*Naturgeschichte*), that is, the explicit recognition that nature changed and developed over time.[4] Actually, *four* fundamental "emergence" problems were driving natural philosophy toward historicization in the eighteenth century: in descending order of scope, (1) the problem of the origin of the regularities observed in the plane and orbital direction of the planets in the solar system; (2) the problem of geological change; (3) the problem of the emergence of life and its differentiation into species; and (4) the problem of the generation of individual living beings.[5] Newton made the first a central issue for natural philosophy with his "General Scholium" of 1713.[6] The second would lead to the establishment of the modern science of geology. The third and fourth became the essential issues of modern life science. Over the course of the eighteenth century, the notion that nature was not constant in its order but rather developmental forced a critical appraisal of the methodological resources science might possess for grasping this.[7]

While Michel Foucault, among others, believed that this notion of historicity was incongruous with the "classical episteme" and could make sense only after 1800, the evidence is clear that a shift began considerably earlier.[8] The movement to a history of nature developed gradually over the eighteenth century. The most obvious locus of this shift to a literal "history of nature" was in the study of the earth.[9] Unquestionably, the discourse of a "history of the earth," starting with the burst of English "theories of the earth" from Thomas Burnet (1635–1715), William Whiston (1667–1752), and John Woodward (1665–1728), was longstanding by 1749.[10] A good proxy for the gestation period of the historicization of nature would be from the composition of *Protogaea* by Leibniz in the late 1680s to its full publication only in 1749, or the similar history of Benôit de Maillet's science-fiction novel, *Telliamed*.[11] The first volumes of Buffon's *Histoire naturelle*, which appeared in 1749, represented a decisive milestone in these developments. An important climax came with his great work of 1779, *Les époques de la nature*.[12] By midcentury the historicization of nature had had a place in the minds of key philosophers and students of nature, and its consequences for other inquiries, particularly in life science, preoccupied the balance of the century.

Martin Rudwick has written that this "new and surprising conception of the natural world[,] . . . the earth itself, . . . as contingent, as unrepeated, and as unpredictable"—was "quite limited" until after 1790.[13] The breakthrough to authentic historical geology ("geohistory") came, he elaborates, with the announcement of a prize competition by the Teyler Foundation in Haarlem in the 1780s. The question posed was "How far can one infer—on indisputable principles, [and] from the known character of fossils, from the beds in which they are found, and from what is known of the past and present condition of the earth's surface—what changes or general revolutions the surface of the globe has undergone; and how many ages must have since elapsed?"[14] The winner of the contest, François-Xavier Burtin (1743–1818), with his *Révolutions générales* (1789), "highlighted the capacity of fossils to act in their own right as *nonhuman* sources of historical evidence."[15] More theoretically, Burtin believed in the possibility that, though the origin of this history "is lost in the immensity of time, admits neither dates nor rigorous calculation, [it does disclose] epochs and a perceptible direction."[16]

Rudwick pursues the history of geology that the Haarlem query *inaugu-*

rated. My concern is how that *question* came to be possible at all: what were the elements of its *gestation*? The language of "epochs" and "revolutions" was firmly established in the mind-set of the moment. Its genesis passed through Buffon and stretched back, however problematically, to Linnaeus, Leibniz, and Woodward. Whatever the "speculative" character of the genre, "theories of the earth" proved the decisive womb in which genuine historicization ("geohistory," as Rudwick calls it) gestated, leading to conceptions of "revolutions" and species extinction ("catastrophism") based on fossils and their stratification.[17]

"The Theory of the Earth is a completely new science; it consists in deducing Phenomena of Nature; the formation of our Globe; the changes it underwent since then and those yet to come. The Ancients have absolutely ignored this science." So wrote Louis Bourguet (1678–1742) in 1729.[18] Bourguet had been in correspondence with Leibniz and was aware of the brief summary of Leibniz's *Protogaea* in the *Acta Eruditorum* (1693).[19] He was also a keen student of British "theory of the earth," particularly Woodward's version, which he and his Zurich friends, the brothers Scheuchzer, especially Johann Jakob (1672–1733), debated with the great Italian naturalist Antonio Vallisneri (1661–1730). Bourguet's own work marked a significant contribution to this "new science." Rhoda Rappaport calls him "the most thoughtful of Continental diluvianists."[20] Typically, retrospective historical study has pounced on the first aspect Bourguet mentions—*deducing* Phenomena of Nature—deeming it a sterile speculative undertaking. That has disregarded the crucial terms "formation of our Globe [and] the changes it underwent since then." For Bourguet, the new science needed to be a science about *historical change* in the earth.

Historical geology arose from the conundrum over the "meaning of fossils."[21] Specifically, seashells appeared atop the highest mountains of Europe. How could that have happened?[22] It was not unreasonable at the time to presume that the Noachian Deluge might be an explanation.[23] Piety alone cannot explain the adherence to the Deluge by geologists from Woodward through Bourguet to Jean-André Deluc (1727–1817).[24] It solved a real scientific problem. But could the one Deluge account for all the separate strata of fossils, and could it really have been universal and not local? Those mountain fossils and all the other anomalies empirically presented in field research became an unbearable weight for the old theories of the earth by the middle of the eighteenth century, and real historicization of nature became inescapable.[25] For a time, some geologists, motivated by religious as well as scientific concerns, sought to find a way to incorporate both extensive duration and the centrality of the Deluge into their theories.

They articulated a theory of early turbulence, punctuated by the Deluge, and succeeded by ongoing stasis through the present. A crucial example is the Genevan geologist Deluc,[26] who became an intermediary between the earlier "geotheorists" and such pioneer paleontologists as Blumenbach and Georges Cuvier.[27]

Central to this whole question of the historicization of nature is Buffon's role. In a provocative and important essay, John Eddy has disputed the substantial efforts by Jacques Roger and Phillip Sloan to uphold Buffon's "historicism."[28] For Roger, the essential locus of historicism in Buffon lay in his more and more intense preoccupation over time with "variation" within species and with "hybridization" among closely related species.[29] Buffon's research, as he composed his natural history of the animal kingdom, forced him to modify his initial definition of species and, by the 1760s, to elaborate a more encompassing notion of closely related species (a "family") among which interbreeding was possible, but only because they shared descent from a common ancestor. This also set limits on the extent of their possible variation and hybridization. This sense of "degeneration of animals," as Buffon entitled his most extensive discussion of this problem in 1766, led him, according to Roger, to a preoccupation with historical questions. Buffon himself wrote of a "far more ancient state [of nature from that of our own day] which we cannot access at all except via inductions and relations which are almost as ephemeral [*fugitifs*] as Time [itself], which appears to have covered over all their traces; nevertheless, we shall attempt to return [*remonter*], via facts and via monuments which still exist, to those first ages of nature and to present the epochs which appear clearly indicated to us."[30]

That, Roger contends persuasively, was a "historical project," indeed, one that would seem to presage Buffon's later *Époques*. Unfortunately, Roger informs us, "the history we find in the *Époques* is of a different order."[31] After 1766, influenced by Dortous de Mairan's study of ice and the establishment of a specific internal heat of the planet itself distinct from solar radiation, Buffon began to concentrate his efforts on a theory of the cooling of the earth as an irreversible geophysical process.[32] The history of living things, in that light, necessarily became ancillary to inexorable physicochemical changes associated with the gradual cooling of the planet. Indeed, that history would end with their utter extinction. In the interval, living things would struggle to adapt to the cooling world by migrating to warmer climes, if possible, but many would simply perish. That model displaced Buffon's earlier concern with "degeneration." While he did not abandon the latter, it no longer seemed so important.

Eddy presses his case aggressively: "Buffon was no more a forerunner of historicism than he was a precursor of evolutionism."[33] In particular, Eddy disputes Sloan's claim to discern in Buffon an "opening to true historicity in the concept of species."[34] Eddy objects: "just because Buffon thought of species as temporal organic continuities (temporal durations) does not mean that he thought of species as historical beings. In fact, . . . Buffon did not think of species as historical in any sense."[35] For Buffon, "species, although manifesting themselves in temporal succession, remain single living wholes, that is, reproductions of prototypes and independent of time."[36] Buffon's "species as organic types were universals, removed from the earth's history . . . [and] not, then, as Sloan believes, only concrete historical lineages; they are also universal types of organic organization."[37] Like Roger in seeing the late Buffon as moving toward geophysical determinism, Eddy concludes that, even in the *Époques*, "Buffon did not develop some new historical vision of living nature."[38] Instead, "Buffon eventually explained all phenomena, living and nonliving, as manifestations of the same Newtonian laws."[39] Rudwick is entirely in agreement: "Superficially, this second geotheory [the *Époques*] might seem to anticipate modern reconstructions of geohistory. But in fact it was profoundly ahistorical, for it postulated a series of changes that had in effect been programmed into the system from the start, and that could be extended into the future with the same degree of confidence."[40]

All such contentions notwithstanding, rather than ask whether Buffon satisfies *our* conceptions of historicity, it proves more germane to consider how Buffon was *received* in his own time. Indeed, whatever Buffon *himself* intended—either in the opening volume of the *Histoire naturelle* (1749) or the later *Époques de la nature* (1779)—the *reception* of his work by the close of the eighteenth century took it to signify historicization of nature.

KANT AND THE CONCEPTUAL CRYSTALLIZATION OF "HISTORY OF NATURE"

One of the key eighteenth-century interpreters of Buffon who unequivocally took him to be a proponent of the historicization of nature was Immanuel Kant.[41] Over the second half of the century, Kant would play a crucial role in defining *Naturgeschichte* to mean quite literally a developmental and generative reconstruction of nature's past, an "archaeology of nature."[42] Despite the reservations he would develop after his "critical" turn—reservations especially spurred after 1784 by the radically historicist reception of Buffon's work by Johann Gottfried Herder—Kant was taken by the wider community in Germany as the key exponent of this usage for the balance of the century.[43] And others took up the impulse to create a developmental theory of *Naturgeschichte* notwithstanding his later resistance.[44] Johann Friedrich Blumenbach would become significantly caught up in this aspect of the field.

The earliest appearance of the term *Naturgeschichte* in Kant's writings came in 1754, at the close of his little essay "Whether the Earth in Its Rotation . . . ," announcing his forthcoming work, *Universal Natural History and Theory of the Heavens* (1755). He proposed "an experiment with a natural history of the heavens . . . in which the earliest condition of nature [*erste Zustand der Natur*], the gestation of cosmic bodies [*Erzeugung der Weltkörper*], and the causes of their systematic relations would be determined from the traits [*Merkmaalen*] that the [current] relations of the order of the universe in themselves demonstrate [*die Verhältnisse des Weltbaues an sich zeigen*]."[45] The specific wording of this passage is very important for Kant's subsequent usages. First, Kant indicated that this was a (thought) experiment, not anything close to an apodictic knowledge claim. Second, Kant affirmed "actualism," that is, the applicability of current "relations" of the universe in reconstructing earlier natural configurations. Third, the basis for such an extension of knowledge into the past was the availability of "traits" (*Merkmaalen*) that persisted into the present. Finally, and crucially, Kant concerned himself with *original condition* (*erste Zustand*) and with *gestation* (*Erzeugung*), not simply development (*Auswicklung*). These are crucial methodological posits of a historicization of nature, and it is important to note that Kant embraced all of them already in 1754. Equally important is Kant's elaboration that his undertaking was simply "to attempt in the large, or better said in the infinite, what the history [*Historie*] of the earth entails in a smaller scope," and hence carried the same warrant as the latter inquiry, which was "in our days the object of considerable efforts at construction."[46] Kant was clearly aware of Buffon's work (and its title) and in all likelihood also of Leibniz's *Protogaea*, both of which appeared in print in 1749.[47]

Kant, in the passage cited from 1754, referred to the genre as "*history* of the earth," not "theory of the earth," and *Historie* meant, literally, formation and development, not description. This was exactly what Louis Bourguet, with whose work Kant was quite familiar, signified as the essence of this *new* science.[48] More immediately, the German translation of Buffon's work, supervised by the important German natural philosopher A. G. Kästner, began appearing in 1750/51 with a title, *Allgemeine Historie der Natur*, emphasizing the new, literally historical sense of Buffon's phrase *histoire naturelle*. Strikingly, in Kant's book of 1755, *Naturgeschichte* was used only in title headings, without conceptual elucidation.[49] In the later 1750s, however, Kant created a pioneering new course on "physical geography" which pursued the questions that animated his little essay of 1754 and also responded to the considerable interest engendered by the terrible Lisbon earthquake of 1755.[50] In the now-published formulation of the initial course in physical geography, Kant devoted "Part Eight" specifically to the "Account [*Geschichte*] of the great transformations which the earth has suffered earlier and still suffers now."[51] Given his frequent teaching of the physical geography course (forty-eight times over his career) and its distinct pedagogical importance for him (in preparing his students for their lives in the "world" of affairs), we have reason to believe that Kant remained very attentive to the literature on "theory of the earth" and on "natural history" over the ensuing decades.[52]

There was a twenty-year gap, nonetheless, before the term *Naturgeschichte* again appeared in a publication by Kant. That reappearance, on the other hand, proves decisive. In 1775 Kant explicitly sought to redefine *Naturgeschichte* away from its traditional descriptive signification into an actual historical reconstruction of geological and biological orders. "We take the terms *natural description* and *natural history* typically as having the same meaning. But it is clear that the knowledge of natural things as they *now are* still leaves wanting the knowledge of what they earlier *were*, and what sequence of changes they underwent in order to come in each locality to their current condition [*durch welche Reihe von Veränderungen sie durchgegangen, um an jedem Orte in ihren gegenwärtigen Zustand zu gelangen*]."[53] His decisive inspiration, as he made clear in his essay, was Buffon.[54] Kant explicated precisely what a history of nature would involve:

> The history of nature, of which we presently have very little, would teach us about changes in the shape of the earth, and also the changes that the creatures of the earth (plants and animals) have undergone through natural migrations, and thereby about the degenerations [*Abartungen*] from the original form [*Urbilde*] of the stem genus [*Stammgattung*]. It would in all likelihood reinterpret a large number of apparently distinct types [*Arten*] into races of the same species [*Gattung*] and transform the currently very diffuse system of academic natural description into a physical system for the understanding [i.e., a science].[55]

That is, in place of logical classes (*Schulgattungen*), one could discern actual natural genealogies (*Naturgattungen*). "The divisions of the schools have to do with *classes* based on *similarities*; the divisions of nature, however, concern lineages [*Stämme*] that discriminate animals in terms of *consanguinity* [*Verwandtschaften*] in terms of their generation."[56] Sloan aptly notes that

Kant's formulation entailed "historical alteration, and it also is concerned with issues of origins, or at least with changes from an original state."[57] In 1775 Kant did not believe that any of this was beyond the reach of science. He was, in fact, inspired by what Buffon was undertaking in geological history and the history of organic life.

The specific language of Kant's passage should be correlated with the formulations in Buffon's crucial "Dégéneration des animaux" essay in *Histoire naturelle* from 1766. Not only does this betoken Kant's study of Buffon's ongoing publication, but it suggests as well Kant's own sense that the "cutting edge" issue was to conceptualize the "stem genus" and to coordinate "varieties" within it in a more rigorous theory. This was precisely what Buffon had come to accentuate in his own work. In the words of Eddy:

> [I]n 1766 he admitted that environmental factors can alter the organic forms of the original prototypes to the point that different but similar species seem to have stemmed from the same source. . . . All the species in a genus, whatever the alteration, stemmed from the same prototype and are presumed to be reproductively compatible. In 1764 and 1766 Buffon identified genera in the same manner that he had earlier, in 1753, identified species: genera are groups of species (the species that descends directly from the original stock plus all altered collateral species) capable of interbreeding with one another and stemming from a common organic source.[58]

Kant's essay on human races constituted a self-conscious elaboration of just this interpretive problem. He believed in the possibility and the future of this new historical approach to earth science. The essay clearly enunciated this agenda and identified his own work (as armchair scientist) as part of this undertaking:

> Natural description (the state of nature in our current moment) is by far insufficient to establish the basis for the multiplicity of variations [*Mannigfaltigkeit der Abartungen*]. It is necessary, however much one might also and correctly oppose insolent opinions [*auch und zwar mit Recht der Frechheit der Meinungen feind ist*], to dare a *history* of nature, which is a separate science [*eine abgesonderte Wissenschaft*] that will in all likelihood [*wohl*] advance from [mere] opinions to [actual] insights.[59]

Indeed, in his essays on race (1775/77, 1785, 1788), Kant considered himself to be actively participating in this emergent natural inquiry.[60]

But what induced Kant to mention "insolent opinions" in this context? To grasp this, we must register that the pursuit of natural history had for some time been decisively embroiled in "physicotheology." Here, the crucial figure was Reimarus, but another important influence, especially for Kant, was Maupertuis.[61] What Kant dreaded was materialism, or, as he phrased it, "hylozoism."[62] The irony was, Kant had personally trained the most formidable German hylozoist, Johann Gottfried Herder, and he would have to deal with that for the balance of his career.

HERDER AND THE MAKING OF A BUFFONIAN HISTORICISM

There can be little question that "Buffon's *Histoire naturelle* had an essential influence on Herder's conception of natural history and his way of thinking in general."[63] We can grasp this more clearly, however, if we begin with Herder's relation with Buffon's philosophical expositor, Denis Diderot. In November 1769 Herder arrived in Paris, after having holed up in Normandy for many months getting up his nerve (and his French) for the venture.[64] One of the claims he made about his visit to Paris was that he met Diderot.[65] Imagine what it would have meant had Herder had the linguistic facility and the intellectual opportunity to have discussed with Diderot the latter's remarkable achievement of the prior summer—the *Rêve de d'Alembert*.[66] Diderot and Herder enunciated some of the most powerful systematic conjectures about nature as a self-contained and infinitely creative system (*natura naturans*) that the eighteenth century would witness.[67] What these two exceptional thinkers shared, even if they could not discuss it together in 1769, was a "Spinozism" that entailed a specific and creative *misreading* of the seventeenth-century philosopher as a theoretical resource for the articulation of a more subtle and dynamic materialism.[68] Herder had a sharp sense for the emergence of the new paradigm of vital materialism that sought to integrate problems of organic life with problems of active principles in the physical sciences. As Elias Palti has noted: "the study of the natural sciences of his time clarifies fundamental aspects of Herder's historical view, and, conversely, the analysis of Herder's philosophy allows us to better understand [the developments in natural science]."[69]

The Diderot-Herder connection is important in the reconstruction of the impetus of *natura naturans* across the eighteenth century. Diderot and Herder shared an insistence on *continuity*—from plant to animal, to be sure, but more radically from inorganic to organic and from animal to man. Diderot reached the same conclusion that Herder would reach: that all the complex forces discerned by "experimental physics" needed to be interpreted in terms of a *single* universal force.[70] Yet there were differences between them. If Spinoza was a vehicle for Diderot to dispense with God, that is, to see *Deus sive Natura* as atheism, for Herder Spinoza was a vehicle to reconstruct God, that is, to see *Deus sive Natura* as pan(en)theism, an immanent, nonanthropomorphic theology.[71] Diderot systematically sought to remove all trace of God. Herder, while consistently pursuing the immanent operations of nature, always retained a theological concern. But as both worked through the concrete issues of interpreting nature, they ended up theorizing a *vital materialism*. On the other hand, Diderot never embraced the possibility of a *history of nature* in the sense of what we would call development. He remained committed to a radically Lucretian sense of chance and emergence.[72] Herder, by contrast, had a very historicist mind. Nature was for him unquestionably historical, though he hesitated—for theological reasons, one suspects—before embracing species change.[73] For Diderot, vitalism reduced to a stark form of "spontaneous generation."[74] For Herder, "formation [*Bildung*]" had an inherently *developmental* character, whether in man or in nature. It was this that drew him from Diderot to Buffon. But it also figured in Herder's seminal reconstruction of Spinoza.

Herder began seriously to work through Spinoza's ideas in 1769.[75] In his last years in Riga he devoted himself to an intense study of Leibniz's *Nouveaux essais* (published in 1765), along with some of his earlier writings, and to the study of Spinoza in a Leibnizian context.[76] Herder read Leibniz through Spinoza and Spinoza through Leibniz to find a philosophical mode for articulating his consistently naturalist insight.[77] Force and dynamism were essential for the new natural philosophy, but it was equally essential that these be seen as *immanent* in nature.[78] Leibniz was important for Herder because he dynamized the natural order. On the other hand, Leibniz robbed the material world of metaphysical reality, whereas for Herder, "Nature had the most exalted reality."[79] This insistence on the actual world brought Herder far closer to the viewpoint of "physical influx" than Leibniz had at all been prepared to go.[80] But this also shows how the project of "hylozoism" needed to appropriate Spinoza. Herder sought to revise Leibnizian dynamism from a transcendent to an immanent monadology.[81] Just this inspired his reinterpretation of Spinoza to balance Leibniz. One found dynamic polarity, Herder wrote,

> spread throughout the whole world order. Everywhere two forces set against one another that nonetheless must work together and in which only by the combined and appropriate influence of both emerges the higher reality of a wise order, development, organization, life. All life arose in such a manner from death, out of the death of lesser forces, all wholes of order and of design from light and shadow, out of diverging, mutually opposing forces. . . . Mathematics, physics, chemistry, physiology of living beings all seem to me to provide evidence for this everywhere.[82]

Herder had developed this ubiquitous vitalism fully by the mid-1770s. He wrote these lines in "Über die dem Menschen angeborene Lüge," which can be dated to those years.

In 1787 Herder claimed his reworking of Spinozism provided a coherent interpretation that would bring

> an end to all the objectionable expressions of how God, according to this or that system, may work on and through dead matter. It is not dead but lives. For in it and conforming to its outer and inner organs, a thousand living, manifold forces are at work. The more we learn about matter, the more forces we discover in it, so that the empty conception of a dead extension completely disappears. Just in recent times, what numerous and different forces have been discovered in the atmosphere! How many different forces of attraction, union, dissolution and repulsion, has not modern chemistry already found in bodies?[83]

Herder's vitalist materialism thus invoked the most important recent developments in the natural sciences, especially in the fields of electricity, chemistry, and physiology.

Central to Herder's magnum opus, the *Ideen zur Philosophie der Geschichte der Menschheit*, was the conviction that there could be no categorical divide between nature and (human) history.[84] Hans-Dietrich Irmscher notes, "It is striking that Herder makes absolutely no effort to bridge [the] gaps [between nature and culture] with reference to the freedom of God and those made in his image. Instead, he calls for a continuous, purely immanent historical transition and coherence."[85] Kant, by contrast, wished to dissociate nature and culture to the highest degree possible without contradiction.[86] As Martin Bollacher has noted, "the main thrust of Kant's critique of the *Ideas* was aimed, not at Herder's argument from concepts that were no longer drawn from sensible experience, but at the pantheistically grounded perspective of a genetic relation and natural-historical development of the 'species.'"[87] Reinhard Brandt grasps the largest sense of difference: Kant could not tolerate the idea of continuity from the inorganic to the organic.[88]

Herder asserted across the board a "natural history" whose cosmological dimension he had learned from Kant and whose biological idea he took from Buffon and Caspar Friedrich Wolff.[89] He saw increasing complexity and differentiation as an immanent principle of natural development, as an intrinsically *historical* character/tendency of the entire physical world. "Everything in nature is connected: one state pushes forward and prepares another."[90] "Nothing in nature stands still."[91] "Inferior powers ascend to the more subtle forms of vitality."[92] Herder sought for a conceptual structure of transition. He found it in the notion of "forces" (*Kräfte*).[93] The result was a theory of the world as composed primarily of forces organized hierarchically. "The one organic principle of nature . . . that we here term *plastic*, there *impulsive*, here *sensitive*, there *artful* . . . is at bottom but one and the same organic power."[94] Thus, Herder proposed to discern morphological universals: "in marine life, plants, and even inanimate things, as they are called, one and the same groundwork may prevail, though infinitely more rude and confused."[95] "The new discoveries that have been made respecting heat, light, fire, and their various effects on the composition, dissolution, and constituent parts of terrestrial substances, the simpler principles to which the electric matter, and in some measure the magnetic, are reduced, appear to me . . . at least considerable advances that will in time enable some happy genius . . . to explain our geogony on principles as simple as those to which Kepler and Newton have reduced the solar system."[96] Herder's specific contribution was to elaborate on Kant's new sense of *Naturgeschichte*, the history of nature, accounting for the immanent changes in physical nature discerned from the empirical record. "Before our air, our water, our earth could be produced, various reciprocally dissolving and precipitating forces were necessary. . . . How many solutions and conversions of one into another do the multifarious species of earths, stones, and crystallizations and of organization in shells, plants, animals, and, lastly, in man presuppose!"[97] And further: "Various combinations of water, air, and light must have taken place before the seeds of the first vegetable organization, of moss, perhaps, would have appeared. Many plants must have sprung up and died before organized animals were produced."[98] Another passage advanced this speculation more concretely: "From rocks to crystals, from crystals to metals, from metals to the world of plants, from plants to animals and finally to man, we saw the form of organization ascend and the powers and instincts of creatures simultaneously become more diverse and finally come together in the human figure (insofar as this could encompass them)."[99]

Herder elaborated into a general interpretive principle the idea of *epigenesis*, as it had come to prominence in contemporary generation theory and embryology.[100] Herder referred explicitly to the work of William Harvey and Caspar Friedrich Wolff:

> How must the man have been astonished, who first saw the wonders of the creation of a living being! Globules, with fluids shooting between them, become a living point; and from this point an animal forms itself. The heart soon becomes visible and, weak and imperfect as it is, begins to beat. . . . What would he who saw this wonder for the first time call it? There, he would say, is a living organic power: I know not whence it came or what it intrinsically is, but that it

> is there, that it lives, that it has acquired itself organic parts out of the chaos of homogeneous matter, I see; this is incontestable.[101]

> If we contemplate these changes, these living operations, as well in the egg of the bird as in the womb of the viviparous quadruped, then, it seems to me, one is not being forthright [*spricht man uneigentlich*] if one talks of germs [*Keimen*] that are only developed or of an *epigenesis* according to which the members would accrete externally [*die Glieder von aussen zuwüchsen*]. It is a matter of *formation* [*Bildung*] (genesis), an effect of internal forces for which Nature has prepared the raw materials [*Masse*] that they incorporate into themselves in order to make themselves visible.[102]

That is, there exists an "internal nature [that] becomes visible in a mass appertaining to it and that must have the prototype of its appearance in itself, whence or wherever it may be."[103] Herder denied individual preformation (*das Evolutionssystem—emboitement*) and merely external, mechanical change (Epicureanism) as sufficient explanations and argued that there had to be an immanent force behind such variation. Herder called it "genetic force."

What Herder was suggesting as the proper sense of epigenesis was a fertile and unpredictable creativity in nature, a sweeping notion of its pervasive fundamental "genetic force." He was not rejecting but rather radicalizing the idea of epigenesis. Kant understood exactly: "As the reviewer understands it, the sense in which the author uses this expression [i.e., *genetische Kraft*] is as follows. He wishes to reject the system of evolution, on the one hand, but also the purely mechanical influence of external causes, on the other, as worthless explanations. He assumes that the cause of such differences is the vital principle [*Lebensprinzip*] that modifies *itself* from within in accordance with variations in external circumstances and in a manner appropriate to these."[104] This was far too open-ended for Kant, and he insisted that this "genetic force" was not unlimited and could not lead to a mutation of species, and that the proper terms for it should be "germs [*Keime*]" or "original endowments [*natürliche Anlagen*]." Such an interpretation, Kant added, should simply accept this capacity for variation as a given incapable of further determinate elucidation.[105] Otherwise, the very notion of species distinction was at risk. To be sure, Herder was careful to repudiate such a speculation explicitly: "In truth, ape and man were never one and the same species."[106] But that could not disguise for Kant the radical potential latent in Herder's text; indeed, this was its historical impact for "attentive and adept readers," as Heinz Stolpe has acutely observed.[107] Charlotte von Stein wrote very aptly to this effect in a letter to Karl Ludwig von

Knebel.[108] H. B. Nisbet, too, emphasizes that it was Kant's recognition of the potential in Herder's type theory for a *transmutation* theory that led to his critique.[109]

Herder deliberately set about erasing the borders Kant had so carefully drawn not only between life and matter but also, and even more grievously for Kant, between animal and man.[110] Drawing extensively from the analogy of botanical and zoological forms with those of humanity, he identified erect posture as man's decisive physiological difference from the rest. In what was the most imaginative segment of the work, Herder sought to correlate all man's distinctive cultural attributes with this essential physical attribute of erect posture.[111] Even in those aspects of humanity that were most authentically spiritual, Herder continued to uphold immanence of transition. Thus, he proposed to read even *reason* as a natural emergent: "Reason is not innate, but acquired. . . . Theoretically and practically, reason is nothing but something *received*, an acquired proportionality and direction of ideas and faculties, to which the human being is formed by its organization and way of life. . . . He acquires reason from infancy, being formed to it, to freedom, and to human speech through art, as he is to his ingenious mode of movement."[112] The issue of the continuity of man with nature was the touchstone of the epoch for which Herder composed *Ideas*.[113] For Goethe, enthusiastically following Herder's composition of the *Ideas*, this was Herder's whole point: in Goethe's words, "nothing [physiologically] specific could be found to differentiate between man and animals."[114] This was Herder's salience for late eighteenth-century German science: no one articulated with the same breadth and vivacity as Herder the prospect of confirming that continuity.

CHAPTER SEVEN

Johann Friedrich Blumenbach and the Life Sciences in Germany: His Rise to Eminence from the 1770s

Johann Friedrich Blumenbach must become our Peter Camper!

J. G. ZIMMERMAN TO BLUMENBACH, September 15, 1779[1]

If Albrecht von Haller dominated German life science from 1740 to his death in 1777, another Göttingen professor, Johann Friedrich Blumenbach (1752–1840), came to prominence in the remainder of the century and helped crystallize the emergent new discipline of biology.[2] His long and distinguished career linked him as teacher, colleague, and mentor to virtually all the important figures in the field, so that Blumenbach came to be the patriarch of German life sciences well into the nineteenth century.[3] And yet, as Tanya van Hoorn observes aptly, "no comprehensive, ground-laying study of Blumenbach has yet been presented." She elaborates that the absence of a full-fledged monograph on Blumenbach "is all the more incomprehensible since Blumenbach's importance for the development of science is always being highlighted."[4] My treatment here makes a modest contribution in light of this gap.[5]

DEFINING "NATURAL HISTORY": FROM THE GREAT APES TO PHYSICAL ANTHROPOLOGY

As John Gascoigne aptly points out, "it was in Germany particularly that natural history began to be reconceptualized. . . . Blumenbach sought to elevate the conceptual significance of natural history. Such views eventually set the scene for the emergence of the discipline of biology."[6] By shifting natural history toward comparative zoology, Blumenbach moved away from medicine as it was traditionally oriented (concerned with humans and their health) toward a relatively autonomous research program. We have to begin from this shift in the semantic field of *Naturgeschichte*. In German eighteenth-century discourse, there was a clear distinction between *Naturlehre*, the general term used for natural philosophy or physical science, and *Naturgeschichte*, taken still in its traditional sense of the description of nature.[7] A classic illustration of this was the composition of two key textbooks by the Göttingen scholar Johann Christian Polycarp Erxleben (1744–77), the short-lived prodigy whose textbook in physical science dominated German academic instruction for the last quarter of the eighteenth century.[8] In addition to his famous *Anfangsgründe der Naturlehre* (1st ed., 1772), he also published *Anfangsgründe der Naturgeschichte* (1768).[9] The terminology, as reflected in Erxleben's titles, set apart the still-descriptive research domain (*Fachgebiet*) of natural history from that of the physical sciences: the first was content merely to collect accurate data, while the second sought the *causes* behind phenomena.

Erxleben's two textbooks illustrate how natural inquiry in Germany articulated its semantic field in the third quarter of the eighteenth century. Of even greater significance are some emergent mutations in this field.[10] An early exemplar was the inaugural address delivered in Jena in 1762 by the professor of medicine Ernst Gottfried Baldinger (1738–1804), later to move to Göttingen and become Blumenbach's academic supervisor (*Doktorvater*).[11] In his address, *Grenzen der Naturlehre*, Baldinger drew three distinctions in terminology.[12] First, he distinguished *Naturlehre* from applied mathematics by virtue of its empirical, as opposed to a priori, element. Second, he distinguished *Naturlehre* from more specialized fields of investigation, like chemistry and mineralogy, by virtue of its universal scope. Finally, he distinguished it from traditional *Naturgeschichte*, because it had causal and explanatory ambitions, not just descriptive ones. Baldinger's statement reflected the moment of transition of the semantic field of German natural science. What he still termed *Naturlehre* would come to be called *Naturwissenschaft*, but, as in his account, it would stress the empirical as against the mathematical in its ambition to extend its scope to embrace *all* the emergent domains of natural-scientific inquiry.[13] And, of greatest relevance here, a new conception of *Naturgeschichte* would seek recognition as *one such* explanatory natural science, not merely description.[14] This aspiration had been clearly articulated by Buffon in the opening statement of his *Histoire naturelle*, and the case for a nonmathematical conception of "experimental physics" got taken up with enthusiasm by Diderot and even by Maupertuis (first in France and then in Berlin).

Over the second half of the eighteenth century, "natural history" came to be problematized and reformulated along crisscrossing dimensions. The first and most salient was the line of tension between a concern with *classification* (taxonomy), associated most eminently with Linnaeus, and a concern with *development* (the literal *history* of nature), which the eighteenth century associated with Buffon. The career path of the great Dutch naturalist Peter Camper (1722–89) was paradigmatic here. He began as an enthusiastic Linnaean, but by the end of his career, as he committed himself to developmental zoology, he called Linnaeus the most "superficial naturalist" he knew.[15] A second polarization has already come under scrutiny in this study, namely, that between the "party" of Réaumur (Trembley, Bonnet, Sennebier), ostentatiously committed to "observation," and the "party" of Buffon (Maupertuis, Diderot, et al.), who advocated "interpretation." Réaumur's partisans imputed the "spirit of system" to their opponents, insisting that the business of inquiry should be restricted to the observation and collection of data, with interpretation to be deferred indefinitely. They lambasted Buffon for "speculation," for "fiction" and "romance" in place of rigorous inquiry. But theoretical interpretation could not be deferred indefinitely, even by such ardent "observers" as Bonnet, who proved altogether a "theorist" in his own right.[16] Buffon's effort to raise the level of natural history to a form of natural philosophy ended up triumphing by the close of the century. Camper, again, is illustrative in his condemnation of *bad* natural history in which one could find "nothing to do with the internal structures, nothing to do with the senses, nothing with the growth of offspring."[17]

In the course of the eighteenth century, the attention of natural history shifted away from the traditional tripartite *division* (animal, plant, and mineral) to the *boundary* between the animate and the inanimate and to the question of *transition* or *continuity* between domains—that is, the nature and the origins of life, the border between plants and animals, and especially that between animals and man. Boundaries needed to be *explained*, not just described, if there were to be *unity* in the sciences of nature. In just that measure natural history took on far more ambitious *theoretical* aspirations. As one instance of this general turn, the self-formation of crystals assumed great salience in construing the transition between inert matter and organic form.[18] Yet even the crystal's demonstration of "formative forces" in nature could not fully capture the uniqueness of organic form, which did not simply accrue at the margin of the entity (*per appositionem*) but integrated new material throughout the growing form (*per intussusceptionem*).[19]

A final, more concrete dimension of the reorientation of natural history had to do with its research focus with respect to botany and zoology. There had long been an emphasis on botany in the medical school curriculum, oriented, of course, to pharmacology.[20] Thus, no medical school was without its botanist, and in the era of Linnaeus botany was no minor pursuit. He made botany the exemplary domain in achieving a rigorous taxonomical approach to natural history.[21] Still, by the second half of the eighteenth century, *zoology* had become the most exciting domain of inquiry.[22] Buffon's great *Histoire naturelle* began with many volumes treating the animal kingdom. This became the prime concern of Camper, then Blumenbach.

Zoological interest concentrated overwhelmingly on the *vertebrates* until the close of the century when Lamarck's great work revealed the enormous theoretical significance of the character and variety of the invertebrate realm.[23] Indeed, interest focused still more narrowly for a time on what had come to be classed as "mammals."[24] Two major bodies of new evidence galvanized this fascination with vertebrates and more specifically with mammals. The first was the discovery of fossilized remains of remarkable creatures that were at once similar to living forms and yet at the same time strikingly different.[25] In Siberia, Peter Simon Pallas (1741–1811) raised to theoretical salience the remains of elephants (mammoths) and also of rhinoceroses whose dimensions vastly exceeded currently existent species.[26] In Ohio, the remains of a creature equally large but quite problematic—hence its original name: the "Ohio-unknown" (eventually, thanks to Cuvier, baptized a mastodon)—became an international topic of debate.[27] In Europe itself, discoveries in caves of the bones of a huge creature eventually identified as a bear added to this fascination.[28] What these finds all indicated was a very substantial shift in the habitat and in the dimensions of creatures very similar to current species and hence, plausibly, evidence of significant recent geological changes and especially of animal *extinction*.[29]

The second sort of new evidence that caught the imagination of natural historians was the introduction into Europe of extensive travel reports and eventually of actual specimens, first dead, then living, of anthropoid apes.[30] Many living specimens died very shortly after arrival; an even vaster number failed to survive sea transport from their origin to Europe.[31] The first chimpanzee arrived in Europe around 1630.[32] Dead specimens of orangutans arrived around 1770, and the first live specimen later that decade.[33] In addition, these were almost all *juvenile* specimens, because they were more easily captured, and this complicated interpretation still further.[34] The most provocative reception of this new evidence came in Linnaeus's *System of Nature*. First, in 1735, Linnaeus audaciously included humans within the taxonomy of animals, in a group termed Anthropomorpha, which included initially not only humans with geographically varying ("racial") features but also apes, sloths, and bats.[35] More dramatic still was his advance in the tenth edition of the *System of Nature* (1758) to the characterization of a new order of *Primates*, from which the sloth and bat were excluded but within which Linnaeus now included *two* types of humans—*Homo sapiens (diurnus)* and *Homo troglodytus (nocturnus)*—and segregated these from the *simians* (lesser apes and monkeys).[36] Under *Homo troglodytus* he classed not only "orang-outangs" (a *generic* term for the anthropoid apes) but also what we would call albinos.[37] In a note to the twelfth edition, Linnaeus even considered a *third* type of human—one with a tail! This combination of the reception of the most recent zoological evidence from the non-European world with a host of bizarre received tales of monstrous human proxies carried over from ancient and medieval European sources indicated not simply the credulity of Linnaeus, or his desire to include every sort of "attested" creature in his scheme, but the genuine and pervasive bafflement of natural historians as to the variety of humanoid forms and their implication for (European) human self-understanding.[38] There was no clear theory of teratology—of monstrous deformity—though it was a topic of considerable interest.[39] Any given individual aberrance might accordingly be the manifestation of a variant *kind*.

Most of these issues had already come under careful scrutiny in a work that, over the course of the eighteenth century, became the touchstone of serious advance in the field: *Orang-Outang, sive Homo Sylvestris: or, The Anatomy of a Pygmie compared with that of a Monkey, an Ape, and A Man. To which is added a Philological Essay concerning the Pygmies, the Cynocephali, the Satyrs, and Spinges of the Ancients, wherein it will appear that they are all either Apes or Monkeys, and not Men, as formerly pretended* (1699), by Edward Tyson (1651–1708).[40] The full title indicates two decisive features of the work. First, it undertook a deliberate project of *comparative anatomy*, aiming to distinguish, by empirical evidence, the specific traits of the specimen under examination (whose multiple names will be taken up below) from others already known—that is, the monkey, the human, and the Barbary ape (familiar since antiquity and in all likelihood the object of Galen's pioneering anatomical investigations).[41] Second, it undertook, equally deliberately, to discriminate a project in natural history (i.e., to establish, by actual physical-anatomical investigation, the variety of Anthropomorpha) from the accumulation of centuries of mythical lore that only encumbered that investigation with delusions.[42] In this philological debunking, we can identify Tyson's work with a longer-standing endeavor, going back to Francis Bacon and John Browne, earlier in the century, to purge natural history of its accretions of fanciful errors. Tyson's contemporary, the leading naturalist John Ray (1627–1705), recognized and welcomed this aspect of Tyson's work. But Tyson was adding a dimension to natural history that even Ray did not fully realize: that the discrimination of varieties could not rely simply on external features but had to probe into the interior of the organisms to establish similarity and difference. Thus, natural history in zoology had to entail *comparative anatomy*.[43] This would be a principle even Buffon could not fully embrace, leading to tensions with his collaborator, the anatomist Louis-Jean-Marie Daubenton (1716–1800), that would ultimately rupture their partnership. But Tyson's way would prevail in the generation after Buffon: in Daubenton's disciple Félix Vicq d'Azyr (1748–94) in France; in Peter Camper in Holland; and in Johann Friedrich Blumenbach in Germany.

The connection linking Tyson to Blumenbach, however, was hardly a straight line but rather a sprawling web, with many a wayward strand. Tyson's title gives us a key to that, as well, in the variety of names Tyson offered for the creature he examined: first, and most importantly, *Orang-Outang*, which he identified with the Latin *Homo sylvestris* (man of the woods, i.e., the human *savage*), and second, most prominently for Tyson, *pygmie*, a term, as becomes clear from his philological appendix, with a long and misleading cultural heritage.[44] Nomenclature became a crucial problem in the conceptualization of apes over the eighteenth century, prominently in Linnaeus but also in Buffon's two crucial essays "Nomenclature of the Apes" and "Orang-outang, or the Pongo and the Jocko."[45] Tyson dissected an African chimpanzee, but he called it "orang-outang" for good reason, since this name had already been (mis)applied to the African apes by his predecessors, notably Nicolaes Tulp (1593–1674).[46] Moreover, the notion of a "man of the woods," which Tulp offered as the translation of the Malay term *orang-outang*, set the decisive conceptual issue: namely, what had this creature to do with *man*? Interestingly, it seems that "chimpanzee" had a similar meaning in the original African language: "man of the woods."[47] In both native contexts, that is, apes were taken to be (deviant) *humans*.[48] The Latin *Homo sylvestris* made that salient in a European cultural context, bringing to the forefront of attention the relation of humankind to all the other mythical and actual life-forms that resembled or dissembled (deranged) that kind.[49] In this context, the fanciful illustrations and tall tales that accompanied reports of the anthropoid apes, with their persistently accentuated affinities to human form, added to the damaging confusion, and even the finest naturalists were not above the use of these dubious humanizing illustrations and the still more dubious (and salubrious) tales.[50]

The question, bluntly put, was this: could comparative anatomy contribute to the consolidation of human distinction? The danger, of course, was that any merely physical enterprise appeared to undercut the most prized conceptualization of that distinction, namely, that the human differed from the animal by virtue of a *rational*—and, for Christians, an *immortal*, God-given, and God-*like—soul*.[51] In the traditional European view, humans partook in a totally different, *spiritual* order of being. That Linnaeus dared to include humans among the animals in 1735 provoked scandal and opprobrium.[52] Tyson was already attuned to this parameter for inquiry in 1699. Descartes had inflamed the issue with his "animal-machine" hypothesis. The "dignity of man"—that is, his radical superiority to all other living things—appears to have been a rather frail thing in the dawning Enlightenment, in need of conservative retrenchment or progressive reconfiguration.[53] But the question of a spiritual human nature was only *one* of the great anxieties that would intensify across the century. Another was the question of the *unity* of the human species: the challenge of polygenism and racism.[54] The meandering path that led from the comparative anatomy of apes to physical anthropology to the conception of a "scientific" racism is the essential historical context for the career of Johann Friedrich Blumenbach.

Robert Wokler, in a series of essays, has highlighted the "ape debate"—"controversy about apes, man and language[,] . . . the distinction between the natural and cultural determinants of human behavior"—as a central feature of Enlightenment discourse.[55] Similarly, Miriam Meijer has retrieved the notion of a "century of the orangutan."[56] Something major was unleashed in the problem that had come to the attention of Tyson. Certainly it had to do with the "chain of being" with which Tyson was clearly preoccupied, and more specifically with the question of the *gap* between man and the rest of the animal kingdom.[57] In his specimen, Tyson explicitly offered a creature "more resembling a Man, than any other animal, . . . in this *Chain* of the *Creation* . . . an intermediate Link between an *Ape* and a *Man*."[58]

What, then, did this link-creature offer on the question of the specific difference of the human species? Justin Smith has recently offered a very illuminating account. Two differentiae had been salient in the discourse up to Tyson's moment: the "bodily" one of erect gait, with attendant behavioral functions, and the "spiritual" one of spoken language as the talisman of rationality.[59] Obviously, the physiology of the human organism had to accommodate this spoken language, so the anatomy of the larynx and other physical features involved in speech production became centrally important in the physical discrimination of humans from other animals. Tyson for that reason meticulously studied these features in his chimpanzee specimen. Yet there, precisely, lay the paradox, as Tyson saw it. He was under the misapprehension that his creature naturally and normally walked erect. He had a brief experience with the juvenile male chimp before it died, but it was already quite ill, and Tyson presumed that it was only ill health that led it to lean upon its knuckles as it moved. Moreover, his creature certainly had a larynx—and, presumably, all the other physical accoutrements for language—identical to that of humans, yet it could not speak.[60] Tyson came to believe that anatomical endowment did not necessarily entail functional use, a discord that put enormous pressure on the generally teleological notion of natural order.[61] But in this particular case, Tyson argued, it underscored that a grasp of human distinctness *had to* turn to spiritual endowments.[62] The capacity for language had to come from some *immaterial* intervention or supplement (the rational soul); it could not emerge or be explained adequately from physiology alone.

This incongruity between anatomical endowment and cultural capacity became the obsessive question of the "ape debates." Linnaeus made clear that he could discern no characteristic in the physical nature of things that could discriminate the human species from that of the ape.[63] Buffon made Tyson famous throughout Europe by citing him on this precise point and concluding, resoundingly, that this was the preeminent proof that there could be no materialist account of human capacities, thereby securing the privilege of the human rational soul.[64] As a consequence, neither Linnaeus nor Buffon—nor Bonnet nor Haller—could countenance counting apes and humans as of one kind. This was particularly important in the face of circulating notions that such apes were the product of the "degeneration" of humans or of the interbreeding of humans with simians.[65] A disciple of Linnaeus, the German Johann Christian Fabricius (1745–1808), became well known for suggesting that black Africans were the product of "degenerate" interbreeding of (white) humans with apes.[66]

An interesting feature of this discourse—exhibited by both native informants and their eager European interpreters—was the recurrent contention that male apes had an overweening lust for human females.[67] This evoked association with the satyrs of ancient European mythology—which frequently carried over into the very nomenclature attached to apes. Not only sexuality but also aesthetics figured in these concerns about human difference, as frequently it was contended that some of these anthropoid apes seemed more comely than some humans, particularly the "Hottentots" and the "Kalmucks," both taken to be particularly hideous by European standards of taste.[68] Clearly, the discourse of *race* became thoroughly interfused in the discourse of *apes*, and by the end of the eighteenth century, Robert Wokler has contended, the first discourse simply displaced the second as the prime concern of physical anthropology.[69]

Notably, no one imagined that humanity *ascended* from the ape—and for a variety of reasons.[70] By contrast, the notion of *degeneration* was particularly prominent in Buffon, for whom the term betokened not simply *descent* from a common ancestor but *decay* of original vigor. Buffon made "degeneration" a central element in his account of the domestication of animals, and it carried over into his conviction regarding the pervasive inferiority of New World life-forms to those of the Old World.[71] That view captured, both for him and for his avid disciple Cornelius De Pauw (1739–99), the degenerate character of the indigenous human population of the New World, the Amerindian "race."[72] Buffon inaugurated the consideration of the human *species* as an object of natural history.[73] "Man became simply a part of natural processes, developing like all other natural creatures from an elemental mud," and humankind "could be fruitfully compared to the lowest creatures and placed in the context of the unfolding of matter and motion in an evolutionary process."[74] "The location of human beings among the animals was combined with a radical historicizing and naturalizing of the human species that would pursue zoogeographical analysis of humanity in connection with a gradually developing schema of a naturalized account of cosmological and geological history."[75] Still, Buffon refused to erase the boundary between man and animals. What physiology could not define, he was prepared to stipulate in terms of reason and language as irrefutable evidences of a spiritual nature in man irreducible to natural elements. Buffon insisted that the "ape was no mediating link between the human and the animal orders of nature, but 'in truth just a plain animal.'"[76] Indeed, anatomical similarity only reinforced the claim that the difference that constituted humankind had to be sought in a separate, spiritual dispensation. Reason and language belonged to a divine intervention: that was the line Buffon took up alongside many other thinkers of the time.[77]

If Buffon and Linnaeus were content to adopt Tyson's posture in dealing with the disconnect between anatomy and function with regard to speech (and gait), others in the eighteenth century would not be. There were two lines of dissent: first, on the question of language and, second, on the question of erect posture. La Mettrie inaugurated the first line of dissent, insisting that it was inconceivable that there could even be such a disconnect between form and function, and *therefore*, it had to be a mistake to believe the orangutan—that is, generically, the higher apes—incapable of speech and hence a distinct species. On the contrary, La Mettrie contended, the orangutan *could* be taught to speak, and it was, therefore, conjecturally human.[78] Jean-Jacques Rousseau, in his famous tenth note to the *Discourse on the Origins of Inequality*, made the case even more compellingly, suggesting that humans in their natural state were quite similar to orangutans and especially like them in their lack of speech.[79] It could not be established that they were separate species without experiments in interbreeding, which Rousseau pronounced impracticable and unseemly.[80] The most extravagant case along these lines was proposed by James Burnet, Lord Monboddo (1714–99). In 1773, and especially in his revisions of 1774, he took up all Rousseau's arguments and pushed them still further in asserting the commonality of orangutans and humans.[81] All that distinguished the latter from the former was progress in cultivation of mental agility via socialization, and this could not be denied a priori for the former case, given proper circumstances.

The revisions that Monboddo undertook between 1773 and 1774 arose from the critical intervention of his friends, notably John Pringle, president of the Royal Society of London, pointing out his lack of familiarity with the work of Tyson and his consequent misunderstanding of the arguments of Buffon. Monboddo thereupon studied these materials, incorporating them in a still-unpublished draft essay entitled "Of the Orang-outang and Whether he be of the Human Species" (1773–74) and adapting much of this material into revisions in the second edition of his *Origins of Language*.[82] Pringle, among many others, found the new version no more credible than the earlier one, and he invited his Dutch correspondent, Peter Camper, who was becoming widely known as the leading anatomist of the orangutan, to intervene. This resulted in the first publication by Camper specifically dealing with the true orangutan of Indonesia. In the *Philosophical Transactions*, 1779, he argued, against Tyson on anatomy and against Monboddo more generally, that the orangutan did *not* have the physiological equipment for human utterance.[83]

Language was one avenue of argument. The other was erect posture. Tyson had believed that his specimen was a natural biped. The question of the erect gait of the anthropoid apes became a very important issue in the eighteenth century. Indeed, the erect posture of *humans themselves* came into dispute. In 1771 the Italian Pietro Moscati argued that humans were by nature quadrupeds; their original constitution entailed walking on all fours. Hence, it was a *cultural* deviation—and one with significant costs in human health—to assume an upright gait.[84] Kant found this a very appealing line of thought, which he praised in a very rare book review, for it seemed to him to mark the radical intervention of rationality in human development, displacing and overriding the merely animal.[85] For physiologists and medical men, Moscati was far less persuasive.[86] Blumenbach dismissed him with scorn, and he was not alone.[87] The anatomical and physiological evidence that humans were by nature distinctively bipeds was overwhelming.

That case was buttressed substantially by two crucial anatomical interventions. The first came even before Moscati's publication, with a pathbreaking essay by Daubenton on the occipital cavity in humans relative to other animals.[88] By establishing the relation between the carriage of the head and the top of the spinal column, Daubenton demonstrated that the human skeletal frame fit upright posture, but those of animals did not. The second contribution came from a series of dissections of orangutans by Peter Camper over the 1770s, the results of which were disseminated orally long before they came to be published. As Visser notes, "from 1770 to 1777 he examined eight animals in all; five of them were anatomized."[89] This was widely known before any of his results were published. Blumenbach referred to this work in his dissertation of 1775; it was this reputation that induced Pringle to invite the paper for *Philosophical Transactions* in 1779. In that same year, Camper published on the orangutan in Dutch, and in 1781 Samuel Thomas Soemmerring (1755–1830) made that widely known by translating extensive excerpts into German in a key paper, "Etwas vernüftiges über den Orangoutang."[90]

What Camper established is that the higher apes, while they could stand up on their hind limbs, naturally walked on all fours.[91] This upheld his view and that of his successors that upright posture did indeed physically distinguish the human animal from all others. In Germany, Camper's ideas had enormous influence, and both Herder and Blumenbach would claim upright posture as the decisive physical basis for human nature, with which many specifically cultural capacities could be correlated.[92] A final, ancillary feature of Camper's anatomical enterprise with the orangutan had to do with the intermaxillary bone. For many anatomists, this bone was a ubiquitous feature of all animals *except* humans.[93] Camper upheld this view.[94] In France, Vicq d'Azyr established that it was false, but this was not widely known, especially in Germany.[95] A notorious affair ensued in the mid-1780s upon Goethe's proclaimed discovery of an intermaxillary bone in humans.[96]

Camper's self-formation as a naturalist followed a crucial pattern, which I have already identified in Haller, Buffon, and others: commitment to the "experimental Newtonian" tradition. Camper learned this at its heart: the University of Leiden, where he studied, 1734–36, under Van Musschenbroek. He absorbed Dutch experimental Newtonianism from the methodological writings of 'sGravesande as well as the class instruction of Van Musschenbroek.[97] As a student at Leiden, Camper concentrated on animal anatomy.[98] His first major publication came in 1760–62: *Demonstrationem Anatamo-Pathologicarum*.[99] Over his career, Camper worked and taught in medical schools at various universities in the Netherlands.[100] But although he was a medical doctor, and quite capable as a surgeon, he was not primarily interested in clinical practice or even in human health, but rather, from the beginning, he pursued zoology "for its own sake."[101] He was eager to devote himself entirely to his research in comparative anatomy.[102]

In 1782 Camper published the first and only volume of what he envisioned as his major work: *Natural History of the Orang-Outang, the Double Horned Rhinoceros, and the Reindeer*.[103] It established Camper as the foremost comparative anatomist in Europe. "At the basis of Camper's comparative anatomy was the idea that the animals are interrelated through a uniform plan."[104] He had elaborate ideas for publication, but his "rage for research" often induced a tendency to rush from one project to the next, without bringing any to fruition.[105] Nonetheless, Camper swiftly became known as "the indispensable source of information on the orangutan."[106] His only rivals as a comparative anatomist were Vicq d'Azyr in France and John Hunter in England. His greatest successor would be Blumenbach in Germany. It would be one of the foremost sources of pride for Blumenbach that he was the first to offer a course in comparative anatomy at a German university.[107] In the preface to his widely acclaimed and internationally disseminated textbook in comparative anatomy, he wrote of its centrality to the discipline of natural history.[108]

Camper proved decisive in the development of Blumenbach's scientific identity. As the epigraph of this chapter indicates, for his countrymen, Blumenbach's role was to *become* the German Camper.[109] Such analogies are not to be taken altogether lightly, for we are concerned, here, with the gestation of a science, and that must always entail the construction of quite specific scientific identities, modeled after paradigmatic practices. Buffon and Camper did offer such paradigmatic models of natural history for Blumenbach, and his self-formation as a scientist decisively reflected their assimilation. Above all, Camper played this role.

Haller was Blumenbach's self-proclaimed model, but it is noteworthy that Haller died in 1777, just as Blumenbach was coming into his own. Camper lived considerably longer; he and Blumenbach would meet in person and engage in a correspondence. Their correspondence began in August

1778 and they met personally when Camper came to Göttingen in October 1779.[110] Thus, they were colleagues and, in crucial matters, even competitors. As a result, Blumenbach never fully acknowledged the paradigmatic status of Camper for his practice of natural history.[111] It remains, however, that this was palpable to his peers and admirers from very early on. Blumenbach's whole sense of natural history is unimaginable without (Buffon and) Camper.[112]

JOHANN FRIEDRICH BLUMENBACH'S ACADEMIC CAREER

In Jena, where Blumenbach commenced his university studies (1769–72), he got his first exposure to natural history from J. E. I. Walch (1725–88), one of the pioneers in German mineralogy and fossil collecting (though he was a professor of rhetoric!).[113] When Blumenbach transferred his university studies from Jena to Göttingen in 1772, he moved from a relatively respectable medical school to the most exciting university in Germany.[114] To be sure, its guiding light, Albrecht von Haller, had returned to Switzerland in 1753, but he remained a decisive force in the Göttingen Academy, in the *Göttingische Gelehrten Anzeigen*, and even in the university, which kept hoping for his return and which, under the general guidance of Christian Gottlob Heyne (1729–1812), looked to his counsel from afar in most of its important policy considerations.[115] Heyne, who had arrived in 1763, dominated university affairs by 1770, and it was to Heyne that Blumenbach came, bearing references.[116] Heyne took him under his wing, put him in contact with Haller, gave him the management of the newly acquired natural history "cabinet" of the Göttingen medical professor Christian Wilhelm Büttner (1716–1801), and thus opened his way to a splendid career in the rapidly transforming field of natural history.[117] It also helped that Blumenbach's mentor at Jena, Ernst Gottfried Baldinger, was called to Göttingen in 1773, where he would serve as the director of Blumenbach's dissertation, though not as its inspiration, a role that belonged in equal parts to Büttner and to Haller.

At Göttingen, Heyne saw to it that Blumenbach entered into correspondence with Haller as early as February 1775, and Blumenbach acknowledged this connection as seminal for his entire formation as a naturalist.[118] But Blumenbach emphasized as well the "whimsical but remarkable" Büttner, whose collections in natural history furnished the basis for the ethnographic museum of Göttingen.[119] Blumenbach studied natural history with Büttner and was drawn especially to his extensive "quantity of books of voyages and travels" as well as being motivated by Büttner's insistence that humans should be a primary topic in natural history. This was a fundamental impulse in the field, emanating from Buffon. Blumenbach summed up: "It was thus I was led to write as the dissertation for my doctorate, *On the natural variety of mankind*."[120]

What was the study of nature like at Göttingen in Blumenbach's time? At Göttingen, physics, chemistry, mineralogy, and geology were esteemed essential propaedeutics for the study of human medicine, grounding it in wider "experimental physics," the nonmetaphysical and largely nonmathematical approach to natural inquiry emphasizing "observation and experience." This had been the agenda of Albrecht von Haller and it was systematically maintained, especially in the medical faculty of the 1770s when Blumenbach received his training. Subsequently, he was instrumental in perpetuating it over his long tenure.[121] Georg Lichtenberg (1742–99) was Blumenbach's closest colleague in the sciences.[122] Not only did he edit and expand the Erxleben physics textbook four times up through 1794 and manage other important journals in natural science, but Lichtenberg's own work (primarily on electricity) was highly regarded as well.[123] He came to be considered an arbiter in German natural science over the last decades of the century. Kant, for example, wanted Lichtenberg to review his *Metaphysical Foundations of Natural Science* (1786).[124] Lichtenberg was also the figure to whom many turned for counsel in the bitter dispute over Lavoisier and the "chemical revolution."[125] He was an ardent Anglophile and frequently visited England. In that context, one important contact that Lichtenberg brought to Blumenbach was the Swiss geological thinker Jean-André Deluc, who would eventually present Blumenbach with a series of letters on the theory of the earth.[126]

Another of Blumenbach's important colleagues was Samuel Thomas Soemmerring, who enrolled at Göttingen in October 1774 and achieved his doctorate in 1778 with a dissertation of highest distinction on anatomy.[127] His thesis director, Baldinger, already predicted on that basis that Soemmerring would become one of the great masters of anatomy.[128] In the years 1776–78, Soemmerring took courses with Blumenbach, about whom he offered a ringing testimonial in his curriculum vitae: "Exc. Blumenbach was not only my most desirable instructor in general zoology, mineralogy, physiology, pathology, the particular history of man, and in relating the traditions of medicine, but also a distinguished patron, who deigned to treat me as a friend."[129] Having completed his celebrated dissertation, Soemmerring went on extended tour (*peregrinatio medica*) in 1778–79, meeting and becoming closely associated with Peter Camper in Holland and with Georg

Forster (1754–94) in England, relationships that would define his intellectual and personal life.[130] In addition, Soemmerring met the Hunter brothers and he stayed with and learned a great deal from Alexander Monro II in Edinburgh. Yet Soemmerring had problems finding a position when he returned to Germany. There were few chairs in natural history among German medical faculties: they were a luxury in a stringent time, when the energies of the larger faculty were directed at clinical professional advancement. The first defined *chair* in natural history in Germany was specially created for Christian Büttner at Göttingen in 1763, so the university could secure his prized "cabinet" of natural curiosities.[131] Blumenbach happily inherited his mentor's chair upon Büttner's retirement in 1776. Thus, Heyne, a master of university politics, counseled Soemmerring that Blumenbach had been very lucky to find a position in a medical faculty that accommodated his interest and it might be hard for Soemmerring to find something similar. Even though courses were widely offered in natural history, it was very difficult to establish secure academic *lines* in the field.

The narrow group of Blumenbach, Camper, the Forsters (father and son), and Soemmerring (with one more, remote participant, Eberhard A. W. Zimmermann [1743–1815]) constituted the cutting edge of natural history for Germany.[132] Blumenbach's relationships with both Camper and Forster were mediated by Soemmerring, though in each case Blumenbach had made a prior connection. Later in his life, Blumenbach recollected that he met the younger Forster in November 1777.[133] In all likelihood this was really in 1778, while Georg was visiting Göttingen to seek a position for his distinguished father, Johann Reinhold Forster (1729–98), the expedition naturalist on Cook's second voyage to the Pacific (1772–75), on which Georg had accompanied his father.[134] In 1779 Georg was appointed (instead of his father) as natural historian at the Collegium Carolinum in Kassel. The younger Forster and Soemmerring had become intimate friends in London during the latter's stay, and he secured a position at Kassel in anatomy for Soemmerring.[135] They remained the closest of friends over the next two decades. For much of that time, as well, they were frequently in Göttingen and associated with Lichtenberg and Blumenbach.

THE DISSERTATION OF 1775 ON HUMAN VARIETY

The theme of Blumenbach's dissertation was highly topical: the renewed challenge of "polygenism" in man.[136] This matter was "much discussed in these days" in German academic circles (and, of course, more widely in the European discourse of "race").[137] The weakening of Buffon's position through equivocations about fertile hybrids in later volumes of his *Histoire naturelle* had offered opportunity for the revindication of polygenism. In *Sketches on the History of Man* (1774), Henry Home, Lord Kames (1696–1782), ruthlessly demonstrated that Buffon had compromised his monogenism.[138] The work was translated into German in 1775 to considerable notoriety.

For the developing field of natural history the importance of Blumenbach's dissertation lay in the conceptual frame treating the problem of species and variety in zoology generally.[139] Blumenbach elaborated the idea of "total habitus" as the only viable empirical approach to classification in life science.[140] While it was admittedly still "artificial," it came closer to the ideal of a "natural system" than Linnaean classification schemes, without falling into the empirically unmanageable difficulties of Buffon's genealogical approach.[141] In all this, Blumenbach followed dutifully the line of argument of his great Swiss mentor Haller. The opening page of Blumenbach's dissertation cannot say enough of the "immortal labours of the great Haller," of his "profound sagacity."[142] Blumenbach steeped himself in everything Haller wrote; he is widely regarded as the most competent late eighteenth-century commentator on Haller's sprawling opus, and he clearly set out to be heir to Haller's mantle in the field.[143] He would envision his great works, the *Institutes of Physiology* (1787ff.) and above all the pioneering *Handbook of Comparative Anatomy* (1805ff.), as worthy successors to Haller's most popular work, *First Lines of Physiology* (1747), which had defined the field for the second half of the eighteenth century.[144] Blumenbach sought from the outset to ground his inquiry in Haller's theory of preformation. He further followed Haller in doubting Buffon's definition of species. Therefore, Blumenbach thought it essential to address the question of hybrids—that is, "the conjunction of animals of different species."[145] He was highly skeptical of most proposed examples, because of the lack of proportion in the organs of reproduction as well as in the gestation periods of different species.[146] He distrusted even long-suggested instances, notably that of the "jumar," an ostensible cross of horse and cow.[147] Still, Blumenbach wrote: "There is no reason for doubting that hybrids have sprung from the union of the fox and the dog, and those too capable of generation."[148] While he agreed with Buffon that interspecies generation was *unlikely* and that hybrids were so typically infertile that enumerating exceptions was "tiresome," he would not adopt Buffon's principle as a necessary and sufficient criterion for species differentiation.[149]

Like Buffon and Haller, Blumenbach believed that man should be categorically distinguished from the other animals on the basis of "the endowments of the mind."[150] He elaborated: "man alone ought to be held to possess *speech*, or the voice of reason, and beasts only the language of the affections."[151] While he noted that "Linnaeus could discover no [anatomical] point by which man could be distinguished from the ape," he was himself convinced of the importance of some distinctions, in particular the form of the human hand and more generally man's erect posture.[152] Indeed, retrospectively he would take his discussion of erect posture in his 1775 dissertation as establishing his precedence in advocating this trait as the definitive morphological difference distinguishing man from animals (well before Herder's advocacy of this notion in his famous *Ideen* of 1784).[153] Blumenbach disputed as "not quite serious" Pietro Moscati's recent claims that erect posture caused physiological problems for man.[154] Following the latest reports on anatomical dissections of various "animals which are most like man" and relying on his own systematic dissections, Blumenbach was confident he could spell out empirically what morphological differences separated man from animals. He pointed specifically to a series of dissections of apes he performed in the winter of 1774, in which he concentrated on the brain.[155] Drawing on all this work, Blumenbach disputed the assertion by Rousseau and others (Monboddo and La Mettrie, in all likelihood) that orangutans were of the same species as man, making the point bluntly that such writers were "ill-instructed in natural history and anatomy."[156] While the orangutan might be "like man in structure," Blumenbach asserted, it was incapable of speech.[157]

Having carefully followed the research on apes, he discriminated two distinct species: chimpanzee and orangutan.[158] He was already aware in 1775 that Camper had conducted dissections of the latter that indicated substantial differences from Tyson's earlier anatomy of the "pygmie" (i.e., the chimpanzee).[159] He was also beginning to register a radical separation in habitat for the two: Angola versus Borneo. By 1779 his characterizations of the two species in the *Handbuch der Naturgeschichte* would be quite clear and set the terms for future work in the field.[160] Another aspect that Blumenbach noted in his dissertation was the presence of an intermaxillary bone in Camper's anatomical studies of the orangutan.[161] Blumenbach would later be peripherally involved in the disputes over Goethe's claims for the intermaxillary bone in humans (1783ff.), in which Soemmerring and Camper would prove so disappointing to Goethe.[162]

Reaching, finally, the central issue of polygenism in human origins, Blumenbach ascribed the revival of this view to "ill-feeling, negligence, and the love of novelty" rather than any scientific soundness. He specifically identified Voltaire and Kames as guilty here. He charged bluntly: "it was much easier to pronounce [humans] different species than to inquire into the structure of the human body, to consult the numerous anatomical authors and travellers, and carefully to weigh their good faith or carelessness, to compare parallel examples from the universal circuit of natural history, and then at last to come to an opinion, and investigate the causes of the variety."[163] While Blumenbach conceded to the polygenists that differences among humans might seem to warrant considering them "as forming different species of mankind," he insisted nonetheless that this was misguided. Clearly, on the basis of the alternative and appropriate method he accused them of neglecting, Blumenbach came to the opposite conclusion, namely, monogenism.

The problem, then, was how to account for the *variety* among humans, especially since, "when the matter is thoroughly considered, you see that all do so run into one another, and that one variety of mankind does so insensibly pass into the other, that you cannot mark out the limits between them."[164] Variation *within* any population of humans was so rife that to leap to a species discrimination *among* humans, as Kames did, left Blumenbach "astonished."[165] For the purposes of his dissertation, Blumenbach confidently asserted: "even if it be granted that lascivious male apes attack women [an idea that ran sensationally through the travel literature], any idea of progeny resulting cannot be entertained for a moment."[166] Thus, the insinuation of simian origins of African populations, entertained by Voltaire and Fabricius, he flatly dismissed.

In discriminating varieties, Blumenbach invoked "the whole bodily constitution, stature, and colour" first, then "the particular structure and proportion of individual parts."[167] The former group of traits he identified as "owing almost entirely to climate alone."[168] Blumenbach postulated a defining relation between heat, moisture, and bodily form: "That in hot countries bodies become drier and heavier; in cold and wet ones softer, more full of juice and spongy, is easily noticed."[169] He carefully adduced all the comparative anatomical research that had been done to establish the generalization. Stature, too, Blumenbach correlated with heat and cold: "the latter obstructs the increase of organic bodies, whilst the former adds to them and promotes their growth."[170] He brusquely dismissed Kames for having "presumed with the greatest confidence to think otherwise."[171]

While Kant in 1775 and especially in 1777 made skin color his decisive criterion for racial distinction, Blumenbach found this trait equivocal.[172] He saw a continuous spectrum of changes in skin tints whereby "the most distinct and contrary colours so degenerate, that white men may sensibly pass and be changed into black, and the contrary."[173] "There is an almost insensible and indefinable transition from the pure white skin of the German lady through the yellow, the red, and the dark nations, to the Ethiopian of the very deepest black."[174] Thus, two vital points informed Blumenbach's approach to the variety in human skin coloration in 1775. First, it was not only an effect of extended exposure to climatic difference but also *reversible* under exposure to different climatic conditions for similarly extended periods.[175] Second, there were simply no definable discontinuities adequate to constitute varietal boundaries. His conclusion was unequivocal: "from all these cases, this is clearly proved, . . . that colour, whatever be its causes, be it bile, or the influence of the sun, the air, or the climate, is, at all events, an adventitious and easily changeable thing, and can never constitute a diversity of species."[176]

Instead, a more promising standard for diversity seemed the "various shapes of the head."[177] This was a portentous turn in these matters, prompted by Johann Joachim Winckelmann's raptures over the Grecian profile and Johann Caspar Lavater's physiognomy, which culminated in Peter Camper's theory of the "facial angle" as the key to racial classification.[178] Much of Blumenbach's later theory of "race" would be invested in such craniology. The intervention not only of cranial-capacity measurement but of ethnocentric aesthetic judgments of physiognomy would long dominate "race" theory, as Stephen Jay Gould so memorably demonstrated.[179] Camper, following explicitly upon Winckelmann, began to conjecture about the ideal facial angle in lectures of the 1770s at Groningen.[180] The published fruit of this conception came only posthumously, but the oral transmission was swift and potent, and Blumenbach noted at several points in his 1775 dissertation that he had learned key features of Camper's theories from mutual acquaintances.[181] Indeed, Blumenbach showed the influence of this aesthetic orientation: "J. B. Fischer has published a drawing of a Calmuck's skull, and [he maintains that] it is ugly . . . and in many ways testifies to barbarism."[182] But he quickly supplied counterevidence from the Siberian explorer Pallas, who "describes the Calmucks as men of a symmetrical, beautiful and even round appearance."[183] (One presumes the discourse is of skull shape.)

Already in 1775, then, Blumenbach found "the physiognomy and the peculiar lineaments of the whole countenance in different nations" a "very vast and agreeable field," yet it is crucial to note that he believed "almost all the diversity of the form of the head in different nations is to be attributed to the mode of life and to art"—that is, that these were *not* matters of natural endowment.[184] Blumenbach considered a body of evidence concerning the willful or adventitious modifications of head shape based on the manner in which infants were swaddled or carried. He pointed to some evidence of deliberate manipulations of the plastic infant cranium to achieve culturally desired aesthetic standards. He even toyed with the idea that these acquired characteristics might become hereditary.[185] Yet, for all that, clearly Blumenbach was not ready—as he would be in 1795, after amassing and assessing his famous collection of skulls from many of the world's peoples—to assert that there was in fact a "racial face" and that skull shape was the definitive indication of human racial variety.[186]

Blumenbach clearly saw any classification scheme as "very arbitrary indeed both in number and definition."[187] In 1775 he followed Linnaeus in settling upon four "varieties" (he eschewed the term "race" in the entire discussion), of which "the first and most important to us (which is also the primitive one) is that of Europe, Asia this side of the Ganges, and all the country situated to the north of the Amoor, together with that part of North America [in later statements this is more sensibly North *Africa*], which is nearest both in position and character of the inhabitants."[188] This is a striking congeries of peoples, one might observe. Blumenbach acknowledged this: "Though the men of these countries seem to differ very much amongst each other in form and colour, still when they are looked at as a whole they seem to agree in many things with ourselves."[189] The other three varieties he identified geographically with South and East Asia, with Africa, and with America.[190] By 1779 he would find this Linnaean fourfold division too confining, and he added a fifth variety, the "Malay," to recognize the problematic character of the peoples of the vast archipelago including the Dutch East Indies, Australia–New Zealand, the Philippines, and all of the South Pacific.[191] The Pacific was the scene of the most rapid growth in European geographical and ethnographic discovery in Blumenbach's lifetime, owing to the voyages of Cook and others, and he took an avid interest in all of this.[192] Indeed, by the turn of the century, Blumenbach had built up the largest collection of natural-historical materials from the Cook expeditions and generally on the Pacific region to be found in Europe.[193]

What are we to make of this dissertation? It is clear that that he was involved in a quasi-Linnaean project of classification. Blumenbach's approach was one of comparative morphology, applying the idea of the "total habitus," as it had been taught to him by Heyne, to the organism. But because he had no clear criterion for variety, indeed insisted repeatedly on the fluidity and arbitrariness of such classification schemes, his fourfold division seems even less motivated than that of Linnaeus. Blumenbach in 1775, then, appears to have derived monogenism from the indeterminacy of any categorical discrimination among varieties of human. I must dissent from the widely shared view that Blumenbach expressed strong "racialist" bias in his early texts and came only later in his career to the defense of blacks. There is every reason to contend that already in the 1770s he (and Camper as well) stood sturdily against arguments, whether explicitly polygenist or not, that would affirm the radical inferiority of other "races"—and particularly of blacks.

THE *HANDBUCH DER NATURGESCHICHTE* AND *ÜBER DEN BILDUNGSTRIEB* (1779–1782)

Blumenbach devoted a good deal of energy to organizing, cataloging, and expanding the natural-history cabinet that had been entrusted to him. Indeed, when he was appointed extraordinary professor in 1776, he was actually given three distinct assignments: the professorship, to be sure, but also the complete organization of the natural-history cabinet and, at his specific request, the position of "prosector" in the anatomy theater of Professor Heinrich August Wrisberg (1739–1808).[194] That entailed substantial time in the dissecting theater, which Wrisberg, apparently, had allowed to fall into considerable disorder.[195] Blumenbach devoted himself assiduously to all three roles. It was only when Heyne could report that Blumenbach had accomplished a complete catalog of the collection that the administration in Hanover in 1778 approved his promotion to ordinary professor.

This promotion carried with it the obligation to produce and publish a Latin inaugural dissertation. Blumenbach summarized his work, entitled *Prolusio Anatomica de Sinibus Frontalibus*, in an announcement in the *Göttingische Gelehrten Anzeigen* (*GGA*).[196] From this summary we can discern some important impulses in his approach. First, Blumenbach chose to present a dissertation in *anatomy*. Second, dealing with the frontal sinuses, Blumenbach was already showing his interest in the facial structure and the skull, matters that would be central to his life's work. Third, he presented a distinctly *medical* dissertation, dealing with the formation and *de*formation of human sinuses and with their maladies: clinical concerns usually absent from his work. And, finally, but perhaps most importantly, he sought to explain the structure of the frontal sinuses functionally and via comparative anatomy, systematically relating the size and number of sinuses in humans to those of other animals in connection with their physiological role. This was portentous of the major directions in which Blumenbach's whole career would move the field of natural history and comparative zoology. No less a figure than Johann G. Zimmermann recognized the promise of the work, prompting the line to Blumenbach that I have chosen as my epigraph for this chapter, that Blumenbach should become the German Peter Camper, a role that he did come to fill.[197]

Perhaps the most important event in the interval between the publication of his dissertation in 1775 and the publication of the first part of his *Handbuch der Naturgeschichte* at the end of April 1779 was the death of Albrecht von Haller, in early December 1777. Haller was Blumenbach's idol, but he also cast a very deep shadow over the field. Dissenting from his views was highly risky. Blumenbach had long since developed reservations about aspects of Haller's work, some of them quite substantial.[198] The most important of them concerned the theory of generation. His experiments on freshwater hydra had been under way since 1774 at the latest, and it is inconceivable that the evidence from these experiments, published only in 1780, had not already aroused theoretical restiveness that would culminate in the idea of the *Bildungstrieb*. But as long as Haller lived, Blumenbach never challenged his theory of preformation. Conversely, already in 1778 (i.e., shortly after Haller's death), Blumenbach ventured to express skepticism about preformation in an article in a Göttingen popular magazine: "can it be more than mere plausibility [*Wahrscheinlichkeit*] that we have lain as seeds in our mothers from time immemorial and that our fathers, like all the other male animals, in the end are little more than a supplement whose role [*Bestimmung*] is merely to enliven these seeds stuck in eternal sleep and encourage them to develop?"[199] Even when he did come to overthrow Haller's theory, Blumenbach remained extraordinarily deferential, intimating that Haller himself would have come round to this view had he only lived longer and revised his physiology.[200] Toward Bonnet, however, Blumenbach was hardly so considerate.[201]

Other developments marked the years 1777–78, when Blumenbach was writing his *Handbuch*. In London, in 1777, Georg Forster, defying the British Admiralty, published *A Voyage Round the World*, his father's naturalist report of the Cook expedition: a major landmark in the ethnography and natural history of the Pacific.[202] Forster would later see to the publication of a German translation of his father's book in Berlin.[203] Blumenbach's acquaintance with the younger Forster dates exactly to this period, and there is no doubt that he devoured the elder Forster's text with intense interest.[204] Simultaneously, the years 1777 and 1778 saw the publication of Lavater's famous works on physiognomy, which took Germany by storm.[205] Only in 1778 would Georg Lichtenberg begin deflating Lavater's pretensions.[206] Blumenbach was very close to Lichtenberg at this time, and for all the reasons mentioned in connection with his 1775 dissertation, he could not have been indifferent to the question of Lavater's physiognomy.

Even closer to his own bailiwick, in 1777 Immanuel Kant published a revised version of his 1775 Königsberg University course announcement on the problem of human races.[207] This revised version appeared in a very widely circulated popular journal, J. J. Engel's *Philosophie für die Welt*, and Kant's essay (which stood out incongruously in that volume of mostly literary work) gained considerable attention.[208] As evidence of this, the publisher of the journal invited Kant to publish more extensively on the topic, which Kant declined to do.[209] The scientific community clearly noted Kant's contribution. Given its direct parallel with his own work, Blumenbach undoubtedly read it, though there is little evidence that Kant's essay had any immediate impact on his own ideas.[210] Perhaps this was because in 1778 in the preface to a work that had a much more profound impact on the field, Eberhard A. W. Zimmermann sharply criticized Kant's views.[211] Zimmermann was a pioneer in biogeography whose work Blumenbach and other German natural historians considered monumental in the field.[212] In any event, Blumenbach knew of Kant from 1777 forward, and thus, when, in the mid-1780s, Kant entered into dispute with Christian Heyne's close friend Johann Gottfried Herder, and with Blumenbach's own friend Georg Forster, over the issue of human variety, Blumenbach was primed to take a much more decisive interest in his views. Finally, in 1778, the elder Forster published his *Observations on Physical Geography*, with acknowledgments of the contributions of Blumenbach to the field.[213] Thus, Blumenbach found himself situated in a body of scholarship concerning human variety and ethnographic diversity, all sharply inflected in a "natural-historical" direction with which he was himself identified. He was poised to take a leading role in directing this new field, and that was what his *Handbuch der Naturgeschichte* aimed to achieve.

Blumenbach dated the preface to the first volume of his *Handbuch* April 1779. This twelve-page preface introduced a book of some 450 pages, dealing only with the animal kingdom. The continuation in 1780 completed the whole design, covering plants and minerals on pages 449–559.[214] The organization of the work would remain unchanged through all the myriad editions stretching well toward the middle of the next century.[215] It is the material in the first four parts that most warrants attention, in both this original edition and the many succeeding ones. In the 1780 continuation, what is striking is that Blumenbach devoted only part 10 to plants, while aspects of mineralogy took up parts 11–16.[216] While minerals get attention in eighty-six pages, all plant life is covered in a mere twenty-three pages, and this includes an introductory section of four pages, restating the main themes of the whole book. Blumenbach was simply not interested in botany. The introductory part on minerals (part 11) and the concluding one, on fossils (part 16), are particularly important, especially as the editions succeeded one another, for an understanding of the development of Blumenbach's idea of natural history.

The foundation upon which Blumenbach constructed his *Handbuch* was the principle of the total habitus as the key to classification. He argued in the opening section of part 1, "On Natural History Generally," that Linnaeus's "artificial" system was clearly inadequate but that the ideal of a "natural system" was empirically unattainable, and hence, something in between must be found. The most promising avenue was the pursuit of a classification using *all* characteristics accessible for interpretation, the total habitus of the organism. This was the argument he had already made in his 1775 dissertation and in a presentation on classifying mammals delivered to the Göttingen Academy in 1775, but now he was able to demonstrate its utility more fully.[217]

Blumenbach rejected the idea of transitional kinds between plant and animal or between inorganic and organic realms. Such "missing-link" inquiries had in fact been the object of his own research, not only on polyps, which some had suggested were plantlike but which he had established incontrovertibly as animal, but also on freshwater mosses, which seemed to have animal properties but were distinctly plant forms.[218] That led Blumenbach to a most important argument, which he sustained in version after version of the *Handbuch*: namely, that the "chain of being," or the "law of continuity" in the objects of nature, prominently associated with Leibniz in metaphysics and with Bonnet in life science, was only a heuristic analogy, useful for the pursuit of the ideal "natural system" but in itself false to the complexity of nature and the providence of God.[219] Adamantly he insisted that it was folly to regard the orangutan as the link between man and the apes.[220] Blumenbach reiterated and amplified all the arguments whereby man was distinct from the animals. He assigned humans to their own genus, separate from the apes, thus repudiating Linnaeus's inclusive class of primates. The core distinction turned on behavior: man had no instincts, but all the other animals were determined by them.[221] In his terminology and argumentation, Blumenbach clearly drew on Hermann Reimarus, though he would cite him explicitly only in later editions in connection with these arguments. Man was distinct precisely by virtue of his lack of instinct and organs of self-defense. His unique order was "*Inermis*—here taken specifically to betoken the absence of innate weapons, constructive instincts [*Kunsttriebe*, a technical term from Reimarus], cover [*Bedeckungen*, i.e., armor or camouflage]—in short, all those [vulnerabilities] that man is rescued from by reason."[222] In the *Handbuch* Blumenbach explicitly connected the absence of instinct in man with his ability to penetrate all habitats on earth and establish ubiquitous settlement.

Revisiting, as well, the theme of human variety of his 1775 dissertation, Blumenbach firmly reasserted his monogenism: "humans of all times and all climes can have descended from Adam."[223] Indeed, "all the differences flow so indiscernibly together that there can be none but the most arbitrary boundaries established between them."[224] Here, Blumenbach first introduced his *five*-variety classification scheme, as I noted earlier. Of great interest is the language he used in characterizing the first such variety: "The original [*ursprüngliche*] and greatest [*größte*] race [*Race*] comprises first all Europeans, including the Lapps," then Asians west of the Caspian Sea and the Ganges River, plus North Africans and Eskimos. All these peoples were white-skinned and "according to our concepts of beauty the most well formed of humans."[225] By 1779, then, Blumenbach not only advanced the five-variety model but explicitly used the term "race." He still believed in the primordiality of the "European" white race and grounded its grandeur, in some measure at least, in its beauty, albeit acknowledging that this aesthetic judgment was an ethnic preference.

Important as all this was for anthropology and "race" theory in the late eighteenth century, for biology it is secondary to the question of Blumenbach's thoughts on organism and on generation. That is, in the 1779 work what were the criteria by which Blumenbach distinguished the organic from the inorganic orders of nature, and how did he appraise the theories of generation? For Blumenbach, the question of the emergence (*Entstehung*) of organized bodies was, despite all the efforts of modern science, "one of the most difficult topics of physiology." He elaborated: "Whence the *original basic material* [*erste Grundstoff*] of each and every animal and plant arises [*hervorkomme*], and by what sorts of forces this material is subsequently developed [*ausgebildet*], are each problems whose solution hitherto has been shrouded in darkness."[226] There was not even agreement over the roles of the male and the female in propagation. Blumenbach declared his loyalty to Haller's ovism: on the basis of the latter's experiments on chicken embryos, "Herr von Haller has drawn conclusions that make this doctrine considerably more than merely probable [*bei weitem mehr als blos wahrscheinlich*]."[227] In part 2, §11, Blumenbach gave clear definitions of the competing theories of epigenesis and "evolution" (preformation). The weakness of the former theory was "that all sorts of forces were assumed to carry out this process." He identified three forms of the epigenesis idea: outright animism ("The spiritualists have made the soul into the master builder"); Buffon's view (whereby an "inner model" present in the older organism provides the scheme after which "the basic material of the new organism is patterned"); and, finally, Caspar Friedrich Wolff's notion of a *vis essentialis*.[228] Blumenbach then characterized "evolution" (preformation) in the starkest form as "preexistence" or "encapsulation." In §12, he considered the experimental work of Haller in more detail, then offered some reservations. He pointed to hybrid generation, to the inheritance of polydactyly from both sexes over generations, and to species in which male and female demonstrated radically different morphology—all classic difficulties for theories of preformation.[229] Blumenbach was about to make a major intervention in this whole question of generation with his new idea of the *Bildungstrieb*, but in 1779 there was still no hint of that.

Blumenbach devoted considerable energies to a Göttingen journal founded in 1780 and edited by his two colleagues Georg Lichtenberg and Georg Forster, the *Göttingisches Magazin der Wissenschaften und Litteratur*.[230] Blumenbach contributed to the inaugural issue and then provided many more essays over the next two years, including his most seminal scientific paper, "Über den Bildungstrieb (*Nisus formativus*) und seinen Einfluß auf die Generation und Reproduction," the original presentation of his theory of the *Bildungstrieb*. This came in late 1780, in the fifth and last issue of the first volume.[231] The article title itself is important not only for its introduction of his famous term, with its Latin equivalent (*nisus*, not *vis*), but also for its incorporation of propagation (German, *Generation*) and regeneration (German, *Reproduction*) as essential aspects of the phenomenon.

Blumenbach was drawn to the most exciting experimental work in life science: Trembley's famous work on polyps. Indeed, while still a medical student, his own field research brought him to the study of freshwater hydra near Göttingen, and these studies would lead to his greatest theoretical innovation, the idea of the *Bildungstrieb*.[232] Blumenbach explained that his experiments had led him to a new and significant finding: the regenerated parts were always *smaller* than the originals. When, in his subsequent medical practice, he observed that after the healing of a particularly extensive knee injury, the restored tissue permanently manifested a depression (i.e., was smaller in mass), he saw the parallel and inferred a generalization of very wide significance:

> That in all living creatures from man to mite and from cedar down to mold, a distinctive innate drive [*Trieb*] lies that remains actively at work over their entire life span, at the beginning to help them assume their definitive form, then to maintain it, and if that be destroyed, as much as possible to restore it.
>
> A drive (or propensity or endeavor, however one wishes to term it) that is entirely different both from the general properties of all bodies whatsoever and from the other characteristic forces [*Kräfte*] of organized bodies in particular, which appears to be one of the original causes of all propagation [*Generation*], nutrition, and regeneration [*Reproduction*], and which I here, in order to avoid all misinterpretation and in order to distinguish it from other natural forces, provide the name formative drive [*Bildungs-Trieb*] (*nisus formativus*).[233]

The definition would remain virtually unaltered in all subsequent formulations, both in the texts devoted explicitly to the *Bildungstrieb* and in the texts that made use of the concept, for example, later editions of the *Handbuch der Naturgeschichte* and the various editions of his *Institutes of Physiology*.

The term itself deserves some attention. The first part, *Bildung*, is one of the most studied in modern German intellectual history.[234] Its importance for aesthetics and for cultural orientation has been unquestionably established. But it had a powerful role in the life sciences as well, with its important conception of processual development, of *formation*, especially in the organic realm.[235] On the other hand, the *last* component of Blumenbach's famous term has not quite drawn the attention it deserves. The concept *Trieb* played a central role in the developing life sciences of the eighteenth century and one explicitly pertinent to Blumenbach's dissertation: namely, the difference between man and animals, the question of *instinct* versus reason. Here, Blumenbach was clearly very attentive to the pioneering work of Hermann Reimarus. There is also an important contrast between *Trieb* and *Kraft* (*nisus* and *vis*), which mattered both to Blumenbach and to Kant (and, later still, to Goethe and Schelling).

Blumenbach proposed that propagation, nutrition, and regeneration were all fundamentally aspects of one single force, the *Bildungstrieb*. He contended that this new discovery should not be assimilated to older notions, such as that of *vis plastica*, associated with early eighteenth-century discussions of organic life, or the *vis essentialis* of C. F. Wolff. His description of each of these alternative views came in the form of brief footnotes.[236] In later versions he would elaborate, especially as he endeavored to make clear his differences from Wolff. But Blumenbach had a more important argumentative task than distinguishing his position from those of other epigenesists. He had clearly parted company with preformation theory, and he needed to adduce his reasons for doing so, especially since he had formerly embraced Haller's preformation theory in numerous publications.

This was a bold essay, and he swiftly saw the need to elaborate on his argument to make it more convincing. The result was the publication in 1781 of a book-length version: *Über den Bildungstrieb und das Zeugungsgeschäfte*. The core of the eighty-seven-page book is the unaltered repetition of the article, but the supplements are worth attention. The book version opens with a long, clearly anxious kowtow to Haller, lamenting "this split from Haller—the man to whose writings and whose correspondence I owe so immeasurably much."[237] Blumenbach consoled himself that had Haller lived longer he would have himself come round to the new view. Even were that not the case, "under no circumstances would Haller's fame suffer the slightest diminution of its deserved brilliance."[238] Blumenbach claimed his new insight was just luck, not superior labor. The humility is oppressive though presumably genuine.

In the book version, Blumenbach expanded on the difference of his view from earlier forms of epigenesis, *vis plastica* and *vis essentialis*. He expressed a bit of pique at Lazzaro Spallanzani for lumping all such theories together the better to uphold his Hallerian preformationist orientation.[239] In §5 Blumenbach moved his discussion of the difference of his *Bildungstrieb* from Wolff's *vis essentialis* from the original footnote into the body of the text, but the treatment amounted to no more than a citation from Wolff's *Theorie von der Generation* where Wolff gave a characterization of *vis essentialis*.[240] A great deal more would be required to settle the real differences between their views.

More definitive was his characterization of the weaknesses of preformation theory. Now he suggested there were problems in Haller's experimental results as summarized in Haller's famous review of Wolff in the *GGA*. Blumenbach confessed himself to have been convinced for a time, but now he raised objections, claiming (in line with Wolff, incidentally, but without acknowledgment) that in the first days of embryonic development of the fertilized chicken egg there were no signs of blood vessels at all.[241] He dismissed the "spermists" more brusquely, arguing that the various illustrations offered by Leeuwenhoek, Hartzöker, Lieberkühn, and Spallanzani were so mutually incompatible that they made any general claim problematic.[242] In §35 Blumenbach brought to bear for the first time explicitly the decisive experimental work of Joseph Gottlieb Kölreuter (1733–1806), who had demonstrated the possibility of mutating one species of tobacco plant into another and then back over a series of generations, thus dissolving some of the most essential fixities that inspired preformation and eighteenth-century life science altogether.[243] Kölreuter's results, Blumenbach averred, "ought to convert even the most committed defenders of the theory of evolution from their prejudice."[244] Finally, he turned to the notion of "panspermism," the revival of the Lucretian idea that organic germs were dispersed throughout creation and simply came together by chance convergence.[245] For Blumenbach such a theory was formed from "the most adventurously arbitrary presuppositions [*abentheurlichsten willkürlichen Voraussetzungen*]."[246] Having disposed of the rival theories, Blumenbach devoted the rest of his text to the role of the *Bildungstrieb* in propagation, nutrition, and regeneration, elaborating on the different regenerative capacities among animals, especially their limits in warm-blooded ones.

In 1782 Blumenbach issued the second, revised edition of his *Handbuch der Naturgeschichte*, but with virtually no change in the account of animals, with one noteworthy exception: in the discussion of the apes, the 1782 edition introduced the notion that they could be particularly distinguished from man by the presence of an elongated snout caused by the intermaxillary bone, absent in humans.[247] Blumenbach referred explicitly to Camper's dissection of an orangutan proving the presence of this bone in them.[248] The greatest innovation in the 1782 edition was the inclusion of his theory of the *Bildungstrieb*. The concept made its appearance in the 1782 *Handbuch* in the revisions of part 2 ("On Organized Bodies Generally") and in the revisions of part 10 ("On Plants"). Part 2 gives the impression of far greater confidence and concentration in Blumenbach's presentation of the idea of the *Bildungstrieb*, which he defined in essentially unchanged language in §11 of the new edition.[249] Preformation got short shrift, given its "numberless and irresolvable difficulties," and he did not even bother to distinguish his idea from other forms of epigenesis.[250] His argument was that preformation had nothing to say concerning the ongoing organic processes of nutrition and regeneration, whereas his theory of the continued presence of this formative drive took all this directly into consideration.[251] Moreover, the regularity as well as the deviance of birth defects and teratology generally could be explained via the *Bildungstrieb* but not by preformation.[252] Finally, in the section on plants, Blumenbach offered a more extended discussion of Kölreuter's experiments, arguing that preformation was helpless to account for them, while they fit very nicely with his own theory of the *Bildungstrieb*.[253]

CHAPTER EIGHT

Blumenbach, Kant, and the "Daring Adventure" of an "Archaeology of Nature"

> It is commendable to do comparative anatomy and go through the vast creation of organized beings in nature, in order to see if we cannot discover in it something like a system, namely, as regards the principle of their production.... Despite all the variety among these forms, they seem to have been produced according to a common archetype, and this analogy among them reinforces our suspicion that they are actually akin.
>
> KANT, 1790[1]

In the first volume of the first edition (1779) of Johann Friedrich Blumenbach's *Handbuch der Naturgeschichte*, there is a crucial section (§39) in part 3, "Of Animals Generally," that would be removed from all subsequent editions, not because Blumenbach repudiated anything in it, but rather because he would expand upon it in a separate, major component of later editions. That section dealt with the question of extinction. He wrote: "Since we are acquainted with so many animals only in fossil form, and not in [live] nature, some famous men have concluded that probably some species [*Gattungen*]—indeed genera [*Geschlechter*]—may have died out. Against this, to be sure, one might protest that a very large part of the earth is not yet investigated, and that we cannot know what might lie hidden at the bottom of the seas, in the interior of Africa, and other places where natural history has not yet made its way." This was cautious, to be sure, but Blumenbach clearly inclined toward the idea of extinction. He noted that there were so many fossils without contemporary living counterparts that notwithstanding the doubters "still [*doch*] we have to see from all this that our earth over time has suffered very grave catastrophes."[2]

BLUMENBACH AND PALEONTOLOGY

In the second part of this first edition, which appeared in 1780, Blumenbach presented a section specifically dealing with fossils (*Versteinerungen*) that aimed to establish clearly the distinction between organic remains and inorganic minerals.[3] The opening section on mineralogy in general showed no changes between the first two editions (1780, 1782). In §221, Blumenbach presented his overview of the field in crucial language largely unmodified in subsequent editions:

> To start with, something concerning the *origin* [*Ursprung*] of minerals, namely, concerning the main paths [*Hauptwegen*] by which in part in earlier times they emerged all at once [*theils vor Zeiten mit einemal entstanden sind*] and in part [they] have emerged gradually and even now continue to emerge [*nach und nach und noch immerfort entstehen*]. To shed some light on this, we must necessarily go back to the origin of our earth itself, an investigation in which, to be sure, a few daring conjectures must always be allowed, although we do not want to give ourselves over to the flights of clever men who have offered comets and burned-out suns as the basis of their system of the earth[. Instead,] we offer our more modest opinion, to which we have been brought first of all through the investigation of fossils [*Versteinerungen*] and their observed difference from currently existing organized beings, as well as through the comparison of ancient volcanoes [*ehemaligen Vulcane*] and their products with those still burning, etc., and though this remains just another hypothesis, yet it is a hypothesis that nature and what is observable can easily and effectively accommodate.[4]

In the very next section of the 1780 text, Blumenbach made a most radical claim:

> We believe ourselves persuaded, accordingly, that our globe [*Erdkugel*] has experienced at some time at least *one* apocalypse [*wenigstens schon einen Jüngsten Tag einmal erlebt*], and that we have the last judgment that took place then to thank for the current state of the world.[5]

I think the phrase "at least" in this passage needs to be registered as highly significant, though assuredly Blumenbach did concentrate on a single catastrophic event in this account:

> This great catastrophe was, in all likelihood, occasioned by a subterranean fire that presumably thrust the floor of the sea into the heights and at the same time occasioned the dry land all at once to be swamped by the sea. Thus was everything living on earth drowned [*die ganze beseelte Erde ertrunken*], and by the same token the sea animals perished in being thrust out of their element onto

> dry land. Hence the size and the regularity of the strata on the highest peaks, full of the majority of fossilized molluscs, etc., which have not been discovered and are indeed hard to discover in [living] nature.[6]

There is compelling evidence that Blumenbach drew this model of the great catastrophe from the writings of the Genevan geologist Jean-André Deluc. Blumenbach reviewed Deluc's *Lettres physiques et morales sur l'histoire de la terre et de l'homme: Addressées à la reine de la Grande-Bretagne* (1779/80) in *Göttingische Gelehrten Anzeigen* in December 1780.[7] It offered Blumenbach a general theory into which to situate his own issues of natural history and comparative anatomy and thus brought him to the cutting edge of the field of paleontology. Once attuned, Blumenbach recognized that paleontological evidence lay all about him—indeed, as he famously put it, in every paving stone in Göttingen.[8]

Blumenbach discriminated two elements in Deluc's text, a rambling travelogue and a systematic scientific theory, and he concentrated entirely on the second of these, which he took to be a finely articulated *system*.[9] Summarizing, Blumenbach observed that Deluc "divides all of world history into two great periods"—a former world (*Vorwelt*) and a "newer" earth—separated by a great catastrophe:

> Our earth has its current form thanks to the Deluge—but this came to be in the following manner: that the dry land, after it was gradually undermined by subterranean fire, eventually collapsed and sank below the level of the ocean of that time, which accordingly was drawn into these new depths and abandoned its former beds, which in turn, having been left dry, became the new landmasses. This new, still-current dry land therefore remained stable and unchanged in its original fixed stratification, and its previous mountains and valleys had almost all already existed while all this was still at the bottom of the sea. The highest peaks had stood out over the surface of the waters as islands. The rest was covered by the waters.[10]

Since then, Blumenbach continued his summary, the oceans and continents had remained roughly in the same configuration. Most significantly, according to Blumenbach, Deluc maintained that "since this great catastrophe . . . only a few thousand years could have passed" up to the historical present.[11] Deluc drew his evidence, Blumenbach noted, from the "archive" of the mountains to be found on the new continents. Deluc offered a tripartite typology of these mountain formations: primordial, aquatic, and volcanic. Blumenbach rendered these into "our"—that is, the Wernerian geognosical—terminology, with which his German readers would be familiar. What Deluc called "primordial," Blumenbach rendered *Ganggebirge*; what Deluc called "aquatic," Blumenbach termed *Flötzgebirge*.[12] The latter were particularly decked with strata upon strata of fossils (*Versteinerungen*).

Since Deluc posited that in the former world there had been many upheavals thanks to volcanic action, this allowed for the *local* submersion and reemergence of dry land, relative to the seas, hence accounting for the fossil presence of land animals and plants in strata beneath those of sea life, and conversely. The volcanic explosions were outbursts of a subterranean fire and, combined with explosions resulting from penetration of seawater via the resulting cracks, widened cavernous gaps in the earth's crust, gradually spreading laterally until the whole surface was undermined, setting up the great catastrophe. This further explained the overlay of volcanic and sedimentary strata, since there were many such dislocations and resettlements over the vast span of preadamite time. Two outcomes were particularly salient for Deluc in this analysis. First, the age of the *current* continents could be measured by geological observations—the use of "nature's chronometers" such as the accumulation of silts in river deltas like that of the Rhine—to be quite recent. And, second, this all converged gratifyingly, but ostensibly without deliberate refitting, with the biblical narrative of Moses.[13]

Blumenbach's review accomplished a remarkably concise distillation of Deluc's prolix presentation. There can be little doubt from this account that Blumenbach drew heavily on Deluc for his theory of the early history of the planet in 1780. He made this explicit in a footnote to §227 of the *Handbuch* of 1797: "There is no geognosic system known to me (and one could count already in the year 1764 no fewer than forty-nine) that satisfies this demand [of grounding the fundamental knowledge claims of geognosy in a careful test against physics and chemistry] other than that in Mr. Deluc's geological letters which are translated from the French manuscript in Hr. Professor Voigt's *Magazin* [*für den Neueste aus der Physik und Naturgeschichte*]."[14]

Blumenbach's 1780 formulation augmented Deluc's account in its treatment of volcanoes. If fossil remains represented one of the two classes of empirical evidence for the ancestral earth, Blumenbach noted, volcanic remains formed the other:

> In a thousand places, however, the fire broke through the crust of the earth, hence the countless extinct volcanoes that in the most recent times have been recognized as such and that from Göttingen to the banks of the Rhine alone have been counted at about fifty. Perhaps it was through this great catastrophe

> that granite assumed its current appearance and, consequently, like the vast majority of fossils [*Petrefacten*], along with the majority of extinct volcanoes and columnar basalt formations, should be considered the ruins of the former world, that preadamite earth, and accordingly be distinguished from all the other minerals that, gradually or also through similarly violent catastrophes, emerged on the subsequently established earth, after the Creator, in the manner recounted by Moses, gave it new life with its current creatures.[15]

In this passage Blumenbach both embraced the biblical account (Moses) and situated it in a context of preadamite times that was clearly a more recent and nonscriptural conception.[16] In short, Blumenbach arrayed two substantial classes of empirical evidence for a radically different earlier epoch of the earth, and he contemplated the possibility of "similarly violent catastrophes," that is, a plurality of such events. Indeed, in §223, Blumenbach moved from the conception of the unique catastrophe to a notion of general causes of geological disruptions: "subterranean fires and deluges, the two means by which, in our opinion, the former world was annihilated, are still two considerable sources also on our current planet sometimes for the destruction and other times for the transformation and emergence of minerals."[17]

The next edition of Blumenbach's *Handbuch* appeared in 1788, and it showed primarily an expansion of Blumenbach's documentation of fossil finds.[18] He had become quite interested, as museum director, in accumulating a collection of such fossils for Göttingen, and he would build one of the foremost such collections in Europe by 1800. Theoretically, however, he made no revision in the 1788 edition. By the 1791 edition, Blumenbach did make important changes in the mineralogy section, including a clearer formulation of its taxonomy.[19] But in the interval, of course, he had published the first edition of his *Beyträge zur Naturgeschichte* (1790), which proved his most important publication in the field.[20]

The first volume of the first edition of his *Beyträge* was dated April 1790.[21] It was accompanied, later that year, by an article in *Magazin für den Neueste aus der Physik und Naturgeschichte*, with the title "Beyträge zur Naturgeschichte der Vorwelt."[22] Only the first sections of volume 1 of the book dealt directly with the questions of earth history and paleontology; the entire second volume dealt with other issues.[23] The article of late 1790 did take up questions of earth history, since it was motivated by his immediately ensuing critical résumé of James Hutton's "eternalist" theory of the earth.[24] Clearly, Blumenbach preferred the historical-developmental account worked out by Deluc (and others, including the German Wernerians) to the cyclical theory developed by Hutton, especially since the latter made light of species extinction, in stark contrast to Blumenbach's own view. This may have been a central topic in his conversations with Deluc a year later (1791) at Windsor. Their common opposition to Hutton may have been an important motive for Blumenbach's invitation to Deluc to draw up a concise summary of his rival theory for a German audience, which appeared in the same journal where Blumenbach had himself critically appraised Hutton's ideas. A decade later, in 1801, Blumenbach gave an anniversary address to the Göttingen Royal Academy of Sciences, "Specimen Archaeologiae Telluris," which he summarized in a report for the *Göttingische Gelehrten Anzeigen* in that same year, then published in Latin in 1803.[25] A summary of his summary was then appended to the third section of his *Beyträge*, part 1, in the second edition of 1806.[26] The contentions of these three versions, and often the very phrasing, are roughly identical, with the exception that the Latin paper published in 1803 had a more extended treatment of particular fossilized organisms.[27] This body of material is the most direct evidence we have of the development of Blumenbach's ideas on paleontology and the history of the earth.

As he put it in a footnote to §2 of the 1790 edition of his book, "nearly the only, but therefore all the more important, use of the knowledge of fossils [*Versteinerungen*], is the solution that the history of the changes of the earth's surface derives from it."[28] Blumenbach articulated clear methodological constraints, already in 1790 and then more extensively in 1806: "If petrifactions [*Petrefacten*] can be made of regular use for the archaeology and the physical geography of the earth, as the surest documents of the archives of nature for the fruitful history of the catastrophes that have been connected with our planet since its creation, the study of them . . . demands . . . a thorough critical comparison of them with the organized bodies of the present creation, . . . of their different locations, and their geognostical relations."[29] If one observed this methodological propriety, "a wider examination of these differently made fossils, and of these equally various sorts of condition, brings us to a closer conclusion as to the oldest history of the body of this earth . . . [and] the numerous catastrophes . . . through which its crust has acquired its current appearance . . . built out of such great convulsions."[30]

The most important feature in these texts is that Blumenbach insisted upon a plurality of "revolutions"—"the sequence of the totally different catastrophes [earth] has gone through, by which the numerous fossil remains of former organic creations have come to their present positions."[31]

He clearly articulated *three* conceptually distinct phases; moreover, each of these phases in its turn involved *multiple* incidents.[32] Working backward in time, Blumenbach first collocated those fossilized (and semipreserved) remains of creatures that could be identified with currently existing species. Second, he identified remains that were *similar* to current species but not the same and, in addition, whose closest living analogues inhabited distinctly different climatic zones. Here he placed the widely discussed elephant, rhinoceros, and other remains that had been the preoccupation of zoologists over the second half of the eighteenth century, and whose correct reconstruction would become the basis of Georges Cuvier's rise to preeminence in the early nineteenth century.[33] Blumenbach clearly discerned a mass extinction between the first and the second phase: "a total alteration of the climate took place, which occasioned the destruction of the then living generation of these tropical creatures, as of many other genera and species of organized bodies which existed among them . . . such as the unknown of Ohio [*Ohio-incognitum*] among great land-animals. . . . This revolution, which seems to have been merely climatic, must be distinguished from those earlier and much more formidable ones [note plural!], from which we must date the petrifactions of the *third* division, the oldest of all."[34]

In his third and most ancestral phase, Blumenbach placed fossilized remains that were utterly different from any known living forms and that, accordingly, were evidence for a former world (*Vorwelt*) and a discontinuity far more radical than he ascribed to the border between his first and second phases. It was this more ancient and drastic divide that Blumenbach accorded the title of a "total" revolution: "in my opinion, it becomes more than probable that not only one or more species but a whole original organized preadamite creation has disappeared from the face of our planet."[35] That is, "as I have said [already in 1780], . . . our earth has already suffered a complete revolution and experienced one last day."[36] He elaborated: "This general revolution . . . is quite different from the subsequent one" dividing the first and second phases.[37] For *all* these phases his consistent employment of the plural indicated that *many* disruptions, though not all global in scope, took place both before and after the great catastrophe. Significantly, he wrote of that as "the *last* catastrophe" (my emphasis), hence one of many that beset the *Vorwelt*.[38] He insisted that the fossil remains could not all be "explain[ed] . . . by one and the same catastrophe."[39] Clearly, Blumenbach conceived of multiple disruptions in the *Vorwelt*: "these destructive catastrophes themselves were again of more than one sort, and were very far from happening all at the same time."[40] To be sure, Blumenbach was cautious about how much order could be brought to the periodization of the planet's surface changes: "it is scarcely possible at present to determine with any certainty the chronological arrangement of the successive periods[,] . . . to say nothing of the causes of them."[41]

What triggered Blumenbach's thinking was the empirical evidence for extinction, and not simply rare instances of it but massive hordes of no-longer-to-be-found organic forms, illustrated, in his own works, by some two hundred species of ammonites for which no correlative life-forms could be found in the living world.[42] One can hardly argue that Blumenbach was the first to take up the issue of extinction. Already in 1779 he wrote of "a few famous men" who advocated the idea and with whom he chose to align himself.[43] Blumenbach made the question of extinction the leading principle not only for the study of fossils but also for the theory of life-forms in general—a foundation, in conjunction with comparative anatomy, for a true life science. Moreover, in Baron's words, "Blumenbach's periodization of fossils represents one of the earliest efforts to draw a parallel between the history of the earth and the history of organisms."[44] Thus, Blumenbach was among the first to see that the reconstruction of the "individual phases" of the historical process of the development of organisms could be achieved through the study of fossils.[45] The *direction* of this relation is crucial here. In the history of geology, the significance of "biostratigraphy" lay in dating strata sequences, that is, *geological* forms. But strata sequences *also* created effective empirical evidence for periodization in the sequence of *life*-forms—*evolution* in our most general sense of shifts in the basic form and function of living things across the span of earth history. To be sure, Blumenbach knew he was far from a good causal account. Yet, by that very token, such a realization obviously *called* for one.[46]

The single most important conundrum presented by this periodization of earth history and its living inhabitants was how the extinct organisms came to be *replaced* after the great catastrophes, and especially after the most massive one separating the *Vorwelt* from the recognizable configurations of what Blumenbach took to be the current world (*Schöpfung*, literally, "creation").[47] Mass extinction made the idea of a "total revolution" very problematic, as Rudwick notes, since the question of a "new" creation seemed to require literally a deus ex machina.[48] Blumenbach clung to the fact of the matter, even if he could not offer a compelling causal account. "A whole creation of organized bodies has already become extinct, and has been succeeded by a new one," he wrote.[49] "*How* indeed this subsequent creation took place, that I can no more say than how in early times the first spermatic animalcule came into being; that, however, they *were* subsequently created seems to me undeniable, and I lay that to the account of the great mutability of nature."[50]

Blumenbach, to get some perspective on this, insisted that extinctions and new emergences were *continuing* to take place: "Creatures enough die away in a locality, and fresh ones again become naturalized and spread themselves.... So there is nothing contradictory in the idea ... that a species may have become extinct; and on the other hand, a fresh one may likewise be sometimes very easily created [!] subsequently."[51] This was the perspective from which to consider the aftermath of the "last great catastrophe." "If the former world [*Vorwelt*] suffered a total revolution, as seems unmistakable, and *if* this revolution was probably caused by a general conflagration of the earth [*durch einen allgemeinen Erdbrand bewirkt*], afterward there must have been a very long span of time before the newly changed crust of our planet had cooled down and its surface once again became at all ready to be enlivened with a fresh vegetation and vivified with a new animal creation."[52] Blumenbach surmised that "the Creator" would then have "permitted the same natural forces in general to achieve the production [*Hervorbringung*] of a new organic creation" to fulfill the same purposes that the older order of life had served in the prior world.[53] Noteworthy here is not only the *transcendent intervention* but the *teleological construction* of the order of organized life in both worlds. That teleological structure was, moreover, ascribed to *natural forces*, which, even more importantly, remained constant across the catastrophe.

This allowed Blumenbach to have recourse to his own master concept, the *Bildungstrieb*, to grasp the phenomenon. "Just that, after such a total revolution had so changed the materials at hand, the *formative drive* clearly was compelled to take a direction differing more or less from the old one in the production of new species."[54] That is, "the formative drive in these two creations, to be sure, functioned in a similar but not in the identical manner."[55] The language of an "alteration of direction of the formative drive" became a centerpiece of Blumenbach's teaching in the 1790s, as evidenced in its prominence in Christoph Girtanner's representation in his 1796 monograph.[56] In his text of 1790, Blumenbach was at pains to distinguish that mode of explanation from a rival view proposing "*degeneration* (*Degeneration*) acting for a long series of thousands of years."[57] Evoking the contrasting direction of the spiral in fossil seashells relative to living ones, he firmly asserted: "Such a thing is not a consequence of degeneration, but a remodelling [*Umschaffung*] through an altered direction of the formative drive."[58] He considered the possibility, based on the similarity of ancient fossils to currently existing life-forms, that *some* organisms might have persisted across *all* the catastrophes into the present, but he asserted a preference for the hypothesis that these were *only* similarities induced by the formative drive of nature since it was indeed perennial, though it adapted to the new circumstances of the postcatastrophe earth.[59] Thus, Blumenbach proposed "an altered direction of the formative drive [*eine veränderte Richtung des Bildungstriebes*]" as his key. In a footnote added to the second edition, he elaborated: "the formative power of nature in these remodellings [*Umschaffungen*] partly reproduces again creatures of a similar type to those of the old world, which however in by far the greatest number of instances have put on forms more applicable to others in the new order of things, so that in the new creatures the laws of the formative drive have been somewhat modified."[60]

Rather than abandon his commitment to the immutability of species (though not of *varieties* within a stem-line), Blumenbach insisted on the formation of entirely new life-forms, with only structural parallelism to earlier forms due to their common formation by the ubiquitous *Bildungstrieb*. The notion of perennial forces generating different particular forms in differing ecological circumstances has some resonance with Buffon's notion of the persistence of indestructible "organic molecules" and the action of various "interior molds" that allowed for similar emergences of new life-forms across geological change.[61] That assumption of perennial forces, which took into account altered circumstances, seemed to Blumenbach more consistent with "physiology" than the notion of immanent developmental change leading to new species.[62] The functionally constrained (teleological) "animal economy" of each species (or its stem-line) was so finely integrated in terms of the relations of its various organic parts that any significant change would induce drastic dysfunction, in his view.[63] The force of his "comparative anatomy" thus induced conservatism about biological transformism, just as it would in the parallel positions of Cuvier.[64]

KANT'S RETREAT FROM THE "DARING ADVENTURE OF REASON" IN HISTORY OF NATURE

If Kant was one of the most important advocates of history of nature as a new empirical science, by 1790 he also became suspicious of its potential. Kant made a decided shift over the 1780s from *participation* in actual life science (to be sure, from his armchair) to a much more skeptical *critique* of its method.[65] As he waged his bitter disputes over "race" in the 1780s with Herder and Forster, and as he simultaneously evolved his own critical philosophy, epistemological scruples overshadowed Kant's scientific ambitions, undermining the very possibility of a *science* that proposed to account for the genesis or even the organicism of living things in nature.[66] Raphael Lagier traces across the 1780s a "progressive reduction of *empiricism* in [Kant's theory of] the sciences," which at the extreme "seemed to disqualify from the outset" new impulses in late eighteenth-century science, especially life science.[67]

Phillip Sloan notes three contexts for Kant's shift: the controversy with Herder, Kant's return to the problem of race, and Kant's "development of a rational philosophy of science and a classification of the sciences in the *Metaphysical foundations of natural science* of 1786."[68] These three contexts are thoroughly interconnected. Sloan recognizes the key: Herder's "vital, transformative pantheism, accounting both for the origin of the earth and also for its living creatures, and eventually for human beings and for the emergence of reason itself, seems to have drawn Kant abruptly up against the conclusions to which a full-blown 'history' of nature, one even more ambitious than that of Buffon, could be taken."[69] Kant became alarmed over "the possibilities (and the dangers) inherent in a developmental history of nature—the options either of a dynamic pantheism or of an atheistic materialism."[70] The decisive passage in Kant's review of Herder's *Ideen* made clear what troubled him:

> As regards the hierarchy of organisms, . . . its use with reference to the realm of nature here on earth leads nowhere. . . . The minuteness of differences when one compares species according to their *similarity* is, in view of such a great multiplicity [of species,] a necessary consequence of that multiplicity. But a *consanguinity* [*Verwandtschaft*] among them, according to which either one species springs from another and all of them out of one original species, or as it were they originate from one single generative mother-womb, would lead to ideas which are so monstrous that reason shrinks back.[71]

Kant also rejected Herder's idea of a single all-pervasive force as the principle of the organization of nature. "The unity of organic force . . . is an idea which is entirely outside the field of empirical natural science and belongs to merely speculative philosophy."[72] Indeed, Kant suggested that practitioners of *Naturbeschreibung* would do well to draw back: "As to whatever this contribution to comparative anatomy through all the species of animals down to the plants may betoken, those who conduct natural description [*die Naturbeschreibung bearbeiten*] can decide for themselves how useful this suggestion may be for new observations and whether it even has any basis whatever [*ob sie wohl überhaupt einigen Grund habe*]."[73] This is an unmistakably hostile stance toward an enterprise he had himself been central in advancing, though, to be sure, he did not entirely *proscribe* the new *Naturgeschichte*.

In his classification of natural inquiry into a hierarchy of "proper" science in the preface to *Metaphysische Anfangsgründe der Naturwissenschaft* (1786), Kant distinguished (with his epoch) between "historical" and "rational" sciences: that is, those that rested entirely on empirical, a posteriori observations and those that had a rational, or a priori, foundation.[74] For Kant, "proper science" (*eigentliche Wissenschaft*) could belong only to the latter category. He therefore divided natural theory into "historical natural theory [*historische Naturlehre*], which is nothing more than systematized facts about natural things, . . . and (proper) natural science [*Naturwissenschaft*]." Under the merely empirical category of *historische Naturlehre* Kant offered two further subdivisions: "*natural description* [*Naturbeschreibung*], as a system of classification of these natural things according to similarities, and *natural history* [*Naturgeschichte*], as a systematic representation of these things in various times and places."[75]

In his 1785 essay on race, Kant took up the question of the relation of *Naturgeschichte* to *Naturbeschreibung* again. "*Species* [*Art*] and *genus* [*Gattung*] are not in themselves distinguishable for *natural history* (in which what matters is only generation and origin [*es nur um die Erzeugung und den Abstamm zu tun ist*]). Only in *natural description*, since it is just a matter of the comparison of traits [*Vergleichung der Merkmale*], does this distinction arise."[76] Thus, for Kant, Linnaean classification sufficed to define life-forms according to a "*nominal genus* [*Nominalgattung*] (in order to classify them according to certain similarities), but never a *real genus* [*Realgattung*], for which the absolute minimum requirement would be the possibility of descent from a single common set of parents [*durchaus wenigstens die Möglichkeit der Abstammung von einem einzigen Paar erfordert wird*]."[77] It was this characterization that provoked criticism from the famous naturalist Georg Forster.[78]

In his crucial essay of 1788, "Über den Gebrauch teleologischer Principien in der Philosophie," the response to Forster's criticisms, Kant distinguished a project to explain "the original emergence [*Entstehen*] of plants and animals"—which he agreed with Forster "would be a science for gods," that is, something "to which human reason [*Vernunft*] cannot extend"—from history of nature. The latter "would, by contrast, concern itself with investigating the connection between certain present properties of the things of nature and their causes in an earlier time in accordance with causal laws that we do not invent but rather derive from the forces [*Kräften*] of nature as they present themselves to us, pursued back, however, only so far as permitted by analogy." Kant insisted that such an enterprise was "not only possible, but one which is attempted frequently enough, as, for example, in the theories of the earth formulated by careful natural scientists (among which the theories of the famous Linnaeus also find their place)."[79]

Kant maintained that *Naturbeschreibung* and *Naturgeschichte* were "thoroughly heterogeneous," and "if the first (natural description) appears as a science in the complete splendor of a great system, [while] the other (natural history) can only offer fragments or faltering hypotheses," still he defended upholding *Naturgeschichte* as a distinct science:

> Even if for now (perhaps even forever) [this undertaking] can be conducted more in outlines [*Schattenrisse*] than in a thoroughly elaborated science (in which for most questions a "still open" [*Vacat*] will be found to be indicated), nevertheless I hope to be able to see to it that an insight not be accorded to the one [vantage] that actually belongs only to the other and to show how one might more precisely understand the extent of actual knowledge in natural history (for we do have some such knowledge), while at the same time [understand] its limits, which lie in [the nature of] reason itself, together with the principles in accordance with which it might be extended in the best possible manner.[80]

This conceptual discrimination was in itself a valuable contribution to science, Kant insisted, for it kept off "the recklessness [*Sorglosigkeit*] which allows the proper boundaries of the sciences to be transgressed."[81] He went on:

> the greatest difficulty in this proposed innovation lies simply in the names. The term *history* [*Geschichte*], as with the Greek *historia* (storytelling, description), has a meaning that is already so thoroughly and so long a matter of custom that one could not easily permit assigning it another meaning, which would be that of the natural inquiry into origins [*Naturforschung des Ursprungs*], while it is also not without difficulty to find for it in this last sense another, appropriate technical expression.* (*I would like to suggest for natural description the term *physiography*; for natural history, by contrast, *physiogony*.) In any event, the linguistic difficulty of discrimination cannot remove the difference in fact.[82]

Kant distinguished between "*Abartung* (progenies *classifica*)" and "*Ausartung* (degeneratio s. progenies *specifica*)"—that is, variation *within*, versus mutation *into another*, species, or what in §80 of the *Kritik der Urteilskraft* he distinguished as *generatio homonyma* and *generatio heteronyma*, then added in a footnote: "The terminology of *classes* and *origines* expresses quite unequivocally a merely *logical* discrimination, which *reason* makes among its concepts for the sake of strict *comparison*; *genera* and *species*, however, can also express the *physical* discrimination which *nature* herself makes among her creatures in terms of their *generation* [*Erzeugung*]."[83] A few pages later, Kant reasserted the distinction between *Naturgattung* and *Schulgattung*.[84] The first was a category of *Naturgeschichte*, and the second, a category of *Naturbeschreibung*.

It was the task of *Naturgeschichte*, not *Naturbeschreibung*, to account for both the fixity of species and the varieties and races within them that had emerged over the course of natural history.[85] But there were rational limitations: "the physical first beginning of organic entities cannot be established by either of us [himself or Georg Forster] or by human reason in principle, just as little as the half-bred inheritance of traits in propagation [between races]."[86] Only teleology could work here; to attempt otherwise, "to supplant [teleological judgments] by physical ones in considering organisms in terms of that which has to do with the fixity of their species, cannot even be thought, and this form of explanation imposes no new burden on natural inquiry to go beyond what it in any event can never get past, namely here to follow merely the *principle of purposes*."[87] This led Kant to the epistemological posture that grounded the "Critique of Teleological Judgment" two years later:

> Because the concept of an organized being carries with it that it is a material [entity] in which all [elements] stand in mutual interaction as ends and means, and this can only be thought of, moreover, as a *system of final ends*, and its possibility accordingly only teleologically—never, however, in terms of a physical-mechanical mode of explanation at least for *human* reason—thus, it may not be investigated in physics whence all organization itself has sprung. The answer to this question, if it is at all accessible to us, would lie *beyond* natural science, in *metaphysics*. I for my part derive all the organization in an organized being (via generation) and subsequent forms (of this sort of natural being) according to laws of gradual development from *original endowments* (of which one can often come upon instances in the dissemination of plants), which are to be traced to the organization of its stock. To account for how this [original] stock itself *arose*, this obligation lies entirely beyond the bounds of all physics possible for humans, within which I, however, believe I must contain myself.[88]

Sloan argues that by 1788 Kant drastically restricted any "science of the historical genesis of nature in the Buffon-Herder tradition."[89] While he still recognized the *possibility* of a history of nature, he starkly limited its *prospects*. Indeed, already in his reviews of Herder, he wrote of science "shrinking back" from monstrous hypotheses of genesis and transformation.[90] To justify to Forster this shrinking back, Kant appealed to the constraint of "original endowments," which predetermined the "line of descent" and which could be triggered but not created by environmental conditions.[91]

In the *Kritik der Urteilskraft*, Kant observed:

> If the name *natural history* that has been adopted for the description of nature [*Naturbeschreibung*] is to remain in use, then one can call that which it literally means, namely, a representation of the *ancient* condition of the earth—about which, even though there is no hope for certainty, there is reasonable ground for making conjectures—the *archaeology of nature*, in contrast to that of art. To the former belong fossils, just as to the latter belong carved stones, etc. For since we are really constantly, if also, as is fitting, slowly, working on such an archaeology (under the name of a theory of the earth), this name would be given, not to a merely imaginary branch of research into nature, but to one to which nature itself invites and summons us.[92]

This is a crucial passage, in which Kant gave up trying to extract *Naturgeschichte* from its longstanding confluence with *Naturbeschreibung* in ordinary language and instead offered a new technical term for it. He clearly saw that such a historical science of nature had been pursued and that it was a natural desideratum. He allowed there could be some knowledge, though not much certainty, about such a topic.

In the main text (to which this observation was attached merely as a footnote), Kant described the evidence that the earth had undergone a series of catastrophes, with drastic effects on organic life. He suggested that this offered little prospect of a simple, teleological account of the development of earth (i.e., that it was *designed* to support life and especially human life). He wrote: "Now if the habitat, the maternal soil (the land) and the maternal womb (the sea) for all these creatures, yields no signs of anything except an entirely unintentional mechanism for their generation, how and with what right could we demand and assert another origin for those products?"[93]

At the very least we are entitled to find Kant quite equivocal about this new science of the historicization of nature. Some years after the publication of the *Kritik der Urteilskraft*, in Rink's edition of Kant's physical geography lectures, we find the following passage:

> If one presents the characteristics of nature as a whole . . . as it took form across all times, then—and only then—would one be presenting a properly so-titled natural history. Whatever changes arose [within any given organism] by vir-

> tue of the differences of the land, the climate, dissemination [*Fortpflanzung*], etc. through all times—that would be a natural history—and one that could be made for every particular part of nature. However, there is this difficulty: that one can only offer hunches based on experiment more than one can be in a position to provide an exact report of all this [*daß man sie mehr durch Experimente errathen müßte, als daß man eine genaue Nachricht von allem zu geben im Stande sein sollte*]. For natural history is hardly more recent than the world itself, but we can't even attest to the certainty of our reports from the time the art of writing emerged. If one, then, undertook accordingly to go through the conditions of nature so that one observed what changes these underwent through all epochs, then this undertaking would provide an authentic natural history.[94]

Kant, by the 1790s, while not surrendering his conceptual articulation of a historical science of nature as a separate science with determinate ambitions, nevertheless believed that its actual harvest of empirical knowledge would remain quite limited. A final, very revealing piece of evidence in this vein comes from Kant's lectures in physical geography from 1786, where, for the only time we have documented, Kant commented on Buffon's great work of 1779, *Les époques de la nature*. Kant offered very left-handed praise: "The only work that deals with real history of nature [*Naturgeschichte*] is Buffon, *Epochs of Nature*. But Buffon let loose the reins of his imagination too much and therefore composed far more a novel of nature than a true history of nature."[95] Neither Buffon nor Herder could satisfy Kant; the project itself seemed dubious.

Significantly, in the third *Critique* Kant assigned a "system [of nature] according to teleological concepts" not to natural history but to natural *description* (*Naturbeschreibung*), then added: "But concerning the origin and the inner possibility of these forms [i.e., the project of *natural history* in Kant's new sense], . . . such information is still properly to be attained through theoretical natural science."[96] By "theoretical natural science," Kant must be taken to have meant science according to the mechanistic maxim.[97] But, if so, Kant must also be taken to have meant that these were *inaccessible altogether*, for his whole argument in the third *Critique* was that mechanism could not explicate organic nature even descriptively—hence the need for teleological thinking in the first place. The real point, then, is that Kant made *three* distinctions: first, full-fledged natural science on the mechanistic model; second, strictly "reflexive" teleological description (which allows us at least to organize our thoughts about organic nature); and third, questions of the "origin and the inner possibility" of things—particularly "natural purposes"—which must remain "inscrutable" *in principle*.[98]

Methodologically, Kant upheld a lifelong conviction that mechanism could never explain organic form.[99] Now he added that the only prospect for natural science in this domain was, in Lagier's phrase, "to *limit* the hemorrhage into the supernatural."[100] In the first *Critique* Kant proposed to understand reason by analogy to organic forms ("epigenesis of pure reason").[101] However, in the third *Critique* Kant reversed himself, asserting, as Lagier notes tellingly, "the purposive systematicity of organic form is not properly speaking even thinkable except via the limited implications of an analogy with the *intentional products of the will* (of practical reason), such that the teleological orientation of the products of nature is in principle always *the imposition by the subject himself* of a purpose onto nature, and never an objective property of objective 'life.'"[102] Indeed, Kant's ultimate considerations of teleological judgment turned away from natural history to human self-realization, from "physiological" to "pragmatic" anthropology.[103]

The *Kritik der Urteilskraft* (1790) essentially proposed the reduction of life science to a kind of prescientific descriptivism, doomed *never* to become "proper science," never to have its "Newton of the blade of grass."[104] That for Kant "biology" *cannot* be an empirical science but must draw on "metaphysics" is the open secret of Clark Zumbach's provocative title: *The Transcendent Science*.[105] While I have made this larger argument about Kant and biology elsewhere, what I stress here is a more concrete issue: Kant's view on the prospect of a new empirical science of "natural history" as a developmental-historical approach to geological and biological matters.[106] In the context of the *Kritik der Urteilskraft*, this brings us to the famous and controversial passage in §80 concerning a "daring adventure of reason."

The text forms the second section of what Kant termed "Methodology of the Teleological Power of Judgment." Having, in the prior section, made clear that teleological judgment "provides no information at all about the origination and the inner possibility of [organic] forms" but can at best have "a negative influence on procedure in theoretical natural science," Kant recognized in §80 that this could hardly appear congenial to practicing inquirers in this emergent field of empirical science.[107] He acknowledged their ambition to find an empirical scientific approach to these problems, which for him meant integrating them into "the mechanism of nature, without which there can be no natural science at all," but he insisted that this would never be successful—*not* "because it is impossible *in itself* to find the purposiveness of nature by this route, but only because it is impossible *for us* as humans."[108] That is, Kant's suspicion of biology was a direct consequence of the critical epistemology, the establishment of the limits of human reason, which Kant worked out from 1781 to 1790.

Thus, the balance of the section should be seen as a *counterfactual* exposition: a presentation of what one might wish could be done, but which Kant, from the outset and in principle, denied could be done.[109] Kant alluded to a "commendable" temptation to use "comparative anatomy" to construe an actual historical development of life-forms: "This analogy of forms, insofar as in spite of all the differences it seems to have been generated in accordance with a common prototype [*Urbild*], strengthens the suspicion of a real kinship among them in their generation from a common protomother [*Urmutter*], through the gradual approach of one animal genus to the other, from . . . human beings, down to polyps, and from this even further to mosses and lichens, and finally to the lowest level of nature that we can observe, that of raw matter."[110] Such an "archaeologist of nature," as Kant characterized the empirical scientist involved in this inquiry, would "have the maternal womb of the earth, which has just emerged from a condition of chaos (just like a great animal), initially bear creatures of less purposive form, which in turn bear others that are formed more suitably for their place of origin and their relationships to one another, until this birth mother itself, hardened and ossified, has restricted its offspring to determinate species that will degenerate [*ausarten*] no further, and the variety will remain as it turned out at the end of the operation of that fruitful formative power."[111] Kant postulated that this developmental process had long since *closed*, that the world he contemplated was one of fixed species.[112] He then went on to make a crucial philosophical point about such an empirical speculator: "ultimately he must attribute to this universal mother an organization purposively aimed at all these creatures, for otherwise the possibility of the purposive form of these products of the animal and vegetable kingdoms cannot be conceived at all."[113] That is, Kant was committed to *preformation*—even if "generic" rather than individual—and this meant that there was an ineluctably *metaphysical* foundation for any consideration of organisms and life in the physical world.[114] The "original principle of organization" was "inscrutable"; that is, it lay beyond the reach of human reason.[115] In a footnote to §80, Kant continued his reflection on this empirical-speculative impulse: "One can call a hypothesis of this sort a daring adventure of reason, and there may be few, even among the sharpest researchers into nature, who have not occasionally entertained it."[116] The balance of the footnote made the point that while this was not an *irrational* undertaking, as the hypothesis of "spontaneous generation [*generatio equivoca*]" would be, it simply lacked *any* empirical evidence. All natural history could demonstrate, Kant affirmed, was species fixity, *generatio homonyma*, for "*generatio heteronyma*, so far as our experiential knowledge of nature goes, is nowhere to be found."[117]

In these passages, Kant was hardly *endorsing* this "daring adventure of reason" but in fact suggesting drastic limitations for the empirical pursuit of a historical inquiry into nature. One of the casualties of the critical Kant's constriction of authentic scientific credibility ("proper science") was his own vision of a historical approach to nature. This "daring adventure of reason" had no prospect of becoming "proper science," and little even of becoming a "systematic organization of facts," in the terms of Kant's *Metaphysische Anfangsgründe der Naturwissenschaft*.[118] What remains is to ask whether his warning was taken to heart. In fact, Kant's interdict failed. To demonstrate this, we must assess his relationship with Blumenbach, in whom we have established an unequivocal commitment to the historicization of nature.[119]

BLUMENBACH AND KANT

In defending himself against Forster in his essay of 1788, Kant invoked Blumenbach in a footnote to dismiss what in the *Kritik der Urteilskraft* he would call a "daring adventure of reason"—namely, the transformation of the great chain of being from a taxonomy to a phylogeny.[120] This "widely cherished notion preeminently advanced by Bonnet" had been questioned by Forster, and Kant was happy to report that, under the critical scrutiny of Blumenbach's *Handbuch der Naturgeschichte*, all the weaknesses of that position had been exposed.[121] Then he added the observation: "this insightful man also ascribes the *Bildungstrieb*, through which he has shed so much light on the doctrine of generation, not to inorganic matter but solely to the members of organic being."[122]

More extensively, in the *Kritik der Urteilskraft* (1790), Kant elaborated his estimation of Blumenbach's important contribution: "He makes organic substance the starting point for physical explanation of these formations. For to suppose that crude matter, obeying mechanical laws, was originally its own architect, that life could have sprung up from the nature of what is void of life, and matter have spontaneously adopted the form of a self-maintaining finality, he justly declares to be contrary to reason."[123] This was the essential postulate to which Kant had committed himself in his second essay on race (1785), and the stakes were not small: without some fixity in the power of generation (*Zeugungskraft*), the prospect of the scientific reconstruction of the connection between current and originating species would be altogether dim.[124] Yet it was not simply a *methodological* issue, however dire. There was also an essential *metaphysical* component. All organic form had to be fundamentally distinguished from mere matter. Kant was adamant that the *ultimate* origin of "organization" or of the formative drive (*Bildungstrieb*) required a *metaphysical*, not a physical, account. He invoked Blumenbach for support in these metaphysical stipulations. The leading life scientist of the day seemed to be affirming just the same metaphysical and methodological discriminations that Kant himself demanded.

How justified was Kant in appropriating Blumenbach to his cause? And how important for Blumenbach, conversely, was Kant's philosophy of science? Blumenbach began serious consideration of the philosophy of Kant in the mid-1780s as a direct consequence of the disputes surrounding Kant's reviews of Herder's *Ideen zur Philosophie der Geschichte der Menschheit* and especially Kant's controversy with Georg Forster. Kant's essay appeared in the *Teutsche Merkur* in January and February 1788. How did Blumenbach respond to Kant's appropriation of his ideas? His first subsequent publication came almost immediately: the third edition of the *Handbuch der Naturgeschichte*, whose preface was dated March 1788. Unsurprisingly, it shows no evidence of Kantian ideas.[125] Less than a year later, in January 1789, however, Blumenbach published his revised version of *Über den Bildungstrieb* and sent Kant a copy in acknowledgment of Kant's references to him in the 1788 essay.[126] In this second edition of the *Bildungstrieb*, Blumenbach emphatically repudiated hylozoism, advancing beyond epistemological scruples to an ontological distinction between the general order of nature and the specific order of the organic: "No one could be more totally convinced by something than I am of the mighty abyss that nature has fixed [*befestigt*] between the living and the lifeless creation, between the organized and the unorganized creatures."[127] Indeed, Blumenbach shared Kant's skepticism about a bridge from the inorganic to the organic and about the phylogenetic continuity of life-forms. What bound them most together was their commitment to the fixity of species and their rejection of the reality of the "great chain of being." *Yet Blumenbach drew neither of these commitments from Kant.* They were already expressed with clarity in his dissertation of 1775 and especially the first edition of his *Handbuch* of 1779. These were basic issues for anyone taking up natural history in the eighteenth century. What remains is to consider whether the *reasons* for Blumenbach's commitments were the same as the reasons for Kant's commitments to these positions.

Peter McLaughlin identifies crucial changes that Blumenbach introduced in 1791, after he had absorbed Kant's ideas not only from the 1788 essay but from the *Kritik der Urteilskraft* which Kant sent him.[128] In 1788 Blumenbach found "throughout all nature the most unmistakable traces of a virtually general drive to give matter a determinate form, which already in the inorganic realm is of striking effectiveness."[129] In 1791 Blumenbach pruned the cited line as follows: one finds "in the entirety of organic nature the most unmistakable traces of a generally distributed drive to give matter a determinate form."[130] The final clause from 1788 was eliminated altogether. In 1789 Blumenbach compared the *Bildungstrieb* to "the *terms* attraction and gravity[,] . . . generally recognized natural forces." But in 1797 he changed this to "The term *Bildungstrieb* just like all other life-forces such as sensibility and irritability . . ."[131] The thrust, as McLaughlin notes, was to make a radical distinction between the organic and the inorganic realms and to assign the drive exclusively to the former.

The point, here, is that Blumenbach wished his formative drive to be considered only in comparison with other *life-forces*. Still, that did not mean one could not draw *analogies* from the inorganic to the organic, for, Blumenbach went on, "even in the inorganic realm the traces of formative forces are so unmistakable and so general. Of formative *forces*—but not by far of the formative *drive* (*nisus formativus*) in the sense this term assumes in the current study, for it is a life-force [*Lebenskraft*] and accordingly as such inconceivable in inorganic creation—rather of other formative forces [*Bildungskräfte*], which provide the clearest proof in this inorganic realm of nature of determinate and everywhere regular formations [*Gestaltungen*] shaped out of a previously formless matter."[132] This distinction between the formative *forces* (*Kräfte*) that structure the inorganic realm and the formative *drive* (*Trieb*; note that it is always singular in Blumenbach's usage), which is unique to organic life, and indeed a *Lebenskraft* among others, proved crucial for Kant. This was what Kant found most gratifying in Blumenbach's new book of 1789, as he reported in his letter of acknowledgment to Blumenbach.[133] In the *Kritik der Urteilskraft* he elaborated:

> Blumenbach . . . rightly declares it to be contrary to reason that raw matter should originally have formed itself in accordance with mechanical laws, that life should have arisen from the nature of the lifeless, and that matter should have been able to assemble itself into the form of a self-preserving purposiveness by itself; at the same time, however, he leaves natural mechanism an indeterminable but at the same time also unmistakable role under this inscrutable *principle* of an original *organization*, on account of which he calls the faculty in the matter in an organized body (in distinction from the merely mechanical *formative force* [*Bildungskraft*] that is present in all matter) a *formative drive* [*Bildungstrieb*] (standing, as it were, under the guidance and direction of that former principle).[134]

This passage in the *Kritik der Urteilskraft* makes the distinction between formative force and formative drive prominent, but I submit that the distinction of a formative *drive* creates problems within Kant's system.[135] What is problematic is twofold. First, Kant suggests that the formative forces (of general, physical nature) constrain the formative drive in organized life-forms. This is a plausible scientific claim in itself, but it goes against the metaphysical thrust of his whole argument, which is to suggest that organisms as natural purposes urge us toward the notion that there is a larger purpose in nature as a whole that constrains the physical order (a "supersensible substrate").[136] Some commentators on this key passage have been so motivated by this larger concern that they have inverted the relation in Kant's text. Second, it is not clear how the notion of "drive" (*Trieb*) fits in his philosophy: in what measure is it really different conceptually from "force" (*Kraft*)? Are they not equally "inscrutable," or is there a supplementary inscrutability about *life*-forces? In a word, what is the ontological status of *immanent purposiveness* in Kant's transcendental philosophy? This is a cardinal question, but it has more to do with Kant than with Blumenbach.

Most pertinently in regard to the latter, it is not clear that Blumenbach *ever* considered his formative drive merely a regulative idea, not a constitutive force in nature. Blumenbach and his school understood the *Bildungstrieb* as actual, not speculative. Their project was to specify its *effects* through the mechanisms (*Bildungskräfte*) it set in motion. Kant's regulative/constitutive distinction proved useless for them in that pursuit, though it gave them some metaphysical comfort, especially given the thinness of their analogy to the Newtonian mysteriousness of gravity. Blumenbach, to be sure, found Kant's suggestion that he brought teleological and mechanical explanations together in his scientific practice quite pleasing, but it is not clear that he understood Kant's painstaking argument for their radically different roles in scientific explanation. In short, Blumenbach's affiliation with Kant is best understood as a *misunderstanding*—though an influential one in the constitution of biology as a discipline in the succeeding decades.[137]

There is no doubt that the life scientists of Blumenbach's day reached out to Kantian philosophy as legitimation of their methodology.[138] They sought to evade both mechanistic materialism and animistic vitalism, construing themselves securely within Kant's "Newtonian" paradigm. But this Kantianism was more a convenient ploy than an epistemological commitment. There is perhaps no more widely accepted historical finding about life scientists in Germany in the 1790s—even, or especially, when they invoked Kantian critical terms—than that they slid one and all from a strictly regulative into an unmistakably constitutive use of natural teleology.[139] This was indeed an inevitable consequence of their commitment to the *empirical practice* of a life science, which Kant's philosophy in fact proscribed.

Indeed, as both Robert Richards and Frederick Beiser recognize, these life scientists were closer in many ways to Kant's disparaged former student Johann Gottfried Herder than they were to Kant.[140] It was Herder, not Kant, who offered avenues toward synthesis—perilous as well as inspiring—to shape the natural science of the epoch 1790–1820.[141] Major theorists of natural science in that era could hold Kant and Herder *both* in high esteem.[142] If that is true, then there was something historically misguided about Kant's effort to exclude Herder's views from authentic science. Kant's method *and* Herder's manner were decisive conjointly in constituting the scientific imagination of the age.[143] And that can be grasped, historically, only as a defeat of Kant's program. What was most innovative in Herder's uptake of Spinoza (vital materialist monism) emerged clearly in and through Kant's own boundary work and helped constitute the culture of science of the turn of the century. Goethe found even Kant himself in the *Kritik der Urteilskraft* equivocating between constitutive and regulative uses of teleology.[144] Kant's own effort to understand specific organisms under the regulative rubric simply created more problems than it solved, and at the very least made it *necessary* to resort to the idea of the teleology of nature as a whole.[145] If Kant could not hold this line, it can hardly be surprising when the leading naturalists of his day, even in invoking his theory, found it impossible in practice to observe it.

What role *did* Kant play? Christoph Girtanner's *Über das Kantische Prinzip für die Naturgeschichte* (1796) offers a very useful vantage for assessing how Kant was understood by Blumenbach and the German naturalists of the 1790s.[146] The essential point was Kant's new research program for *Naturgeschichte*.[147] It would ask, in Girtanner's words, "what the primal form of each ancestral species of animals and plants originally consisted of, and how the species gradually devolved from their ancestral species."[148] It would explore and explain how environmental changes on the earth—indeed "violent revolutions in nature"—occasioned dramatic changes in life-forms. Yet however dramatic, these were not *chaotic* changes; rather, the variation in observed traits in current species emerged always under the guidance of a "natural law" requiring that "in all of organic creation, species remain unaltered."[149] Kant's great achievement, in Girtanner's eyes, was his connection of this law to a more determinate "natural law" (proposed by Buffon) to explain this process: namely, that "all animals or plants that produce fertile offspring belong to the same physical [i.e., real] species," notwithstanding considerable variation in observed traits.[150] That is, these organisms *must* have "derived from one and the same stem [*Stamme*]."[151] While there could be hereditary variations (*Abartungen*) *within the confines of the governing stem*, there could not be "degenerations" (*Ausartungen*), that is, permanently heritable departures from the fundamental traits of the ancestral stem.[152] *Races* constituted decisive evidence for this theory, because their crosses always showed perfect proportion in the offspring: *Halbschlachtigkeit* (half-breeding).

To account for these internal variations within species, Kant offered the view that "the ancestral stem of each species of organic life contained a quantity of different germs [*Keime*] and natural potentialities [*natürliche Anlagen*]."[153] Girtanner followed Kant literally in identifying *Keime* with the source of changes in the *parts* (organs) of a life-form, while *natürliche Anlagen* occasioned changes only in the *size or proportion* of such parts. Kant used winter feathers in birds to exemplify the first, and thickness in the husk of grain to exemplify the second, and Girtanner replicated these examples. To help explicate the *process* of variation, Girtanner turned to his teacher Blumenbach. It was "through different directions of the *Bildungstrieb* [that] now these and now those [germs or natural potentialities] developed, while the others remained inert."[154] Only climate acting on organisms over an extended time could educe such variation, such shifts in the "direction of the *Bildungstrieb*," and thus permanently alter "the primal forces of organic development and movement."[155] Moreover, once such shifts in direction took place, once certain germs or natural potentialities were triggered into actualization, the rest atrophied and the process proved irreversible.[156] This claim represented one of Kant's decisive interventions, separating him sharply from Buffon, for example.[157]

Girtanner was acutely aware of the way in which Kant's "natural history" interpenetrated with his theory of organic form. Not only did Kant require a specific theory of generic transmission, but he needed a theory of organic life in which to cast it. The only form of generation that had been empirically observed, Girtanner parroted Kant, was *generatio homonyma* (*Abartung*), the persistence of species, though *generatio heteronyma* (*Ausartung*) was not impossible (against reason) but only unheard of (against experience). The essential point was that these both contrasted with *generatio aequivoca* (spontaneous generation). "That by mechanism organized beings should emerge from unorganized matter . . . contradicts reason as well as experience."[158] That is, "it contradicts all known laws of experience that matter which is not organized should have by itself, without the intervention of other, organized matter, organized itself."[159] *Antihylozoism*, then, was the essential posit of Kant's theory of organic form. Girtanner stressed Kant's idea of organism: not only was it "not a machine" in consequence of the mutuality of cause and effect, of parts and whole, but neither was it fully the "analogue of art," for "organized Nature organizes itself."[160] Because life-forms showed characteristics—reproduction, growth through nourishment and assimilation, regeneration of lost organs and self-healing generally—that could not be assimilated to the mechanistic model of natural science, they represented anomalies requiring recourse to teleological judgment, the analogy of "purposiveness."

If Girtanner replicated Kant's presentation of the *perplexity* of natural purpose, he did nothing to advance its *resolution*. On the other hand, he clearly did not find the regulative/constitutive distinction of any use in the science he proposed to elaborate. Girtanner understood Blumenbach's *Bildungstrieb* not as a regulative ideal type but as an actuality in the physical world—namely, "that force by virtue of which the chemical and physical laws are subordinated under the laws of organization."[161] Yet by clinging to the antihylozoism and species essentialism Kant and Blumenbach shared, Girtanner, publishing in 1796, in fact missed the decisive shift of the new generation toward a fully historicized natural history, transformism.

BLUMENBACH'S "GÖTTINGEN SCHOOL": A PALEONTOLOGICAL REINTERPRETATION

Crucial for the generation of the 1790s at Göttingen, I suggest, was Blumenbach's specific interest in the historicization of nature. Paleontology, as it was reflected in the problem of extinctions and geological "revolutions," had become a central research interest of Blumenbach's by the 1790s. He had amassed one of the largest collections of fossils in Europe, and he was beginning the crucial process of correlating geological strata with organic life-forms in a historical sequencing. That was profoundly important to his key students—Heinrich Friedrich Link (1767–1851), Alexander von Humboldt (1769–1859), Ernst von Schlotheim (1764–1832), and Gottfried Reinhold Treviranus (1776–1837). They were among the first in Germany to opt for an explicitly "transformist" (evolutionary, in the modern sense) theory of life. They steeped themselves in historical geology in order to undertake historical life science. It is no coincidence that Humboldt and Schlotheim chose to continue their studies at the Freiberg School of Mines with the great geologist Abraham Gottlob Werner.[162] One of Link's first publications was on mineralogy. Treviranus would make this the centerpiece of his six-volume conspectus of the emergent science. By 1800, through the influence of Göttingen and specifically of Blumenbach, the historical reconstruction of life-forms in terms of geological change had become a widely accepted research program in German life science. One can even wonder whether the infusion of Blumenbach students into the Freiberg research community might not have played some role in its later inflection toward *Flötzgebirge* and fossils, away from its earlier, virtually exclusive concern with primary formations (*Ganggebirge*) and mineralogy.

The core exemplar of this historicizing turn of the "Göttingen School" was Treviranus. Until very recently, the only commentators interested in Treviranus were local historians of his hometown of Bremen.[163] His place as one of the notable propagators of the term "biology" around 1800 earned at most a mention in the wider histories of life science; his six-volume study was dutifully listed in bibliographies, but its content and its place in the epoch were passed over largely in silence. Treviranus was typically considered yet another *Naturphilosoph*, with all the dismissal that entailed. A first effort to take Treviranus seriously in mainstream history of science came with an essay by Brigitte Hoppe in 1971, in the context of a colloquium on his contemporary, Lamarck.[164] Timothy Lenoir highlighted Treviranus in one of his essays on teleomechanism and the Göttingen School.[165] But then even his interest faded. Not until around 1990 did two substantial dissertations attempt a fuller consideration of Treviranus's work and impact.[166] I will build on their contribution to situate Treviranus at the core of the crystallization of life science in the Germany of his time.[167]

In his curriculum vitae, composed at the close of his course of studies in Göttingen in 1796, Treviranus recognized that he would need to make a career as a practicing physician and only "utilize my leisure time for the promotion of physiology."[168] As Trevor DeJager aptly puts it, "Treviranus travelled the road of *Wissenschaft*, even while he had to earn a living from his medical practice."[169] Thus, although he set about composing his massive monograph, *Biologie*, immediately upon receiving his degree at Göttingen, it would be eight years before the first volume was completed and published, and it would take him twenty years more to bring the full conception to completion.[170] To be sure, Treviranus was able to make a research trip to Paris in 1810 to meet and work with Cuvier and others, but there was nothing approaching institutional security and support for his research project.

Treviranus was the eldest of a number of children of a struggling Bremen businessman.[171] From early in his childhood he evinced a strong interest in natural science and especially mathematics, and in 1793, at the age of seventeen, his academic merit earned him admittance to the premier university in Germany, Göttingen.[172] That was not only the best university in Germany but also quite expensive for students, and Treviranus needed to work as a tutor for most of his years of study there, just to get by. The overexertion led to his first bout with tuberculosis in 1794.[173] Nevertheless, he made great academic strides. Already in 1794 he read a paper on the history of mathematics to the Societas Physica Goettingensis and won admission to that prestigious group.[174] The aged but eminent mathematician Abraham Gotthelf Kaestner prized the young student and would eventually offer him a teaching position in the field.[175] He studied not only with Kaestner in mathematics but with Friedrich Bouterwek (1766–1828) in philosophy (a course on Kant's philosophy of science in winter semester 1793/94).[176] Though mathematics was his first love, Treviranus needed to be eminently practical, as his father urged, and he determined to earn a medical degree and take up private practice to contribute to his struggling family's upkeep. He concentrated his practical training under the eminent clinician August Gottlieb Richter (1766–1812) for his future career's sake.[177] Intellectually, his ultimate inspiration was Blumenbach.[178] He also took a very serious interest in the experiments with infusorians of Heinrich August Wrisberg. Treviranus resolved to devote his intellectual energies for the balance of his life to the project of life science, though only as he could squeeze out the time from earning a living for himself and his extended family.[179] Already as a dissertation student, he published a paper in Johann Christian Reil's (1759–1813) distinguished journal, *Archiv für die Physiologie*: the first essay not by Reil himself to appear in the journal.[180] Reil's interest in the student's work can be connected not only to the content of the essay but perhaps even more to Treviranus's vision of the state of the field and its needs. He wrote: "our entire medical art can only achieve a scientific form [*Durchbildung*] with the help of a sound physiology, which in that position would prepare a path of certainty for practical medicine."[181] That was precisely Reil's vision. Treviranus's dissertation was reviewed in the scholarly literature with favor as well.[182] His should have been a prominent academic career, but he could not afford it. His father pressed him to come home at once and take up private practice to help support the family, and he returned to Bremen as soon as he had completed his degree, in October 1796.[183]

It helped him when his brother, who earned a medical degree at the University of Jena in 1802, returned to Bremen and joined him in his practice and in his studies. Ludolph Christian Treviranus (1779–1864) was a specialist in botany and an ardent follower of Schelling's *Naturphilosophie*.[184] When the wealthy parents of Gottfried's new wife passed away and left her a considerable inheritance, he was able to leave the bulk of the clinical work to his brother and devote himself to research.[185] As a result, the first three volumes of *Biologie* appeared in 1802, 1803, and 1805. But then political and economic turmoil (the Napoleonic invasions) and illness beset him again, and it would be more than a decade before he could publish the next volume.[186]

Thus, it is the first three volumes of *Biologie* that will concern us in terms of establishing the research field of the time.[187] Volume 1 was a general theoretical overview; volume 2 offered a major contribution in what we would now call biogeography; and volume 3 offered a clearly "transformist" (or, in our sense, evolutionary) statement on paleontology and historical development of species. Each volume, then, achieved important progress toward the consolidation of the *Fachgebiet*. Only with great injustice can this work be dismissed as a "prospectus."[188]

The first volume presented a considered effort to conceptualize life as a natural-scientific problem, taking up the debates then current about "vital forces." It also took on Kant not only epistemologically but also substantively, proposing an alternative theory of the physical forces Kant had elaborated in his *Metaphysische Anfangsgründe* and using these to conceive a developmental hierarchy of forces integrating the inorganic with the organic realm. Treviranus strove for an approach to development and evolution that took ecological factors as central. For him, the essence of life was self-preservation in interaction with a changing and challenging environment.[189]

In an effort to distinguish Treviranus's thinking from speculative *Naturphilosophie*, scholars have attempted to assimilate him to Kantianism.[190] DeJager makes a very effective case that this cannot work. To be sure, from the beginning Treviranus had to come to terms with Kant. But the upshot was that he substantially rejected or revised many of the crucial Kantian claims in both method and substance regarding natural science in general and life science specifically. As DeJager correctly claims, "Kant actually eliminates the possibility of any true novelty to emerge in nature."[191] Life was an impenetrable mystery: "As far as Kant was concerned, nature has no creative capacity when it comes to the production of living beings."[192] By contrast, Treviranus shared the fundamental intuition of his age that "nothing in nature is fixed and permanent. . . . One phenomenon is transformed into another by changes in the balance of forces."[193] Thus, for him, "[n]ature could only be constructed as a system if it were understood historically."[194] DeJager summarizes the sense of the *Fachgebiet* around 1800: "The *Naturforscher* were now being called upon to investigate the process of nature, the course of change itself."[195] Hence, "a history of the living world was only possible if one were prepared to drop Kant's restrictions."[196]

Treviranus devoted himself to the enterprise of a real developmental history of nature, "the problem of reconstructing the course of nature through time."[197] In the second volume of *Biologie*, Treviranus wrote: "Any investigation into the influence of nature as a whole on the living world must begin with the principle that all living forms are the products of physical influence, occurring now in time, changing only in degree or duration."[198] This was a methodological affirmation of "actualism," but it was also a theoretical commitment to physicalism and to evolution. He shared the conviction of Kant, Blumenbach, and the entire research community that life-forces were irreducible to the mechanical laws of physics.[199] But he rejected the view articulated by Kant and affirmed in a measure by Blumenbach that there was an utter discontinuity between the inorganic and the organic. Above all, Treviranus would have nothing to do with Kant's regulative/constitutive distinction and its implications for the scientific status of research into organic forms.[200] Thus, he found less of interest in Kant's *Kritik der Urteilskraft* than in his *Metaphysische Anfangsgründe der Naturwissenschaft*, and less of interest in the latter work's stipulations about "proper" science than in its "dynamic physics" of forces.[201] His originating query was, in DeJager's terms, "is there a single fundamental force in nature of which all the various forces, which are assigned different names, are simply modifications," and then, "what is the cause of their modification into distinct forces?"[202] He never accepted Kant's view that there could be no natural science of organic forms. In particular, he rejected Kant's views on spontaneous generation.[203]

The second volume elaborated this ecological frame by developing extensively and seminally the ideas of biogeography that had been initiated by Buffon and especially by E. A. W. Zimmermann.[204] As DeJager aptly puts it, a theory that would construe the development of life-forms in terms of geological changes must begin by relating the current life-forms to their environmental niches. That was a prospect that Buffon had already contemplated, especially in his theories of animal migrations caused by changes in climate, but it was developed in a far more complete and complex form by the noted naturalist Zimmermann in works published at the end of the 1770s.[205] Treviranus was the most thorough assimilator of Zimmermann's initiatives in the development of a general biogeography.[206] Together, they are regarded as founders of the field.

The third volume proved beyond question the prominence and thoroughness of a "transformist" approach to the history of nature in German natural science around 1800.[207] Having studied the literature on invertebrates and conducted considerable research of his own on infusorians and the life-forms in the gray zone between plants and animals, Treviranus argued that these simpler organisms arose first in geological time, that they were far more immediately responsive to environmental changes, and that all more complex organisms evolved from these over the course of geological time: "not, as is usually assumed, that it was great catastrophes of the earth that extinguished the animals of ancient times, but rather that many of these survived and that they have vanished from current nature because the species to which they belonged completed the circuit of their being and have transformed into other species [*in andere Gattungen übergegangen sind*]."[208] It was his view that "the succession of fossils in the strata of rocks of different ages demonstrates clearly the relation that exists between the organisms that underwent transformations and the lawfully regulated development of these changes."[209]

Ernst Mayr famously dismissed Treviranus (among others) for offering only a "prospectus" for the science of biology, which had to await the "modern synthesis" to achieve realization. But instead of reading Treviranus as the anticipation of something long in the future, the balance of this study will consider him as the summation of a far more compelling project of consolidation of the *Fachgebiet*. That means we have to consider the lineage from Blumenbach through Carl Friedrich Kielmeyer, Johann Wolfgang von Goethe, and Friedrich Schelling and the long-despised context of Idealist *Naturphilosophie*.

CHAPTER NINE

Carl Friedrich Kielmeyer and "an Entirely New Epoch of Natural History"

> Without a doubt future ages will date . . . the *Address* of Professor Kielmeyer . . . already in the year 1793 . . . as the onset of an entirely new epoch of natural history.
>
> SCHELLING, 1798[1]

In many ways, *force* was the concept that animated revision in the natural philosophy of the eighteenth century.[2] While new forces in physics—such as gravity, electricity, magnetism—became objects of intensifying experimental and theoretical inquiry, so did new forces in *chemistry*: heat, affinity, eventually oxidation.[3] These *supplemented*—and *could not be reduced* to—particulate mass or the physical forces. In the same way, by the end of the eighteenth century, "it was felt necessary to distinguish a concept of force that applied for organic nature alone."[4] The critical frontier between the physical sciences and medical physiology was open-ended in both directions.[5] Thus, the research problem which galvanized experimental physiology was what Brigitte Lohff has charmingly called "a banquet of vital forces."[6] That banquet first served up Haller's two forces of irritability and sensibility in 1753.[7] Then it presented Caspar Friedrich Wolff's *vis essentialis* in 1759 and Blumenbach's *nisus formativus* or *Bildungstrieb* in 1781.[8] A few years later Blumenbach elaborated *five* vital forces in his widely influential *Institutiones Physiologiae* (Institutes of physiology; 1787).[9] In 1793 Carl Friedrich Kielmeyer (1765–1844) worked these up into another complex menu that, in my view, proved decisive for the gestation of biology.[10] The generation of the 1790s would struggle to digest the ensemble.[11] Vital forces constituted the basis for the autonomy of biology, as Gottfried Reinhold Treviranus asserted in adopting the term in 1802.[12]

To reconstruct the development of this "banquet of forces," a useful point of entry is with the first text to use the crucial term *Lebenskraft* as its title, Friedrich Casimir Medicus's lecture of 1774, *Von der Lebenskraft*. His aim was to trace the leading figures and issues in German life science up to his moment and articulate where the frontier of debate stood in their wake. Notably, there appears to be a striking omission (at least overtly) in his catalog of relevant figures and ideas: namely, Caspar Friedrich Wolff and his notion of *vis essentialis*. That will demand reflection. Very early we tumble upon a pervasive anxiety that runs through Medicus's remarks: his concern with "materialists." Medicus constantly sought to rebut "our contemporary materialists."[13] But who were they? He mentioned no names. Now, Medicus was writing in 1774, and the most salient articulation of materialism in that moment was unquestionably Baron d'Holbach's *System of Nature* (1770), which occasioned a huge stir all over Europe and notably in Germany.[14] Yet materialism was no sudden intrusion. French materialism made substantial inroads into Germany earlier in the century, as we have established. It carried from La Mettrie, Buffon, Diderot, and Maupertuis through the Berlin Academy to the work of Caspar Friedrich Wolff. Medicus, I suggest, gives evidence of the persistence of that challenge in German life science into the 1770s. Thus, the suspicion arises that the *absence* of Wolff from Medicus's consideration may be linked to his aspersions against "materialists."[15] If so, Medicus both underestimated the serviceability of Wolff's *vis essentialis* for the sort of argument that he wanted to make and yet, perhaps, showed a telling sensitivity to the potentially "materialist" implications of his work. Wolff was assuredly no materialist. He was simply a *naturalist*, an empirical scientist of the physical world, for whom transcendent interventions were not relevant, yet for an age and culture still anxious about its pieties, this could *seem* quite "materialist."

In any event, clearly Medicus wanted to dispute the unnamed "materialists." He insisted that since matter was *essentially* inert, it could by definition have no *intrinsic* force or capacity for self-movement.[16] The fundamental error of the "materialists," he claimed, was to ascribe activity to matter in itself. "The clever constructions of the materialists collapse in the face of this one principle, the inertness of matter."[17] Moreover, it would not do to salvage this idea of self-movement for *some* matter by adding the modifier "organized," Medicus went on. An organized body—that is, a living organism—was "an artificial configuration of matter," producing "harmony and balance" by a regulated relation among the parts.[18] But this regulation could not come from the material itself. "Organization accordingly provides the body with no force."[19] A dead body, Medicus argued, had all the organic matter, but none of the life. That required something additional, and this was the decisive consideration.

Taking up human nature, Medicus rolled out all the doctrines of the Leibniz-Wolff school. The soul was essentially thought and will. "The soul is a simple substance[,] . . . immortal and self-conscious. . . . It cannot be material. For from inert matter no thinking substance can arise."[20] Some of "our contemporary philosophers" suggested that "humans have a rational soul, while animals have a sensible [*sinnlich-empfindende*] soul."[21] For Medicus, this was not quite plausible. Given the definition of soul, there seemed to him a clear contradiction in the idea of an animal soul. He acknowledged that animals appeared to sense and to feel, but that did not suffice for the term *soul,* in his view. This became clearer as he turned to the perennial issue of mind-body interaction. "Philosophical physicians" were divided in how to deal with this issue, he explained.[22] One school wished to hold the soul thoroughly capable of causing all the motions of the body, voluntary and involuntary. This was Stahl's position, and it had been revived, Medicus noted, by Sauvage in Montpellier and Whytt in Edinburgh. The other school favored interpretation according to strictly mechanical laws. Medicus referred in this context to Friedrich Hoffmann and Herman Boerhaave. According to Medicus, this school abjured "metaphysics" and professed contentment with "hydraulic, hydrostatic, mechanical laws."[23] Then Haller, Boerhaave's student, muddied the waters. He proposed an intrinsic force in (muscular) matter, *Reizbarkeit,* thus reopening the door, Medicus would seem to be suggesting, to the "materialist" error.

Medicus argued that Stahl's position should not be taken lightly. The latter had understood properly the significance of the inertness of matter. But he then overextended the notion of soul to save the phenomena. Soul, Medicus repeated, was simply thinking and willing. That was irreconcilable with involuntary processes. Soul essentially required volition and self-conscious agency, yet most life processes (*die Geschäfte des Lebens*) took place involuntarily. That was the strong point of Boerhaave's alternative notion of a *Lebensgeist* mechanically propagated via the nerves. While there remained problems in the physiology of nervous tissue and the transmission of nervous force (Medicus appears to have been unfamiliar with Unzer's crucial work of 1771), this approach seemed superior, in his view, to Haller's notion of an irritability *intrinsic* to muscle tissue. As far as Medicus could see things, muscular motion was incited by nerve action.[24]

If matter was inert, what made it move? "Unbiased men in recent times have . . . proclaim[ed] that the life processes cannot be explained by either physical or mechanical laws."[25] Medicus posited that divine agency ultimately accounted for all motion, but he still sought the *instrumentality* of this divine intervention.[26] The Stahlians, again, seemed closer to the point: "a simple substance that is not material [is required] to enliven and move organic matter."[27] Since by definition the soul could not serve this role, Medicus concluded (here was his "innovation") there had to be *another* immaterial substance/force that could do it, an "enlivening force [*belebende Kraft*]," or *Lebenskraft*, the *Triebfeder* (driving impetus) of organic life.[28]

Adding this second immaterial force to the account of human (and all organic) nature required differentiating it from the soul. This Medicus did by three distinctions. The soul's demands on the body induced discernible resistance and weariness; those of the *Lebenskraft* did not. The soul had to learn its functions and develop its skills in order to perform them aptly; *Lebenskraft* was immediately efficacious. Finally, all the actions of the soul were self-conscious, whereas none of the actions of the *Lebenskraft* were conscious at all.[29] Medicus recognized that his conception had some affinities to the notion of "innate capacity [*angeborene Fertigkeit*]," or instinct, as developed in the work of Reimarus. Indeed, *Kunsttrieb* seemed to Medicus to be a very good analogue of his notion.[30] By contrast, "animal soul" did not. Thus, he was very happy to recognize a recent, major achievement in "metaphysics," Herder's *Abhandlung über den Ursprung der Sprache* (Treatise on the origin of language; 1772), which established beyond doubt, in Medicus's view, that only humans had language and the thought it expressed; hence, animals could have no soul because they could have no thought.[31] This demolished all efforts to imagine apes with linguistic capabilities, Medicus noted, and it made absurd Linnaeus's inclusion of apes with humans in the same taxonomical order.[32] While there was a measure of blur at the boundary between plants and animals, he averred, there could be none at the boundary between humans and animals.[33] And that, he concluded, should put the "materialists" out of business.

What Medicus believed he could resolve with a few dogmatic fiats and his simple inference to a new immaterial force in organisms proved the subject of earnest philosophical struggles, based on the empirical evidence of neurology and psychology, by figures like Herder and Tetens over the balance of the 1770s. Moreover, it would come under a priori scrutiny in Kant's transcendental philosophy, as he worked it out from the *Kritik der reinen Vernunft* in 1781 to the *Kritik der Urteilskraft* in 1790. These philosophers all understood far more profoundly than Medicus the enormity of the issues he believed he could resolve offhand. Similarly, medical anthropologists like Ernst Plattner and neurophysiologists like Johann Unzer worked on these matters as well, trying to formulate a "physiological psychology," an account of immanent teleology in human involuntary processes (with significant implications for wider circles of "sentient" life).[34]

The essential step within the life sciences themselves came with Blumenbach's articulation of the notion of a *Bildungstrieb*. His ambition was to displace the preformation theory propounded by his great model, Haller, in the direction of epigenesis, but what he in fact triggered was a far more fundamental revisionism regarding Haller's most important idea, *irritability*, itself. Stephane Schmitt has offered us a very useful rubric for conceptualizing this. He writes of the "dismemberment" of Haller's irritability into a proliferation of life-forces across the later eighteenth century.[35] Jörg Jantzen, Brigitte Lohff, and Eve-Marie Engels have also offered penetrating reconstructions of the burgeoning discourse of *Lebenskraft* over the later eighteenth century.[36] What marked this proliferating discourse was *pluralization* of life-forces. Yet, as Lohff perceptively notes, this was all really for the sake of retrieving a *unified* sense of the organism in the wake of Haller's all-too-dualistic formulation of irritability and sensibility.[37] In that light, Blumenbach's notion of the *Bildungstrieb* provided many integrating dimensions, to be sure; he believed it encompassed not only generation but also nutrition and regeneration, yet this was still not enough to integrate all the life-forces Blumenbach himself came to recognize. In his crucial textbook, *Institutes of Physiology* (1787), he in fact discriminated *five* life-forces. Moreover, for thinkers of the 1790s it would not be so much these specific forces that proved most important. Rather, *Bildungstrieb* became the general term for the distinctiveness of living things, synonymous with *Lebenskraft* itself, in a manner that vastly exceeded the specificity of its creator's intentions.[38]

Attending Blumenbach's lectures in just these years when he first published his physiological textbook, and therefore certain to have heard all about these five life-forces and the whole protracted debate behind them, a brilliant young zoologist began to conceive his own construction of the problem. His name was Carl Friedrich Kielmeyer. In 1798, famously, Friedrich Schelling proclaimed that Kielmeyer "without a doubt" has opened "an entirely new epoch of natural history."[39] That "new epoch," it is my contention, saw the dawn of what would be called "biology" by several key figures in the very years Schelling was writing.[40] How did Kielmeyer come to inaugurate this new epoch? In 1799, in another key text, Schelling traced a filiation of ideas that led to this crucial breakthrough: "The idea of a comparative physiology is already found in Blumenbach's *Specimen phisiologiae comparatae inter animalia calidi et frigidi sanguinis*, and further explicated in the discourse on the relations of the organic forces by Mr. Kielmeyer, whose major idea is taken from Herder's *Ideas for the Philosophy of the History of Humanity*, first part, pp. 117–126; namely, that in the series of organisms, sensibility is displaced by irritability, and as Blumenbach and Sömmerring have proven, by the force of reproduction."[41] While I will reserve to a later chapter the consideration of Schelling himself, this historical reconstruction must occupy us here. Like Schelling, our historical point of departure must be Blumenbach (and, behind him, Haller) but, like Schelling again, the supplement we will consider, in construing Kielmeyer and his legacy, will be Herder.

CARL FRIEDRICH KIELMEYER'S "SYSTEM PROGRAM OF BIOLOGY"

In one of the most important passages in his *Ideen zur Philosophie der Geschichte der Menschheit* (Ideas for a philosophy of the history of mankind), Johann Gottfried Herder suggested that a skilled life scientist might develop the full potential of the idea of epigenesis for a comprehensive—genetic as well as organic—theory of life on earth:

> Who would not be delighted should a philosophical anatomist [*philosophischer Zergliederer*] undertake a comparative physiology of several animals, particularly those close to man, in order to provide in accordance with this empirical experience the distinctive and definitive [*vestgestellten*] forces in relation to the whole organization of the creature? Nature presents us with her work: from the exterior in disguised form [*verhüllte Gestalt*], a covered-over relation of inner forces [*Behältnis innerer Kräfte*]. We perceive its mode of life; we infer from the physiognomy of its face and from the relation of its parts perhaps something of what takes place inside; but it is there on the inside that the mechanisms [*Werkzeuge*] and the masses of organic forces themselves are laid before us, and the closer to man, the more we have a basis for comparison. I dare, though I am no anatomist, to draw a few examples from the observations of great anatomists: they prepare us for the structure [*Bau*] and for the physiological nature of man.[42]

In many ways, Kielmeyer can be taken to have fashioned his life's work in answer to this Herderian call.[43] Herder was a decisive inspiration for Kielmeyer's project, as thinkers from his own time to ours have stressed. We have evidence that Herder himself noted this with satisfaction.[44] Schelling was perhaps the first to point out the connection.[45] Hegel and Schopenhauer, too, saw this lineage.[46] In recent scholarship, it has been taken up by Owsei Temkin, William Coleman, Wolfgang Pross, and Thomas Bach.[47] While it is not clear exactly when Kielmeyer absorbed Herder's work, it is well known that Jacob Friedrich Abel (1751–1829) at the Karlsschule was very enthusiastic about Herder's program for physiological psychology and also for his grand natural history of the human understanding, and Abel propagated these ideas in his philosophy courses in the mid-1780s.[48] It is likely that the uptake of Herder's *Ideen* was also one of the features of the milieu at Göttingen from 1786 to 1788, when Kielmeyer studied there with Blumenbach, among others.[49]

One of the central historical considerations of this study has been the self-formation of naturalists—that is, how they identified themselves and their projects. That illuminates the differentiation of research life science from medical practice. Like Haller, Blumenbach, and C. F. Wolff before him, Kielmeyer fits this model precisely. In the words of the scholar who has done the most to advance Kielmeyer studies, Kai Torsten Kanz, "Kielmeyer . . . was, to be sure, a trained doctor, but he never practiced medicine, devoting himself entirely rather to zoological (and chemical) research."[50] Such dedication to natural-history research was still quite unusual: "taking up natural history in those days was for most merely a 'hobby,' since very few could earn a living by it."[51]

There are two historical-biographical questions from which we must, then, take our point of departure. First, how did Kielmeyer come to this self-conception and project, and second, integrally related, what were his sources and resources? If Herder inspired Kielmeyer, it does not follow that it was Herder alone or even preponderantly who created the matrix of ideas for Kielmeyer's science. We clearly need "a reconstruction of Kielmeyer's research matrix."[52] Kielmeyer was not generous in his citations; Kanz could say only that he drew on "a number of the best sources."[53] Here, some crucial names besides Herder are Haller, Wolff, Blumenbach, E. A. W. Zimmermann, Tetens, and Kant. Kielmeyer's theory of organic forces caught up the essential impetus of the life sciences since Haller, influenced not only by physiologists like Wolff and Blumenbach but also by two crucial philosophical constructions of this work, by Tetens and Herder.[54] From Tetens came the impetus to focus on the *relation* of vital forces, especially the two crucial ones Haller had introduced—irritability and sensibility. As Bach reconstructs it, Herder added three things: first, that *more* forces needed to be identified; second, that nonetheless *all* these forces were the expression of a *single* ultimate force; and, finally, that in their expression these forces fell under a sort of "law of compensation"—that is, proliferation of any one came at the expense of a reduction in some or all of the others.[55] Kielmeyer brought these philosophical reflections back into an empirical context with an explicit set of hypotheses, above all an explicit "law of compensation" for the "animal economy," both in individual organisms and across the whole living world.[56]

Kielmeyer proved extremely chary of publishing his ideas. The result is that for the most part we have to work from student lecture notes still preserved in archival form as well as from a staggering plagiarism whereby one of Kielmeyer's courses from 1807 was published by a mid-nineteenth-century author as his own work.[57] Kanz has been the major bibliographer of Kielmeyer. He stresses two points: a vast amount of material from Kielmeyer remains in archives, without a critical edition; and most of the secondary literature simply regurgitates the points made by earlier scholarship. He adds a third observation: Kielmeyer appears never to have sustained the interest of scholars; he is always taken up as a way station to a different research goal.[58] This study cannot escape most of these quandaries, but it will attempt, at least, to make the case for Kielmeyer's centrality in the emergence of German biology.

His training paved the way to this status. Instruction in the natural sciences was in fact better at the Karlsschule in Stuttgart than at many universities in the German lands because the prime motivation for its creation was to promote economic life and practical application in the duchy.[59] That is, "occupational connections, the closeness to questions of active life, differentiated the school from the universities."[60] Kielmeyer's father was a forestry official in the ducal government, and Kielmeyer's initial admission to the Karlsschule envisioned him following in his father's footsteps. But Kielmeyer gradually distinguished himself in natural science and earned a medical degree, the "practical" professional credential appropriate in such a Karlsschule education.

Philosophy also had an important place at the Karlsschule, above all in the teachings of Abel, "the primary definer of philosophy at the Karlsschule."[61] In his "Philosophy of Common Sense," Abel articulated sharp criticism of aspects of Kantian philosophy.[62] Moreover, Abel's orientation to the question of psychology stressed the physiological, as revealed in the various dissertations that Friedrich Schiller composed, primarily under his direction.[63] Schiller's effort to go beyond Haller got him in trouble, we know, but it was not idiosyncratic insouciance that motivated his effort; Schiller was working toward physiological psychology: a naturalist, if not a materialist, theory of the relation of mind and brain, something both fertile for the development of life science in the balance of the century and anathema to the Kantian "critical" philosophy he would later come to embrace.[64]

Instruction in natural history was part of the curriculum from the Karlsschule's inception in the early 1770s. The appointment of Karl Heinrich Köstlin (1756–83) to the first chair in natural history in 1780 placed the Karlsschule in a very small company of institutions of higher learning: it was only the sixth institution in the Germanies to create a chair in the field.[65] In his natural-history courses, Köstlin used Linnaeus as his textbook for botany and the second edition (1782) of Blumenbach's natural-history handbook for zoology.[66] He brought the latest work in life science into his classes, threw in ideas of his own, and presented it all with such liveliness that it inspired students to take up the field.[67] "His influence on the young Carl Friedrich Kielmeyer is said to have been very great."[68] Köstlin died only three years into his appointment, and his chair went unfilled until Kielmeyer himself was appointed in 1790.[69] Yet Köstlin's three years sufficed to inspire Kielmeyer and leave a legacy that was strong enough for Georges Cuvier and Christoph Heinrich Pfaff later to create a student association for natural-historical research, though neither experienced the teaching of Köstlin directly.[70] As his curriculum vitae suggests, Kielmeyer soaked up as much natural science as the school offered. There was no formal instruction in chemistry, so he had to learn that on his own.[71] This was no minor pursuit; it became the topic of his medical dissertation, completed in 1786, and he would be recognized as a leading chemist throughout his career, notably with his appointment to the chair in chemistry at the University of Tübingen in 1796.[72] At graduation from the Karlsschule in 1786, Kielmeyer had distinguished himself enough in his scientific pursuits to earn the permission and financial support of the duke to go on to Göttingen to continue his studies. He observed that in contrast with his Stuttgart experience, where he was largely thrust upon his own resources, at Göttingen there was excellent support for studying natural science.[73] His studies, 1786–88, entailed a series of courses with the chemist Gmelin (a fellow Swabian), with the physicist Lichtenberg (who became his closest academic mentor and later referee in numerous job applications), and, of course, with Blumenbach.

The central historical-biographical issue of the Göttingen years is Kielmeyer's relation to Blumenbach. What exactly is at issue here? Clearly, he had studied Blumenbach's work already, and he took several courses with Blumenbach at Göttingen. We cannot doubt that he was fully cognizant of Blumenbach's contributions and importance. Evidence from his letters home intimated that he admired Blumenbach and saw him outside the classroom on occasion, but without any established intimacy. In one letter to his parents, Kielmeyer characterized Blumenbach as similar to a teacher he knew back at the Karlsschule, only Blumenbach was "far more discriminating, indeed beyond compare more discriminating [*viel gescheuder, ja ganz ohne vergleich gescheuder*]."[74] Blumenbach was, he wrote in another letter, "a man of outstandingly good character." But Kielmeyer enrolled in his course in natural history "not foremost to learn natural history but far more in order to be able to make better use of the natural-history cabinet here, since he is the man in charge of it."[75] He already knew Blumenbach's ideas from Köstlin's courses at the Karlsschule; more, he had ideas of his own, and he was mainly interested in expanding his database. For similar reasons, Kielmeyer went on an extended field trip through the neighboring countryside gathering specimens with his classmate Heinrich Friedrich Link in 1787.[76] After he finished his course work at Göttingen, Kielmeyer made a similar, more extended field trip across northern Germany.

The question is really driven by Timothy Lenoir's contentions about a "Göttingen School" and the suggestion that Kielmeyer should be understood as inspired by and affiliated with the "teleomechanism" of Blumenbach and Kant.[77] The answer to *that* question is unequivocal in the recent scholarship: no such inspiration/affiliation can be established. Dorothea Kuhn, Kai Torsten Kanz, and Frank Dougherty have each made a very strong case that Kielmeyer should not be viewed as derivative of Blumenbach and Kant, since his ideas were formed independently. At best, his ideas can be seen as consistent with, but not adopted from, their views on teleomechanism.[78] The case has been made most aggressively by Frank Dougherty: Kielmeyer was of course a student of Blumenbach's, and he developed a personal acquaintance, but there is no evidence for any simple derivation of his ideas or methods from Blumenbach.[79] Dougherty argues that Kielmeyer never used Blumenbach's key concept of the *Bildungstrieb* and that his conception of the organic forces, especially the "reproductive force," did not tally with that of Blumenbach.[80] Moreover, Dougherty suggests that Blumenbach's sense of the *Bildungstrieb* had a far stronger *philosophical* (metaphysical and epistemological) orientation than Kielmeyer's corresponding *empirical* approach to organic forces. Indeed, the latter aimed to connect his vital forces with organic chemistry in an empirical, "materialist-mechanist" approach that was in fact closer to the ideas of C. F. Wolff and, we might add, Johann Christian Reil.[81] That suggests that the standpoint for inquiry that Kielmeyer brought with him from the Karlsschule was one that Blumenbach might not have felt entirely comfortable with.

Kielmeyer incorporated Blumenbach's teachings, there can be no doubt. But he started from a different perspective and he advanced it in directions that were clearly his own. That was apparent from his earliest research and teaching, after he returned to Stuttgart and took up a post at his old school. Kielmeyer became the professor for zoology at the Karlsschule in 1790 and he taught there till the school closed in 1794. His courses drew students of both agricultural management and medicine. This was how the young medical student Christoph Heinrich Pfaff began studying with Kielmeyer, eventually attending every course he offered.[82] The first course, summer semester 1790, covered general natural history and was based on the textbook that the first instructor at the Karlsschule to deal with this material, Gottlieb Konrad Christian Storr (1749–1821), had developed before he left in 1774.[83] A partial manuscript for Kielmeyer's course has been published under the title "Über Naturgeschichte" (On natural history) in the collection of Kielmeyer's writings edited by Holler. Already in that first course Kielmeyer advocated a dramatic enhancement of the field of natural history from its traditional descriptive role to a true history of nature: "The history of the phenomena that our earth as a whole manifests must, in accordance with the concept of natural history, articulate not merely the question about its current condition but also about the conditions preceding and perhaps following from it: thus, how it is, how it was, and how it will be."[84]

This was the last time that Kielmeyer offered a course in *general* natural history, however. From that point forward he directed his teaching far more concretely to zoology.[85] In the winter semester of 1790/91, he offered a course on "zoology according to Blumenbach," but from the following summer semester he announced that his lectures would follow his own texts.[86] Ingrid Schumacher lists Kielmeyer's course notes that began circulating across Germany from this period forward: *Vorlesungen über vergleichende Zoologie, allgemeine und spezielle Zoologie* (1790–92); *Vorlesungen über allgemeine Zoologie insbesondere* (1790–92), with supplements through 1816; *Vorlesungen über vergleichende Anatomie der Tiere* (1790–92), with supplements through 1816.[87] Pfaff's notes from all these lecture courses are preserved in the archives, and another set of notes (from 1807) served as the basis for the mid-nineteenth-century plagiarism by Münter.[88] Kielmeyer himself obtained such student note sets and annotated them for his own use.[89] At Goethe's request, Kielmeyer obtained, annotated lavishly, and sent on to Goethe in Weimar two sets of these notes. He also annotated two sets of Pfaff's notes to be relayed to their mutual friend, Cuvier.[90]

The central thrust of these materials was Kielmeyer's theoretical shift from *Naturgeschichte* to the "developmental history of animals [*Entwicklungsgeschichte der Thiere*]."[91] That would set the course of the *Fachgebiet* for the next several decades.[92] Thus, as Schumacher puts it, "already in his lectures at the Hohe Karlsschule Kielmeyer linked anatomy and physiology with the natural history of animals."[93] Like his French contemporary Vicq d'Azyr, Schumacher argues, Kielmeyer construed "comparative anatomy as a physiological, functionally determined domain."[94] More generally, "as sciences, comparative anatomy and comparative developmental history were closely related and interacted with one another."[95] In an early draft essay entitled "Idea for a More Comprehensive History and Theory of the Developmental Manifestations of Organisms [*Entwicklungs-Erscheinungen der Organisationen*]," Kielmeyer drew explicitly on the notion of a *Hauptform* from Herder to conceptualize a historical pattern of development based on a "fundamental form that comprehended in its simplicity all the possible forms of manifestation of organisms and accordingly included all the developmental stages of organisms," in the paraphrase of Schumacher.[96]

Three aspects of comparative anatomy seemed central to Kielmeyer at this point. First, it should determine the actual structures in animals and correlate these with functions. Second, it should develop a "physics of life"—that is, laws of the processes that constituted living systems. Finally, it should develop a coherent view of the entire zoological order.[97] What is crucial is that for Kielmeyer a "developmental history of animals" implied a "physics of life," and conversely: history of nature was a theoretical science of developing life-forms, and the "organic forces" at work had to be related (not "reduced") to the other forces in the physical world.[98] Continuity of the inorganic with the organic was the condition for the coherence of natural science.[99] Thus, Kielmeyer took a strong professional interest in the *chemistry* of living forms and in microscopic observations of their formation.[100] That same interest made him one of the earliest to take up experiments in Germany in response to Luigi Galvani's proposals concerning "animal electricity."[101]

Kielmeyer's address *Über die Verhältniße der organischen Kräfte unter einander in der Reihe der verschiedenen Organisationen, die Gesetze und Folgen dieser Verhältniße* (On the interrelations of the organic forces in the series of different organizations [i.e., life-forms], the laws and consequences of these [inter]relations), delivered at the Karlsschule in Stuttgart on February 11, 1793, and published shortly thereafter, had a massive impact on the German philosophical and scientific scene. The address of 1793, according to Kielmeyer's own characterization, was only a fragmentary and high-level summary of his lecture program of the prior three years and required more systematic elaboration in a publication that he promised would soon be forthcoming.[102] That publication became eagerly awaited, but it was never to appear.[103] It is, indeed, a pity that Kielmeyer never published a systematized text, but at least we still have the address itself, and for all its brevity, it remains a masterpiece of life science for its day. As a historian of biology of the 1950s, Theodor Ballauf, put it succinctly, Kielmeyer's address constituted for the life sciences "the summation of the eighteenth century and the point of departure for all subsequent articulations."[104] Similarly, Reinhard Mocek notes that it "presented to the educated public a synthesis of all the available knowledge to that point concerning the fundamental forces of life."[105] Kristian Köchy has termed it "a milestone in the history of biology" that "introduced the step from an overwhelmingly descriptive natural history to an explanatory natural-scientific biology," capturing "the burning questions in the metatheoretical discourses of philosophy and natural science of its epoch."[106] Thus, Thomas Bach has come up with the striking formulation that Kielmeyer's address was the "first system program of biology," paralleling the contemporary and more famous "Earliest System Program of German Idealism."[107] A decade before Treviranus named it "biology," Kielmeyer issued its "declaration of independence," the claim that a new science was called for, with its own domain of inquiry and based on its own laws.[108] Kielmeyer explicitly proposed that his considerations pointed toward the "foundation of a general natural science of life or of the organism."[109]

One of Kielmeyer's favorite metaphors, "the great machine of the organic world," in fact problematized the conventional sense of "mechanism." Bach makes the perceptive suggestion: "what seem to be mechanistic notions in Kielmeyer would be better construed as systemic."[110] When Kielmeyer wrote of "the great machine of the organic world," he meant to stress systemic holism, a holism that embraced even the inorganic in a "world-organism." Thus, Bach observes, "Already in Kielmeyer nature is conceived as an organism," the defining feature of Romantic *Naturphilosophie*.[111] Kielmeyer's invocation of the great machine of nature betokened that *systematicity* of the natural world could best be conceived as *organic*. That was the crucial sense of "vital materialism" as it dominated the sciences of the late eighteenth century. In his important letter to Cuvier, December 1807, explaining his relation to *Naturphilosophie*, Kielmeyer made the crucial point that it was through philosophical arguments that the age had "become more accustomed" to the idea of "regarding nature as a whole and in its immensity [*im ganzen und im großen*] as an organism and alive in all its effects."[112] Just because Kielmeyer stressed the *systematicity* of nature, he denied that *natural* explanations required a *spiritual* (transcendent) supplement (as has been the standard—and crude—reconstruction of "vitalism").[113] Instead, he believed that "upon the physical forces of attraction and the chemical [ones] of affinity a further force was superimposed here in the domain of the organic."[114] It was not that the laws of physical and chemical forces did not apply, it was that they were *modified* in the context of living forms, and this betokened a supplementary force, something distinctly organic, to be investigated in its own right for its own governing principles.

Kielmeyer insisted that natural science sought regularities (laws) that governed process. In that sense, he was committed to an explanatory approach. Still, for him the organic world's laws were not already available in physics or chemistry: rather, a specific science was required to explicate *life* forces in their higher-order manifestation.[115] Kielmeyer sought laws according to which a developmental history of life-forms emerged. His model was not the *cycle* of identical recurrence typical of the inorganic world but rather the *spiral* of patterned irreversibility that characterized the historicity of life-forms.[116] To achieve a comprehensive *theoretical* closure, one had to work from *historical-developmental* changes. Kielmeyer envisioned a *historical* science, in parallel with emerging geology.

In the address of 1793, Kielmeyer undertook what Arthur Lovejoy famously termed the "temporalization" of the "great chain of being."[117] By natural history, Kielmeyer clearly understood history of nature, and he more concretely believed "that classification of natural history should be based on consanguinity [or descent, *Verwandtschaft*] and not on artificial criteria."[118] His goal was explicit: "a developmental history of the animal kingdom in relation to the epochs of our earth."[119] Most saliently, Kielmeyer not only moved from a synchronic to a diachronic conception of the order required to make sense of the phenomena but also moved from a metaphysical to an empirical footing. Whereas Leibniz made metaphysical arguments for his famous principle of continuity, Kielmeyer was content to stress the boundless abundance of different life-forms, the prodigality of distinct individual exemplars of each, the multiplicity of organs in each such organism, and so on.[120] The empirically observable vastness sufficed to render the theoretical question of a coherent *ordering* critical.

Kielmeyer started from Haller's two forces, sensibility and irritability, and deliberately omitted evocation of either Wolff or Blumenbach. Then he went on to proliferate the number of organic forces on his own account, adding three general organic forces: *Reproduktionskraft*, *Sekretionskraft*, and *Propulsionskraft*. The last two forces were his own constructions. By "secretion," he meant the capacity to assimilate new material into the organism and elaborate it and integrate it into the existing tissues, aspects we might associate with metabolism, generally, and which for the eighteenth century Bourguet had elaborated under the important name *intussusception* adopted from Réaumur.[121] By "propulsion" Kielmeyer had something quite concrete in mind, namely, the capacity to drive fluids through the organism, a feature he associated primarily with plant life, invoking the earlier work of figures in Britain (presumably Hales, *Vegetable Statics*).

Notably, Kielmeyer insisted that in animals the heart and circulatory system could not be exhaustively explained by irritability, as Haller had maintained. He expressed skepticism about the scope of irritability, scoffing at the "*schlumpfsinnige Wiz* [retarded judgment]" of Haller's myriad epigones, who wanted to explain everything as irritability. For himself, Kielmeyer asserted in a note, "the manifestations of irritability are nothing else but manifestations of elasticity."[122] While Kielmeyer did not go so far as to disparage Haller personally, his treatment of the epigones and his own reduction of irritability to elasticity suggest a drastic challenge to the preeminence of Haller in the field. In any event, Kielmeyer did not choose to elaborate any further on secretion or propulsion in the balance of his address. He concentrated instead on the relations among the first three organic forces and the laws that expressed their divergence across the chain of living things.

What interested Kielmeyer was not simply diversity but order: despite the enormous multiplicity of the forces in interaction, nature appeared to go serenely forward as a systemic unity.[123] Yet this unity involved development, just as the unity of mankind involved historical change. Thus, Kielmeyer observed, "the great machine of the living world . . . appears to stride forward along a path of development [*scheint in einer Entwicklungsbahn fortzuschreiten*]."[124] For Kielmeyer, not only did each individual organism seem to have a life course, a developmental path (*Entwicklungsbahn*), but so did the entire organic world. How living things accomplished this was the really interesting scientific question he wished to explore.[125] He posed three questions in his address to illuminate this historicization of the natural world. First, what organic forces could be found across the hierarchy of living things? Second, and most prominently, "what are the relations of these forces to one another in the different forms of organized life, and in accordance with what laws do these relations change in the hierarchy of different organizations [life-forms]?"[126] Finally, what *explanation* could be surmised for these relations and the descriptive laws expressing them?

Starting with sensibility as an organic force, Kielmeyer noted the simplification of dedicated sense organs as one worked downward in the hierarchy of living things from the human and higher animals to lower forms of animal life. Certain animals had weaker vision, others seemed to lack hearing; in lower animals there were no dedicated sense organs. In organisms without a nervous system, indeed, stimuli appeared to be registered generally throughout the tissue structure, as in the response of Trembley's famous polyps to light. What Kielmeyer observed, decisively, was that with the diminution or elimination of a determinate sense facility, it appeared that the remaining senses *compensated* by expanding their range and acuity.[127] This was the key insight that drove his entire argument. He acknowledged that this compensation was only approximate—that, indeed, the reduction in the multiplicity of sensory receptivity was only partially compensated by the increase in the remaining facilities.[128] Nonetheless, as a general rule, this pattern of compensation deserved consideration, in his view, as the first "law" of the relation of organic forces across the hierarchy of living things.[129]

Turning to irritability, Kielmeyer concentrated on the vast wave of experiments with severed animal parts in the wake of Haller, noting a dramatic difference in the persistence of irritability between warm-blooded and cold-blooded animals.[130] While the former swiftly lost all irritability once severed from their brains or hearts, the latter could survive an amazingly long time. Kielmeyer observed similar persistence of irritability in tissue in the still-lower organisms. Thus, he came to his second law of compensation: the persistence of irritability was inversely proportional to the hierarchical placement of animals in the scale of living things according to sensibility.[131]

Finally, he turned to the force of reproduction, which he considered the most common across the hierarchy of living things and also the force that consumed most of their lives.[132] He discerned here again an inverse relation—this time between the number of offspring and the rank of the organisms in the chain of living things. Higher organisms took longer to develop in the embryo stage and longer to reach maturity, often requiring substantial parental support; hence, this restricted the number of offspring. The lower in the hierarchy one moved, the more prolific the number of offspring and the less parental involvement. Even more interesting for Kielmeyer was the proliferation of *asexual* reproduction as one moved down the scale of living things.[133] That led him to postulate a third law of relation: the variety of reproductive forms and the number of offspring increased as one moved down the chain of living beings, and, inversely, as one moved up that chain, the complexity of development favored both *sexual* reproduction and a substantial reduction in the number of offspring.[134]

Having considered the patterns within each of the three forces, Kielmeyer was prepared to make his encompassing generalization: "the capacity for sensibility in the hierarchy of living things is gradually displaced by irritability and reproduction, and in the end even irritability gives way to the latter."[135] Kielmeyer clearly formulated this generalization as a rule of the displacement of "higher" forces by "lower" ones, but he recognized that this could just as well be formulated inversely, resulting in a model of development, not simplification. This was clearly the case when he drew the important analogy between these laws of compensation across the hierarchy of living things and the stages of embryonic development in individual organisms.[136] What Kielmeyer noted here would be taken up by Johann Meckel the Younger (1781–1833) and propounded early in the nineteenth century as "recapitulation," then formulated into a slogan later in that century by Ernst Haeckel (1834–1919): the "biogenetic law" that "ontogeny recapitulates phylogeny."[137]

Even more interesting was Kielmeyer's concluding suggestion that "through carefully selected analogies" one might use his laws of compensation to reconstruct "the generation of the first living things on our earth [*ersten Hervorbringung der Organisationen auf unserer Erde*]."[138] That was a decisive gesture toward "transformism" in the biological world, toward an "archaeology of nature," to use Kant's phrase.[139] As Gabrielle Bersier puts it, Kielmeyer created "a system of biodynamic interaction encompassing all living organisms and regulated by the law of compensation."[140] He capped that off with a thought that showed his full and enthusiastic embrace of these altogether Herderian impulses: "what was previously irritability develops in the end into the capacity for representation [i.e., consciousness; *was zuvor Irritabilität war, entwikelt sich am Ende zur Vorstellungsfähigkeit*]."[141] Human consciousness was itself an emergent consequence of the development of organic forces within the living world. This was a radically *naturalist* position.[142] In the compass of his little address, Kielmeyer had gathered up a host of the most urgent issues of the age and thrust them in a coherent and provocative new direction: the historicization of life-forms all the way up through man.

KIELMEYER AND PALEONTOLOGY

Kielmeyer's letter to Cuvier from 1801 contains such an explicit statement of his interest in paleontology that it is worth reprinting in translation:

Since I believe that the differences between fossilized animals and those that can be found still living upon our earth are not so much—at least not always—to be ascribed to the extinction of species, but rather far more to a transformation in the formative forces of the majority of still-existing species running parallel with the revolutions of our earth (since even the fossil species manifest so many similarities with those to be found still living), I would be very delighted to see you bring your own work on this matter to completion. Perhaps from it one might then be able to achieve some illumination not only concerning the revolutions of our earth but above all concerning the laws according to which the transformation in the formative forces of animals takes place, always in accordance with the varying conditions of our earth. Perhaps these laws are the same as the laws according to which even now transient and persistent varieties of flowers emerge, or the laws of the variation of the formative force in tulips and carnations. What above all fortifies my conviction in thinking of a transformation in the formative type of the species (for the majority of cases) is, in addition to the similarity between the still-living animals and the organic formations of earlier nature, the observation that I have been able to draw quite generally from all the data I have encountered and come to know that the fossilized species are considerably larger than the similar and corresponding species among the still-present animals. This very observation permits me to believe in the possibility that one may find laws for the transformation in the formative force of animals each in accordance with the varying circumstances of our earth. Only, for such a purpose it would be necessary for our inquiry to take up not only investigation and comparison of fossilized quadrupeds with living ones but also more generally to extend to all the remains of the earlier organic world, in particular plants and crustacea as well. In addition, it will be necessary to pay particular attention to the times in which these remains occurred, since it is highly likely these were very different. To do this last bit would admittedly be difficult, because the dates are not included as they are with buried coins! Still, the geological stratification and other circumstances in which the fossilized animals are found in the earth could often suffice to establish what came first and what came later and provide us with a semiotic for the span of ancient time. A hindrance in the execution of this whole consideration of laws of formative variation according to the circumstances of the earth is clearly this: that we know next to nothing about the circumstances of our earth or its history and that a part of the revolutions was violent and sudden; still it seems to me that we need to use these violent revolutions at most to explain only the amassing of such fossilized animals in a few small locations, but not their appearance in different climes. This latter seems to be based upon more regular, more developmental, slower revolutions of the earth's surface. For the knowledge of this last there remains hope via chemistry and a comparative consideration of heavenly bodies. And were the sequence of changing

> circumstances that our earth has undergone to be clarified, then it would also be possible . . . maybe . . . maybe . . . to determine the influence of these changing circumstances on living formations, just as the influence of a certain garden dung permits the determination of plant formations. But for dreams we have the night, not the daylight in which I am certainly writing.[143]

There is substantial evidence that Kielmeyer was pursuing these ideas already—and extensively—in his years at the Karlsschule between 1790 and 1794.

Had he given up on that project by the mid-1800s? Crucial evidence seems to come from his correspondence with Carl Windischmann (1775–1839). Kielmeyer had been asked by Windischmann for his appraisal of the question of "the developmental history of the earth and its life-forms." In his letter of November 25, 1804, Kielmeyer replied:

> In all events, I consider praiseworthy [*beifallswürdig*] the idea of a closely connected *developmental history of the earth and in addition of the sequence* [*Reihe*] *of organic bodies* in which each would illuminate the other *reciprocally* [*wechselsweise*]. The reason I think this idea is right is this: because I take *the force* through which at one time on our earth the *sequence* of organic bodies was brought forth as, in its essence and lawfulness, *one and the same* as the force through which still now in every organic *individual* the sequence of its developmental conditions, which is *similar* for every sequence of organization, gets brought forth. If, however, the force in both cases is the same and if it were in the latter case to be considered *analogical to magnetism*, as I attempted in a lecture in the year 1792 to demonstrate analytically, *based on experience*, then it must be assumed that this was also true in the first case, as something that is analogical with the *magnetism of our earth* and somehow connected *with it*. Since, now, *changes of our earth will also affect its magnetism*, and therefore, its strength and orientation, etc. must be assumed to change, so too must the effects of such a force alter in proportion to such changes, hence the produced organisms, the children of the earth—. These produced organisms, however, demonstrated a certain regularity among themselves in the sequence of their formation and beyond this a similarity to the developmental circumstances in an individual organism; hence, it can be retroactively inferred that the changes that our earth has undergone as well, and specifically its magnetism, were *regular, i.e., developmental changes*, and thus in the end it can be concluded that the *developments of our earth* and accordingly in the sequence of organic beings among one another *hang together precisely* and just for that reason share a common history.[144]

Kielmeyer noted that he had developed these ideas extensively in his courses at the Karlsschule in the early 1790s on the *Theorie der Entwicklungs-*

erscheinungen der Organisationen (theory of the developmental phenomena of organized beings).[145] Yet, even as he agreed with Windischmann about the ambition to formulate such a history of earth and its life-forms, he now admitted he doubted "the possibility of carrying out such a history."[146] He made this judgment, he elaborated, not only because there was insufficient evidence to formulate the histories that would determine the continuously interacting relation between earth, its magnetism, and the forms of organic development, but because he no longer had confidence that such a continuity *ever* existed.[147] Instead, he now suggested, there were always *intervals* between the life-forms; each stage in the development of an individual organism, he now observed, demonstrated discontinuities with the succeeding one, and so it must have been with the developmental line of organisms in the history of earth. More emphatically, Kielmeyer insisted that there was too little knowledge of the developmental history of the earth or of the laws of animal development or of the hierarchy of currently existing organisms. Since he could find neither a convincing law of gradation in animal development nor a compelling theory of magnetism, Kielmeyer admitted to having gradually given up on publishing his theory.[148]

He presented yet another reason for his doubt. Now he believed that the way in which organisms developed differed at different times over the history of the earth. "Some species of organism probably emerged out of others just as now the butterfly develops from the caterpillar, the flower parts from the leaf or the rest of the plant. Others, however, are *original children* of the fertile earth. Maybe all these primitive ancestors have thoroughly died out."[149] Faced with such diversity of process, he despaired of finding a uniform law of development. Finally, Kielmeyer suggested that *retrogression* was a possibility that further complicated any coherent historical account.[150] As to extinction, Kielmeyer professed himself unpersuaded, declaring that he leaned to Lamarck's view that changes in the direction of formative forces, aligned with changes in the earth, had occasioned the gaps between current and earlier species.[151]

Clearly, Kielmeyer had become even more skeptical on this subject than the whimsical close of his earlier letter to Cuvier intimated. Yet there remains the fact that Kielmeyer had advanced his theory in the early 1790s, that he thought enough of it to recapitulate it in some detail to Windischmann in 1804, and that it contained both the specific hypothesis and the reasons for skepticism that he had already presented in his far more positive letter to Cuvier of 1801. There is, in my view, no reason to believe either that Kielmeyer misrepresented the enterprise when he called it "praiseworthy," or that the wider community would have taken him, for the intervening decade and a half since he began teaching it, to have been misguided. On the contrary, they may—like Windischmann himself—have wished to pursue the project more boldly than Kielmeyer by 1804 thought warranted.

My conclusion is that by 1804, in his letter to Windischmann, and by 1807–8, in his letters to Cuvier about *Naturphilosphie*, Kielmeyer may well have begun to rue the impulses he had helped set loose in life science, paleontology, and philosophy of nature, and that some of the endeavors of his enthusiastic admirers may have come to seem to him highly uncongenial. But that does not change the fact that he *did* provide an enormous impetus to this whole line of thinking. Even if *he* did not believe it could pay off, within twenty years of his letter to Windischmann, a detailed biostratigraphy *was* fully established (by "stratigraphy Smith" in England), and by the mid-nineteenth century Darwin would present a theory of evolution. Kielmeyer was at once bold and reticent. We should not lose sight of the former in taking into account the latter. Above all, we must not allow ourselves to condescend about his hunches concerning magnetism so much that we lose sense of his extraordinary project of correlating the development of life with the history of earth, the core notion of evolution. Kielmeyer taught it to a whole generation. Like Lamarck, whom he cited favorably at the end of his letter to Windischmann, Kielmeyer may not have had the *mechanism* of evolution that we now warrant, but it was an idea of evolution nonetheless, a spectacular, pathbreaking idea for the modern life sciences.

THE IMPACT OF KIELMEYER: THE RECEPTION, 1793–1820 AND BEYOND

Kielmeyer's view of natural science is bracing in its sense of the enterprise of inquiry: "in seeking general laws, which is what we are about here, it is not so much a matter of piling up a huge number of facts as it is the evaluation of a few less, and the natural capacity for insight or a feeling for what's right drawn from practice, that is decisive. Discovery typically arises from a lucky leap whose danger only becomes apparent in retrospect; securing that possession is what then permits the slow and sure advance along the chain of MANY experiences."[152] One of his most noted students, Karl Eschenmayer, recollected in 1832 the importance of Kielmeyer's methodological pronouncements: "The first lasting sense for natural science I derived from the penetrating lectures of State Counselor Kielmeyer. . . . From him I derived the fundamental idea of the continually varying proportions of the three fundamental life-forces in living nature, from plants to man. . . . His lectures caught me at an age that was most receptive for the ingenious formulation of analogies and of induction."[153] This sense of the calling of science galvanized a generation.

The response to Kielmeyer's address was immediate and enthusiastic. His former teachers Jacob Abel and Friedrich Bouterwek sent him letters of congratulation. Notices appeared in key journals. The Göttingen chemist Gmelin wrote of the address in *GGA,* June 3, 1793; and the philosopher G. K. Storr, formerly of the Karlsschule, published a highly favorable notice in the *Tübingen Gelehrten Anzeigen,* June 20, 1793. Reviews appeared in the following year, notably by his former classmate Heinrich Friedrich Link, in *Neues Allgemeine deutsche Bibliothek.* The address came almost immediately to the attention of Goethe, who relished it and who immediately noted the affinities to Herder's approach. He wrote Herder a letter sometime in 1793 or 1794 asking if Herder was aware of the address and assuring him that he would be pleased. Everyone took Kielmeyer at his word that a full publication would soon flesh out the extraordinary outline he had provided. When that failed to happen, instead student lecture notes stepped in. A veritable industry developed in Stuttgart producing notes from each of Kielmeyer's courses at the Karlsschule. These began rapidly and pervasively to circulate through German academic circles. With them, Kielmeyer's reputation soared still further.

The peak of Kielmeyer's celebrity came in 1798, when Blumenbach himself acknowledged the importance of the address in the second edition of his *Institutes of Physiology,* and most famously, Schelling hailed him in *Von der Weltseele* for launching a whole new epoch in natural history. If his career hit a brief lull with the closing of the Karlsschule in 1794, he took advantage of it to conduct extensive research into invertebrates on the north coast of Germany and then to renew his contacts at Göttingen in pursuit of a new position. That came with his appointment to the chair in chemistry at the University of Tübingen in 1796. He gave his inaugural lecture there on the question of the gestation of infusorians, reopening the question of spontaneous generation. His interest in organic chemistry and animal electricity as the basis of living form carried him in the ensuing years more deeply into plant chemistry, leading to an inaugural lecture in 1801 that, once again, captured the imagination of his contemporaries.[154] He received calls to chairs at Göttingen and Halle—even, eventually, to Berlin—but he declined them all. He retired from teaching in 1816 as a living legend and spent the rest of his life in government service.

Though Alexander von Humboldt had taken issue with some of Kielmeyer's findings in his wide-ranging *Versuche* of 1797, in 1806 he dedicated his *Beobachtungen aus der Zoologie und vergleichenden Anatomie* to Kielmeyer as "the foremost physiologist in Germany."[155] A whole series of works from the early years of that decade testify to Kielmeyer's pervasive influence: Henrik Steffens, *Beiträge zur innern Naturgeschichte der Erde* (1801); Stefan Wickelmann, *Einleitung in die dynamischen Physiologie* (1803); J. F. Fries, *Reinhold, Fichte, Schelling* (1803); Ignaz Troxler, *Versuch in der organischen Physik* (1804). In 1806, perhaps most influentially, Johann Meckel identified in Kielmeyer's address the origins of the idea of "recapitulation," and this imputation was seized upon with enthusiasm by Samuel Thomas Soemmerring in his review of Meckel's book.[156]

The leading figure of the next generation in life science, Johannes Müller (1801–58), observed in 1826 that "Germans can be proud that Kielmeyer was the first to recognize the approach to comparative anatomy" based on "following Nature in generation, in the living process of its production."[157] In his famous address "Goethe as a Natural Scientist" in 1861, the leader of a later generation, Rudolf Virchow, noted: "Still in the days I was a student, Kielmeyer's teachings were laid out in the lecture courses on physiology, so deep and lasting was his influence, even if it was transmitted almost exclusively in the traditional manner, from mouth to mouth."[158]

In one of the testimonials at Kielmeyer's passing, his friend Karl Friedrich Philipp von Martius (1794–1868) made the observation that while Kielmeyer was no *Naturphilosoph*, he was a philosopher of nature.[159] That discrimination has three senses. First, of course, it denies Kielmeyer's affiliation with *Naturphilosophie*. But with the term "philosophy of nature" two other senses come to the fore. On the one hand, it recalls to our historical attention the persistence of the idea of "natural philosophy" as the general name for theoretical natural science in the early modern period, from Newton through Kant. Martius meant that Kielmeyer was heir to that tradition. But, on the other hand, it also suggests something distinctive about Kielmeyer: his genius for theoretical synthesis.

We can see all this even more clearly in Christoph Heinrich Pfaff's characterization of his teacher and friend: "Kielmeyer is more than just a systematizer; he is a truly philosophical mind—though one far from allowing general metaphysical ideas to serve as the foundation for the elaboration of natural science, or from considering the elaboration of such ideas as the highest goal of natural science."[160] Pfaff clearly wished to distinguish Kielmeyer from *Naturphilosophie* and to situate him alongside their mutual friend, Georges Cuvier, as a natural scientist.[161] He stressed Kielmeyer's empirical method of observation and inductive inference, aligning it with the tradition of Bacon and Kant. But Pfaff then went on: "Without being a natural scientist in the rigorous sense of the term, that is, without adding to the inventory of facts and objects by his own observations and experiments, Kielmeyer instead gathered with inexhaustible energy the data available, appraised it and organized it, though not according to what were essentially external similarities and differences but rather according to their most essential, deeply internal relations, which only a theoretical [*geistiger*] view, such as his, could have discerned."[162] This retrospect was composed in 1845 and it betokens both the culture of science of that moment—the adamantly "experimentalist" commitment in German physiology and the equally adamant positivism of its philosophy of science (a blend of Kant and Comte)—and Pfaff's concern to make his mentor and friend respectable in that culture.[163] Hence, Pfaff insisted that Kielmeyer belonged to the "Kantian school" and should not be aligned with the now-despised "black death" of science in Romantic *Naturphilosophie*.[164]

Perhaps the best that can be said here is that we can register in Pfaff's retrospect the definitive split of a modern (albeit positivistic) idea of "natural science" from the old idea of "natural philosophy."[165] In fact, what Pfaff was calling "philosophy" in this passage would be far more reasonably rendered in our current discourse simply as *theory*. Moreover, Kielmeyer's theory, as Pfaff himself attested, built methodologically from an empirical database via inference—something even nineteenth-century positivism more or less embraced. In short, philosophy of science inflects (if not in*fects*) history of science with the attitudes of the day, and *our* philosophy of science (or "science studies" as a wider frame) offers a far more sympathetic basis for the historical reconstruction of Kielmeyer—and, beyond him, of German *Naturphilosophie* generally—than the dogmas of fading positivism.[166]

"ANIMAL ELECTRICITY" AND VITAL FORCE

Significantly, Kielmeyer's only other publication in the 1790s dealt with "so-called animal electricity."[167] This reflected the enormous impact that Luigi Galvani had in 1791 on the German research community.[168] Galvanism was one of *three* theoretical interventions into physiology in the 1790s in the Germanies. The second was the new "antiphlogistic" chemistry of French provenance.[169] "In the closing decade of the eighteenth century, German chemists were engaged in a bitter rearguard polemic over Lavoisier's challenge to the phlogiston theory."[170] The major journals of German chemistry, Lorenz von Crell's *Chemische Annalen* and Friedrich Gren's *Neues Journal der Physik*, had fought hard against this "French chemistry," and their defeat ruined the prestige of the editors and their journals. "By the autumn of 1793, the debate had run its course," but this set the stage for the reception of galvanism: "In the pages of Gren's *Journal*, the subject of phlogiston quickly gave way to exuberant reports of experiments in animal electricity"; that is, "galvanism provided an opportunity to shore up the wounds of the immediate past."[171] Christoph Girtanner drew the new chemistry directly into the context of physiology with his first essays of the 1790s.[172] Chemistry proved central, as well, in the work of Johann Christian Reil and his "school" at Halle over the 1790s.[173] Theirs was a concern to grasp *chemical*, rather than merely *mechanical*, explanations of life processes. The third intervention was the medical theory of John Brown (1736–88)—which Girtanner introduced into Germany as well, though without proper acknowledgment. The German craze for Brownian medicine is best reserved for its role in Schelling's *Naturphilosophie*. But "antiphlogistic" chemistry and galvanism triggered a notable explosion of works on *Lebenskraft* in the mid-1790s, *before* Schelling's works even appeared or the Brownian movement in medicine took shape in Germany.

In a lively and persuasive monograph, Marcello Pera has illuminated the "Galvani-Volta Controversy" in terms of their divergent research goals.[174] "Galvani observed his phenomena with the eye of the physiologist, while seeking to make their interpretations compatible with the contemporary state of knowledge in electrical (electrostatic) science."[175] By contrast, "in Volta's interpretation . . . one should always begin by seeking *physical* explanations, without resorting to other kinds (for example, biological or—a kind he scarcely envisaged—chemical)."[176] Was "galvanic fluid" distinctively *organismic* or simply a feature of the physical world generally, that is, a force (or property) "that pervaded all bodies?"[177] Alessandro Volta succeeded in winning over the bulk of scientific opinion when he appeared to demonstrate that electrical phenomena could be generated without any animal tissue; hence, "animal" became redundant for this phenomenon known as "electricity."[178] But he never proved that there was *no* electricity in animals or, more importantly, that animals could not *generate* such electricity within their own bodies rather than react to it as an external stimulus.[179]

Naum Kipnis makes what I take to be a compelling case for redirecting attention from the "victory" of Volta over Galvani to the question of the importance of "animal electricity" for German *physiology*.[180] "Contrary to physicists, physiologists were interested not so much in explaining galvanic phenomena as in finding something useful for physiology."[181] Concretely, the issue was whether the new "galvanic fluid" might be identical with the longstanding notion of a "nervous fluid." The established fact of electrical charge in the electric eel and the Torpedo-fish offered empirical precedent for the new discourse, which sought to generalize the phenomenon across *all* animal life.[182] At the close of the eighteenth century, physiologists needed a *theory*—and even more they needed experimental confirmation of the *actuality*—of "nervous fluid." Galvanism promised "a breakthrough in the experimental study of the nervous fluid," a "new opportunity to study the nervous act and perhaps to solve the mystery of life."[183]

The enthusiasm for galvanism did not last: the "anticipated revolution in neurology had aborted" by 1800.[184] Thus, galvanism was an intense but short-lived phenomenon, from the initial reception of Galvani's paper in the spring of 1792 just up to the turn of the century.[185] Yet its impact was enormous in that short interval. In Germany, Galvani's theory fed directly into the program of the 1790s to establish a theoretical and experimental basis for a new science of life through a theory of vital forces. For a time, "animal electricity" seemed a prime prospect.[186] This was what drew Pfaff, Humboldt, and Johann Wilhelm Ritter to the topic.

Late eighteenth-century physiology wished to draw upon resources in the physical and chemical sciences of the day while not allowing itself to be reduced to or absorbed by them.[187] Thus, Ritter understood galvanism as "a more complex concept than those of electricity and chemistry," which "incorporates in itself electrical and chemical action."[188] In the state of disciplinary indeterminacy in the sciences in Europe at that moment, a given scholar could well be simultaneously a physicist and a physiologist (to say nothing of a chemist and a physician).[189] Since the frontiers (and, indeed, the interior architectonic) of these respective domains for scientific inquiry had not been clearly established, *boundary work* was important, especially for those seeking to carve out a new and legitimate special science of their own.[190] Thus, it was not uncommon for naturalists to criticize one another for improper border crossings. Pfaff, as Kipnis notes, "severely criticized Humboldt for the improper, in his opinion, use of chemistry in physiology."[191] That would prove central in the controversies of Johann Christian Reil with theorists of *Lebenskräfte* starting in the mid-1790s and get taken up directly into Schelling's crucial texts on *Naturphilosophie* at the close of the decade.

Christoph Girtanner's work marked a crucial moment in the attunement of the German life-science community to the core issues of "vital force." Indeed, some scholars believe he was the single most important trigger for the proliferation of debate over this topic in the 1790s.[192] His first major publication appeared in French in the *Journal de physique* of 1790, a two-part essay entitled "Mémoires sur l'irritabilité considerée comme principe de vie dans la nature organisée."[193] It was published the following year in German.[194] In 1791 he published two major contributions to the propagation of Antoine-Laurent de Lavoisier's (1743–94) antiphlogistic chemistry in Germany.[195] Girtanner provoked the new debate by dragging into the question of life-forces new vocabularies drawn from the medical theory of John Brown ("excitability") and from the chemical revisionism of Lavoisier, in particular "oxidation" (as a process, though he conflated it thoroughly with oxygen as an element). He even brought in electricity—from a pre-Galvani source, the abbé Bertholon.[196] What was in a European perspective a derivative, indeed plagiaristic work proved nonetheless seminal for the elements it brought into German discourse, made especially provocative by the crude reductionism of its hypotheses. First, Girtanner aimed to bring Haller's two life-forces into unity simply by deriving sensibility from irritability. For him, nerve action was epiphenomenal; indeed, there were organisms, like the polyp, that were excitable throughout their bodies without any trace of nerves.[197] In more complex organisms, he claimed, nerves were themselves triggered by the irritation of some nonnervous tissue. Second, Girtanner argued that irritability itself was simply chemical: *oxygen* was "the principle of irritability." On that basis Girtanner claimed one could establish a "universal physiology."

It is hardly surprising that Girtanner drew down upon himself a host of critics. His plagiarism of Brown was easily exposed.[198] More important, however, were his suggestions about life-force. Immediately, but poorly, Johann Ulrich Gottlieb Schäffer (1753–1826) penned a learned rebuttal.[199] Christoph Pfaff scoffed that it was simply a repetition of Girtanner's material, accompanied by an equally crude negation.[200] Within the Göttingen community, where Girtanner was an active and important presence, his claims were taken up with more discrimination. Göttingen in the mid-1790s was the center for German discourse of physics and physiology engaged with galvanism.[201] The Göttingen figures read Girtanner against the backdrop of the complex "banquet of life-forces" articulated by Blumenbach in 1787 and Kielmeyer in 1793, as well as in response to the new thrill of galvanism after 1792. They produced a plethora of commentaries on *Lebenskraft* around 1795. Pfaff stands out here because of his direct connection with

Kielmeyer. He had come to Göttingen to study electricity with Lichtenberg, and his first essay, "Abhandlung über die sogennante thierische Elektricität," appeared in Gren's *Neues Journal der Physik* (1794), to be followed the next year by his widely celebrated monograph, *Über thierische Elektricität und Reizbarkeit* (1795).[202] Alexander von Humboldt's earliest publications on the topic grew out of his letters to Blumenbach in just that same moment: "Über die gereizte Muskelfaser, aus einem Briefe an Hrn Blumenbach" (1795) and "Neue Versuche über den Metallreitz . . . Aus einem Briefe an den Herrn Blumenbach" (1796).[203] Kielmeyer himself had take up residence in Göttingen in these very years and participated intensely in the discussions involving his former student Pfaff, his former classmate at Göttingen Heinrich Friedrich Link, and others in Blumenbach's inner circle—not only professorial colleagues like Lichtenberg and Gmelin but former students like Girtanner himself, Humboldt, and Joachim Brandis.[204] Thus arose what the distinguished physiologist Ignaz Döllinger (1770–1841) labeled in 1806 an "army of treatises on the life-force that have just recently appeared."[205]

We can begin engaging this "army" by taking on Pfaff's acclaimed *Über thierische Elektricität und Reizbarkeit* (1795).[206] In the first chapter of his treatment of sensibility and irritability, Pfaff described the variety of life-forces currently under discussion, drawing explicitly on Kielmeyer's set of five.[207] He undertook to explore the mechanism of action in each specific force. To contextualize his account he offered a brief history of the "main schools of thought" on the question of life-force in recent physiology.[208] It began with Haller, whom Pfaff celebrated for having initiated a new era "in the history of the cultivation of physiology." Beyond all the merely theoretical physiology of his predecessors, like Boerhaave and Stahl, "his experiments opened a new path."[209] Haller famously distinguished irritability from sensibility, but "in general Haller expresses himself very ambiguously and inconsistently concerning the role of nervous force in the phenomenon of irritability."[210] Haller maintained that the irritability of muscle fibers was sui generis, and this provoked a lively opposition insisting that it was in fact caused by *nervous force*.[211] That polarized physiology after Haller. Girtanner upheld Haller's view. Indeed, "Girtanner goes considerably further [than Haller himself] in subordinating sensible fibers to the irritable ones since he considers the impressions [*Eindrücke*] to be the effect of the latter on the former."[212] In addition, Pfaff noted, Girtanner sought to integrate all that into "antiphlogistic" chemistry.[213]

The opponents of Haller themselves split into two camps: first, those who believed in the agency of a soul in all organic movement; and, second, those who believed that life functions could be explained by nervous force—but as a bodily, not a transcendent, element. Stahl was the inspiration for the first camp, and Whytt its more recent exponent, according to Pfaff.[214] He also placed Ernst Plattner in this camp.[215] In the second camp Pfaff placed Christoph Ludwig Hoffmann, Johann Schäffer, and Johann Christian Reil. He disparaged the originality of the first two, arguing that they had been completely anticipated and surpassed by the "ingenious Unzer."[216] For Pfaff, Johann Unzer had done the pioneering work on neurophysiology and nervous force.[217] Hoffmann and Schäffer were trivial in comparison. By contrast, "the treatment of irritability articulated recently by Prof. Reil is outstandingly elegant [*vorzüglich schön*]."[218] For Pfaff, Reil was of the same stature as Unzer as a neurophysiologist. In addition, Pfaff noted, Alexander von Humboldt had provided considerable experimental evidence that supported the insights of Unzer.[219]

Pfaff devoted a whole second chapter of this part of his study to the critical examination of Girtanner's other hypothesis, concerning oxygen as irritability, or, more generally, concerning the *chemical* explanation of physiological processes, especially metabolism.[220] Far from accepting Girtanner's simplistic account, Pfaff, drawing on Kielmeyer, argued that the action of oxygen needed to be supplemented by the important role of nutrients carried in the bloodstream to the active tissues.[221] Thus, Pfaff argued, to account for the chemical processes involved in physiological functions, one could not follow Girtanner in assuming that the positive *correlation* of oxygen with life-force meant their *identity*.[222]

Alexander von Humboldt's work *Aphorismen über die chemische Physiologie der Pflanzen* (1794) struck Pfaff as a much more sophisticated, experimental consideration of the role of oxidation in physiology.[223] Humboldt had studied with Blumenbach at Göttingen, concentrating on plant physiology and especially the emergent field of plant chemistry. His results appeared in the work that Pfaff found so fruitful. There, as well, Humboldt offered a theory of *Lebenskraft* that drew the attention of a number of figures in the discussion, including Reil, who singled Humboldt's view out for specific commentary. The thrust of Humboldt's contribution drew *Lebenskraft* into close proximity with the views to be developed a bit later in France by Xavier Bichat (1771–1802): life-force holds off the dissolution of organisms into ordinary physicochemical processes.[224] Looking back from the vantage of 1849, in his third edition of *Ansichten der Natur*, Humboldt characterized *Lebenskraft* as he understood it in 1793 as "the unknown cause that hinders the elements from following their original forces of movement."[225]

Humboldt's omnivorous scientific interests carried him from Göttingen to the Freiberg School of Mines and the pursuit of geognosy. He then accepted a position as a mining official, under family pressure, chafing already to undertake those faraway expeditions that would eventually make him world famous. Writing to Blumenbach, he assured his mentor that his new interest in geognosy had not quenched his concern with plant physiology, on which he continued to work.[226] In the meantime, Humboldt became one of the earliest German scholars to take an interest in galvanism.[227] He plunged into extensive experimentation in the field over 1794 and 1795—some three thousand experiments on live and dead animals, a variety of metals and other materials, and even on himself.[228] He even visited Volta in Como in August 1795.[229] In 1795 and 1796 he reported his experimental results in a series of letters to Blumenbach at Göttingen, and in this measure, he was part of the ongoing Göttingen discourse on life-forces.[230] Like many of the other contributions to this Göttingen discourse, Humboldt's findings were published in the key *Neues Journal der Physik* edited by Gren in Halle.[231] In these same years, Humboldt visited with Goethe and Schiller in Weimar and Jena, and he and his brother worked with the Jena anatomist Justus Christian Loder in a number of experiments.[232] For Schiller's new journal, *Die Horen*, he provided a whimsical essay in 1795, offering a literary parable on *Lebenskraft*: "Die Lebenskraft oder der rhodische Genius."[233] But by 1799, at the end of the second volume of his *Versuche*, Humboldt concluded that the concept of *Lebenskraft* had simply not been established.[234] In 1849 Humboldt explained that he had come gradually to abandon it in his own works, without, however, abandoning insistence upon the difference between living and nonliving matter.[235]

There would be an ongoing, critical exchange of views between Plaff and Humboldt over the 1790s on questions of galvanism and *Lebenskraft*.[236] Their level of experimental and theoretical expertise set the standard for German research in the decade. Humboldt interrupted publication of the results of his extensive experiments in book form upon engaging with the work of Pfaff; indeed, it induced him to make extensive revisions of his own study.[237] The first volume of Humboldt's *Versuche über die gereizte Muskel- und Nervenfaser nebst Vermuthungen über den chemischen Process des Lebens in der Thier- und Pflannzenwelt* appeared in 1797, and the second volume in 1799.[238] It was probably the most massive German experimental and textual response to the galvanism controversy.[239] But its very prolixity made it hard to digest.[240] Karl Rothschuh saw Humboldt rambling in this work across many domains, contributing "in part to physics, in part to chemistry, and in part to physiology."[241] But the crucial point, in Rothschuh's estimation, was "that he unshakably stood fast against Volta in affirming the existence of a galvanic fluid in animals."[242] As Maria Jean Trumpler sums up his view, by the end of the first volume, Humboldt was certain that galvanism was "a phenomenon of irritation," while by the end of the second volume, he "had become convinced of the chemical nature of the Galvanic phenomenon."[243] By the time the second volume appeared, in any event, Humboldt was off on his great expedition to South America and cut off from the ongoing discourse of German life science for the better part of a decade.[244]

Heinrich Friedrich Link, another Blumenbach student, steeped as well in plant physiology, took up this same constellation of issues.[245] Like Pfaff, he set out from the proliferation of notions of life-force. In his classification he drew explicitly upon Kielmeyer's set of five, with one important exception: he replaced Kielmeyer's "reproductive force" with *Bildungstriebe*—in the plural (Blumenbach used only the singular for this concept).[246] For Link it was crucial to grasp that a *drive* (*Trieb*) was not a *force* (*Kraft*) in the physical sense but a "purposeful formation [*zweckmäßige Bildung*]."[247] The opening essay of the second volume of his *Beyträge zur Naturgeschichte* endeavored to render terminology more rigorous. He criticized himself for an earlier publication in which he had not recognized the difference between a force and a drive, and he set about making careful discriminations between the physical sense of force—what induced motion—and the sorts of phenomena the various terms for life-forces aimed to capture.[248] Thus, he was comfortable with the idea that force had been used in the proper sense when it came to the stimulus and motor response in Haller's original notion of muscular irritability or contractility. Still, it was impossible to achieve a mathematical formula for this mechanism, as was conventionally the case in physical mechanics. Hence, he concluded, it was perhaps better to construe the phenomenon more as a "faculty [*Vermögen*] than a force [*Kraft*]."[249] Alternatively, he suggested, the term "capacity [*Fähigkeit*]" might be even more accurate. Even in the domain of irritability, Link noted two additional issues. First, external stimulus could trigger response that was both involuntary (muscle contraction) and also sensitive (pain). What marked this difference in physiological response? Could simply dichotomizing these in terms of particular tissues, as Haller had done, suffice? Second, there was the case of voluntary motion, in which *internal*—particularly *mental*—stimulus occasioned motor action. What made the body responsive to such interventions? Link was convinced that "force" could not make sense of that.[250] When he turned to "secretion" as a force, he was persuaded that chemistry could have some role, but he was clear that it had not been worked out as yet.[251]

All this persuaded Link to divide life-force into two classes: one that could *never* be assimilated to the current notion of physical force and another that *might* be, though it had not yet effectively been theorized. In the first class, Link placed not only the *Bildungsvermögen*, by virtue of their purposive character, but also *sensibility*, associated both with receptivity (*Empfindungsfähigkeit*), whereby the body transmitted impulses to consciousness (soul), and with voluntary motion (*Reizungsvermögen*), whereby consciousness (soul) induced physical motion. Such communication between mind and body, what in philosophy had been termed "physical influx," lay beyond explanation in physical terms.[252] In the second class Link placed all those life-forces that appeared to have *some* prospect of correlation with physicochemical explanation: the contractility of muscle fibers, the impetus of the "propulsive force" whereby fluids were circulated through organic forms, and the chemical process of the "secretive force." None of these, he was clear, had received an adequate physicochemical account, but they were at least conceptually amenable to such an account. On the other hand, he warned, there was no reason to assume that *one* account would cover all the cases, that these life-forces could be reduced to a singular life-force.[253]

Link became one of the foremost plant physiologists of his generation, and it is not surprising that he devoted particular attention to the question of the life-force in plants.[254] He was altogether opposed to the notion—widespread in the late eighteenth century—that plants should be explained entirely mechanically, that is, excluded from the living world.[255] He insisted that plants were completely eligible for inclusion under the conception of *Bildungstrieb*, as, indeed, Blumenbach would have agreed.[256] Moreover, he construed the "secretive" and "propulsive" forces as thoroughly consistent with plant physiology.[257] He even considered aspects of *motion* in plants—for example, the turn toward sunlight and the closing of leaves or petals with nightfall—as potentially eligible for consideration in terms of "irritability."[258] But he drew the line at sensibility. For this, he argued, there was no evidence whatsoever.[259]

Most significant in Link's essay was the effort to take into account Kielmeyer's pioneering extension of the relation of life-forces from the individual organism to the hierarchical array of all organismic form. Link recognized the centrality of Kielmeyer's notion of compensation and believed it offered a "law of transformations [*Übergänge*]," though one that would need careful modulation in terms of a complicating "law of multiplicity" when applied to actual cases.[260] Considering each of the life-forces in turn, Link asked what correlations could be discerned between their presence or intensity in a class of organisms and the rank of that class in the hierarchy of living forms. Link had been one of the immediate reviewers of Kielmeyer's published address, and we can discern from his analysis here in his 1795 essay a thorough understanding of Kielmeyer's arguments and evidence.[261] That makes all the more important, in my view, the degree to which Link confirmed the impulses in Kielmeyer's developmental history.

Link concluded his essay with the question whether *Bildungstriebe* could be discerned in the inorganic world. After all, he observed, even in inorganic materials there appeared to be a "drive . . . to take on determinate form."[262] To keep some clarity in such considerations, Link argued, it was useful to enforce the distinction between *regular form*, which was fully demonstrable in the inorganic world, and *purposive form*, which seemed to him distinctly organic. That, in his view, was the crucial insight in Blumenbach's concept of *Bildungstrieb*. It might be useful for scholars who had difficulties accepting this notion, he added, that Immanuel Kant had enthusiastically endorsed it in 1790.[263] As a conclusion, of sorts, he wrote: "True life-forces, life faculties, or life capacities, however one wants to label them, consist merely in the capacity of the material to form a living body, and in the relations of the body to the soul. . . . Soul is that in which consciousness is grounded, which, were it also material, could nonetheless be satisfactorily distinguished from all other materials."[264]

In an extended work, Joachim Brandis undertook to uphold the notion that all life-forces could indeed be integrated into a *single* one and clarified further by a consideration of the "phlogistic process of the organic machine."[265] Brandis had still another driving ambition in his monograph, namely, to dispute the notion that life-forces—indeed forces in all physics—should be conceptualized as material *bodies*: "whether it is appropriate to consider as real bodies [*würklich als Körper*] several substances like light, electricity, magnetic force, etc."[266] The hypothesis of imponderable fluids as material bodies, he insisted, could not be experimentally confirmed.[267] Above all, he wished to dispute the notion of "nervous force" (variously, *Nervenkraft, Nervensaft, Nervengeist*) as the actual existence of a "subtle fluid [*feiner Flüssigkeit*]."[268]

The locus of consideration of chemical physiology had to be at the level of the "simple fiber," Brandis contended, and this posed experimental problems, since a single fiber was not observable as such.[269] Even inorganic chemistry had to work by *analogy* from the macrolevel to the microlevel in reconstructing reactions. This fundamental methodological premise, that one *could* infer from the macrolevel to the microlevel, was a decisive component of the experimental Newtonian tradition, starting from Newton himself. What Brandis contended was that this methodological practice entailed an inescapable epistemological uncertainty. Thus, the issue for science was to recognize the boundary between sound hypotheses and mere speculation.[270] For Brandis, the ostensible materiality of imponderable fluids constituted a cardinal instance. This had thoroughly compromised the conception of life-forces, in his view. The effort to relate organismic functions to physicochemical laws had involved positing "mediating bodies [*Mittelkörper*]"—one might think of the "plastic nature" of Ralph Cudworth or the *Archaeus* of Jan Baptist van Helmont—but these were "utterly imperfect offspring of our phantasy [*ganz unvollkommne Kinder unserer Phantasie*]."[271] The practice of multiplying entities in explanation had long since been debunked, Brandis maintained, and he wondered whether anyone really believed that gravity or magnetism could be explained as particles of matter, and even whether light or electricity could be.[272] Forces, in his view, were not material bodies but causal principles.

It was this vantage he wished to bring to bear upon the problem of *Lebenskraft*. Life entailed a force unlike any already established in the physical world.[273] Thus, the concept really designated an unknown.[274] Starting from his crucial argument about the immateriality of force (i.e., its nonparticulate nature), he argued that "this [life-]force acts immediately in organic matter [and] is not a consequence of the form of the matter, or its organization."[275] This was a pivotal contention, for Brandis insisted that life-force *created* organic matter out of inorganic fluids; it *preceded* organic matter and *gave* it its organization.[276] In doing so, this particular force *modified or mitigated* the physicochemical laws prevalent in the wider world, allowing a coherence in the organism that resisted dissolution into the inorganic.[277] This was the significance of *death*, Brandis argued, taking up a line that was soon made famous by the French physiologist Bichat.[278] The moment death took hold, organic bodies became fully subject to "fermentation [*Gährung*]" and rot, but as long as life was present, these processes were held at bay.[279]

Brandis believed that theories of a material (particulate) form for life-forces, however "subtle," did violence to this extraordinary phenomenon, rendering a "divine instrumentality [*göttlichen Werkzeuge*]" into a "hydraulic machine."[280] Stahl, Brandis noted, had already "brought forward the relevant reasons against the existence of a *body* that might be termed nervous spirit [*Nervengeist*]."[281] To be sure, against Stahl, Haller's arguments that the soul could not cause all animal motions were "irrefutable."[282] More recent discussions had simply struggled to fill the gap with another "imponderable," as we have seen in Medicus. A good summary of the issue, Brandis noted, was offered in the work of Plattner.[283] Brandis urged that it was nonetheless possible to conceive of a single life-force that could account for all the distinctive phenomena in organisms that had occasioned the recent proliferation of specific forces in physiological theory.[284] The basic idea was simple: organisms responded to stimulation. Stimulus and response (*Reiz* and *Reizbarkeit*) were the essential elements, which had divergent effects depending on the specific form of tissue or organ involved.[285] Even nervous action derived from the same process of stimulus/response discerned in muscle contractility, Brandis argued.[286] The "purity" of the effect of life-force on tissue was thus a function of the "refinement" of the tissue involved, a matter of the levels of "perfection" in tissue organization that Brandis claimed he derived from Goethe. Thus, a flower was a more "refined" form than a leaf or a root, nerves were more "refined" than bones, and accordingly the action of the life-force would be more "perfect" in these more "refined" forms.[287]

The most interesting problem for Brandis was the explanation of *Reproduktionskraft*, in the sense of the restoration of tissue by organisms.[288] For him, this was keyed directly to the question of nutrition and metabolism in the augmentation of tissues. Brandis took this to be the core of the *Bildungstrieb* unique to living forms. He praised Blumenbach explicitly for two major contributions along this line.[289] Brandis himself proposed that the whole matter could be construed in terms of a *chemical* process—that is, "phlogiston in the organic machine."[290] "It is thus an almost *certain* presumption that in the living organic machine a similar phlogistic process takes place as in the burning of other bodies."[291] It was to this hypothesis that he devoted the entire second half of his monograph.[292] What Brandis aimed at, with somewhat more clarity than had Girtanner, was *oxidation* as a fundamental chemical *process*, not simply oxygen as an *element*, a material *body*, present in these reactions.[293] For him there were two crucial issues to clarify.[294] First, *where* did this "phlogistic process" take place? Brandis insisted it was in the tissues themselves, at the level of the simple fiber, rather than, as some physiologists had claimed, in the lungs or the circulatory system.[295] Second, what was the *mechanism* involved? For Brandis, analysis of *Reproduktionskraft* made it "more than probable" that life-force really was electricity.[296] Hence, he found Galvani's theory of "animal electricity" very promising for the prompt clarification of the life process.

In his penetrating review of Brandis, Johann Christian Reil recognized both the creative contributions and the difficulties in the monograph. Reil "admire[d] the penetration with which the author claim[ed] that the phlogistic process takes place immediately in the solid parts [of the organism] themselves" and more generally "share[d] the opinion of the author that in all solid and fluid parts of the body, and particularly in the solid parts, there are constantly ongoing changes in composition [*Mischungsveränderungen*—chemical changes] that we c[ould] for the present call *phlogistic processes*."[297] However, Reil declared himself unconvinced about the argument for the singularity of life-force: "A single principle as the immediate cause of this contraction seems to be an unnecessary assumption. It remains to be studied what kinds of materials (whether carbon or oxygen, etc.) are involved in organic bodies in the actions that divide and unify their instrumentalities [*Werkzeuge*—organs or tissues, one presumes], where this takes place, and in what manner."[298] To be sure, Reil noted, Brandis had made an important contribution in insisting that the function of metabolism at the level of organs and tissues had not been sufficiently clarified.[299] But Brandis had gone too far in trying to collapse nerve response into the same class as muscle contraction. No such contraction could be observed in nerves or in the spinal cord or brain.[300] Most fundamentally, Reil objected to the overarching claim that forces could not be material: "However: can matter not of itself, by virtue of its physical and chemical properties, in particular through its changes in composition [*Mischungsveränderungen*], bring forth movement without an external and distinct cause?"[301] Reil would build his own alternative to the theory of *Lebenskraft* on precisely that foundation.

Most scholars take Reil's extensive essay, *Von der Lebenskraft*, as the most important contribution to the discourse on vital forces of the decade of the 1790s.[302] As Lee Ann Hansen aptly characterizes his ambitions, "Reil intended in the *Lebenskraft* essay to apply [a] mixture of Kantian philosophy and antiphlogistic chemistry to the current debates over the nature of the vital force and Blumenbach's concept of *Bildungstrieb* or formative drive."[303] The terms *Lebenskraft* and *Bildungstrieb* became virtually synonymous in his usage, and thus, Reil inflated Blumenbach's "carefully limited concept . . . until it became the foundation of all manifestations of life."[304] For Reil, Blumenbach merely "drew a circle around a blank spot on the map of biology and gave this a scientifically usable name—*Bildungstrieb*."[305] Reil's treatment was part of a general dissemination of the term in German discourse of the 1790s, which became particularly vivid in the texts of Schelling and his followers.[306]

The challenge for physiology as a research program was to find some coherence in its methods compatible with the standards of "proper science [*eigentiliche Wissenschaft*]" that philosophy seemed to demand.[307] In the opening dedication of his new journal, 1795, Reil put the matter plainly: "In fact, philosophy could do medicine a great service if it could put the concepts of physicians into good order, offer them efficient methods for investigation, inform them on how to develop specific rules for establishing inference from fact and general laws from individual observations, demonstrate to them the boundaries beyond which human inquiry dare not venture, and show them the way out of the realm of metaphysics by which they so readily stray back into the domain of physics."[308] This formulation bespeaks the general epistemological crisis of physiology and medicine of which Brigitte Lohff has given us a thorough reconstruction.[309]

Reil wanted to rescue physiology, and medicine in general, from dire quandaries.[310] He assessed the state of physiology as dismal: "Among all the sciences, if we leave out anatomy, physiology has made the least progress, and for the most part contains nothing other than a wasteland [*Wust*] of partly unfounded, partly senseless hypotheses."[311] The problem lay in the guiding orientation of the field: "we seek the foundation of animal phenomena in a supersensible substrate, in a soul, in a universal world spirit, in a life-force that we think of as immaterial."[312] That is, living things were taken to be inexplicable by mechanism, or indeed, on any simply material basis, so a metaphysical supplement seemed indispensable. This was the conviction of the entire eighteenth century, contested only by materialists. But Reil challenged it. Using quite Baconian rhetoric, he denounced "fanciful conjectures [*Vernunfteleien*] and hypotheses," demanding instead "experiments in accordance with logical rules, through which discovered results could be used to derive general laws."[313] The "backward state of physiology" needed to be overcome not so much by more clinical effectiveness as by a more coherent methodological and theoretical framework, he asserted.

Initially, Reil sought to ground his revisionist program for physiology in the "critical" epistemology of Kant.[314] He picked up the attempt of his friend and mentor Marcus Herz "to merge the boundaries of physiology and Kantian philosophy."[315] But "by the time the *Critique of Judgment* was published, in 1790, Herz and Kant had already drifted apart."[316] Hansen makes the very appropriate point that "the *Critique of Pure Reason* yields a very different practical science than the *Critique of Judgment*."[317] Herz showed no enthusiasm for the third *Critique*.[318] For Herz, Kant's philosophy was not easily reconciled with physiological psychology, and his epistemology in fact created enormous obstacles for any scientific psychology at all.[319] In his medical psychology Herz had already "shifted Kant's concepts from epistemological categories to psychological constructs."[320] Thus, Hansen sees Herz very swiftly reduced to mouthing "familiar Kantian pieties" about "the limits of reason."[321] The more Herz, then Reil, sought to find empirical, medically grounded approaches to the brain and mind, the less they could keep faith with the essential tenets of "critical" philosophy.

In short, Reil was invoking Kant's transcendental philosophy to establish his empirical realism.[322] He was interested only in blocking transcendent hypotheses, not in problematizing empirical findings. For Reil, the crucial point was to stick to the phenomenal/empirical level: "I therefore will consider the representations [*Vorstellungen*] in the natural science [*Naturlehre*] of animals as phenomena of their own sort and as forces in the chain of natural forces for which we have no experience of a further absolute ground."[323] There could be no place in such inquiry for spirit (*Geist*) in the traditional sense, whether as substance or soul. Philosophers had offered metaphysical posits, not experiential actualities. What Reil was rejecting was *animism*: "Life force is in no way a force that animates matter, but simply the 'force of matter which characterizes the plant and animal kingdom.'"[324] Reil advocated a program of relentless naturalism, indeed, materialism.[325] In this exploration "physics and chemistry must be the torchbearers on the path of investigation."[326] The object of physiological inquiry, for Reil, should be to "achieve a better acquaintance with animal matter and with its various forces, relations, and modifications."[327] Thus, the "tenor of life-force [*Stimmung der Lebenskraft*]" is permeated by the material components of the organic being [*organisches Wesen*], and "a change of the materials causes a change of all of its forces [*eine Veränderung der Materie verursacht eine Veränderung ihrer sämtlichen Kräfte*]."[328] Reil proposed to reject the conventional sense of *Lebenskraft*, ironically, for the sake of a *vital* materialism. As he wrote in his important review of Joachim Brandis, *Versuch über die Lebenskraft* (1795), "Cannot matter through itself alone, through its physical and chemical properties, especially through its alterations in combination, produce movement without a cause external and heterogeneous to itself?"[329] "*Naturlehre* is the science of the properties of things in the phenomenal world and of the appearances that depend upon these properties."[330] Reil aimed at a thoroughgoing materialization of the questions, taking *Lebenskraft* as "the meeting ground for mind and body" and hypothesizing that nerve physiology fell within the bailiwick of the "subtle fluids" of "experimental physics"—like magnetism and electricity.[331] His real point was that *Lebenskraft* would be grasped rigorously in physiology only if it were grounded in *Mischung und Form*, that is, in chemistry.[332] All in all, in a manner utterly un-Kantian, in *Von der Lebenskraft* Reil "abolished the distinction between the organic and the inorganic . . . [and] included reason itself as an organic force; the highest force, it is true, but rooted in the chemical properties of matter like all the others."[333]

Kant himself made no comment on this heresy, but his epigones did. Perhaps the most extensive such critique came in Reil's own journal: Johannes Köllner's massive "Prüfung der neuesten Bemühungen und Untersuchungen in der Bestimmung der organischen Kräfte, nach Grundsätze der kritischen Philosophie" (Examination of the most recent exertions and investigations into the definition of organic forces according to the tenets of the critical philosophy).[334] The fiercest critic of Reil on Kantian grounds turned out to be Friedrich Schelling! In *Von der Weltseele* (1798), Schelling attacked Reil explicitly: "it is the height of *un*philosophy to maintain that life is a property of matter."[335] Such hylozoism was utterly unbearable in a Kantian frame. As Schelling put it in his reproach to Reil, "matter is the product of life," the physical world was the expression of an immanent, self-actualizing spirit, and therefore, all nature was an organism articulating this self-organization. Of course, what Schelling proposed was equally abominable to Kant: *constitutive*, rather than regulative, invocation of a "supersensible substrate."[336]

At first, Reil would resist this solution; eventually, he got on board. Indeed, over the first decade of the nineteenth century, Reil converted to Schelling's *Naturphilosophie*, leaving Kantianism definitively in his wake. That is extremely significant for the reconstruction of the relationship between emergent biology and *Naturphilosophie*.[337] Hansen claims: "In the years between 1795 and 1803 Reil moved steadily away from the Kantian program for physiology that he had proposed in the *Lebenskraft* essay, and in the direction of a full-blown reliance on Schelling's *Naturphilosophie*."[338] This claim about the deviations of Reil from Kantianism has been contested by Reinhard Mocek.[339] While Mocek may be correct that Reil took somewhat longer to make the transition to *Naturphilosophie*—thus, not in 1802/3, as Hansen argues, but in 1807—the essential point is that Reil *made* that transition and, in doing so, exemplified a movement across all the medical sciences in Germany. The best evidence for that, as both Hansen and Mocek agree, is the development of Reil's key journal, *Archiv für die Physiologie*.

Archiv für die Physiologie was the first specialized journal for physiology in Europe, signifying a moment of crystallization for the research field. Reil founded the journal in 1795 to advance the argument for the primacy of the "theoretical" in medical training, "redefining physiology's traditional concerns in terms of more fundamental questions about the general possibility of scientific knowledge."[340] Especially after his appointment to a prestigious chair in medicine at the new University of Berlin in 1810, he became noted for his clinical commitment and his aspiration to keep all the medical sciences, including physiology, unified.[341] In this, he stands as the exception to the historical rule I have tried to build about the crystallization of research identities in an emergent science of biology. While there is no question about his commitment to experimental natural science—and especially organic chemistry—as the foundation of medicine, it is also clear that he was deeply committed, as well, to clinical practice. Indeed, his central goal was to maintain the unity of these two commitments through reform of medical education at the university. And yet his actions in fact undermined these intentions, and he became a central figure in constituting physiology as a special research science after 1795. In this, Reil stands as a bookend to Georg Ernst Stahl. Both turned to experimental physiology to buttress clinical medical training, yet both ended up fomenting specialization in research in physiology that would lead to the calving of physiology from general medicine into a separate life science.

Although he may have aimed to keep physiology and medicine together, Reil in fact helped bring the fracture of physiology from clinical medicine to full realization with his demand that the new physiology be able to ground itself in physics and especially chemistry.[342] The idea for a *journal* and not a monograph, Thomas Broman suggests, lay in Reil's efforts to mobilize a community with a shared "research program in physiology" that would transform the medical profession by raising its scientific rigor.[343] "It was precisely to find a new mooring for physiology that Reil founded the *Archiv*."[344] The backwardness of physiology would not be salvaged by more clinical research; it required a more coherent theoretical framework, he asserted. "Reil attempted to place physiology at the base of a unified science of medicine."[345] Many of the publications in the journal elaborated advances in experimental research without any particular concern for medical practice. The content of the *Archiv* "attempted to advance Reil's program," and he "drew heavily upon dissertations [mainly written by his own students or by Reil himself for them] as a model."[346] An early dissertation from 1794 may well have been at the basis of Reil's own essay *Von der Lebenskraft*. Certainly, the unfinished dissertation of his student D. von Madai, given prominent place in *Archiv für die Physiologie*, represented Reil's own further developments of his chemically grounded theory of vital forces.[347]

But "the *Archiv* began to evolve in a direction quite distinct from the one envisioned by its founder."[348] By 1805 "the disintegration of Reil's program for a unified science of medicine appeared complete."[349] Instead, the *Archiv* proved to be a journal dedicated to the special research program of physiology, apart from medical application. "Physiology had indeed become a new science, though clearly not the one Reil had originally intended."[350] After a two-year hiatus in the publication of the journal, caused by the Napoleonic invasions, this became even more explicit when, with a new coeditor, Johann Heinrich Autenrieth (1772–1835), a student and Tübingen colleague of Kielmeyer's, the journal took on "a new program: morphology, the study of animal form."[351] That is, "the *Archiv* largely became a journal of anatomy and morphology."[352] For Broman of even greater significance is the fact that with this transition it was clear as well that "*Naturphilosophie* had arrived in the *Archiv*."[353]

After 1810 Broman recognizes the consolidation of "animal morphology as the research program of a self-conscious community of largely university-based researchers."[354] When Johann Friedrich Meckel the Younger took over Reil's *Archiv für die Physiologie* in 1815, renaming it *Deutsches Archiv für die Physiologie,* he made it the flagship for "the existence of a self-conscious disciplinary community."[355] That is, "the *Deutsches Archiv* invoked a community of scientists engaged with a common set of problems and, most importantly, a community of judges of each other's work . . . [and] the qualifications for participation consisted of mastery of research practices and knowledge of the discipline's language and theoretical concerns."[356] Broman concludes: "During the next decade the links between physiology and other areas of medicine, and to the larger university environment, withered as German physiologists defined a set of problems internal to their own subject, and asserted their right to do research without paying heed to its potential clinical ramifications."[357] Lynn Nyhart has drawn the appropriate conclusion in her aptly titled *Biology Takes Form*—a special science had emerged in Germany. What changed decisively was personal professional orientation: "by mid-century, medical students like Carl Gegenbauer, Ernst Haeckel and Emil Du Bois-Reymond could take an M.D. without seriously intending to make a career as practitioners."[358] They were research biologists, not practicing physicians. Biology had unequivocally become a discipline.

CHAPTER TEN

Polarität und Steigerung: The Self-Organization of Nature and the Actualization of Life

It is an old folly to believe that organization and life cannot be explained through natural principles. If that is as much as to say that the *first* origin of organic nature cannot be investigated *physically*, then this *unproven* claim accomplishes nothing more than to crush the ambition of the researcher.

SCHELLING, 1798[1]

There was a crucial intermediary between the deliberations over *Lebenskraft* and the impact of Schelling's *Naturphilosophie* on German life science: Johann Wolfgang von Goethe. Goethe was a major figure in the constitution of German life science as a patron and personal interlocutor *before* 1817, the date at which a systematic body of his morphological writings began to be available to the public.[2] Timothy Lenoir has written that when G. R. Treviranus invoked the term "biology" for a new science in 1802, "he was in fact consciously synthesizing discussions that had been going on for at least a decade in Germany involving such persons as Johann Friedrich Blumenbach, Karl Friedrich Kielmeyer, Heinrich Friedrich Link, and the von Humboldt brothers. But one of the most distinguished co-workers in this enterprise was . . . Johann Wolfgang von Goethe."[3] In the incipient phases of a science, such discussions are not inconsiderable contributions. Thus, what we might call Goethe's *oral culture of science* proved inspiring and constitutive for the emergent life sciences.[4] I will try to reconstruct both *what* Goethe's sense of a biological science was in the 1790s and *how* it was transmitted to the practicing community of scientists. We can trace Goethe's impact in terms of his interactions with several key figures. The famous conversation with Friedrich Schiller of July 1794 and the subsequent discussions regarding Goethe's reception of Kant's third *Critique* provide the most obvious context. In addition, Goethe worked directly with the brothers Humboldt on various issues in natural science, especially around 1795.[5] Over the 1790s, Goethe's contacts extended widely. He was in active correspondence with—among a host of others—Merck, Blumenbach, and Soemmerring. For my purposes, Goethe's engagements with Kielmeyer and Schelling prove most important.

Like others involved in the 1790s in the crucial shift from comparative anatomy to comparative physiology, Goethe was well aware of the epistemological challenges of physiological inference from anatomical investigations. Like Johann Christian Reil, from whom he received a personal copy of the journal issue of *Archiv für die Physiologie* of 1795 containing the key essay *Von der Lebenskraft*, he recognized that "it is easy to understand why physiology had to lag behind for so long and why it keeps lagging behind."[6] Morphology was Goethe's paradigmatic answer to this epistemological challenge. Gabrielle Bersier notes that he first coined the term in that context, on September 25, 1796.[7]

That Goethe turned his thoughts "towards a new science of morphology," Nicholas Boyle suggests, was at least "partly" the result of the "whirlwind passage of Alexander von Humboldt through Jena and Weimar, from March to May" of 1795, still "full of his most recent *opus*, a study of muscular irritability based on 4,000 experiments." Goethe met with him almost daily in Jena during March and then in Weimar in April. "Their conversations on electrical polarity were probably Goethe's first systematic introduction to [the] concept."[8] With Humboldt, Goethe wrestled over the question of the coherence of inorganic and organic forces, with reference to Kant's *Metaphysische Anfangsgründe der Naturwissenschaft*, and "before the combined authority of Kant and Humboldt Goethe . . . consented to eliminate from his Morphology the inorganic sciences."[9] This was the moment when Goethe began to compose "an essay, attempting to define the field of 'morphology' and its relation to the other sciences. . . . It was to be one of the many 'servants' of that all-embracing science of life—here given the name 'physiology'—which did not yet exist, and perhaps could never be attained."[10] Humboldt was a congenial, as well as encyclopedic, discussion partner, with an insistently experimental orientation. Indeed, Boyle maintains that Goethe "had more in common with Alexander von Humboldt than with Baader or Schelling."[11] With Humboldt, Goethe became caught up in the experimental frenzy at the University of Jena concerning galvanism and animal forces. From 1795 to 1797 Goethe not only spent a lot of time in Jena but spent it on his morphological studies.

JOHANN WOLFGANG VON GOETHE AND THE EMERGENCE OF MORPHOLOGY

Asked in the late 1820s to reflect upon a panegyric to Nature that had recently come to light and seemed to be in his own hand, Goethe declared that he could not recollect having written it (in fact, it was composed by Georg Christoph Tobler in the early 1780s), but he affirmed that he found the text to "reflect accurately the ideas to which my understanding had then attained."[12] That was the high tide of the Sturm und Drang, and as Goethe noted, the "tendency toward a form of pantheism" in the document tallied with his own Spinozistic views of the time.[13] Goethe sought to introduce two important distances between his earlier sentiments and his current views in 1828. The first retraced the historical course of his own development. "During the years in which this essay probably falls I was largely occupied with comparative anatomy." There followed his revelatory experiences during the journey to Italy, culminating in his elaboration of morphology. The second distance was the conceptual clarity with which he could now formulate that paradigm: "The missing capstone is the perception of the two great driving forces in all nature: the concepts of *polarity* and *intensification*. . . . Polarity is a state of constant attraction and repulsion, while intensification is a state of ever-striving ascent."[14] The two concepts of *Polarität* and *Steigerung* have come down to us as the quintessence of Goethe's vision of morphology and, more than that, the fountainhead of much that we associate with *Naturphilosophie*.

While he had ostensibly journeyed to Italy to renew himself as poet and artist, upon his return Goethe could scarcely take up any poetic project.[15] Instead, he wanted to pursue a different life course, as a natural scientist.[16] "To Knebel, who of all his old acquaintances was the one now most concerned with the natural sciences, he wrote . . . as he left for Silesia [in 1791] that he was 'beginning a new career . . . my inclinations drive me more than ever into natural science.'"[17] Goethe had written to Charlotte von Stein from Italy on June 8, 1787: "I am very close to the secret of plant organization. . . . With this model and the knowledge how to use it one could go on forever inventing plants that would be consistent and would have an inner necessity and truth."[18] Still earlier, on July 10, 1786, Goethe wrote to von Stein: "I am beginning to grow aware of the essential form with which, as it were, Nature always plays, and from which she produces her great variety."[19] Thus, Goethe had clearly arrived at a sophisticated conception—perhaps, visualization—of morphology in the course of his Italian experience. It applied, of course, to the plant, but what was clear from his letters to von Stein was that he believed he had come upon a universal principle that extended "to all the realms of nature—the whole realm."[20] His language of *Urphänomene* and cognate constructions throughout his later writings indicate that this model remained at the core of his thought. By the time he returned from Italy, the terms "type" and "metamorphosis" dominated his writings not only on botany but on zoology.[21] They would be complemented, especially via his engagement with Schelling, by the notions of polarity (*Polarität*) and intensification (*Steigerung*).[22]

Before we can take up his interactions with Schelling, we have to consider Goethe's interactions with Schiller and Kant. Goethe published his *Versuch die Metamorphose der Pflanzen zu erklären* just weeks before Kant's *Kritik der Urteilskraft* appeared in 1790. He read Kant's book by October of that year, and Kant's work would preoccupy Goethe for some time thereafter.[23] His encounter with Schiller came in the wake of a presentation at the Jena Natural History Society in July 1794. Both were put off by the particular presentation, which seemed too narrow and uninspired.[24] As they left the session together, they fell into conversation and Goethe announced he could have done a better job, offering to share with Schiller some conceptions he believed could considerably advance natural history. As Goethe recollected years later, he presented "a different approach" to scientific inquiry, "not . . . concentrating on separate and isolated elements of nature but . . . portraying it as alive and active, with its efforts directed from the whole to the parts."[25] It may well be that the language Goethe used here in retrospect was inflected by his engagement with Kant. But that would underestimate the persistence and depth of this insight in Goethe and its origins much earlier in his thought. In an essay composed in the context of the *Pantheismusstreit* of the later 1780s, Goethe wrote: "The things we call the parts in every living being are so inseparable from the whole that they may be understood only in and with the whole."[26]

When they reached Schiller's dwelling, the conversation continued. Goethe went on to sketch a "symbolic plant," which represented the type for all possible plant development; that is, he laid out the theory he had published in his *Versuch die Metamorphose der Pflanzen zu erklären* and, behind it, the idea that he had earlier called the *Urpflanze*. It is important to note that Goethe *sketched* his "symbolic plant." While in Italy, he had been instructed, by Peter Camper's son, regarding Camper's techniques of drawing the transformation of one vertebrate form into another—for example, of horse to man.[27] This intense visualization of metamorphosis was crucial to Goethe as artist and as empirical scientist. It was something *real* that Goethe believed he was reconstructing, though that reality lay behind and between any actually existing instances.[28] Famously, Schiller insisted that what Goethe called an empirical observation was in fact a rational *idea* in the Kantian sense: "That is no experience, that is an idea!" Goethe was not amused, though he claimed to make light of it by saying he was delighted that he could *see* ideas: "Then I may rejoice that I have ideas without knowing it, and can even see them with my own eyes."[29] It was the beginning of an intense dialogue whereby Schiller gradually convinced Goethe to reformulate his insights in more Kantian language while Schiller himself came to appreciate the vivacity of Goethe's intuitive attunement to nature.[30] Schiller was close to the key circles at Jena propagating the philosophy of Kant, so that Goethe's discussions with him reverberated through wider circles at the University of Jena.[31] There, of course, Goethe already had important interlocutors in anatomy (Loder), botany (Batsch), and geology (Voigt), to say nothing of his administrative presence in the university.[32] It was, as Goethe recalled, "a long struggle" between "a cultivated Kantian" and his own "stubborn realism," occasioning "the gradual development of my aptitude for philosophy (insofar as such an aptitude lay in my nature)."[33] That parenthetical qualification is crucial, because in fact Goethe did not—*could not*—assimilate Kant wholesale into his "nature."[34] The residual difference was decisive for his own thought and for the epoch.

Goethe admitted that, up to the encounter with Schiller, "I had no sense for philosophy in the real meaning of the word." His encounter with Kant's *Kritik der reinen Vernunft* had been fruitless. He learned that he operated "with unconscious naiveté; I truly believed that my eyes beheld what my mind thought true."[35] And "then the *Critique of Judgment* fell into my hands," delighting Goethe because in this work "products of art and nature were dealt with alike. . . . The products of these two infinitely vast worlds were shown to exist for their own sake." Thus, "I was glad to find poetry and comparative science related so closely: both are subject to the same faculty of judgment." Goethe now "found my most disparate interests brought together," occasioning a "wonderful period . . . in my life."[36] The narrative here seems heartfelt, but the author was a master of irony, and we should not be too lulled by his ostensible enthusiasm. He gave fair warning: "occasionally something seemed to be missing" even if "the main ideas in the book were completely analogous to my earlier work and thought."[37] Analogy, as Kant would be the first to point out, is not identity; it always entails difference.[38] If, indeed, these ideas were "*completely* analogous," how could Goethe's earlier thought have labored under "unconscious *naiveté*?" And, how, indeed, could he discern "*something missing*"?

From his engagement with Kant, Goethe came away with a "passionate enthusias[m to] pursue my own paths."[39] Notwithstanding his ostensible conversion experience, "the what and how of my discoveries met with little approval among the Kantians." While "the Kantians . . . listened to me, [they] were unable to respond or help in any way." Thus, "one or the other of them would admit with a bemused smile: this is indeed an analogue to Kantian thought, but a peculiar one."[40] Schiller must be counted among these reluctant Kantians with whom Goethe could never come to accord. In his characterization of their famous conversation to his friend Johann Daniel Falk, Goethe made that clear: "For really from the first moment of our acquaintance we were never able to come to a complete agreement about a single point."[41] Goethe returned to Kant's text repeatedly, taking it up in his "peculiar" and "analogical" manner. "I slowly grew accustomed to a language which had been totally foreign to me . . . [since] it encouraged a higher level of thinking about art and science."[42]

Goethe was never completely converted to Kantianism, however fashionably he retooled his language, however firmly he allied himself with Schiller in the project of Weimar Classicism. We need to recur to that "something missing." Goethe balked when "the philosopher . . . asserts that no idea is fully congruent with experience." He elaborated, crucially: "Our intellect cannot think of something as united when the senses present it as separate, and thus the conflict between what is grasped as experience and what is formed as idea remains forever unresolved."[43] Goethe could not submit for long to such Kantian strictures, because "this difficulty in uniting idea and experience presents obstacles in all scientific research."[44]

In another essay, Goethe brought his ironic sense to exegesis of the Kantian corpus: "In seeking to penetrate Kant's philosophy, . . . I often got the impression that this good man had a roguishly ironic way of working: at times he seemed determined to put the narrowest limits on our ability to know things, and at times, with a casual gesture, he pointed beyond the limits he himself had set."[45] Registering Kant's distinction between the constitutive and the regulative, Goethe turned to the famous passage at §77 of the *Kritik der Urteilskraft* where Kant conjectured about the *intellectus archetypus*.[46] He understood the restrictive use Kant wished to make of the notion: "the author seems to point to divine reason." But Goethe argued that once the portal had been opened, even if only for our "practical use" in moral freedom, "why should it not also hold true in the intellectual area that through an intuitive perception of eternally creative nature we may become worthy of participating spiritually in its creative processes?"[47] Famously, Goethe claimed that, despite Kant's reservations, and inspired by Kant himself, "there was nothing further to prevent me from boldly embarking on this 'adventure of reason.'"[48] This is, in my view, a most dramatic instance of the fundamental pattern of the 1790s (and beyond) in the reception of Kant: namely, that his warning was ignored and the possibilities that had provoked it became the driving concerns of the age.[49] That set the tone for what Goethe called "the most important period, the final decade of the [eighteenth] century," when he shared the great post-Kantian adventure with "Fichte, Schelling, Hegel, the Humboldt brothers, and Schlegel."[50]

GOETHE'S MORPHOLOGICAL PARADIGM

One of the essential components of Goethe's thinking was the *type*, for which all actually existing instances of the organism were strictly *tokens*, none of them attaining to every aspect of the type, yet each incontrovertibly identifiable with it. This allowed Goethe to conceive of *species* as an empirical *actuality* behind and within every individual instance.[51] In this, Goethe embodied the impulse toward "objectivity" that Lorraine Daston and Peter Galison have discerned in the view of science that dominated the eighteenth and early nineteenth centuries, for which it was the goal of science not to note every idiosyncracy of the specific empirical observation but to see through the observation to the principle it expressed.[52] Goethe put it clearly: "Phenomena, which others of us may call facts, are certain and definite by nature, but often uncertain and fluctuating in appearance. . . . There are many empirical fractions which must be discarded if we are to arrive at a pure, constant phenomenon . . . [to] establish a type of ideal."[53] Thus, Goethe conceived the idea of *simultaneity* of form in the type.[54] "After observing a certain degree of constancy and consistency in phenomena, I derive an empirical law from my observation and expect to find it in later phenomena."[55]

A parallel insight, notably working from within aesthetics, is Kant's "aesthetic normal idea," articulated in the first part of his third *Critique*.[56] Working from such classic aesthetic models as "Myron's cow," Kant identified a capacity of the human mind to discriminate the typical and ideal form implicit in a myriad of instances, not via some extended mathematical computation reducing deviations to a mean, but as an immediate, complete, and compelling realization. This aspect of Kant, rather than Schiller's invocation of the Kantian *rational idea*—namely, that which exceeded in principle all intuitive schematization—better captures Goethe's insight.[57] Kant's "aesthetic normal idea" speaks both to Goethe's artistic practice and insight and to his notion of empirical practice in science as expressed in the notion of "aesthetic intuition [*anschauende Urteilskraft*]."[58] The mind was active, for Goethe, but it was imaginative and interpretive, not simply discursive, as Kant conceived it.[59] As Thomas Pfau concludes, "Goethe . . . seeks to distill—not *infer*—through a series of precise empirical observations and descriptions."[60] That is the "delicate empiricism [*zarte Empirie*]" that Goethe advocated.[61]

"A natural process . . . must be conceived of in idea as both simultaneous and sequential."[62] Goethe was not concerned merely with synchronous or simultaneous form but also and above all with *sequential* or developmental form, expressed initially in his notion of *metamorphosis*.[63] For Goethe, it was never the *particular* manifestations, no matter how many one observed, that defined a form but rather some principle *behind* these instantiations. He drew a crucial distinction between the German term *Gestalt*, which meant for him a discrete, static shape, and *Bildung*, which always contained within its connotative range not only *form* but the process of *formation*, a temporal-developmental course toward that form.[64] This teleology was *immanent*, not external. As Ronald Brady aptly summarizes: "The forms of a graded series have the peculiar property of appearing to be arrested stages—we might call them 'snapshots'—of continuous 'movement.'" Crucially, one must "begin . . . study of the series *from the progression itself* rather than from a single form."[65] Any transient state is derivative, an "arrested stage" of what is a continuous process of formation. Just as, synchronically, every instance exemplified its type, so, diachronically, each momentary manifestation exemplified its normal developmental path: the quest was for a principle or law of developmental form.

In the words of Dalia Nassar, "The archetypal plant is not a thing or a completed product, but productivity. Thus, it cannot be made equivalent with any one of its products."[66] In short, at its core we must recognize "the 'process' nature of Goethe's thought."[67] As Eckart Förster recapitulates Goethe's methodology, "we must follow a natural process completely, from beginning to end. Second, this process must then be held together, as it were, and viewed as a whole, as a single phenomenon."[68] As Goethe put it, "What higher synthesis is there than a living organism? Why would we submit ourselves to the torments of anatomy, physiology, and psychology if not to reach some concept of the whole?"[69] Förster articulates Goethe's insight thus: "I have to recreate in thought the formative forces that are active as it were *between* the leaves. What matters are the transitions, the *Übergänge*, not so much the formed parts or products, and these transitions I experience only in the observation of my own formative (*nachbildend*) thinking. . . . [I]t can be experienced only in thought, in the mental participation of the formative motions of the plant."[70] Förster cites Goethe's articulation: "When we are able to survey an object in every detail [*in all seinen Teilen*], grasp it correctly, and produce it again in our mind [*im Geiste wieder hervorbringen können*], we can say that we intuit it in a real and higher sense [*im eigentlichen und höhern Sinne*]."[71]

Goethe made plain that the sensibilities associated with the aesthetic had a crucial role to play in the imaginative and interpretive practice of empirical science.[72] He insisted: "if we would succeed, to some degree, to a living view of Nature, we must attempt to remain as active and as plastic as the example she sets for us."[73] As he wrote: "our full attention must be focused on the task of listening to nature to overhear the secret of her process."[74] In his famous reflection on the influence of Kant, Goethe summed all this up pointedly: "In the end, the phenomena must form a series, or rather, overlap; thus they give the scientist a picture of some organization by which the inner life of the phenomenon become[s] manifest as a whole."[75] This is simply to transgress Kant's transcendental strictures on human knowledge, and Goethe knew that. It is similarly to transgress the Kant-inspired insistence that "proper" scientific knowledge must proceed from parts to whole, in unidirectional temporality, the core maxim of mechanism.[76]

In Goethe's approach to life science two crucial notions converged: form or structure (*Bau*) and design or telos (*Plan*). The paradigm that Goethe propounded to his age, morphology, was the empirical science of reconstructing—from observations *and integration through and beyond observations*—the *Bauplan*, the principle behind the development in actual organic life-forms.[77] The essential point about life-forms, for Goethe, was their developmental directionality. This was lawful, but it was concrete. Each instantiation was unique yet utterly expressive of its type. Moreover, the developmental plan could itself change, altering the course of the whole sequence.[78] In this, biological regularities showed a decisive divergence from inorganic, physical regularities. Life changed its expressions; physical matter simply reiterated its manifestations with ceaseless uniformity.[79]

Inspired by Kant's positive notice of Blumenbach in the *Kritik der Urteilskraft*, Goethe returned to Blumenbach's work with new interest and through him came upon the whole line of argument about life-forces I have traced: "I discovered that Caspar Friedrich Wolff formed a link between Haller and Bonnet on one side and Blumenbach on the other."[80] Goethe immediately criticized the "materialism" of Wolff's *vis essentialis*. "This type of terminology proved untenable," he wrote. "Basically the word 'force' means something purely physical, even mechanical." That led him to contend (rather uncritically) that Blumenbach had "achieved the ultimate refinement" in "anthropomorphiz[ing] the phrasing" by substituting *Trieb* for *Kraft*.[81] For Goethe, neither the entrenched school of "evolution" (preformation) nor the increasingly assertive school of "epigenesis" had an adequate model for organismic development.[82] "I will go so far as to assert . . . we cannot grasp the unity and freedom of [an organism's] formative impulse [*Bildungstrieb*] without the concept of metamorphosis."[83] From the epigenesists, Goethe embraced the notion of developmental change. From the preformationists, he embraced the notion of lawful regularity. Morphology was intended by Goethe as what in Hegelian terms we might call an *Aufhebung*, a dialectical synthesis: the discernment of the lawful principle of formation that lay behind—constraining as much as enabling—the creative change of actual developmental paths in organic life. "Morphology should comprise the doctrine of the form and of the formation and transformation of organisms," or as he put it in the manuscript version, "morphology is mainly concerned with organic forms, their differences, their formation and transformation."[84]

Given that there were developmental regularities in the ontogeny of each individual organism that bespoke its species type, this insight could be extended, crucially, to the discernment that there were developmental regularities *across* species types. Here Goethe's thought converged with that of Kielmeyer, perhaps drawing on their common source in Herder and, beyond Herder, Buffon.[85] Such developmental regularities, it should be noted, did not, for Goethe or for most morphologists in his wake, entail explicitly the historical descent of one species from another, which the age termed "transformism."[86] Yet it would be doctrinaire to deny that there was—starting in Buffon, and much more energetically in Herder and Kielmeyer—a *propensity* toward historicization, in which sequence suggested genetic succession.[87] The "daring adventure of reason" that Kant had described in his *Kritik der Urteilskraft*, and that Goethe embraced notwithstanding Kant's reservations, had been couched in explicitly historical-genetic terms.[88]

GOETHE AND KIELMEYER

Very shortly after Kiemeyer's address was published in 1793, it came into Goethe's hands. Kielmeyer's thought remained very vivid in Goethe's consciousness, so that when the occasion presented itself some six years later in the course of a trip to Switzerland, Goethe stopped to visit with Kielmeyer in Tübingen. His diary entry describing their conversation, while compressed, is very indicative:

> Tübingen, the 10th September 1797: Early with Professor Kielmeyer, who visited with me, [went over] various topics concerning anatomy and physiology of organic nature. His program in relation to his lectures will very soon be in print. He laid several ideas out for me, concerning how he is inclined to connect the laws of organic nature to general physical laws, for example, polarity, the interactive tenor [*Stimmung*] and correlation of extremes, the power of diffusion of expansive fluids....

> Concerning the idea that in their development higher organic natures have advanced several stages, leaving the others behind. Concerning the important consideration of . . . simultaneous and successive development.[89]

The themes of this conversation touched upon crucial issues that were galvanizing the life sciences in the 1790s, and what became obvious through this brief encounter was a striking convergence of views. Goethe had been working for some time on his own ideas of type and metamorphosis with regard to both plants and vertebrates, and even earlier he had digested the implications of stratigraphical geology and paleontology. He could see the full implications of Kielmeyer's sketch, and like the rest of Germany, he was eager for the monograph in which Kielmeyer promised to flesh out his schemes. When that did not emerge, Goethe prevailed upon Kielmeyer to provide him with two personally annotated copies of student lecture notes from his zoology courses. Goethe returned to Kielmeyer's address in 1806 for a new reading, as he was preparing a comprehensive account of morphology with a view to publication.[90] He returned to Kielmeyer's lecture notes in 1813, as once again he prepared to publish his morphological paradigm.[91] Thus, Goethe fully recognized the epochal character of Kielmeyer's work and its dramatic convergence with his own.

Bersier suggests an even stronger case. She argues that Kielmeyer's address of 1793 occasioned a "turning point in Goethe's scientific development."[92] She elaborates: "The impact of Kielmeyer's physiological model on Goethe's scientific approach can best be ascertained if one juxtaposes his osteological essay comparing human and animal form written in 1790–91 with the works on comparative anatomy composed in 1795–96 and published in the morphological notebooks of 1820."[93] She maintains that "in the post-Kielmeyer essays . . . attention shifts from a structural to a morphological viewpoint, from the apparent bone shape to the inner physiological functions revealed through anatomical forms."[94] Goethe himself retrospectively attached great significance to the year 1795: "That year I hardly lost sight of anatomy and physiology."[95]

GOETHE AND SCHELLING: GOETHE'S PATRONAGE AND PARTICIPATION

Goethe made systematic breakthroughs in his paradigm of morphology in 1795–96, many of which he discussed with Kielmeyer in September 1797. Nevertheless, he published none of the key papers at the time. He was still deeply troubled by his lack of reception by the professional scientific community, but he never gave up wanting to convey his new vision for science, and he decided to reach out to a wider public via a genre in which he felt unquestioned competence, poetry. That led him to consider the question of how to convey theory in poetry, the generic issue of "didactic" poetry.

The grand exemplar of such poetry was the ancient work of Lucretius, *De rerum natura*. This poem was not just a *formal* model, however, for it contained the sort of comprehensive natural philosophy—cosmogony and theory of emergent life—that Goethe wanted to reformulate in modern terms.[96] Of course, Lucretius was associated with materialism, and the French celebration of this connection made the German reception problematic. "Interest in the poem in Germany did not reach its height until the last two decades of the [eighteenth] century."[97] Kant, for example, went to great lengths in the introduction to his *Allgemeine Naturgeschichte* (1755) to distance himself from such materialism.[98] He still repudiated it, explicitly, in his *Kritik der Urteilskraft* of 1790.[99] By contrast, not surprisingly, Herder found it attractive.[100] Goethe himself made no secret, in the late 1780s, of his sympathies for Lucretius and Spinoza, even if the sort of materialism that attracted him was of a more vitalistic turn, as in Herder, and not the dreary form he discerned in d'Holbach.[101] Thus, Lucretius became a central concern for Goethe in the last years of the century, especially since his close friend Karl Ludwig von Knebel was involved in the first full poetic translation of Lucretius into German.[102] Goethe "constantly discussed Lucretius" with Knebel in these years.[103]

The question remained: was didactic poetry possible in the modern age, and could it carry the burden of *modern* scientific ideas? The possibilities of didactic poetry became a major theoretical topic in his discussions of poetic artifice with Schiller, who expressed considerable reservations about the strength and flexibility of didactic poetry in the modern age.[104] Both were profoundly dissatisfied with Erasmus Darwin's efforts along these lines.[105] While Schiller did not rule it out altogether, he posed sufficient *formal* concerns, and Knebel was experiencing sufficient *actual* difficulties in his Lucretius project, that Goethe had to recognize that the challenge was formidable. His initial poetic recourse was to a genre that he had already used effectively, the elegy.[106] He formulated the key doctrines of his *Versuch die Metamorphose der Pflanzen zu erklären* in an elegy, incorporating all the key ideas of his morphological theory of plant formation into a love poem.[107] It was composed in June 1798 as his first experiment in didactic verse for the sake of his new vision of science. As poetry it proved more welcome than his essay had, though not among professional naturalists.

That was not, however, his last or grandest endeavor. Much earlier, when he first began to study earth science, Goethe came up with the notion of composing a "novel about the history of earth [*Roman über das Weltall*]."[108] The notion of a "novel" was itself ironic, for, as he well knew, the term "novel [*Roman*]" had been used disparagingly against Buffon's great work, *Époques de la nature* (1779), and more generally against *any* theoretical system-thinking.[109] To turn the tables on such shortsighted resistance to new ideas and present this material explicitly in a novel would be a suitable rejoinder, Goethe believed at the time.[110] He did not stay with that project very long, but his ideas for a grand literary synthesis as well as his sentiments against leaden scientific reception lingered. The new circumstances of the 1790s rekindled both Goethe's ambition and his animus. The result was his decision to undertake a didactic poem on the model of Lucretius to put forth his vision of a modern science, especially his morphology. As Nisbet contends, we can take his draft poem, *Metamorphose der Tieren* (1799?), as a component of this larger project of a Lucretian poem for the modern age.[111] While the large project quickly came to naught, Goethe's various drafts and fragments eventuated much later in a very important collection of his poems, *Gott und Welt* (1821–22).[112]

Strikingly, it was in immediate proximity to Goethe's composition of his elegy *Metamorphose der Pflanzen*, and in the context of Goethe's efforts to organize resources for the larger project, that Goethe received Schelling's new work, *Von der Weltseele* (1798), and they met for the first time. Their first meeting was highly orchestrated: "it was thanks to the effective staging of [Schelling's Württemberg] countrymen Schiller and Niethammer that a personal meeting was arranged in Jena on May 28, 1798, at Schiller's home."[113] Schelling had earlier hoped that with the publication *Ideen zu einer Philosophie der Natur* (1797), he would win Goethe's attention and an appointment to the University of Jena, but Goethe had his reservations about the book, which seemed to him too much caught up in transcendental philosophy, and not open enough to the objective grandeur of nature.[114] In his new work, *Von der Weltseele*, Schelling made a decided effort to address himself to Goethe's concerns with life science and even to cite from Goethe's *Versuch die Metamorphose der Pflanzen zu erklären* with approval. None of this was lost on Goethe. Still, at the meeting he wanted to test Schelling not just for his philosophical tenor but also for his politics. In the light of the fiasco over Johann Fichte (1762–1814), the last thing Goethe wanted at Jena was another "sans-culotte spouter [*Sansculotten Tournure*]." He was pleased to find the young man "a very clear, energetic mind, organized according to the latest fashion."[115] As a final initiation rite, Goethe invited Schelling to participate over the next several days in his optical experiments, and Schelling demonstrated the appropriate enthusiasm for Goethe's color theory and the equally important skepticism toward Newton.[116] Goethe reversed himself and recommended Schelling's appointment to Jena, which went forward without a hitch.[117] Strikingly, in his recommendation of Schelling for a position at Jena in his letter to Carl Voigt, Goethe expressed the hope that Schelling "would be introduced to experience and experimentation and an assiduous study of nature," that is, weaned away from whatever residual "transcendental philosophy" he still carried with him.[118]

Goethe sat down and read Schelling's *Von der Weltseele* on June 7–8, 1798.[119] Less than two weeks later (June 17–18), Goethe composed his elegy.[120] For the next two years, 1799–1800, their dialogue was intense and fruitful. Goethe read the *Erster Entwurf eines Systems der Naturphilosophie* (1799) in page proofs and helped Schelling revise the *Einleitung zu dem Entwurf eines Systems der Naturphilosophie* (1799), the bases for Schelling's course offerings at Jena and the defining texts of German *Naturphilosophie*. "In the fall of 1799 Goethe and Schelling met every day for a week, to go through and edit what came to be Schelling's most path-breaking work."[121] Schelling himself noted he was at Goethe's place "daily and had to read and work through my text on the philosophy of nature with him. What a growth of ideas these conversations were for me."[122] Goethe and Schelling reinforced each other in the development of their respective viewpoints. From the exchange each gained clarity and intensity of articulation.[123]

Many scholars, including most recently Dalia Nassar, believe there is a sharp break between Schelling's writings prior to his encounter with Goethe and after it, thus between the *Weltseele* text of 1798 and the *Erster Entwurf* and *Einleitung* of 1799. The key difference was in the theory of a self-fashioning nature, which Nassar argues Schelling derived from Goethe's notions of metamorphosis.[124] Thus, Nassar concludes: "The most significant difference in Schelling's conception of nature in the *Einleitung* is that nature is a self-producing organic whole. This means that nature has within itself a capacity which Schelling had previously only identified with the self. Radically, this implies that self-production is not limited to a self-conscious being—self-production is no longer identified with the act of reflection."[125] The transfer of agency from the mind to nature as well—and indeed more fundamentally—was the driving impulse in *Naturphilosophie*.[126] Nassar is arguing that Schelling got this notion from Goethe. They shared what was ultimately an *ontological* intuition: "Schelling, like Goethe, argues that metamorphosis is not simply a description of nature's development, but the very source or ground of nature, the archetype that underlies and constitutes nature's parts and their relations."[127]

Their period of maximal association (1799–1800) embraced Goethe's project of a modern Lucretian epic of nature. Goethe invited Schelling to join him in this very project, and when Goethe gave up on it, in 1800, he explicitly transferred it to Schelling.[128] If we are to find the point of inflection between Goethe's project and that of Schelling with *Naturphilosophie*, this might be the most pertinent locus. In 1901 Margarethe Plath already discerned this and, behind it, what drew the two figures together: "If we consider . . . Goethe's *God and World* and the writings of Schelling on *Naturphilosophie*, it would appear that the foundation of the world conception of both men found its classic expression in that poetic form of Spinozism that Herder, under the influence of Leibniz, created in his *God* and in the *Ideas for a Philosophy of the History of Mankind*."[129] Up to their meeting, Plath contends, Schelling did not draw that directly on Goethe's ideas in the formation of his vision of world order. Rather, they shared a set of common sources, most prominently, as she noted, Herder, but including Bruno, Spinoza, Hemsterhuis, Kielmeyer—"ultimately the whole natural-scientific–philosophical current" of the age.[130] What all these figures shared, with Herder at the forefront, was a rejection of Haller's interdict: "No created spirit can penetrate into the interior of nature."[131] Goethe would compose a poem for his crucial collection *Gott und Welt* (1820) contesting this idea:

> "Into the interior of nature"—oh, you philistine!—
> "No created spirit can penetrate"
> "Happy he who can know the mere outer shell!"
> I've heard that repeated for sixty years now.[132]

His deepest conviction was that nature hid no secrets from a trained sensibility.

Kant had drawn similarly sharp boundaries for natural inquiry, not unlike Haller. But Herder transgressed them inveterately. What Herder conveyed to Goethe and to Schelling (and, indeed, to Kielmeyer) was that "life, which for Kant represented an inexplicable boundary concept for a mechanistic explanation of nature, became the central concept of the universe."[133] All these figures adopted Herder's intuition that the forces in the physical world were all expressions of one organizing force, one integrating principle. "The whole of Nature should be regarded as a self-sustaining organism—even though that would be a totalizing step of precisely the kind which Kant's critical philosophy was designed to prohibit—a single, universal spiritual principle or 'world soul.'"[134] Kant repudiated this notion with a vengeance.[135] Thus, we must reckon that what drew Goethe and Schelling together was that "something missing" in Kant, something that Herder—especially via Kielmeyer—made central to the agenda of the epoch. Here lie some of the grounds for arguing that Herder must be taken to be as important as Kant in shaping the scientific horizon of 1800. Notwithstanding Schelling's deliberate suppression of the influence of Herder for the sake of authenticating his own "Kantianism," only these two streams from Kant and Herder combined could feed the enormous intellectual surge of *Naturphilosophie* in both its speculative and its empirical expression.[136]

As with Kielmeyer, and with hints already even in Herder, *magnetism* seemed to Goethe and to Schelling the most likely candidate for the primal principle in the natural world. Schelling had stressed magnetism and polarity in his 1797 work, *Ideen*.[137] Goethe began magnetic experiments in the wake of reading that work. The day before he began composing his elegy, Goethe wrote to Knebel about his hopes for writing "a poem concerning magnetic force."[138] Magnetism would henceforth constitute for him "an urphenomenon which one need only articulate to explicate it; thus it becomes the symbol for all else."[139] In the summer of 1798 Kielmeyer's student Karl von Eschenmayer (1768–1852) published a text entitled *Versuch, die Gesetze magnetischer Erscheinungen aus Sätzen der Naturmetaphysik, mithin a priori zu entwickeln*. This would be one of the texts upon which Schelling would draw for his own *Naturphilosophie* and, with Eschenmayer and others, build it into a movement. The text came to the attention of Goethe almost immediately upon its publication.[140] What attracted all these thinkers to magnetism was its *essentially* polar structure. Polarity had its literal seat, they believed, in magnetism. That made magnetism the *Urphänomen*, from which polarity extended by analogy into such realms as electricity, chemistry, organic formation, and sexuality. What is most striking about this intuition about magnetism and polarity as the *Urphänomen* is that it seemed thoroughly compatible with the utterly metaphysical conviction that the natural world had to be grasped as *alive*, that is, as agential and creative, and in just that sense *spirit*, not just inert matter.[141] The link concept was *force*.[142] The key shift was from product to process: from *natura naturata* to *natura naturans*. Spinozism, in Herder's vitalist reformulation, galvanized the age.

SCHELLING AND *NATURPHILOSOPHIE*

Friedrich Schelling was a philosophical prodigy, leaping to the forefront of German Idealism while still a student to become Johann Fichte's foremost disciple.[143] Yet almost as swiftly, he went beyond Fichte to pursue his own system building, demonstrating a mercurial character that persisted across his entire career. Schelling hopped from system to system with a half-life scarcely exceeding a half-decade. Thus, his monumental impact upon *Naturphilosophie*, a position that is all but identified with his name, came in a very brief interval in his meteoric career, spanning the years 1797–1806. By that last year, he had become exasperated with the idea, and he disengaged from it entirely, plunging first into philosophy of art, then surging on to philosophy of religion.[144] Thus, we need to assess both the enormous enthusiasm and the evanescence of Schelling's engagement with *Naturphilosophie* in these years, and why it struck such a powerful resonance in his age.

When Schelling published the first edition of his *Ideas for a Philosophy of Nature* in 1797, two crucial motivations appeared to drive him, as he affirmed explicitly in the second edition of 1803.[145] Not only did he propose to engage philosophically with the latest developments in the empirical natural sciences, but he also proposed to move beyond Kant's positions on the philosophy of natural science.[146] The need for a system of nature was ineluctable in human cognition: "to explain this necessity is a major problem of all philosophy."[147] Kant had made similar claims: first, in the "Transcendental Dialectic" of the *Kritik der reinen Vernunft*, concerning the need of reason to achieve systematicity as a regulative ideal; then, in the *Kritik der Urteilskraft*, concerning the need of empirical inquiry to establish a system of coherent laws as a matter of reflective judgment.[148] While Kant presented the first need as an authoritative *mandate* of reason, the second need appeared more as a requisite psychological *reassurance* to sustain inquiry faced with the inscrutability of the natural world.[149] In any event, Kant made clear that philosophy had a role vis-à-vis natural-scientific inquiry: the role of fostering, guiding, and evaluating the actual practice of that inquiry.[150] But it was clear to Kant and equally to Schelling that philosophy could not *replace* that empirical inquiry. Allegedly, Schelling disregarded empirical science to concoct "deductive" science from his armchair. In fact, he undertook extensive surveys of the latest scientific work in all his books on *Naturphilosophie*.[151] He was explicit, moreover, that *Naturphilosophie* was intended to complement empirical science, and he expressed enormous respect for experimental research, maintaining that all knowledge arose initially through experience. Schelling observed: "the philosophy of nature has nothing further to do than recognize the unconditionally empirical. . . . Empiricism extended to include unconditionedness is precisely philosophy of nature."[152]

There was more than epistemology in the Kantian impulse. His *Metaphysische Anfangsgründe der Naturwissenschaft* had provided not simply Kant's conception of "proper" science but an innovative *physical theory*—a *dynamic*, as opposed to *atomistic*, theory of matter and force.[153] This (quasi-empirical) theory of the "movable in space" in terms of attractive and repulsive forces captured the interest of naturalists in fields from rational mechanics through chemistry to physiology, as well as occasioning Schelling's own construction of "speculative physics," or *Naturphilosophie*. Almost all the interesting work on galvanism in the 1790s in Germany made use of Kant's "dynamic theory of matter" in trying to establish physical, chemical, and physiological facts about electricity. A dynamic approach offered the prospect for a theory of chemical bonding, either as elective affinity or more concretely as oxidation, and even for conceiving physiological forces or drives. This swiftly extended to magnetism, which seemed to many the most concrete—because intrinsically *polar*—phenomenon of nature, hence the potential ground or model for all the rest.[154] Ultimately, the dynamic approach might even achieve a unified theory of nature. For the most audacious researchers, and a fortiori for Schelling as a philosopher concerned primarily with this universal level, the issue was whether Kant's critical philosophy could accommodate these ambitions.[155]

While not replacing empirical science, "a philosophy of nature *ought* to deduce the possibility of Nature, that is of the all-inclusive world of experience, from first principles," Schelling affirmed.[156] Philosophy derived its ultimate warrant to undertake this grounding enterprise because empirical inquiry by itself could *never* "explain the possibility of a world system."[157] One could never achieve a whole by the aggregation of particulars through the enterprise of induction.[158] Instead, systemic unity had to be *prior* in the transcendental sense.[159] Yet Schelling was not satisfied with Kant's solution, the so-called "Copernican Revolution," for in his view all that Kant had achieved across the whole trajectory of the critical system was to affirm a *subjective* necessity for coherence or systematicity, one exclusively *for the subject*.[160] But what natural science aspired to, what human knowledge *needed*, was to find *objective* necessity in the order of nature.[161] To Kant's subjective coherence (and Fichte's in his wake), Schelling insisted there needed to be added an *objective* coherence, and he pointed with boldness to the "dogmatic" metaphysicians whom Kantians believed they had forever overthrown: Spinoza and Leibniz.[162]

As Schelling saw it, Kant made the valid point, in the *Kritik der Urteilskraft*, that "organization as such [i.e., purposiveness] is conceivable only in relation to a *mind*."[163] Yet Kant also recognized that the purposiveness of organisms was *intrinsic*.[164] That was the dilemma in which Kant left the prospect not only of understanding organisms but of developing a unified order of nature.[165] To resolve it, Schelling argued, philosophy had to go beyond Kant's "regulative" approach and retrieve an insight that Leibniz had already achieved: "I cannot think otherwise than that Leibniz understood by substantial form a mind *inhering in* and regulating the organized being."[166] Thus, in a sense radically different from Kant's, for Leibniz, too, "a *concept* lies at the basis of every organization."[167] Leibniz found a bridge across the divide between organic and inorganic nature that blocked Kant and all the other mechanistic philosophers of science. For Leibniz, "even in mere organized matter there is *life*, but a life of a more restricted kind."[168] Hence, Schelling asserted, "the time has come when his philosophy can be re-established."[169] That required abandoning the mechanistic natural science to which Kant remained committed. "Mechanism alone is far from being what constitutes Nature. For as soon as we enter the realm of *organic nature*, all mechanical linkage of cause and effect ceases for us. . . . The organic . . . produces *itself*, arises *out of itself*."[170] Kant was correct: "the *origin* of an organism, as such, can no more be explained mechanistically than the origin of matter itself."[171] But where Kant balked, Schelling resolved to carry forward.

From the marvel of particular organisms, Kant himself had been "reflectively" drawn to the judgment that all of nature needed to be thought purposive, as a *Technik der Natur* on the model of intelligent design, though without literally divine creation.[172] Schelling now urged that Kant had come to the threshold of the philosophical thought par excellence, without which neither transcendental philosophy nor empirical natural science could establish systemic coherence: "Nature should be Mind made visible, Mind the invisible Nature . . . [yielding] absolute identity of Mind *in us* and Nature *outside us*."[173] This was Schelling's great claim. "The system of Nature is at the same time the system of our mind. . . . But this system does not yet exist."[174] Spinoza had conceived the former system (as substance); Kant (and Fichte) had conceived the latter system (as subject). What remained was to bring them to synthesis. This was Schelling's project.

In his specific articulation of *Naturphilosophie*, Schelling's aim was "to reintegrate the transcendental 'I' into nature, to take it outside its self-sufficient noumenal realm and to show how its reason is the expression and manifestation of the rationality inherent in nature itself."[175] That is, his "*Naturphilosophie* attempted to know the fundamental forces of nature, the infinite productive powers of Spinoza's *natura naturans*."[176] Schelling offered "a synthesis of the vitalism of Leibniz with the monism of Spinoza."[177]

But his emphasis on *dynamism* transcended even their fixation upon substance. "A substance or thing is never as basic as an activity because it is only the result or product of it." In his full-blown *Naturphilosophie*, Schelling conceived of all nature as one living organism, a vast continuum of variously developed levels of organization and forces. But, "strictly speaking, . . . nature is not even an organism, since that would be again only the product of its activity (*natura naturata*). Instead, nature is nothing less than living activity or productivity itself (*natura naturans*)."[178] Thus, Spinoza's *natura naturans* "ceases to be dead and static but becomes alive and dynamic."[179] Here, the role of Herder proved substantial: "the younger generation followed Herder in *vitalizing* Spinoza's concept of substance, which now becomes nothing less than the single cosmic force."[180]

This is, indeed, *heady* metaphysics. But Schelling offered two justifications: first, metaphysics arose out of a blatant aporia in the very project of empirical science itself—its need for, yet incapacity to achieve, systematicity; and second, this metaphysics could succeed only if the *ideal* system it postulated could be *confirmed* in empirical inquiry. In the face of recalcitrant evidence, the philosophical system failed. This was a crucial component of the philosophy of nature as Schelling conceived of it. Moreover, such a conception of philosophy of nature entailed a revision of the philosophy of mind—indeed, of philosophy altogether: "Philosophy . . . is nothing other than a *natural history of our mind*. . . . We consider the system of our ideas, not in its being, but in its *becoming*. Philosophy becomes *genetic*."[181] If nature could be grasped only *in process*, as *becoming*, so too, knowledge could be grasped only as *learning*, as *development*. The dynamic of knowing and the dynamic of being had to be strictly *homologous*.

In its detailed exposition, the *Ideas for a Philosophy of Nature* (1797) offered an exploration of the relation between matter and force, as it had preoccupied early modern physics from Galileo, Descartes, and Newton all the way to Kant. The method of the work was to advance from empirical findings to some higher principle of order that informed their possibility. There was one overarching insight in *Ideas*, upon which all the subsequent system would be grounded: the principle of *contesting forces*. Nature "has admitted . . . no force which is not limited by an opposing one, and finds its continuance only in this conflict."[182] This insight was the key to the "great artifice of Nature."[183] Force and matter were *mutually constitutive*, Schelling contended: "neither force without matter nor matter without force can be conceived."[184] But, he went on, this should not be construed in the Kantian sense of explaining matter empirically from given forces working on given masses. Indeed, already in 1797 Schelling had slipped the mooring of Kant's "dynamic theory of matter" and proceeded to a "dynamical philosophy" that derived all the empirical forces—even Kant's—from something more primordial, a "higher principle." That was simply to see *process* (*natura naturans*) as the essential character of nature, and all *products* (*natura naturata*) as transitory states along its path. But process itself could propagate or develop only through contesting forces.

The most important dimension of empirical science for Schelling in 1797 was what has come to be called the "chemical revolution," especially in the highly contested form it assumed in Germany over the 1790s.[185] "The new system of chemistry . . . may very well develop into the universal system of nature," Schelling speculated, for it "spreads its influence ever more widely over the other branches of natural science."[186] Oxygen seemed more than an element; it seemed the key to all chemical process, a "higher principle" that ordered the whole physical world, especially when considered in alignment with the ethereal fluids—heat and light.[187] Taken in this sense, oxygen could serve as "a leading principle for the investigation of nature, as soon as this discovery ceases to be the exclusive possession of chemistry alone."[188]

But the *Ideas* left things at the level of "a series of individual discussions," because Schelling was not ready yet to bring them all together into a synthesis.[189] His "speculative physics" was just getting started.[190] In 1797 Schelling was not ready to deal with *organic* nature (or with the even grander prospect of unifying the organic with the inorganic into a unified order of nature). To be sure, he already had an *intuition* of unity, and even gestured to the ancient name for it, the *world soul*, in 1797.[191] Yet it would only be in his new work of 1798, which bore this term in its title, that Schelling would make the decisive step forward.

The question is: how over the brief span from 1797 to 1798 did Schelling achieve the new position articulated in *Von der Weltseele: Eine Hypothese der höheren Physik zur Erklärung des allgemeinen Organismus* (1798)?[192] From the internal evidence, he was prompted by the confrontation of two rival positions in life science of the day: *chemical physiology*, propounded most aggressively by Johann Christian Reil, versus the various formulations of *Lebenskraft*, most prominently that of Joachim Brandis. Two texts of 1795 above all embodied this opposition for Schelling: Brandis's *Versuch über die Lebenskraft*, advocating the idea of *Lebenskraft*, and Reil's *Von der Lebenskraft*, urging its displacement by a chemical physiology.[193] Schelling's *Von der Weltseele* sought to transcend this opposition. But it had a further immediate catalyst: Christoph Girtanner's *Über das Kantische Prinzip für die Naturgeschichte* of 1796.[194]

It was necessary, Schelling contended, to get beyond "the opposition between mechanism and organism that has long enough obstructed the progress of natural science."[195] The impasse had arisen because of the ubiquitous presumption that organism must be explained by mechanism, but, Schelling suggested, why not invert this conception? Why not begin from the vantage of the *more complex* and see the less complex as a restriction of the former? "Thus, in the end the *world* is—an *organization*, and a *universal organism* [is] itself the *condition for* (and in that measure the *positive in*) *mechanism*."[196] As he reduced it to a phrase in his later *Erster Entwurf*, "Nature implies a *universal* organism."[197]

Schelling proposed to begin from the result of 1797. The *Ideas* had demonstrated "that, in all nature, divided, really-opposed principles are active."[198] All nature was driven by "first principles of a universal dualism of nature."[199] The first force was *expansive* (positive); uncontested, it would sweep to infinity so instantaneously that it would leave no empirical traces.[200] Thus, it needed the complement of a *retarding* (negative) force to generate empirical manifestations, which alone constituted nature as experience, and thus enabled empirical natural science.[201] All particular entities had the *positive* force in common; it was *restriction* (the retarding force) that generated individuality and multiplicity.[202] "All multiplicity in the world arises originally through the distinctive *constraints* [*Schranken*] with which the positive [force] operates." In that sense, "the positive force first awakens the negative."[203] Together, the expansive and the retarding forces constituted a system, a whole world, which the ancients identified imagistically (*in dichterischen Vorstellungen*) as a world soul.[204] Thus, "it is the first principle of a philosophical doctrine of science [*philosophischer Naturlehre*] *to set out to grasp all of nature on the basis of polarity and dualism*."[205] Schelling elaborated: "The same dynamic sequence of stages prevails in universal and anorganic nature as in organic nature."[206] "If the origin of the organism is one with the origin of nature itself, then it is evident a priori that in anorganic or, rather, in universal Nature, something analogous must become evident."[207] The project would be to establish "that *one and the same principle* binds inorganic and organic nature."[208]

In book 2 of his *Von der Weltseele*, "On the Origin of Universal Organism," Schelling's objective was twofold: first, to discriminate the character of organism or life itself; and, second, to link this characterization of organism with a parallel characterization of inorganic nature to form a total system of nature based on common principles. There were only three strategies to explain the character of living things, he contended. First, one might try to explain them from a *material* principle: the approach of *chemical phys-*

iology. This view maintained that "all functions of the organism follow from chemical laws of matter, and life itself is a *chemical process*."[209] For Schelling, this was physiological *materialism*, with all the anxious associations that word conjured up in European culture.[210] The "chief defender of this chemical perspective," in Schelling's estimation, was Johann Christian Reil.[211] Schelling left little doubt of his scorn for the position Reil embraced. Still, Schelling struggled to give the chemical physiologists their due: "One must allow the chemical physiologists the fame of having been the first, though with obscure awareness, to have raised themselves above mechanical physiology and at least to have progressed as far as they could with their dead chemistry."[212] In *Von der Weltseele*, Schelling appeared even more convinced of the importance of the "new chemistry" than he had been in 1797, yet he remained highly critical. Schelling targeted for criticism not only Reil but also Girtanner, who in his first major publication had argued that irritability could be explained by the action of oxygen.[213] Schelling explicitly criticized Girtanner's thesis.[214] While oxygen clearly played a role, it was only the *negative* factor, he urged. It needed to be complemented by a positive force to achieve the appropriate synthesis. Schelling suggested this was the thrust of Christoph Pfaff's theory of "animal electricity."[215] Schelling also cited Alexander von Humboldt's experiments with irritability and electrical stimulus, claiming that Humboldt had rescued from the criticisms of Alessandro Volta all that was valuable for physiology in Galvani's work.[216]

All that empirical science had developed thus far, he claimed, not only in "mechanical" but even in "chemical" physiology, were the *negative* conditions of life, not its positive principle.[217] By seeking for this positive principle, Schelling observed, "defenders of life-force . . . are far more advanced than the chemical physiologists."[218] This second approach resorted to an *immaterial* principle, some vital force (*Lebenskraft*) beyond mere matter. Schelling made the same point in his *Erster Entwurf*: "advocates of vital force . . . to the extent that they persistently view life as something *sublime, beyond the chemical*[,] . . . infinitely tower over the chemical physiologists."[219] Yet generally Schelling proved as impatient with the life-force advocates as with the advocates of chemical physiology. The former flirted with something utterly *un*scientific, in his view. Already in the *Ideas*, Schelling called *Lebenskraft* "an altogether contradictory concept."[220] His argument in that text was that the postulation of any singular force, unopposed, could not account for the complexity of actual determinations. In addition, he harped on a performative contradiction: however *immaterial* its nature, advocates of *Lebenskraft* always proposed it in "the hope of allowing that principle to work according to physical laws."[221] Yet, as was notoriously the problem with all substance dualisms going back to Descartes, it remained inconceivable how the pure spirituality of the principle could work upon the strict materiality of the physical world. In *Von der Weltseele* Schelling mocked theories of *Lebenskraft* as the invocation of "a magical capacity that suspends, in living things, all the effects of the laws of nature."[222] It helped little to invent "dark qualities" and "unknown principles."[223] In the *Erster Entwurf*, he noted: "to accept a fantasy is good neither for physics nor philosophy."[224]

Schelling undertook to demonstrate that neither approach could, by itself, resolve the matter. A third, integrative or synthetic approach, based on the idea of polar forces in dualistic interaction, would be Schelling's own proposal. Somehow, Schelling had to bring together the *freedom* he discerned in life processes with the *lawfulness* he affirmed in nature, and especially the inorganic physical world.[225] The only concept, Schelling contended, that could support the dualistic process-thinking he proposed for his philosophy of nature was the term *Trieb*.[226] It was necessary to start with a *Bildungstrieb* that pertained to nature as a whole. *Naturtriebe* were prior to and higher than the chemical, metabolical processes that instantiated them in particular organisms. "In organic matter an original formative drive [*ein ursprünglicher Bildungstrieb*] is at work, through which alone it is able to assume a particular form, to maintain it, and ever after to reproduce [*wiederherstelle*] it."[227]

His prime contention was that "the *purposeful formation* of animal matter . . . can be explained only by a principle that lies beyond the sphere of chemical process."[228] It would never be possible to explain *life* by chemical processes alone; rather, chemical processes themselves might best be understood from the vantage of the higher-order process of *life*. Thus, Schelling insisted, it was "time to leave behind dead concepts."[229] Chemical processes should be theorized as "*incomplete processes of organization*."[230] From that "higher" vantage, it could be observed that "the universal formative drive [*Bildungstrieb*] of nature finally *dies out* in dead products [*in todten Produkten erstirbt*]."[231] Living organisms *mirrored*, for Schelling, a far vaster, universal life—*Bildungstrieb*—in nature altogether, of which chemical process was simply one, restrictive expression. Thus, Schelling undertook to lift the term *Bildungstrieb* out of its specific usage and render it a "higher principle" in his *Naturphilosophie*.[232]

Schelling celebrated Blumenbach for having developed the concept, for it opened a path "beyond the limits of mechanical philosophy of nature."[233] But Schelling was perfectly clear that he had appropriated Blumenbach's notion and that the latter would not have been comfortable with the dramatic extension Schelling was introducing.[234] Blumenbach did not recognize that his term was "only a [form of] *expression* [*Ausdruck*], . . . not, however, a *basis of explanation* [*Erklärungsgrund*]." It was a concept alien to natural science and could not be used constitutively but only as "a pathway for investigative reason [*Schlagbaum für die forschende Vernunft*]," that is, merely as a reflective heuristic, in Kantian terms.[235] Moreover, in Blumenbach's own usage "this concept presupposes organic material already, for this drive should and can only take effect in organic material. This principle therefore cannot represent a *cause* of organization, *far more this concept of the formative drive itself presupposes a higher cause of organization*."[236] The concept stood only as a placeholder for a yet-to-be-discerned causal principle. Ultimately, the source of organism was not even to be found in organic material itself but had to be traced infinitely higher into a principle that transcended the organic as such.[237] For Schelling, both Kant and Blumenbach were stymied at this point. He took it that Kant perceived Blumenbach's results as a *boundary* for natural science.[238] But Schelling affirmed: "I am fully convinced that it is possible to explain the natural process of organization from natural principles."[239]

While Kant, like Blumenbach, discriminated terminologically between *Bildungskraft* and *Bildungstrieb*, neither fully explicated that verbal distinction. "The question is: how does the universal formative force of matter carry over [*übergehe*] into formative drive?"[240] Schelling found this crucial. "The formative *force* turns into formative *drive* as soon as to the dead effect of the former something contingent, like the disturbing influence of a foreign principle, is added."[241] Thus, the essence of life appeared to be a "free play of forces . . . under an external influence."[242] But the trick, Schelling insisted, was to see this influence *not* as external but rather as *immanent*, which was the sense that *Trieb* added. "The principle of life does not enter organic material from without (as through an infusion)—an uninspired yet widely shared view—but just the opposite: this principle has *formed* the organic material *in itself* [*dieses Princip hat sich die organische Materie angebildet*]."[243] Thus, Schelling concluded, there were no distinctively organic forces but only the organic instantiation of material forces at the behest of a higher principle, the "free play of nature."[244] Schelling asserted a "universal formative drive of nature"—*drive*, not *force*, because it was *alive* (agential), not *dead* (inertial).[245] The proper term for this, the ancients proposed, was "world soul."[246] It was the principle that *drove* all matter, *used* all the "dead forces" *for purposes of its own*.[247] It constituted "a general continuity of all natural causes."[248] With the term *Bildungstrieb* Schelling moved closer to the ideas of vital force (*Lebenskraft*) that were circulating in German discourse at the time.[249]

Still, this was strictly a philosophical claim, an "a priori" construction to which he had been led; Schelling was not sure that he could demonstrate it empirically.[250] His general idea was that the specific features of the organic world mirrored the ultimate form of nature as a whole. Schelling turned, accordingly, to a consideration of the various empirical forces that had been discerned in organic life, starting with Haller's key forces of irritability and sensibility. "It would be at least *one* step toward such an [encompassing] explanation," he suggested, "were one to be able to show that it is one and the same organization that has formed for itself, by gradual development, the hierarchy of all organisms [*organischen Wesen*]."[251] Carl Friedrich Kielmeyer had suggested such a theory in his great lecture of 1793. The core insight moved from the *hierarchy* of forces in organic life to a *continuity* among them and on to the decisive idea of a *single force*, one principle of nature that caused all life.[252] Schelling proclaimed that Kielmeyer "without a doubt has opened a completely new epoch in natural history."[253]

One might permit oneself the judgment that Schelling's thought in *Von der Weltseele* seems to have circled entirely back to the crossfire between Stahl and Leibniz, if in terms derived from Haller and Reimarus. Nonetheless, this publication achieved what the publication of *Ideas* had not; it won the enthusiastic approval of Goethe and secured Schelling a professorship at the University of Jena. His next publication, accordingly, proved to be a textbook for his new course on *Naturphilosophie*. The initial concern of the *Erster Entwurf eines Systems der Naturphilosophie* (1799) was to relate the project of *Naturphilosophie* to the transcendental philosophy of Kant and Fichte, especially for a student population at the University of Jena steeped in Kantianism and for whom Fichte had been a charismatic teacher for a number of years.[254] Schelling wished to argue that philosophy of nature was a separate but parallel enterprise relative to transcendental philosophy, yet like the latter a *philosophical* project, a matter of logical construction. Thus, Schelling argued that philosophy of nature sought "the point from which nature can be posited into *becoming*."[255] The key notion here was *positing*: a philosophical undertaking. Thus, Schelling urged, one must assert "the first *postulate* of all philosophy of nature," namely, the movement of nature into becoming.[256] Nature was essentially "absolute productivity."[257] The importance of *productivity* in Schelling's approach (*natura naturans* rather than *natura naturata*) is already familiar. The important new term here is "absolute": the talisman of Schelling's distinctive idealism. By "absolute" Schelling meant here what was *aboriginal* in nature: "we have to think the most original state of Nature as a state of universal identity and homogeneity (as a universal sleep of Nature, so to speak)."[258] That presented the "supreme problem of the philosophy of nature: What cause brings forth the first duplicity . . . out of the universal identity of Nature?"[259] For transcendental philosophy "everything that exists is a construction of spirit."[260] Moreover, "every science that is a *science* at all has its unconditioned."[261] For philosophy of nature, the unconditioned principle could only be "*the constructing itself*."[262] Thus, "*being itself* is only *activity*[,] . . . nothing other than a *continually operative natural activity*."[263]

Accordingly, from the vantage of philosophy of nature, "we do not know *nature as product*. We know Nature only as *active*. . . . To philosophize about nature means to heave it out of the dead mechanism to which it seems predisposed, to quicken it with freedom and to set it into its own free development."[264] Thus, "for us the product must disappear behind the productivity."[265] "Nature gives itself its sphere of activity[,] . . . all of its laws are immanent, or Nature is its own legislator (autonomy of nature)." Nature, hence, is "a whole, self-organizing, and organized by itself."[266] In that light, "the chief problem of the philosophy of nature is not to explain the *active* in Nature, but the *resting, permanent*."[267] The solution was to recognize that a *product* "is only *apparently individual*."[268] Thus, "every product that now appears *fixed* in Nature would exist only for a moment, gripped in continuous evolution, always changeable."[269] This evolution left as its empirical trace the "continuity of the dynamic graduated series of stages of Nature."[270] This made the "fundamental task of all natural philosophy: to *derive the dynamic graduated sequence of stages of Nature*."[271]

To conceive of nature in its absolute productivity as the dynamic sequence of stages of nature was to think of the "formative drive" as a universal principle of individuation, a process instantiated across a myriad of transitory products. For Schelling the language of a universal "formative drive" was "the most genuine designation that was possible for the state of physics at the time."[272] In the concept were combined notions of freedom and of lawfulness.[273] "Nature is only one activity. . . . Through the individual products it seeks to present just *one*—the absolute product."[274] Hence, "all individual products of Nature can only be seen as abortive attempts to represent the Absolute."[275] Still, as Schelling would put it in his *Einleitung zu dem Entwurf eines Systems der Naturphilosophie*, "every individual is an expression of the whole of Nature."[276] And "in Nature, in so far as it is real, there can be no more productivity without a product than a product without productivity."[277] Concretely, "organisms overall are to be seen as only *one* organism inhibited at various stages of development."[278]

The question latent in a hierarchy of organic forms, of "stages" in the dynamic of organic nature, is ultimately whether this is strictly *ideal* (a *logical* construction) or also *actual* (i.e., an empirically traceable *phenomenon*—a *historical* development, or evolution in the modern sense). This was the fundamental issue of the "daring adventure of reason," in Kant's famous phrase.[279] What was the relation, in Schelling's view, between "logical construction" and empirical reconstruction? Schelling observed: "One must not allow oneself to be led astray by the appearance of a lack of continuity [in the empirical record]. These interruptions of Nature's stages only exist with respect to the products, for reflection, not with respect to the productivity, for intuition. The productivity of Nature is absolute continuity. For this reason we will present that graduated view of organisms not mechanically, but rather dynamically: that is, not as a graduated series of products, but as a graduated series of productivity. *It is but one product that lives in all products*."[280] The crucial discrimination is between *reflection*, which is a posteriori and inductive, working only from observed *products*, and *intuition*, which grasps the whole pattern and process, the *productivity* behind the instances, concerned not with the discrete but with the continuous. Where did Schelling stand on this, vis-à-vis Kant? Schelling explored the stages of development in organic nature under important rubrics with which we have become very familiar: epigenesis and natural history. Thus, he followed Kant very closely—or, rather, the synthesis of Kant with Blumenbach that had just been elaborated by Girtanner in *Über das Kantische Prinzip für die Naturgeschichte*. Schelling began by affirming "the general principle that no individual preformation, but only *dynamic* preformation exists in organic nature[;] . . . organic formation [*Bildung*] is not evolution, but the epigenesis of individual parts.—Various organs, parts, etc., signify nothing but different *directions* [*Richtungen*] of the formative drive; these directions are predetermined, but the individual parts themselves are not."[281] The language of "directions of the formative drive" stemmed originally from Blumenbach but played a far more elaborate role in the fusion of Blumenbach with Kant by Girtanner.[282] There are strong indications throughout Schelling's text that this was the immediate source of his commentary.

Schelling adopted the whole line of thought that carried from Buffon's definition of species through Kant's elaboration of *Keime* and *natürliche Anlagen* to Blumenbach's *Bildungstrieb* and Girtanner's synthesis. "Since in *natural history* (in the authentic sense of the word [!]), . . . one must assume that in the first individuals of each species [determinate] directions of the formative drive were not yet indicated, for otherwise they would not have been *free*[,] . . . every first individual of its species, although it would incompletely express the concept of its genus, would have been itself again *genus* in relation to the individuals produced later."[283] The formative drive was *free* in the stem organism, with respect to those directions, "because they were *all equally possible*." Thus, when any *one* got expressed, this required an *additional* triggering factor: "an external influence . . . in order to determine the organism toward one of these directions."[284] This was the role of *environment*—over extended time spans—in bringing about the expression of all the original potentialities in the stem genus in a set of determinate subspecies (or "races," as Kant called them).[285]

Schelling observed that neither environmental determinism nor complete organic plasticity could account for the diversity *and* the regularity of organismic forms. Only concrete interaction of determinate environment with determinate potentialities triggered the specific expressions of the organism. He concluded: "Now that which is *developed* (but not, on that account *brought forth*), through external influence is called *germ* [*Keim*] or *natural predisposition* [*natürliche Anlage*]. Those determinations of the formative drive . . . are able to be presented as *original natural predispositions* or *germs* . . . all united in the primal individual . . . [and] the prior development of the one makes the development of the others impossible."[286] This is an excellent distillation of Girtanner's exposition of the Blumenbach-Kant synthesis of natural history. Schelling replaced Kant's "generic preformation" with his own term, "dynamic preformation."[287] Following Blumenbach, Schelling rejected *individual* preformation. While at the origin of each species there was a "multiplicity of tendencies" *in potentia*, there were no actual preexistent forms in miniature, no "*preformed seeds*, for whose existence there is not a shadow of proof."[288] Empirically observable eggs or seeds were "themselves already products of the formative drive."[289] Thus, Schelling pronounced: "We are agreed with Blumenbach in that there is no individual preformation in organic Nature, but only a generic kind. . . . [T]here is no mechanical evolution, but only a dynamical one, and thus . . . only a dynamical preformation."[290] Thus, "*all formation occurs through epigenesis*"; the stages of development in nature were driven by the "directions in which the formative drive . . . operate[s]."[291]

All this, as with Kant and Blumenbach, had to do with development *within* a species or genus line. But what about the question of continuity *across* species? This was the ultimate question for natural history. "The leap from polyp to man appears gargantuan to be sure. . . . The polyp is the simplest animal and the stem, as it were, out of which all the other organisms have branched."[292] This was the direct issue Kant had addressed in his admonitions about the "daring adventure of reason." Schelling recapitulated the prospect: "The hope which so many natural scientists seem to have cherished—to be able to present the origin of all organisms as successive, and indeed as the gradual development of one and the same original organism."[293] Schelling asserted that his new philosophy of nature *transcended* this vantage, leaving it behind while saving its essential concern.[294] In a footnote, he elaborated: "All organisms, as different as they may be, are surely, in terms of their physical origins, only various stages of development of one and the same organism; they may be presented *as if* they had arisen through the inhibition of one and the same product at various stages of development. However, what holds for diverse organisms in terms of a *physical* origin, cannot hold good when transferred to the *historical* origin."[295] By "physical" in this passage, it would appear that Schelling referred to the material constitution of any given organism (its synchronous composition of different organizing forces and elements). By "historical," accordingly, he meant the diachronic continuity from one organism to another. This seems a complete *rejection* of "transformism" as an empirical-historical theory of organismic development. What was Schelling's reason? "The assumption that different organisms have really formed themselves from one another through gradual development is a misunderstanding of an idea which actually does lie in reason."[296] That is, "the productivity was thus one, but not the product. It was just not *one* already fixed and present *product* that developed itself into various organisms."[297] Accordingly, as Schilling put it in his *Introduction to the Outline*, "a true system of natural history, which has for its object not the *products* of Nature, but *Nature itself*, follows the *one* productivity."[298]

To establish an *empirical* science of natural history along developmental lines, Schelling argued, it would be necessary to advance far beyond comparative anatomy into "comparative physiology, . . . a science not yet attempted."[299] Were this to be carried through, "what was *formerly* called *natural history* would be raised to a *system of Nature*."[300] Schelling understood exactly what Kant had envisioned along these lines: "*Natural history* has been, until now, really the *description of Nature*, as Kant has very correctly remarked. He himself uses the name 'natural history' for a particular branch of natural science, namely, the knowledge of the gradual alterations which the various organisms of the Earth have suffered through the influence of external nature, through migrations from one climate to another, and so forth."[301] To establish empirically "a dynamically graded series of stages" would give natural history "a higher meaning, for then there would actually be a *history* of Nature itself."[302]

For all Schelling's seeming Kantianism here, he was *not* denying the empirical-scientific project. In the preface to his *Von der Weltseele* Schelling stood forth vigorously against the Kantian prohibition: "It is an old folly to believe that organization and life cannot be explained through natural principles. If that is as much as to say that the *first* origin of organic nature cannot be investigated *physically*, then this *unproven* claim accomplishes nothing more than to crush the ambition of the researcher."[303] The arguments about discontinuity in observed organisms, central for both Kant and Blumenbach, Schelling dismissed as insufficient: "That our experience has shown us no transformations of nature, no transitions from one genus or species to another, . . . is no proof against this possibility, for . . . the changes that organic nature, as much as inorganic, undergo can . . . take place in ever longer periods of time, for which our short duration . . . offers no proper measure. . . . [They] are so vast that up till now no [human] experience has been able to encompass their full course."[304] Moreover, despite this difficulty, empirical life science *had* made some steps in the direction of historical reconstruction. In a memorable footnote to the *Erster Entwurf*, Schelling traced the crucial line of this theorizing: "The idea of a comparative physiology is already found in Blumenbach's *Specimen phisiologiae comparatae inter animalia calidi et frigidi sanguinis*, and further explicated in the discourse on the relations of the organic forces by Mr. Kielmeyer, whose major idea is taken from Herder's *Ideas for the Philosophy of the History of Humanity*, first part, pp. 117–126; namely, that in the series of organisms, sensibility is displaced by irritability, and as Blumenbach and Soemmerring have proven, by the force of reproduction."[305] Schelling discerned as the key to a history of nature the incorporation of Kielmeyer's theory of the development from one organic force to another in the hierarchy of organisms. In addition, Schelling went on to point out: "Blumenbach and Soemmerring have proven that only those parts that are independent of the brain, and *all* parts only of such animals as do not even have a brain at all, or a very imperfect one, *regenerate themselves*."[306] Earlier, Schelling had affirmed Soemmerring's "law" of the relation between brain mass and sensibility. To repeat the essential point, Schelling proclaimed himself "fully convinced that it is possible to explain the natural process of organization from natural principles."[307]

What Schelling in fact suggested was that the possibility of a scientific history of nature lay in a more complex conceptualization of physiology: not only comparative but *developmental*. "The individual [organism] is only a visible expression of a determinate proportion between sensibility, irritability, and force of [re]production."[308] This allowed for Schelling's culminating synthesis regarding the dynamic stages of nature expressed in organic form:

> If there is a gradation of forces in the organism, if sensibility presents itself in irritability, if irritability presents itself in the force of reproduction, and if the lower force is only the phenomenon of the higher, *then there will be as many stages* of organization in Nature overall as there are various stages of the appearance of that single force. . . .
>
> THEREFORE, THERE IS ONE ORGANISM THAT IS GRADUALLY ATTENUATED THROUGH ALL OF THE STAGES DOWN TO THE PLANTS, AND ONE CAUSE ACTING UNINTERRUPTEDLY WHICH FADES FROM THE SENSIBILITY OF THE FIRST (i.e., highest) ANIMAL DOWN TO THE REPRODUCTIVE FORCE OF THE LAST (i.e., lowest) PLANT.
>
> . . . [N]ow we have *a unity of* FORCE of production throughout the whole of organic nature. It is indeed not one product, but still ONE force, that we observe to be inhibited at various stages of appearance.[309]

While this was assuredly an *ideal*, a *philosophical* construction, it could also serve as the basis for (and stimulate) empirical theory in the life sciences, especially as it directly linked to Kielmeyer's model of natural history and the medical reception of Brownian medicine. That was what made Schelling so important for the scientific community around him. Schelling's reception needs to be situated between that of Kant and that of Goethe. In practicing their science, Robert Richards observes, "after Kant, and especially because of the influence of Goethe and Schelling, biologists came to hold the teleological structure of nature not simply *as if* but as intrinsic. . . . [T]hey conceived nature in a Spinozistic fashion—it was *Deus sive natura*"[310] Thus, "Schelling and Goethe—and those biologists following their lead—[believed] that if archetypes proved a necessary methodological assumption . . . then there was no reason . . . to argue that nature was not intrinsically archetypal." Goethe himself embraced a "Schellingian Spinozism: God, nature, and intellect are one."[311]

CHAPTER ELEVEN

Naturphilosophie and Physiology

> The history of science needs to cast off the legacy of positivism—especially that lurking under Kantian guise—and to realize that Naturphilosophie was nothing less than the normal science of its day.
>
> BEISER, *German Idealism*[1]

In 1799 Schelling moved from relative marginality to the cockpit of the German intellectual universe, the professorship in philosophy at the University of Jena.[2] He was all of twenty-four years old! As he took up his post in Jena, all eyes, not all of them sympathetic, turned to Schelling. In the Jena *Allgemeine Litteratur Zeitung* (*ALZ*), his enemies made mockery of all he had produced. To defend himself, Schelling recruited key allies in the German scientific world—Johann Wilhelm Ritter, Henrik Steffens, Adolph Karl August von Eschenmayer, and Andreas Röschlaub—for a new journal devoted to his *Naturphilosophie: Zeitschrift für spekulative Physik*. There the first defense of Schelling's ideas would be published: Steffens's review of Schelling's latest works.[3] There, too, Eschenmayer would offer a provocative review of the whole theory, relating it to his own work, from which Schelling had explicitly drawn.[4] In short, from 1799 to 1800 *Naturphilosophie* became a *movement* in German thought.

The important question for the history of science is why Schelling's *Naturphilosophie* came so swiftly and so deeply to shape the mentality of the research community around 1800.[5] Naturalists were seeking something in Schelling that would be constructive for their own endeavors. The generation around 1800 saw in *Naturphilosophie* the way to "transform the results of natural-scientific research into a philosophical biology," in the words of Paul Diepgen. Through *Naturphilosophie*, "the interest of these men in the idea of development [could be] kept alive and promoted, but the goal and methodology arose from the more practical requirements for natural-scientific clarity in the results from dissection and the experimental bench."[6] German naturalists sought to grasp the implications of *philosophy of science* for their practices; more specifically, they sought rules for integrating concrete empirical results into a warranted theory. We can distinguish at least three levels of articulation: first, philosophy of *science,* as general epistemological warrant of scientific claims; second, philosophy of *nature,* as the *systemic unity of nature* in its empirical lawfulness; and finally, determinate *theoretical frames* in which specific research fields (*Fachgebiete*) formulated their experimental-observational results.

KANT, SCHELLING, AND THE "DARING ADVENTURE OF REASON"

The "essentially contested" issue between Kant and *Naturphilosophie* was the propriety of what Kant had styled a "daring adventure of reason."[7] His characterization reasserted all his reservations against Herder. Kant admonished those who would undertake the "daring adventure of reason" to observe his distinction between regulative and constitutive principles.[8] That did not happen.[9] If some of Kant's successors "read Kant through [a] Herderian lens, it was to misread him," Phillip Sloan argues.[10] Indeed, Goethe knowingly undertook this strong misreading in order to embark upon the "adventure of reason."[11] Even for Sloan, readers like Goethe "took from this 'adventure of reason' the warrant for drawing from Kant a program of developmental transcendental morphology and even a form of evolutionism."[12] I believe not only that they did so but that this was a scientifically and philosophically *fruitful* misreading. To suggest there might be *something* wrong with Kant distancing himself from the "adventure of reason" opens the way to reconsideration of the starkly hostile historical verdict that has found *everything* wrong with *Naturphilosophie*. The starting point must be with the reception of Kant's philosophy (of science), that is, how Kant's notion of "proper science [*eigentliche Wissenschaft*]" related to the actual practices of science—in particular, *life* science—in Germany at the turn from the eighteenth to the nineteenth century.[13]

In 1795 Samuel Thomas Soemmerring, the preeminent German anatomist of his generation, asked Kant for feedback on a little essay he was planning to publish, dealing with the vexed question of the "seat of the soul" in the body.[14] Soemmerring specialized in neurophysiology, and he proposed to advance the study of the brain by a more accurate assessment of the physiological interaction of the various nerve endings in the brain with the *fluid* in the cerebral cavities, to which he assigned a crucial function as the locus of integration of nerve impulses, a *sensorium commune*.[15] That was the bulk of his little volume, but it was supplemented by a speculative second part that sought to connect this physiological hypothesis to a metaphysical one: namely, that this *sensorium commune* was also, and essentially, the "organ of the soul." In proposing so direct a bodily situation of the soul, Soemmerring knew he was treading on highly controversial ground, which he designated "transcendental physiology."[16] It was concerning this metaphysical adventure that Soemmerring wished most specifically for Kant's comment. His admiration for Kant was flamboyant and flattering, and Kant found himself drawn for this reason among others to take up Soemmerring's invitation. The result was received with such enthusiasm by Soemmerring that he appended it to his text for publication, and indeed, he dedicated the whole work to Kant, "the pride of our age."[17] But it is clear from Kant's text—and from the lengthy drafts that he composed leading up to the version he finally sent to Soemmerring—that Kant was hardly disposed to endorse without reservation either Soemmerring's project in general or his particular hypothesis more narrowly. Yet how Kant responded reveals a great deal about his sense of the relation between the philosophical and the medical faculties, especially in this region of uncertainty between the body and the mind, and also about Kant's sense of his own competence as "someone not altogether unacquainted with natural science [*Naturkunde*]."[18]

From the far longer first draft of his response to Soemmerring we can detect a much sharper sense in Kant that this appeal by a total stranger for a comment on what was primarily a physiological research report might in fact entrap him in an adventure outside his bailiwick.[19] Georg Forster had already made Kant uncomfortable along these lines regarding his essays on race, and he was clearly not disposed to incur similar criticism, especially in connection with one of Forster's closest friends.[20] Thus, this invitation sparked Kant's sense that philosophy as a faculty found itself in a state of conflict not just with the theological or legal faculties but even with the medical faculty. In his response to Soemmerring already in 1795 Kant came to articulate his general idea of a *Streit der Facultäten*, which would be published only in 1798.[21]

Kant simply overrode Soemmerring's proposal about the need for a "transcendental physiology" as a hopelessly confused sense of the boundary in question. Physiology had nothing to say about the transcendental. Above all, the longstanding concern for a "seat of the soul" was a futile and contradictory misadventure of physics in metaphysics, a "subreption" that sought to materialize what was in essence immaterial, to spatialize (i.e., locate in "outer sense") what was accessible only in time (i.e., in "inner sense").[22] Thus, Kant dismissed *any* "transcendental" aspect to Soemmerring's project. Instead, in one draft he asserted the authority of the "critique of pure reason" to adjudicate the proper boundaries between empirical research and a priori knowledge. Then, however, Kant turned the tables, offering to supplement and confirm the strictly *physiological* elements in Soemmerring's study. If the latter had no warrant to meddle in metaphysics, Kant presumed the warrant, as "someone not unacquainted with natural science," to offer a *scientific hypothesis* concerning the core issue that had led Soemmerring to his misguided appeal to "transcendental physiology": namely, could a fluid be "organized" or "animated"? He proposed that the fluid in the cerebral cavity—which he took to be mere water—could not be in itself organized, because that required a stable purposive structure that was not consistent with the physics of liquids. However, he suggested that water need not be understood merely "mechanically" as extended mass, but it could also be understood "dynamically," in terms not only of recent "antiphlogistic chemistry," which had analyzed it into its two component gases, but also of the various theories of ethereal forces—light, heat, electricity, and so on. Thus, the seeming homogeneity of the fluid could accommodate all sorts of alterations of *qualitative* state; that is, the fluid could be transiently "organized" by the interjection of outside forces, yet retain its overall consistency, and return to its prior state when the stimulation dissipated. Were one to consider that these stimulations might be differentiated by the originating nerve-ending stimulus, then this fluid could harbor and transmit and in this sense aggregate and integrate nervous impulses and sense data and thus serve as a *sensorium commune* in a strictly material, physiological sense. It could then serve as a material substrate for the synthesis of intuitive consciousness. But this would not be a *literal materialization* or localization of the soul (*anima*), but only a *virtual* context for consideration by consciousness (*animus*), with no metaphysical stipulations about substance or interaction.

Setting these specifics in a larger context, in his pioneering essay on Kant's commentary on Soemmerring, Peter McLaughlin makes a decisive observation. He argues that Soemmerring suffered from a fundamental "misunderstanding of natural-scientific theory-construction as metaphysics." In Soemmerring, McLaughlin suggests, "a part of [medicine's] own field was transferred to metaphysics, because he confused the construction of theory or hypothesis with metaphysics. The part of natural science that was not simply descriptive he took as already transcendental." That indicated, McLaughlin rightly observed, "a remarkable uncertainty and incapacity [*Unbeholfenheit*] in the domain of theory."[23] But what McLaughlin identifies as the personal weakness of Soemmerring was, in fact, a characteristic "uncertainty and incapacity" of the whole faculty of medicine, for which Kant proved not a recourse but an exacerbation. To be sure, Kant was invoked—ubiquitously—but there was nothing unequivocal about that invocation. In fact, it is not at all clear that the generation of 1790–1820 could come to a clear determination of exactly what Kant's philosophy *meant*. As Brigitte Lohff so well states, "Kant's influence is more frequently assumed than specified," in the secondary literature on the period.[24] His own publications contributed to the confusion. In 1800 the Jäsche *Logic* appeared, offering to a wider readership Kant's general introduction to the field of philosophy as he had presented it to students at Königsberg University for a generation.[25] But this text was an amalgam of precritical and critical formulations, not a guide to the critical philosophy in its finished form. Almost simultaneously, three new editions of Kant's *Allgemeine Naturgeschichte* (orig. 1755) appeared in 1797, 1798, and 1808, to say nothing of its inclusion in several collected editions of Kant's work that were brought to market at the very end of the eighteenth century.[26] The *Allgemeine Naturgeschichte* was Kant's boldest effort toward a history of nature (cosmogony), but it also contained his most imaginative speculations about life on other planets, and more. Here "method" and "manner" were blurred in a way that hardly befitted Kant's "critically" policed natural science.[27] Yet it was widely and enthusiastically received. A striking bit of evidence in this regard is a letter that Wilhelm von Humboldt wrote to Friedrich Schiller, September 28, 1795, in reaction to Kant's commentary on Soemmerring: "This letter is extremely original and contains, besides a very well applied corrective of the peculiarity of searching for a seat of the soul, a hypothesis about how water might act upon the nerves, in which Kant appears exactly as he did in his theory of the heavens, and which has not been seen from him in the many years since then [*in der Kant ganz so, wie in seiner Theorie des Himmels erscheint, und wie man seit vielen Jahren ihn nicht wieder auftreten sah*]."[28]

Even the rigorously "critical" Kant seemed to point in a myriad of directions.[29] McLaughlin has made the very important point that the message of *Metaphysische Anfangsgründe der Naturwissenschaft* was not the same as the message of *Kritik der Urteilskraft*.[30] It was the so-called dynamic physics of the former, though *not* its rigid stipulation for "proper" science, that was most widely taken up by natural scientists of the 1790s.[31] With regard to *Kritik der Urteilskraft*, Kant's account of organism, as McLaughlin correctly notes, posed uncongenial objections to the inquiry into life-forces that was the driving concern of physiological investigation in the era.[32] Kant proved a bewilderment, not a panacea. That is why, as Dieter Henrich observed with harsh accuracy, Kantianism dissolved almost instantly into a host of "post-Kantianisms" over the decade of the 1790s.[33] The consequence of that historical ambiguity in "post-Kantianism," I contend, is that we have to be extremely careful in crediting so-called "transcendental *Naturforschung*" as authentically "Kantian" in any sense that would make Schelling (or Reinhold or Fichte or Hegel) *not* Kantian.[34]

The goal of physiology was not to describe any particular organism but rather to grasp the life process that it instantiated. Process needed to be theorized behind its observed particular objectifications. Simple observation was not sufficient; experiment presupposed hypothesis; evaluating experimental results always involved rational reconstruction, and the knower was always implicated in both construction and interpretation of the experimental system.[35] Results were always partial, even if the aspiration was to a totality. For the era after 1795 those who theorized about medicine as a science had to wrestle with Kant's "critical" characterization of the relationship between the rational and the empirical, between the a priori and the a posteriori, and his particular construal of the limits of human understanding concerning nature—including human nature—as they applied to medical knowledge. It was clear that physiology had to be an empirical science, but what would raise it to the level of valid knowledge? I agree with McLaughlin that we must discriminate between what a late eighteenth-century life scientist took as the object of inquiry and what Kant, as a philosopher of science, interjected at a metalevel. As McLaughlin notes, a life scientist *must* work with the actual, while a philosopher may well question the very possibility of access to it.[36] That is crucial to a reconstruction of what was happening in the life sciences of the epoch 1790–1820. Kant articulated a prescriptive notion of "proper science" that seemed to undermine, rather than enable, empirical science—particularly to the physiological and wider medical community. Lohff documents the bewilderment of a generation of thinkers grappling with Kant's wisdom. His rigorous stipulations of systematicity for authentic science left physiologists and medical professionals generally feeling inadequate—caught, in the words of one of the key spokesmen of the profession, "between system fever and clueless empiricism [*zwischen Systemsucht und zügelloser Empirie*]."[37]

The challenge of Kant's rigorous notion of "science" discomfited medical professionals concerning both their public authority and their fundamental mission: "Should the ultimate goal of the physician be therapy or natural research?"[38] If there were a way to restructure medicine into a "pure science," this potential dilemma would be averted. Thus, Kant "was explicitly and frequently invoked in the physiological texts," but it was not clear whether his point was that there could be no knowledge without experience or that the only authentic knowledge was a priori.[39] Was the "critical" philosophy *cautionary* for the practical pursuit of medical knowledge, testing it against ideals of systematicity and certainty, or was the "critical" philosophy—in Moses Mendelssohn's striking phrase—"*alles-zermalmende*"; that is, did it crush the very possibility of "scientific" medicine?[40] Again, it was not that natural scientists, philosophers, even theologians did not *want* to be Kantian: *everyone* did. The problem was that Kantianism could not be specified in a manner that they could first of all comprehend and thereupon integrate effectively into their practices. Thus, the massive three-volume effort of the theologian Carl Christian Schmid, an early Kant enthusiast, to instruct physiologists on how to do their science under the auspices of Kantian philosophy, while duly noted, had no substantive impact on the *Fachgebiet*.[41]

This gives us some insight into why Kantianism was so swiftly overshadowed by Schelling.[42] Neither Schelling's charisma nor his celebration of living nature by themselves could have made this possible.[43] It was rather that Kantianism itself proved starkly polyvalent.[44] Carl Friedrich Kielmeyer put it well, in his effort to distance himself from Schelling's *Naturphilosophie* in response to Cuvier's query and tacit accusation: not only was Schelling doing something original, but he was responding to *real problems* in Kant's own formulations.[45] Frederick Gregory, no enthusiast for this turn of events, identifies three factors: "that in Kant nature seemed somehow less real than mind, that Kant's scientific description of nature had to be restricted to mechanistic interaction alone, and [more generally] the confusion that reigned about the status of scientific theory and the relation of science to religion."[46]

To establish why life science turned from Kant to *Naturphilosophie*, consider Lohff's detailed reconstruction of the reception of Kant's philosophy by the German physiological community at the close of the eighteenth century.[47] That coincided with and exacerbated a "foundational crisis" concerning the scientific status of medicine.[48] Symptomatic was the immediate uptake of a work by the French ideologue Pierre Cabanis that questioned the "certainty" of medical knowledge.[49] This played into a domestic controversy launched in 1795 by an anonymous article, "Über die Medizin," in one of the key journals of the popular-philosophical Enlightenment, the

Teutsche Merkur.[50] The protagonist, "Arkesilas" (actually Johann Benjamin Erhard [1766–1827]), blasted the pretensions of medicine to science, denouncing "the 'uncertainty' of medical knowledge and its failure to measure up to the criteria of a philosophical *Wissenschaft*."[51] Arkesilas/Erhard invoked Kant for a philosophical standard of "proper" science.[52] He was answered by no lesser figure than Christoph Wilhelm Hufeland (1762–1836), the most important defender of the reputation of clinical medicine in the era.[53] The crisis threatened professional medical authority, and it swept up physicians into debates for the next decade and more. One response was the famous enthusiasm for Brownian medical theory in Germany. Within physiology it pushed even more forcefully toward a scientific identity grounded in experimental research, not practical outcomes.

Schelling's *Naturphilosophie* came to be taken up simultaneously with Kant's philosophy, and often as the resolution of the quandaries in which the latter had left interpreters.[54] What did *Naturphilosophie* have to offer researchers in life science? I take my cue from Schelling. "It is an old folly," he observed in 1798, "to believe that organization and life cannot be explained through natural principles. If that is as much as to say: the *first* origin of organic nature cannot be investigated *physically*, then this *unproven* claim accomplishes nothing more than to crush the ambition of the researcher."[55] This was the decisive point that linked his program in *Naturphilosophie* to the ambitions of emergent life science. And there is no question that the target of Schelling's criticism was Kant. He saw himself undertaking a philosophical rescue operation for a natural science in epistemological crisis, on the one hand, and on the cusp of theoretical breakthroughs, on the other. Lee Ann Hansen Le Roy captures this sense by evoking one of Schelling's earliest and strongest adherents, the young Danish research scholar Henrik Steffens (1773–1845): "As Steffens saw it at the time, Kant had made a philosophy of nature impossible. Schelling had saved the day."[56]

This is the decisive significance of an interview that took place in Jena in 1799 between the eminent "eclectic" professor of medicine Christoph Hufeland and the young Steffens, to discuss reviewing Schelling's new texts—from the *Von der Weltseele* of 1798 through the *Erster Entwurf* and the *Einleitung* of 1799—for the Jena *ALZ*, especially in the wake of Schelling's vehement protests over the hostile reviews that journal had published of his *Ideas* of 1797.[57] Hufeland represented the editorial board of the *ALZ*, which would remain utterly hostile to all aspects of Schelling's *Naturphilosophie* for the balance of its career; he was feeling Steffens out as a potential reviewer who would support the line of the journal, soi-disant "orthodox"

Kantianism.[58] Steffens reported the decisive moment in their conversation in his autobiography, *Was ich erlebte*:

> "You are still convinced," Hufeland said, "that one can go no further in the philosophy of nature than Kant has gone in his *Metaphysics of Natural Science*, on the one hand, and in his *Critique of Judgment*, on the other."
>
> "By no means," I answered very decisively, for I clearly perceived the intent of this question. "The boundaries that Kant saw as insurmountable, on the contrary, make a philosophy of nature impossible, and Schelling has rightly gone beyond them."[59]

That Schelling had addressed decisive aporia in the philosophy of science of Kant, which cried out for an alternative construction, was the crucial realization in the scientific community around 1800 that induced them to take up *Naturphilosophie*. Essentially, *Naturphilosophie* recentered rationality in nature as a whole, deriving human reason as a (preeminent) part of this larger whole.[60] It sought to reformulate the relation between matter and force in the physical sciences, to reanimate the physical world. *Emergence* and *process* became central to the idea of nature in itself. It became inherently creative. Self-production in nature moved from the simpler to the more complex; it took on historical-developmental form. Accordingly, science needed to shift its attention from a set of determinate *products* to the immanent *processes* that generated them—in philosophical terms, from *natura naturata* to *natura naturans*.[61]

Steffens noted in another passage of his autobiography that when he went to meet Schelling at the University of Jena in 1799, proposing to attend his lectures on *Naturphilosophie*, he was greeted by the younger man with real *joy*, for he was the first natural scientist who had approached Schelling in a positive fashion.[62] Thus, Steffens proved a pioneer in the reception of Schelling for natural science. His review of Schelling's works of 1798–99—not in the *ALZ* but in Schelling's new journal, *Zeitschrift für spekulative Physik* (1800)—represented the first major published appreciation of Schelling's system. Meanwhile others were drawn to Schelling's ideas: Goethe and Schiller, as we have already noted; also Johann Ritter, the young experimental physicist at Jena. Ritter was already deeply embroiled in his own research project linking galvanism to chemical and physiological processes and laying the groundwork for a theory of electromagnetism that Hans Christian Oersted and Michael Faraday would carry forward.[63] His interactions with Schelling from 1799 onward proved enriching for both of their programs.[64] From Tübingen, Eschenmayer had already reached out to Schelling in correspondence, sending him his pioneering dissertation of 1796 on a dynamic metaphysics of nature.[65] Schelling praised that work explicitly in his first book on *Naturphilosophie* in 1797.[66] A student of Kielmeyer's at the Karlsschule from 1783 to 1794, Eschenmayer published a set of crucial contributions of his own to the emergence of *Naturphilosophie*: first, *Sätze aus der Naturmetaphysik auf chemische und medizinische Gegenstände angewandt* (1797), a revised version of his dissertation; second, *Versuch, die Gesetze magnetischer Erscheinungen aus Sätzen der Naturmetaphysik, mithin a priori zu entwickeln* (1798); and finally, "Deduktion des lebenden Organism," in *Magazin zur Vervollkommnung der theoretischen und praktischen Heilkunde* (1799), the key journal of Brownian medicine edited by Andreas Röschlaub. Eschenmayer and Schelling would engage in a very important exchange of views on *Naturphilosophie* in the second volume of Schelling's *Zeitschrift für spekulative Physik* in 1801.[67]

Notably, the last segment of Eschenmayer's 1797 *Sätze* had taken up with enthusiasm the new Brownian theory of medicine elaborated by Röschlaub, and this became an important element in his exchange with Schelling in 1801.[68] Already by then Schelling was in direct interaction with Röschlaub, having discerned convergences between his own *Naturphilosophie* and Röschlaub's elaboration of Brownian "excitabilty [*Erregbarkeit*]." In correspondence and then in their published works, Schelling and Röschlaub elaborated that convergence.[69] This became the most important interaction between Schelling and the natural sciences, triggering what some have termed "Romantic medicine."[70]

HENRIK STEFFENS AND THE SCIENTIFIC EMBRACE OF *NATURPHILOSOPHIE*

Steffens stands at the head of this chain of developments. He had studied mineralogy and botany at the University of Copenhagen before continuing his studies at the University of Kiel from 1796 onward.[71] Already exposed, in Copenhagen, to the ideas of Kant, Herder, and Goethe, among others, as well as to Lavoisierian chemistry, at Kiel Steffens read Schelling's first two books of *Naturphilosophie* and experienced what he later characterized as virtually a religious conversion. In the fall of 1798 Steffens went to Jena to attend Fichte's lectures in philosophy. Thus, he was present when Schelling arrived to deliver the inaugural lectures on *Naturphilosophie* in 1799. Steffens attended these and became personally acquainted with Schelling, and they proved lifelong friends. Schelling became enthusiastic about Steffens's project of a dynamic geology, and he published in his journal in 1800 a preliminary sketch that Steffens had delivered to the Jena Natural History Society.[72] Moreover, Schelling urged him to go on to Freiberg to study with the great master of geognosy, Abraham Gottlob Werner, as Steffens did in 1801. In Freiberg Steffens then published his first major work, *Beyträge zur inneren Naturgeschichte der Erde*, aimed at the integration of chemistry and geognosy in a developmental history of the earth.[73] It was one of the first works of natural history explicitly to invoke Schelling's inspiration, though it was dedicated, significantly, to Schelling's patron and intellectual partner in those days, Goethe.

What in Schelling's work inspired Steffens? His review of 1800 is the clearest basis for establishing this. The two-part review commenced with the issue of the general relation between philosophy and empirical inquiry. Steffens took Schelling's point to be that experiments always required a framework from which to design inquiry and to appraise outcomes; thus, a prior theoretical vision was indispensable. Moreover, the goal of natural science was a general system of laws that would establish the *unity* of nature, and this system could simply never be reached from the accumulation of details.[74] It required conceptual work, which is what Schelling meant by speculation.[75] Schelling's whole project of a "speculative physics" needs to be grasped in terms of this complex dialectic between theory and experiment, process and product, particularity and theoretical integration. What empirical inquiry could generate would always be tentative hunches ("hypotheses"), but a true science would need a higher warrant. Ultimately, the goal of knowledge was *necessity*, the "incontrovertible stamp of truth," and such necessity was a philosophical, not an empirical, achievement.[76] Not only did such a view reflect the classic tradition of *scientia* in natural philosophy, but it was also the explicit standard Kant had enunciated in the critical philosophy, the most current and compelling philosophy of science on offer in Germany at the close of the eighteenth century.

Steffens noted that Schelling's position needed to be distinguished on two fronts from other positions of the day. First, within philosophy, transcendental philosophy (in Kant and Fichte) privileged the subject, but *Naturphilosophie* emphasized the self-constitution of objective nature.[77] Second, relative to empirical inquiry, Steffens noted that the physical sciences presumed "dead" nature as the foundation of the living, but this proved a fruitless approach.[78] Schelling's *Naturphilosophie* turned that upside down, envisioning nature as intrinsically productive (alive): "Nature is originally organic, that is, its products are productive."[79] Nature had to be understood as *process*, not simply as *product*. This was the essence of the Spinozist distinction of *natura naturans* from *natura naturata*. Yet Schelling recognized that without determinate articulation into products, this process character would be empirically inaccessible: absolute self-constitution would be instantly completed. Nature, as the universal positive force of productivity, required a negative, or "retarding," force to achieve concrete instantiations in its products.[80] Thus, for Schelling, all determinate products in nature were the results of the interaction of a positive and a negative force. From their interaction arose all concrete qualities in the physical world. Moreover, each such concrete actualization was itself intrinsically changing: every product itself began to produce novelties. Nature was always becoming, never static.

What in Kant was a specific, quasi-empirical matter theory became in Schelling's hands a universal theory of natural self-construction. In all nature, Steffens noted, Schelling saw development from the formless (fluid) to the structured (*Gestaltung*; solidity).[81] Thus, behind all "fixed form [*starre Gestaltung*]," Schelling evoked the fundamental forces that moved in, through, and beyond it.[82] Steffens explained that Schelling elaborated a theory of the interaction of the forces of gravity, magnetism, electricity, heat, and light, culminating in chemical processes and ultimately in organized life-forms. Organized life was not simply the end product, however; it was the originating principle of all nature, so that in organic life, nature was only returning to itself in a more elaborate form. Thus, "organic life is nothing other than concentrated nature itself."[83] All the processes of nature, in the organic as in the inorganic world, constituted a unity: "only *one* force courses through all of organized nature."[84] In his *Von der Weltseele*, Schelling proceeded to this analogy inductively from the results of the empirical sciences, but in the *Erster Entwurf*, he proceeded constructively. That is, he deduced life from the principle of productivity of nature itself.[85]

At the close of his review, Steffens stated that he would utilize these ideas in his own approach to earth science, by relating the forces of interaction on earth (magnetism and chemical process) to the influence of the sun (gravity and light), reciprocal influences that had led to the emergence of organic forms. He promised to present his findings in a monograph, his *Beyträge zur inneren Naturgeschichte der Erde*.[86] That work appeared in the following year, based not only on Schelling's ideas but on intense study with Werner in Freiberg to achieve a better understanding of geognosy.[87] In *Beyträge* Steffens endeavored to array all the earths into two series—the silaceous (*kieselige*) and the calcareous (*kalkige*).[88] In this way, he hoped to synthesize two fields of research—laboratory chemistry and natural history (specifically geognosy).[89] By the same token, he attempted to show that merely chemical analysis could never explain organic life.[90] Thus:

> Those materials that have been elevated to the level of organization are released from the laws of chemical relations during the course of life, just as the same materials in the course of chemical process are released from the laws of gravity; that higher potency of dynamic process in general loses itself in the lower chemical [level] with the moment of the disappearance of life, just as chemical material must fall back under the laws of gravity the moment the chemical process is completed. The merely chemical laws of relation are thus just as inadequate to explain vegetative or animal process as the mechanical laws of gravity to explain higher chemical processes.[91]

With his "inner natural history of the earth," Steffens tried to demonstrate the progression from the inorganic (silaceous and calcareous earths) to the organic (vegetative and animal process), expressing the continuity between the history of nature and the current findings of plant chemistry.[92] Following Kielmeyer's scheme, Steffens contended that the lower the level of the animal organism, the more irritability preponderated over sensibility, then yielded to reproductive force, which completely dominated plant life. The same ratios could be found in animal and plant chemistry: decline in nitrogen compounds and increase in carbon compounds.

Steffens suggested that the mineralogical composition of mountain formations established by geognosy could be correlated to the emergence of the life processes. Drawing extensively from Blumenbach's paleontology, Steffens concluded: "Thus, in the *oldest* mountains we find fossils of the lowest animal levels; gradually in the more recent mountains there appear remains of *higher* levels; and only in the *most recent* do we find the remnants of mammals."[93] For Steffens, the key result was that in nature itself there was an immanent propensity to generate vegetative and animal forms.[94] Using Schelling's *Naturphilosophie* as a framework, Steffens argued that he could follow the "formation of the earth from one stage to the next. . . . This progression in formation examined scientifically constitutes the true *history of the earth*. . . . In order to distinguish this science from so-called natural history, which occupies itself merely with making distinctions in the external appearances of things, I call this *inner history of the earth*."[95] While this could never be completely articulated, Steffens conceded, "we can demonstrate nonetheless the main epochs in the formation of the earth and prove these from experience."[96]

Like Kielmeyer and others, Steffens insisted that the repudiation of "spontaneous generation" had been premature, and that if one considered the lowest forms of life—infusorians and those forms that seemed ambiguous as to their animal or vegetative nature (zoophytes)—the prospect of a continuity from the inorganic to the organic was not only tangible but indispensable.[97] Moreover, the lower orders of animal life also represented the oldest fossil remains. But the process of individualization in the animal order was too complex and diverse to be arrayed on a single linear scale, as Bonnet endeavored, or even in a network. Only a functional approach could make sense of this diversity, as the comparative physiology proposed by Kielmeyer promised.[98] In any event, all this allowed the conclusion that nature actively pursued more and more organized form. Also, in terms of the senses and sensibility generally, the higher the organism, the more differentiated the senses and the more complex the general sentience—culminating, of course, in man and reason.[99]

For Steffens, all this was in the spirit of Schelling's *Naturphilosophie*. Moreover, it drew heavily and explicitly not only on Werner's geognosy but on the new "French" chemistry, on Blumenbach's paleontology and on Kielmeyer's address on the developmental relation of the vital forces. In all these ways, Steffens's monograph documents the mutual fruitfulness of *Naturphilosophie* and all the elements of emerging historical life science that have been traced in this study. By 1806, now a professor at Halle, Steffens had developed a lecture textbook, *Grundzüge der philosophischen Naturwissenschaft*, in which he elaborated a philosophy of science and of nature on an explicitly Schellingian model.[100] In that work, Steffens took on the role of custodian of the movement, defending not only against external enemies but especially against internal deviants from what he understood to be orthodox *Naturphilosophie*. At Halle, too, Steffens undertook the recruitment of Johann Christian Reil for the movement.[101] Although it was not until after 1806, when the university was closed by the Napoleonic incursions, that Reil made the full transition, it is clear that Steffens played a significant role in his conversion.

Another self-appointed curator of Schellingian *Naturphilosophie* was the Göttingen-trained plant physiologist Franz Joseph Schelver (1778–1832).[102] He began his studies at the University of Jena in 1796, taking courses from the whole battery of natural-scientific eminences at the university, as well as philosophy courses from Fichte. Then, in 1797, he switched to Göttingen, where he concentrated on medical courses and particularly attached himself to Blumenbach. Under the latter he completed his MD with a dissertation, *De Irritabilitate*, in 1798. Thereafter, he moved to Osnabrück, where he practiced medicine for a time and also gave lectures, on the basis of which he issued his first publications: "Erster Beitrag zur Begründung eines zoologisches Systemes" (1799) and *Elementarlehre der organischen Natur, erster Teil: Organomie* (1800; not continued), dedicated to his teacher Blumenbach. By 1800 Schelver had converted to Schelling's system, incorporating it into his own work. From 1801 to 1803 he taught courses at Halle in *Naturphilosophie* after Schelling.

In 1802 he launched a journal for whose short life he was the only contributor, *Zeitschrift für organische Physik*. The journal title was obviously drawn from Schelling's own *Zeitschrift für spekulative Physik*. The entire second (and final) issue was devoted to Schelver's "Erste Darstellung des Systems der Physiologie des Menschen," which Bach terms a "potpourri" from different bits of Schelling.[103] It would have significant influence in the field. Bach conjectures that Schelver suspended the journal upon his appointment as director of the Botanical Gardens at Jena in 1803 at the invitation of Goethe. Schelver turned to botanical courses at that point.

In the opening essay of the journal, Schelver recognized the importance of philosophy of science and of nature for the concrete practice of empirical science. He repudiated the demand that all knowledge have immediate practical benefit, and he denounced equally the crude empiricism of mere fact without theoretical foundation, or of mere descriptive classification based on external properties. Against this blind empiricism, Schelver proposed an "organic physics"—more cogently rephrased: a natural science of the organism. His journal proposed to engage the "wonderful and colorful chaos" of the various research fields of medicine, zoology, pharmacology, botany, and mineralogy, seeking to bring them to some level of coherence.[104] Unifying these disciplines was his ambitious, if not entirely attainable, goal.

While Schelver clearly wrote the whole journal himself, he used the term "we" to betoken his membership in the community of Schellingian *Naturphilosophen*. Bach identifies three components in Schelver's advocacy of *Naturphilosophie* in these years. First, Schelver insisted upon the need for philosophy of nature as a basis for natural science. Second, he took on the role of "multiplier"—that is, propagator—of Schelling's system. Finally, he developed a more concrete application of Schelling's system to physiology.[105] For Schelver, with respect to the first point, *Naturphilosophie* needed to be understood as a framework within which natural-scientific inquiry could locate the gaps in empirical knowledge and offer general laws that would guide the concrete investigations of empirical researchers. The third aspect of his advocacy was carried forward in the journal's second issue, with the presentation of his system of physiology.

Of greatest interest is the second component, which found most salient expression outside his own journal. Schelver published a series of reviews for the *Erlanger Litteratur Zeitung* in which he criticized Andreas Röschlaub's journal, *Magazin zur Vervollkommnung der theoretischen und praktischen Heilkunde*, as well as Röschlaub's integration of his Brownian theory of excitability with Schelling's *Naturphilosophie* in his *Lehrbuch der Nosologie* (1801). While Schelling appreciated any advocacy of his system in the German cultural sphere, he was not pleased with Schelver's criticism of Röschlaub, whom he at that point esteemed very highly as an eminence in German medicine far superior to Schelver. He also found fault with Schelver's own understanding of the *Naturphilosophie* that he felt called upon to arbitrate in others. Nonetheless, Bach notes, Schelling supported Schelver's appointment to the directorship of the Botanical Gardens in Jena and remained in correspondence with him for years thereafter. What motivated Schelling's criticism of Schelver was the clear sense that a strategic alliance with Röschlaub would be far more efficacious in establishing his *Naturphilosophie* across the Germanies.

BAMBERG AND WÜRZBURG: THE EFFORT TO SHAPE MEDICINE, 1800–1806

The most salient confluence of these years for *Naturphilosophie* and life science came in the affiliation of Schelling's philosophy with the sudden and explosive interest in the medical system of John Brown in Germany after 1795.[106] Brown had published his work some time earlier, without any German uptake, and he had died in the interval.[107] In 1791 Christoph Girtanner published some of Brown's ideas as his own, and his plagiarism was denounced a year later, but without generating any strong interest in Brown's own work.[108] Then, while still a medical student, Andreas Röschlaub (1768–1835) became acquainted with an Italian edition of Brown's works and sent this on to the popular writer Adam Weikard (1742–1803). Weikard produced a translation of Brown's text into German in 1795. Within a year, a second translation by Christoph Pfaff appeared. More importantly, Brown's ideas were taken up and reinterpreted by key German medical theorists, preeminently Röschlaub. "It was Röschlaub's interpretation which made Brown's principle acceptable" in Germany, Nelly Tsouyopoulos has noted.[109] By 1798 Brownian medicine had established a major beachhead at the Bamberg General Hospital under the leadership of Adalbert Friedrich Marcus (1753–1816), and Brownian medicine began to be discussed everywhere. In 1799 the *ALZ* published a critical review of the literature on Brownianism in Germany, recognizing with distaste its new prominence in the medical world.[110] Simultaneously, Pfaff published a second edition of his translation, accompanied by a highly critical commentary, and Girtanner published a new monograph criticizing Brownian medicine.[111] Röschlaub, who had created a journal to propagate the new ideas in that same year, used that venue to respond, defending his reformulation of the Brownian ideas.

Brownian medical theory seemed a possible answer to the conundrums of a German medical profession deeply anxious about its "scientific" status.[112] "In the 'dogmas' of Brown the post-Kantian philosophers, doctors, and poets saw the first step towards a philosophical treatment of medicine[,] . . . deducing it from the principles of pure reason."[113] At the same time, this seemed a way to answer the challenge posed by the emergent field of organic chemistry, with its doctrines of "*Form und Mischung*" as a basis for organic life, propagated by figures like Reil. Both Brownianism, with its theory of "excitation," and chemical dynamism (Lavoisier's chemistry) articulated theories of forces that connected the inorganic and the organic, environment and organism. Of course, that was why this community was drawn to Schelling's *Naturphilosophie* in these same years. As Lohff characterizes it, "for physiology it was primarily Brownianism and chemical dynamism that had to be brought into consistency with the Kantian and the Schellingian foundations for physiology as science."[114]

Brownian medicine entered into the earliest of Schelling's writings on *Naturphilosophie*. He learned a great deal about Brown's system in his lively conversations with Pfaff in Leipzig in 1796, for Pfaff had just published his own translation of Brown's *Elementa* and Schelling was particularly interested in getting up to speed on life sciences and medicine.[115] But, as Schelling understood Brown through these discussions, the account of disease seemed to assert the passivity of the organism in the face of environmental stimulus.[116] Schelling preferred a more active conception of the organism. He made his skepticism of Brownian medicine explicit in his *Von der Weltseele* of 1798. But then he began to reconsider. The decisive influence came from the work of Röschlaub, the first volume of whose *Untersuchungen* appeared in 1798.[117] An additional source was Eschenmayer's *Sätze zur Naturmetaphysik* of 1797, the second part of which was an enthusiastic discussion of Brownian medicine.[118] A more positive assessment of Brown, with explicit acknowledgment of Röschlaub, appeared in Schelling's *Erster*

Entwurf of 1799.[119] Correspondence between the two figures may have begun in that same year, if not before, though the surviving correspondence dates only from 1800.[120] By early 1800, in any event, Schelling was quite cognizant of the crucial role of the Bamberg General Hospital in the clinical testing of Brownian medicine, led by Adalbert Marcus with the theoretical collaboration of Röschlaub.[121] At that point he resolved to visit Bamberg and get to know all of this—and with it, cutting-edge medical theory and practice—firsthand. He requested and received leave to remain at Bamberg for a semester, and he taught his *Naturphilosophie* at the Bamberg university even as he observed and studied medicine at the Bamberg hospital with Marcus and Röschlaub.

Schelling believed it was crucial to intervene philosophically in medicine.[122] Personal circumstances intruded dramatically to intensify his concern. A series of tragic illnesses involving his beloved Caroline Schlegel and then her daughter Auguste in the spring and summer of 1800, culminating in the death of Auguste, elicited scandalous reports, fomented by the ever-hostile *ALZ*, that Schelling's interventions applying Brownian therapy had led to Auguste's death.[123] Schelling went to Goethe in the hope that the latter would use his office to suppress these scandalous accusations, but Goethe did not feel he could do so. In protest Schelling resigned his position at Jena, much to Goethe's regret.[124]

This was also the moment in which Fichte finally recognized the extent of the departure of Schelling's *Naturphilosophie* from his own system, and the two philosophers fell out into harsh opposition.[125] Haunted by all this, Schelling entrenched himself in Bamberg and intensified his commitment to the study of Brownian medical theory and its integration into his *Naturphilosophie*. Röschlaub was a willing partner, incorporating Schelling's reinterpretation of "excitability" as a dialectical process of irritability and sensibility from the *Erster Entwurf* into his own new synthesis, *Lehrbuch der Nosologie* (1801).[126] Schelling worked closely with Röschlaub and Marcus and their circle at Bamberg to develop the highly desired "scientific" basis for medicine through *Naturphilosophie*. This intense collaboration drew the fire of the *ALZ* in 1802, with a blistering condemnation of a set of dissertations produced in the Bamberg medical school for spouting nonsensical "Schelling-Röschlaubian *Naturphilosophie*."[127] Ever a polemicist and armed with a journal of his own, Röschlaub responded equally fiercely in his *Magazin*. For the German academic public, Schelling and Röschlaub appeared firm allies in the merger of Brownian medicine and *Naturphilosophie*, and their correspondence confirms close and cordial relations between the two figures.

When in 1802 Röschlaub was called to a chair at Landshut, where the old University of Ingolstadt was relocating, Schelling wanted to join him, but Röschlaub was unable to engineer a position for him. With Röschlaub gone, Marcus recruited some new associates for the Bamberg General Hospital, most prominently the surgeons Philipp Franz von Walther (1782–1849) and Konrad Joseph Kilian (1771–1811). But this cluster of figures around Marcus grew envious of Schelling's closeness with Röschlaub and jealous of the latter's prominence in the general reception of Brownian medicine in Germany. As Tsouyopoulos reconstructs it, they set about smearing Röschlaub's reputation and undermining Schelling's confidence in him.[128] In 1802, shortly after Röschlaub departed, a scurrilous pamphlet was published, ostensibly by an anonymous friend, purporting to be a public confession by Röschlaub of the banality and incoherence of his understanding of medicine and philosophy.[129] At the height of his career, Röschlaub did not take this seriously. More dangerous for him was a publication by Marcus's new ally Kilian, *Differenz der echten und unechten Erregungstheorie* (1803), which clearly sought to split Schelling from Röschlaub.[130]

When the University of Bamberg closed in 1803 in the wake of territorial shuffling associated with the Napoleonic conquests, the new ruling authority, the Kingdom of Bavaria, determined to invest all its energies in the overhaul of the University of Würzburg. Marcus was able to obtain a chair for Schelling there. Walther moved on from Bamberg shortly thereafter to Landshut, where he set about challenging Röschlaub's authority in the field. When Röschlaub appealed to Schelling for support, the latter aligned himself more openly with the opposition.[131] A new journal, *Jahrbücher der Medizin als Wissenschaft*, began appearing in 1805 under joint editorship of Schelling and Marcus, with the clear aim of discrediting Röschlaub's interpretation of Brownian medicine and undercutting his position in the German medical and philosophical community.[132] This was simultaneously the high point of Schelling's integration of medicine and *Naturphilosophie* and his final engagement with the latter.

In the pages of the *Jahrbücher der Medicin als Wissenschaft* Schelling laid out in the most accessible form his latest conception of *Naturphilosophie* and tried to link it to the needs and prospects of the medical field. That began with the foreword Schelling provided to the first issue of the new journal, originally intended to appear in January 1805 but only appearing in the fall of that year. The foreword celebrated the priesthood of all naturalists and highlighted the investigation of life as the highest calling: "The science of medicine is the crown and the flower of the natural sciences, just as the organism in general and the human in particular constitute the crown and flower of the world."[133] Because of this sovereign status, "philosophers and naturalists of every sort, chemists and anatomists, zoologists and physicians, unite in the common endeavor to elevate the science of the organism and therewith medicine to the peak that it should occupy, and gradually to move it forward."[134] What he hoped was that "the holy bond that unites all the things of nature without suppressing any [should be] possible among scholars" as well, for as nature was one, so too must the "fundamental vision and perspective of the mind ... in science and art" be "one universal, boundless vision of beauty and truth united."[135] This high Romantic rhetoric had its core in philosophy. "The vantage of philosophy ... is this: the absolute is the original or primal form [*Urbild*]; philosophy as the work of humans is a copy or afterimage [*Nachbild*]; the soul moves between the two to appraise the resemblance of the latter to the former."[136]

To establish the proper relation between philosophy, theory, and experience was the crux of Schelling's philosophical endeavor. In the empirical sciences, individual observations—"facts" without theoretical context—could not constitute any viable insight. Theory was inescapable. But how was it to be grounded? Kant had made the question of the warrant of science a general philosophical problem, aiming at once to enable and to delimit the possibility of science. Yet he also argued that philosophy had to establish its own authority transcendentally, without recourse to experience. This led to attacks from two sides. Empirical researchers insisted that observations had to be the basis for natural science, that "speculation" was empty, echoing the rhetoric of the Réaumur-Bonnet school of natural history and the even earlier radical "Baconianism" of the Royal Society of London in the era of Newton. Transcendental philosophers, on the other side, claimed that philosophy needed and could have no confirmation in the natural order, because the latter was entirely the product of transcendental constitution. Nature was but an epiphenomenon of spirit. Subject contained and posited phenomenal substance. This was the position of Fichte, and it was echoed in the "nature metaphysics" of Eschenmayer in his controversy with Schelling in the pages of the *Zeitschrift fur spekulative Physik* in 1802, as Schelling clearly understood.[137] Nonetheless, Schelling wished to bring his system of "speculative physics," his *Naturphilosophie* in its latest incarnation, into direct connection with medicine. That is, he proposed that the general insights into nature as a whole that were developed in *Naturphilosophie* would have determinate consequences for medical practice.

It was to this that Schelling turned in his most important contribution to the *Jahrbücher*, the essay "Preliminary Characterization of the Standpoint of Medicine according to the Principle of Natural Philosophy."[138] The essay began with a sharp repudiation of Brownian theory in medicine because it approached the organism externally rather than recognizing its inner activity. The latter was what *Naturphilosophie* provided. "The basis of the organism as such cannot lie in any particular principle of nature. For only absolute nature, infinite substance in itself, carries within itself the particulars of the world as eternal fruits [*gewächse*], just as the organic whole, as long as it lasts, forms its members in a fixed form."[139] The one was infinite, the other particular, but they were in essence and originally one and the same. "Every organic energy [*Wirksamkeit*] arises from the essential and innermost character of nature and cannot be explained empirically any more than can gravity."[140] Brown's whole theory of excitability floundered in a confusion between the absolute and the concrete. It situated liveliness on a linear scale of more or less, a strictly quantitative way of thinking that revealed "the complete emptiness of the concepts of *sthenia* and *asthenia* in relation to the phenomena of life."[141] How could there be talk of excess or deficiency without a governing *norm*? But Brownian theory could not provide this norm, only assume it. The "scientific" elaboration of Brownian theory in Germany (i.e., by Röschlaub) ultimately faced this decisive limitation since it considered only one aspect of the organism (its outward relation) and missed entirely its intrinsic self-creation. In Schelling's view, the "best thing written so far expressing the perspective of *Naturphilosophie* on actual medicine" was Ignaz Troxler's *Ideen zur Grundlage der Nosologie und Therapie* (1803), a work explicitly repudiating Röschlaub and Brownian excitability theory in medicine.[142] In endorsing Troxler and in his own thorough repudiation of Brownian excitability, Schelling was publicly breaking with Röschlaub.

Schelling argued that the whole theory of excitability should now be replaced by a dimensional theory of the relation of principles of natural development with concrete developments in the material and organic world. Here he drew especially on Steffens. In this new formulation, gravity, the consolidation of a particular material thing, was the "authentic earth principle." This was the first dimension, associated with magnetism among the fundamental principles of natural processes and with carbon among the chemical elements. In living things it was expressed in reproductive force, in *Bildung* or *Gestaltung*. Light was the second dimension, dissipating or breaking up all the consolidations of the first dimension into freer forms; in the array of life-forces, it was associated with irritability; in chemistry, it was associated with hydrogen. The third dimension was not merely a synthesis but equiprimordial with the others, the source of all animation and life, productivity itself. Among the life-forces, this was sensibility. Chemically it was associated with oxygen. This grand new system needed to be taken up into medical theory, he asserted. It wasn't.

There were grievous consequences of this split between the Brownians and the *Naturphilosophen* in medicine. Round Schelling gathered foremost the physiologists; round Röschlaub, the pathologists.[143] That is, theoretically oriented physiology became estranged from clinical practice. Röschlaub's whole program for Brownian medicine was compromised. But that was not all: "The conflict simultaneously and perhaps more essentially damaged the movement of *Naturphilosophie*[, which] spun off in a speculative line and lost relevance for medicine." In the end, "the real winners in the conflict between Röschlaub and Schelling . . . were the traditionalists, the eclectics" in medicine, like Hufeland.[144] The philosophical interest within clinical medicine, spawned by its moment of epistemological crisis, lapsed, and it resumed its preponderant pursuit of clinical success and public respect under the aegis of Hufeland, especially after his move to the new University of Berlin in 1812.

Naturphilosophie remained important to many figures, especially the inner circle Schelling had formed, including Steffens and Schelver from earlier, supplemented during his Würzburg years by Carl Windischmann, Ignaz Troxler, his own brother, Karl, and Ignaz Döllinger. And, by the close of the decade, *Naturphilosophie* found an important convert in Johann Christian Reil. But, strikingly, by 1807 it was of no further interest to Schelling himself! "By 1804 Schelling's interest had shifted," and he "stopped writing on *Naturphilosophie* by the time he was 28."[145] As Tsouyopoulos puts it, "Schelling himself very swiftly lost all interest in medicine and turned to quite different domains of philosophy. The *Jahrbücher* shut down in 1808."[146]

Lorenz Oken (1779–1851) took up the organizational leadership of *Naturphilosophie* among the natural scientists after Schelling's loss of interest. Oken had become connected with Schelling's circle at Würzburg, made his way to Göttingen, ostensibly to study with Blumenbach but actually to work out his own system of embryological development, and then published a number of early papers linking developmental life science with Schelling's *Naturphilosophie*.[147] Under Oken's leadership, especially after he took up a position at Jena in 1807, *Naturphilosophie* remained a dominant force in the natural sciences until the early 1820s. He established his central position theoretically with *Lehrbuch der Naturphilosophie* and institutionally with his *Gesellschaft deutscher Naturforscher und Ärzte* and with his journal, *Isis, oder Encyclopädische Zeitschrift* (vol. 1, 1817). But it was also against Oken's flamboyantly speculative style of *Naturphilosophie* that the tide of critical revisionism gathered.[148] It is noteworthy that one of the most discerning treatments of Oken suggests that his was not a faithful continuation of Schelling's original *Naturphilosophie* but rather "strongly influenced by older pseudoscientific traditions, especially alchemy and numerology as they had been presented by Robert Fludd."[149] He provoked repudiations of the whole impulse both in his own time—by figures like Heinrich Friedrich Link and J. F. Fries (1773–1843)—and more importantly a generation later, when the whole of *Naturphilosophie* would be spurned as the "black death" of natural science.[150] A consequence has been the reluctance to find any fruitful connection between *Naturphilosophie* and emergent biology.[151] Yet there was a more fruitful connection between the generation of 1800 and the generation of the 1830s. We can find it in the person of Ignaz Döllinger.

IGNAZ DÖLLINGER AND THE LEGACY OF THE EIGHTEENTH CENTURY

An exemplary case of the nexus between emergent biology and Schelling's *Naturphilosophie* was the career of Ignaz Döllinger. To start paradoxically from the conclusion of that story, Döllinger became the esteemed mentor of several of the undisputed titans of German life science of the early nineteenth century, most notably Karl Ernst von Baer.[152] Thus, Döllinger embodied the decisive continuity from Blumenbach and Kielmeyer through Pfaff, Humboldt, Treviranus, and Reil, through Döllinger himself to the generation of von Baer, Johannes Müller (1801–58), and the discoverers of cell theory, Matthias Schleiden (1804–81) and Theodor Schwann (1810–82), with whom no one can doubt that biology as a special science had taken form.[153]

Döllinger was an unequivocal advocate of *Naturphilosophie,* but he was foremost an empirical research scientist in comparative physiology. He clearly drew his philosophical orientation from the *Naturphilosophen,* but "he zealously carried forward physiology after the fashion of Haller"—that is, *experimental* physiology.[154] Indeed, he was decisive in consolidating the experimental physiology of the eighteenth century to lay the foundations for the disciplinary practices of the nineteenth, as one of his most notable publications makes clear in its very title: *Von den Fortschritten, welche die Physiologie seit Haller gemacht hat* (On the progress physiology has made since Haller; 1824).[155] It was certainly in that tradition that he wished to situate his own work, as he made clear in *Über den Werth und die Bedeutung der vergleichenden Anatomie* (On the value and meaning of comparative anatomy; 1814).[156] He himself attempted to systematize the field in an extensive monograph, *Grundriss der Naturlehre des menschlichen Organismus* (Outline of the natural science of the human organism; 1805).[157] In his research and teaching practice, as exemplified in his mentorship of von Baer, Döllinger stood at the forefront of his generation in life science, and thus, his conception of the field—and his sense of the philosophy of science that it required—can be taken to capstone the whole development that made *Naturphilosophie* not only congenial but generative for emergent biology.[158]

Eckhard Struck observes: "Döllinger stands among the thinkers of the turn of the century who considered the guiding influence of Schelling's ideas for natural science—in which he included medicine—to be beneficial [*segensreich*], indeed an indispensable foundation [*geradezu als Voraussetzung*] for their own work."[159] Already in 1875 Heinrich Haeser asserted: "The first place among the nature-philosophers and physiologists who arose from the school of Schelling is indisputably taken by Ignaz Döllinger from Bamberg."[160] Similarly, Neuburger and Pagel ranked his contributions at the same level as Kielmeyer's in the promotion of physiology.[161] Haeser went on to add that Döllinger "held himself freest from the temptation to sacrifice empirical investigation to scientific construction."[162] Along the same line, Albrecht Kölliker observed in 1871 that "Döllinger knew how to keep himself free from the extreme views of nature philosophy."[163] Döllinger put it in his own terms, in the preface to his *Grundriss der Naturlehre des menschlichen Organismus* in 1805: "Now, in my opinion, it is one of the most distinctive features of *Naturphilosophie* to give free rein in the application of its fundamental principles to the perspicuity and investigative spirit [of the individual researcher]."[164] That comment indicated a *hope* more than an actuality (it proved *not* to be the case in his specific instance, as we shall see). In any event, this was the promise he believed *Naturphilosophie* held for empirical science.

Ignaz Döllinger was born in Bamberg, the son of the *Stadtarzt* and professor of medicine at the local university, Johann Ignaz Joseph Döllinger (1721–1800).[165] Bamberg had long been a Jesuit educational fiefdom, but with the abolition of the order in 1773, a new secular university was founded—including, through the efforts of Döllinger-*père*, a medical faculty of which he became senior professor. Ignaz was schooled in Bamberg, attended the university there for a preliminary degree in philosophy in 1787, then advanced into the medical school, where he eventually received his

MD, under his father's tutelage, in 1794. In the course of a medical peregrination, he worked with the eminent physiologist Georg Prochaska (1749–1820) in Vienna and did a stint at the first-rate Italian medical school in Pavia, where the comparative anatomist Antonio Scarpa (1752–1832) sparked his lifelong interest in comparative anatomy and physiology. As soon as he received his MD, he was appointed *ordinary* professor of medicine at Bamberg (at age twenty-four!), and he commenced his academic teaching in 1794–95 with a course of introductory medicine based on Blumenbach's *Institutes of Physiology*.

The gem of Bamberg was its General Hospital, where Adalbert Friedrich Marcus created a cutting-edge facility, with a strong theoretical commitment to the new ideas of Brownianism. Döllinger was in the inner circle of the Bamberg group around Marcus and Röschlaub by 1800, when it became the center of controversy in medical theory in the Germanies. He met Schelling when he arrived at the hospital in 1800. While nothing is known of their initial personal relationship, already in one of his first publications (1803) Döllinger acknowledged his affiliation with Schelling, Goethe, and Steffens (whose "inner natural history of the earth" he explicitly embraced as a model).[166] That is, Döllinger was a partisan of *Naturphilosophie* from the outset of his career.

When the University of Bamberg closed in 1803, Döllinger made the transition to Würzburg alongside Schelling. While in Bamberg he had done some private practice, and he had certainly been active at the General Hospital. But once he took up his position at Würzburg, Döllinger, like so many of the other key figures of this study, abandoned clinical practice altogether for experimental research in comparative physiology.[167] In 1806 he pioneered the fusion of the chair in physiology with the chair in anatomy, setting a pattern that became widespread in German medical schools in the nineteenth century. Döllinger appealed to the arguments of Soemmerring in contending that physiology and anatomy belonged together.[168] Thus, "Döllinger's apology for an independently grounded science of physiology was . . . [that] only a science of physiology grounded in general principles and in possession of secure empirical knowledge would eventually be adequate to developing a science of medical practice."[169] His was a crucial emphasis on the experimental and theoretical elaboration of physiology, for which anatomy was a requisite auxiliary field. At the same time Döllinger dropped the traditional association of anatomy with surgery—that is, clinical application.[170]

We must pursue why Döllinger embraced *Naturphilosophie* and how he embraced it. The main source available for understanding Döllinger's career is the eulogy given by his longtime colleague and friend Philipp Franz von Walther in 1841.[171] Both had been involved in *Naturphilosophie* at the outset of their careers, but Walther was acutely aware, as he memorialized his friend in 1841, that the climate of opinion in German life science had turned sharply hostile to that movement.[172] This accounts for the slant of the eulogy, stressing Döllinger's Kantian affiliations and distancing him from *Naturphilosophie*,a retrospective vindication that did more than a little violence to Döllinger's earlier views—and his own as well.[173] To be sure, Kant was all the rage in their youthful days, and Döllinger was no exception. But when Walther maintained that "the study of philosophy in general and precisely this philosophy in particular" constituted for Döllinger "an important, exciting and decisive influence for his entire life, defining forever the direction of his scientific activities," there are historical reasons for skepticism.[174] Struck argues that there was, besides Kant, another decisive influence on Döllinger's early conception of the field: Goethe.[175] That is, Goethe had "incited the animation of anatomy, which thereby became morphology [*Belebung der Anatomie, die dadurch zur Morphologie wurde*]."[176] Not for nothing does Struck take up, toward the conclusion of his study, "the conflict between Kantianism and Goetheanism."[177] Schelling and his *Naturphilosophie* aimed above all to bridge that divide. Döllinger, Steffens, Schelver—almost all the crucial early adherents of *Naturphilosophie*—took this to heart.

Steffens and Schelver played a direct role in Döllinger's intellectual development. In his earliest publication, in mineralogy, his preface explicitly aligned with Steffens's *Beyträge zur inneren Naturgeschichte der Erde* and expressed the hope of extending some of its findings.[178] When he took up his new position at the University of Würzburg, Döllinger announced as the text for his course in comparative physiology "Schelver's Zeitschrift für organische Physik ersten Bandes zweytes Hefts," winter term, 1803–4, and then again in 1804–5.[179] His own *Grundriss der Naturlehre des menschlichen Organismus*, which appeared in 1805, became the course text thereafter.

The creation of the *Jahrbücher der Medizin als Wissenschaft*, coedited by Marcus and Schelling at Würzburg starting in 1805, proved very important in Döllinger's intellectual development. There he published a two-part consideration of the relation of *Naturphilosophie* to medicine, and there, at Schelling's instigation, Carl Windischmann (1775–1839) published a very negative review of Döllinger's new monograph, *Grundriss der Naturlehre des menschlichen Organismus*.[180] Windischmann's review simply carried out the critique that Schelling's letters asked him to make of Döllinger. Just as

Schelling found fault, earlier, with Schelver's embrace of *Naturphilosophie*, so now Schelling urged Windischmann to chastise Döllinger's work for its residual empiricism.[181] Even as he was himself growing weary of all the squabbles over the *correct* version of *Naturphilosophie*, Schelling never failed to fault his epigones for their inadequacies, and they—his own brother, Karl, not least among them—made every effort to curry his favor and discredit each other in his eyes. It was not altogether unfounded when a reviewer in the *Oberdeutschen allgemeinen Litteraturzeitung* observed in April 1806 "that it was the goal of *Naturphilosophie* to take over medicine; Schelling's system of *Naturphilosophie* intended thereby to establish 'sole dominion [*Alleinherrschaft*]' in the realm of the sciences."[182] Döllinger kept his independence at a cost: Schelling disparaged him privately for the sake of his personal hegemony.

Döllinger's two-part contribution to the *Jahrbücher der Medicin als Wissenschaft* and his monograph *Grundriss der Naturlehre des menschlichen Organismus* together articulated his view of physiology and of *Naturphilosophie* at that crucial moment of 1805. They are entirely consistent in their terminology and appraisal. The first of the *Jahrbücher* contributions offers the clearest point of entry. There, he discriminated three "moments"—he meant impulses—in the current situation of physiology (or medicine, as he retrospectively rephrased the matter).[183] Those three "moments" were, first, the practice of medicine (*Technik der Medicin*), that is, clinical treatment of patients; second, the intervention of physics and chemistry in physiological and medical theory; and, finally, the "influence that *Naturphilosophie* has gained."[184] With these three rubrics, Döllinger outlined the development in medicine up to his moment. His premise was that the primitive art of healing without any theory behind it, an "unruly and unfounded undertaking [*regel- und grundloses Verfahren*]," was not worthy of the name of medicine. Medicine properly began when Hippocrates and later Galen offered a theory—that is, "knowledge as the basis for acting." To be sure, Hippocrates drew his theory directly from bedside observation, but he did try to work up a theory, and it was that tradition that persisted until recent times, even up through Boerhaave.[185] More recently, with figures like Cullen and Brown, however, something more than bedside observations entered into the development of physiological theory. Brown, he averred, tried to develop medicine on the basis of a general theory of organism.[186] The point, he went on, was that one could not understand organisms only on the basis of the condition of illness; one had to grasp them in their healthy condition as well; pathology was only a part of physiology. Indeed, what was required was "a general natural science of the human organism [*allgemeine Naturlehre der menschlichen Organisation*]" in order to provide a sound basis for medicine.[187] That, in turn, required integration with the findings of other fields of natural science. That opened the way for the intervention of the wider natural sciences in physiological theory, Döllinger's second "moment."

Here, Döllinger offered some sharply critical observations about the manner in which chemistry intervened.[188] Stuart Strickland has used these observations to suggest that Döllinger was out of step with the rise to hegemony of chemical research in the German arena of his time.[189] But this is not quite right. By 1805 the high hopes not merely regarding galvanism but regarding organic chemistry as the key to the mysteries of life, and to physiological theory more specifically, had run aground. The greatest advocate of the chemical approach in Germany, Reil, had himself given up on an immediate harvest from this approach by 1805, and for precisely the same reason that Strickland finds unconvincing in Döllinger's articulation.[190] Both Reil and Döllinger realized that the chemistry of dead organic compounds could not explain the chemical activity of living organisms. That was not an inappropriate judgment then, and even today we have to make some important discriminations, especially regarding the problem of the origin of life, however much biochemistry and organic chemistry have advanced. And it is important to note that neither Reil nor Döllinger wanted to *eliminate* chemical approaches to physiology, as Strickland seems to believe in Döllinger's case, but only to *delimit* their sphere. Döllinger wrote: "Now, what could offer a clearer contrast than the manner by which the organism establishes its own qualities and organizes them under itself and the manner in which the chemist decomposes [*auseinanderlegt*] and investigates these again in the dead body?" To be sure, he went on, there was much that physiology could learn from chemistry. "But one should not in the least delude oneself into believing that one could find the basis of life in the component parts that chemistry has been able to decipher."[191] Organic chemistry worked with the detritus of life, not its construction, and Döllinger found little reason to expect great contributions to the physiology of live organisms from this analysis of compounds from dead ones.

Still less, he went on, did he find much prospect in the doctrine of "excitability," which he took to be a merely *mechanical* theory of the influence of the environment on the organism, thus of the inanimate on the animate. This took even less note of the active self-constitution of the organism than the chemical approach. Active self-constitution, he insisted, was the proper domain of physiology, and unidirectional theories that sought to explain life mechanically or chemically could never displace it. "Life is thus no reaction that as such needs to be determined from the outside; it is pure, original, inner activity."[192] "If only the physicists and the chemists made it their object to explain the changes that the presence of organic bodies occasions in inanimate nature, as they assimilate the latter, then they would provide the physiologist real gains for science; then they would work hand in hand."[193]

There is a direct relation between this disillusionment with chemistry and the appeal of *Naturphilosophie*, the transition from Döllinger's second to his third "moment." A proper relation between the physical sciences and physiology would recognize that nature was dynamic, and that organisms had to be understood not as passive products but as active agents: "The insight that the organism, just like all of nature in general, is at one and the same time product and productivity and the knowledge of how this is the case in the organism alone constitute the true uncovering of its otherwise-obscure nature."[194] That required a reinterpretation not only of nature but of natural inquiry: recognizing the precedence of process over product. Here was where *Naturphilosophie* represented the indispensable intervention. Döllinger undertook an account of the conflict between understanding (*Verstand*) and reason (*Vernunft*) in the Kantian critical philosophy, suggesting that understanding only set fixed binaries to contain the natural world, which reason then suspended. He believed that this meant that the understanding offered mechanical explanations of actual empirical phenomena and reason could not derive direct speculative constructions of nature from understanding's fixed binaries. Understanding achieved "relative truth," while reason dealt with the absolute.[195] Hence, "speculative physiology can have no other task than the construction of the organism, that is, the dissolution of all the contradictions that the understanding in its investigation of the latter has set up."[196] He offered several such dichotomies of the understanding—cause and effect, possibility and actuality, subjectivity and objectivity—and insisted that these needed to be transcended; the organism was "subjectobject"—that is, "productivity and product at once."[197] This was Döllinger's understanding of *Naturphilosophie*. He clearly drew on Schelver's reconstruction but also criticized it for hypostatizing process into classes of actual things—plants and animals.[198] Döllinger expressed concern that the "pure active expressions of nature" were being falsely "objectified and materialized." Thus, one spoke "of magnetic, electrical matter." The "way of representing the organism as a mere object is manifest in the chemical and mechanical perspectives that are repeatedly introduced into physiology."[199] He hoped that from *Naturphilosophie* one could receive a clearer framework for understanding nature and the organism concretely in a processual manner. This motivated, one can presume, his creation of his own textbook in physiology to replace Schelver's work.

Döllinger's second contribution to the *Jahrbücher der Medicin als Wissenschaft* investigated the published dissertation of Schelling's brother, Karl Eberhard Schelling (1783–1854), defended at Tübingen in 1803, *Dissertatio Sistens Cogitata Nonnulla de Idea Vitae hujusque Formis Praecipuis*. Döllinger read it as "a first introduction into the nature-philosophical construction of the organism."[200] He set out by contrasting the respective inadequacies of a strictly mechanistic account of organisms, based on chemical principles or heat or "ether," on the one hand, and a supersensual life principle, or soul, along the lines of Stahl, on the other. In the face of these theoretical inadequacies, a number of practitioners of medicine retreated to immediate observations, facts, with a "hateful disposition toward anything theoretical." Döllinger associated with such antitheoretical dispositions a blind notion of linear cause and effect that belied the complexity of the organism. The thrust of *Naturphilosophie*, as Döllinger inferred it from the dissertation, was to overcome all these inadequacies. In "living matter," action and being were synonymous, and this correlated with the character of absolute nature in general. But a concrete living thing was also productivity realized "*in a particular way*," hence a product, a "copy [*Abbild*] of absolute nature."[201] Thus, an organism could be viewed in two ways: either as the activity of absolute nature (productivity) expressing itself or as the result of that activity (product).

In his monograph, Döllinger offered his own synthesis. In the introductory part, he made a fruitful discrimination between *Naturbeschreibung* and *Naturlehre*, arguing that in physiology mere description was not sufficient; a *Naturwissenschaft* entailed knowledge of the essential laws of organic life.[202] Moreover, natural knowledge of the human organism was itself simply a branch of general natural knowledge, which embraced all nature.[203] Hence, comparative anatomy, relating the human to the rest of the life-forms of the natural world, was important.[204] He began his exposition with the pronouncement that while activity was not an essential property in the inorganic world, it was the essence of living things. The contrast between inertness and activity was reformulated into that between product and productivity, but with the proviso that the organism represented both productivity and product; indeed, the product was intrinsically active. What distinguished it from general nature as productivity was just the specificity of its form. But this determinacy was itself the result of the activity of general nature. And the organism itself perpetuated this activity on its own scale. Thus, Döllinger summed up, "the identity and the difference of the product and the productivity in the organism offer the most complete insight into its nature and accordingly . . . it is the principle of the natural science of the organism."[205] The so-called life principle could be construed, he went on, in terms of either the universal principle of productivity working through the particular organism or the productivity of the specific organism itself.[206] The concept of life merely expressed the same idea of self-production that was contained in the notion of the organism: "the expression of an inner, essential activity."[207]

Döllinger then addressed the relation of the organism to its environment, discriminating passive "receptivity" from active response (*Zurückwirken*). The inner capacity to respond to the external world he labeled, after the Brownian fashion, "incitability." His contention was that the determinacy of a particular organism was entirely the consequence of the inner forces of productivity: "the *innate* [*einheimischen*] and therefore *natural* determinations of the form of an organism are alone those which arise immediately from productivity and belong without restriction to its effectiveness [*Wirksamkeit*]."[208] An organism was healthy as long as this relation prevailed; when foreign determinations intruded, it became ill.[209] Thus, pathology was a branch of physiology that considered aberrance from the natural: "illness is the unnatural state of the organism."[210]

Döllinger turned then to the levels of activity in organisms and the laws that differentiated them into levels. The most basic activity of living things was growth and formation (*Bilden*), the expansion of tissue in space. Inwardly, this entailed establishment of a chemical "texture," or qualitative *Mischung*, while outwardly, this tissue growth established a particular shape (*Gestaltung*) or "structure."[211] This was what Blumenbach had identified as the *Bildungstrieb* constitutive of all organic form.[212] This was the preponderant function of plant life, and this vegetative form of [re]production carried forward into animal life, where it was complemented and complicated by the capacity for motion associated with irritability.[213] "What formation is in plant life, in animal life is feeling; there all energy is devoted to the construction of a form . . . , here, everything is about the transformation of the capacity to act in itself."[214] In characterizing irritability, Döllinger minced no words in his criticism of Haller for drawing false conclusions from his observations and in this way deeply confusing physiology.[215]

In the internal, chemical constitution of the organism, productivity began with the fluid or formless and worked toward structured solids. The former were purely qualitative, while the latter had determinate scale. Purely quantitative filling of space he identified with mechanism. The qualitative factors could be associated with the fundamental elements of chemistry—oxygen, hydrogen, carbon, and nitrogen.[216] Following Steffens, Döllinger identified animal life with a preponderance of nitrogen compounds, and in inorganic nature he distinguished between silaceous earths, closely connected with plant life, and calcareous earths, closely connected with animal life.[217] Physiological development of the organism proceeded from fluid forms of chemical activity (*Mischung*) to solid tissues and a distinct shape (*Gestaltung*) via solidification, starting with globules, then tissue (*Zellgewebe*), then fibers, and on to bones, muscles, nerves, and the specific organs. The bulk of his exposition followed this hierarchical emergence of tissues and organs.

In animal life, the synthesis of reproductive force and irritability was "sensibility." It combined the external attunement of the vegetative state with the inward attunement of irritability, feeling with drive, culminating in voluntary motion and self-awareness, "the inner feeling of existence."[218] This inner feeling or awareness of self constituted the distinctive sentience of animal life. Döllinger sharply distinguished this animal capacity for "representation [*Vorstellen*]" from the human capacity for thought.[219] More generally, he found the concept of sensibility profoundly ambiguous. "As widespread as the expression sensibility is, nonetheless it appears that up to now there has been lacking a correct general determination of the *concept* one intends in terms of the *matter* which it is to betoken."[220] In the discourse of medicine, sensibility was construed in three distinct manners: first, as receptivity, that is, responsiveness to stimulation from the environment; second, as "an unnameable, mysterious attribute of the nerves" (what others called *Nervensaft* or *Nervengeist*) with many physiological consequences; and, finally, as the capacity of the nervous system to affect consciousness.[221] These were the perplexities of physiological psychology in connection with neurology, as they presented themselves across the entire eighteenth century in German research from Haller to Unzer to Döllinger's own time, in the work of Pfaff and Reil. The best Döllinger could offer was that "sensibility, responsiveness, better sensation [*Sensibilität, Empfindlichkeit, besser Sinnlichkeit*] represent the unification of animal and plant life into a whole."[222] The step from this animal sensibility to full human consciousness, he averred, was beyond the scope of physiology.

With this work, Döllinger synthesized all the currents of thought in the physiology, natural history, chemistry, and psychology of his time and subordinated them to an overarching schema of *Naturphilosophie*. This provided the foundation for his teaching for the next twenty years at Würzburg, and it shaped a whole generation of brilliant morphological researchers, foremost among them von Baer and Heinz Christian Pander (1794–1865). He supplemented this extraordinary mentorship with two overviews of the field that had great impact. In 1814 he published *On the Value and Meaning of Comparative Anatomy*, and in 1824, *On the Progress That Physiology Has Made since Haller*. We can take these two texts as the contribution of Döllinger's *Naturphilosophie* to the discipline of nineteenth-century biology.

The text of 1814 was a public lecture, and it was very clearly written with an eye to the important concerns of the day, especially the moment of German renewal in the aftermath of the Napoleonic Wars and the reconstitution of the German states.[223] More particularly, it concerned itself with the place of the natural sciences in the social and cultural system that was emerging. Medicine needed to orient itself increasingly to the governmental regulative environment and to the interests of the wider public. Thus, political and utilitarian considerations seemed foremost on the cultural agenda of the moment, and in that context Döllinger wished to make the case for the "value and meaning" of pure research in comparative anatomy. It was "beyond doubt," he allowed, that comparative anatomy "could not be of the status of godly mathematics or noble metaphysics or even praiseworthy history," yet among the branches of natural inquiry, he argued, it was "by far the most meaningful, richest in content, and to be recommended in every sense."[224] This resulted not only from its engagement with the human form, the highest expression of anatomy, but also from its immediate engagement with nature. While physics or chemistry had to extort nature through experiment, and hence their results were always ambiguous as to whether they were artifices of the procedure or truths of nature, comparative anatomy simply contemplated what nature itself presented in all variety directly to the attentive observer.[225] He contrasted descriptive natural history, content merely to characterize the external properties of life-forms, with comparative anatomy, which probed deeper into their organization, seeking to understand the laws of formation (*Bildungsgesetze*). "The laws of formation are what comparative anatomy has first and foremost to establish."[226]

Döllinger argued that nature worked up from a basic, undifferentiated living tissue (*Gallaert*)—exemplified by the polyp—to increasingly differentiated tissues and organ systems. In this process, particular types of organization took form.[227] What the comparative anatomist sought to theorize was how and why this happened and also how to account for the occasional deviations in the natural process. It was, in the end, all about what latent capacities humans might have, he concluded, and all sciences had to contribute to the high goal of the perfection of humanity.[228] Yet to understand the human, it was not enough to know human anatomy; one had to situate it among all the other forms—hence the importance of comparative anatomy.[229] As a justification of his field, or as an invitation to take it up, this was not the most overwhelming rhetorical achievement. But it did indicate how Döllinger understood his field and his own representative status as its proponent.

More penetrating, both historically and analytically, was his address of 1824. In a sweeping overview of medicine from its origins in the West, Döllinger highlighted the achievements of Albrecht von Haller. This "ornament of his age," he elaborated, "achieved the impossible[,] . . . making it known that the most authentic foundation of all medicine in general was physiology."[230] His *Elements of Physiology* was not just the seed awaiting further development but the "whole tree grown to its topmost branches, with its happy, promising flower."[231] And "everything that for about the next fifty years was achieved in the knowledge of physiology derived from this original stem."[232] Thus, that work could stand as the point of departure for the assessment of progress in the field. Haller provided precise observations, but observations needed interpretation in order to assume their proper relevance. This was the task that fell to Haller's successors. The life *process* needed to be theorized. Thus, anatomy needed to be fused with physiology (as they were in his own professorship at Würzburg). Morphology should be linked with "histology"—that is, the microscopic analysis of living matter in its various forms of organization.[233] This linkage would trace the development of an organism across time and explore the emergent relations of the various tissues, organs, and systems as they developed. In particular, Döllinger stressed embryology, grasping life at its beginnings. "The greatest naturalists—Malpighi, Harvey, Wolff, and our Haller have recognized the significance" of working on the embryos of birds to ascertain the developmental course of mammals and men.[234] Döllinger himself had inspired major work in this vein among his students, most eminently Pander.[235] He also celebrated the great period of morphology inaugurated by Goethe.[236] He gestured to the question of generic descent from an original stock into subvarieties and races.[237] He praised the role of animal vivisection but warned that drastic intervention might disrupt more than reveal the intrinsic physiological processes of the organisms investigated, and he warned more generally that presuppositions and procedures could distort experimental research and generate artificial results.[238]

Looking back to the last decade of the eighteenth century as the heyday of *Naturphilosophie,* he conceded that there was considerable play with airy abstractions, but he argued as well that there was a fundamental earnestness about trying to penetrate to the "most hidden mysteries of life" that deserved a more balanced reckoning.[239] Finally, he celebrated the emergence of a coherent discipline: "Biology or the physics of the organic," which "expressed the endeavor to represent living nature in all its scope."[240] Among others he noted Treviranus for this vision.[241] Central to this new discipline, he averred, was comparative anatomy, "that beautiful product of genuine *Naturphilosophie.*"[242] He celebrated Kielmeyer as its great initiator, and Cuvier as its powerful executor, the "ornament of our times."[243] He added, finally, that this new discipline had to be open to the contributions that could come from physics and from chemistry, when used with proper discretion.[244] Fifty years of enormous advances in the natural sciences had passed since Haller, and physiology had developed apace. There was every reason to be confident, he finished with a flourish, that all this would continue energetically into the future. With that, Döllinger, on behalf of his generation, the *Naturphilosophen,* passed on the baton to the new disciplinary biology of the nineteenth century.

Acknowledgments

This culminating and most comprehensive work of my career has taken more than ten years to complete, and over that term I have acquired many debts. I would like first to acknowledge Mr. and Mrs. Bruce Dunlevie for their generous endowment of the John Antony Weir Chair in History, to which I have the honor of being the inaugural appointee. Through this wonderful endowment, I could travel round the world to conferences in my field and carry on my research secure in the knowledge that I had funding throughout. Next, I want to acknowledge my very long and productive history with the University of Chicago Press, now extending over more than a quarter century and involving four books. In all these projects, I have benefited from the editorship of T. David Brent; he saw the current work through board approval before the project was transferred to Karen Merikangas Darling, with whom I have enjoyed a seamless transition and a cordial working relationship. I want to thank the readers of this manuscript for the press: first, Nicolaas Rupke for his enormous generosity and support of the project, and second, the anonymous second reader, whose set of exacting demands for revision have made this a much better book.

The field of eighteenth-century life science is narrow, and I am grateful for the stimulus and insight of that cohort of scholars, especially my good friends Phillip Sloan, Robert Richards, Peter Hanns Reill, Philippe Huneman, Charles Wolfe, Catherine Wilson, and Jennifer Mensch. I have also benefited enormously from the extensive scholarship of Frederick Beiser, François Duchesneau, Justin E. Smith, Ilse Jahn, and Peter McLaughlin. For my ventures into German Idealist *Naturphilosophie* I am grateful for formative conversations with Olaf Breidbach and Thomas Bach in Jena, for the works of Joan Steigerwald and Jocelyn Holland, and for my new and wonderful cadre of colleagues and friends at the University of Sydney—Dalia Nassar, Stephen Gaukroger, Peter Anstey, Anik Waldow, and others. It was through them that I came to meet and discover great convergences with Eckart Förster, to my great pleasure.

I wish to thank the many commentators and peer reviewers, over the decade and more, who provided encouragement or criticism, or most often both, on the various ideas I tried out on them that eventually, through many mutations, made their way into this work. Many of these early versions will be cited in my notes, and I am grateful to all the venues that accommodated their articulation.

Finally, I thank my wife for her endurance and continued commitment across this harrowing epoch of my research. In recompense I offer the finished work to her in token of my devotion.

Notes

INTRODUCTION

1. Michel Foucault, *The Order of Things: An Archaeology of the Human Sciences* (New York: Vintage, 1973), 127–28. Originally published as *Les mots et les choses* (Paris: Gallimard, 1966).

2. Friedrich Schelling, *Von der Weltseele: Eine Hypothese der höheren Physik zur Erklärung des allgemeinen Organismus* (Hamburg, 1798), 298n; Nicholas Jardine, "The Significance of Schelling's 'Epoch of a Wholly New Natural History': An Essay on the Realization of Questions," in *Metaphysics and Philosophy of Science in the Seventeenth and Eighteenth Centuries*, ed. R. S. Woodhouse (Dordrecht: Kluwer, 1988), 327–50.

3. Foucault, *Order of Things*, 127–28 and passim; Ernst Mayr, *The Growth of Biological Thought: Diversity, Evolution, and Inheritance* (Cambridge, MA: Belknap Press, 1982), 108–9.

4. Sydney Ross, "Scientist: The Story of a Word," *Annals of Science* 18 (1962): 65–85.

5. Ogilvie contends that "only in the middle of the sixteenth century did naturalists come to think of themselves as practitioners of a discipline that, though related to medicine and natural philosophy, was distinct from both." Brian Ogilvie, *The Science of Describing: Natural History in Renaissance Europe* (Chicago: University of Chicago Press, 2006), 1. In the eighteenth century, Spary elaborates, "naturalists set about recruiting social and natural-historical credit by fashioning natural history as a distinct program of inquiry, appealing to a number of different audiences." Thus, "defining natural history as a scientific enterprise in its own right, independent of medicine, meant inventing a new expertise over the natural world which could be demonstrated to peers, patrons, and the wider scientific audience." E. C. Spary, *Utopia's Garden: French Natural History from Old Regime to Revolution* (Chicago: University of Chicago Press, 2000), 17.

6. William Wallace, "Traditional Natural Philosophy," in *The Cambridge History of Renaissance Philosophy*, ed. Charles Schmitt, Quentin Skinner, and Eckhard Kessler (Cambridge: Cambridge University Press, 1988), 201–35; Ann Blair, "Natural Philosophy," in *Cambridge History of Science*, vol. 3, *Early Modern Science*, ed. Katherine Park and Lorraine Daston (Cambridge: Cambridge University Press, 2001), 365–406.

7. According to Ogilvie, "from the 1530s through the 1630s, the task of natural history—as a discipline distinct from the mere art of gardening and the lofty science of nature—was describing nature, cataloguing its marvelous and mundane products" (*Science of Describing*, 6). See also Paula Findlen, "Natural History," in Park and Daston, *Cambridge History of Science*, 3:437–68.

8. Phillip Sloan, "Natural History," in *Cambridge History of Eighteenth Century Philosophy*, ed. Knud Haakonssen (Cambridge: Cambridge University Press, 2006), 903–37.

9. All these terms structure my interpretation. I have drawn them from Thierry Hoquet, Rudolf Stichweh, and Peter Hanns Reill. For the notion of "nonmathematical *physique*," see Thierry Hoquet, "History without Time: Buffon's Natural History as a Nonmathematical Physique," *Isis* 101, no. 1 (2010): 30–61. The notion of "semantic field" refers to the language by which a discourse organizes its concepts and their relations. It has been used very effectively in the characterization of the establishment of the discipline of physics in Germany in the eighteenth century by Stichweh, and I will adopt it in my own historical account of life science in that period. See Rudolf Stichweh, *Zur Entstehung des modernen Systems wissenschaftlicher Disziplinen: Physik in Deutschland, 1740–1890* (Frankfurt am Main: Suhrkamp, 1984). For "vital materialism," my main source is Peter Hanns Reill, *Vitalizing Nature in the Enlightenment* (Berkeley: University of California Press, 2005).

10. Thomas Kuhn, *The Structure of Scientific Revolutions*, 2nd ed. (Chicago: University of Chicago Press, 1970).

11. Dirk Stemerding, "Plants, Animals and Formulae: Natural History in the Light of Latour's *Science in Action* and Foucault's *The Order of Things*" (diss., School of Philosophy and Social Science, University of Twente, Netherlands, 1991), preface, n.p.

12. David Depew and Bruce Weber, eds., *Evolution at the Crossroads: The New Biology and the New Philosophy of Science* (Cambridge, MA: MIT Press, 1985); C. Allen, M. Bekoff, and G. Lauder, eds., *Nature's Purposes: Analyses of Function and Design in Biology* (Cambridge, MA: MIT Press, 1998); Bernard Feltz, Marc Commelinck, and Philippe Goujon, eds., *Self-Organization and Emergence in the Life Sciences* (Dordrecht: Springer, 2006); Manfred Laubichler and Jane Maienschein, eds., *Form and Function in Developmental Evolution* (Cambridge: Cambridge University Press, 2009). Cf. E. S. Russell, *Form and Function: A Contribution to the History of Animal Morphology* (London, 1916).

13. On the notion of a "special science," see J. Fodor, "Special Sciences and the Disunity of Science as a Working Hypothesis," *Synthese* 28 (1974): 97–115. I juxtapose *presentism* and *historicism*, not as mutually exclusive, but as mutually dependent approaches to history of philosophy. Of course, we appropriate earlier ideas for our present purposes; but we should do so self-consciously cognizant that this is precisely appropriation, not legitimate historical ascription—at least not until we have established the latter by historical reconstruction. Historicism, by the same token, is no simple antiquarianism: it must be motivated by present problems and interests. See my review essay "Reconstructing German Idealism and Romanticism: Historicism and Presentism," review of *German Idealism: The Struggle against Subjectivism, 1781–1801*, by Frederick Beiser, and *The Romantic Conception of Life: Science and Philosophy in the Age of Goethe*, by Robert Richards, *Modern Intellectual History* 1, no. 4 (2004): 427–38.

14. The importance of the new and simultaneous invocation of the term "biology" around 1800 has been most pertinently articulated by Bach: "That in this context [of the simultaneous use of the term "biology" by various authors around 1800] a general trend was actually manifesting itself can be further discerned in that it was not only simultaneously but also from the most diverse vantages that the problematic of a comprehensive science of life under the rubric biology came to formulation, e.g. from the vantage of theories of life-force, from that of clinical medicine and from that of physiology." Thomas Bach, *Biologie und Philosophie bei C. F. Kielmeyer und F. W. J. Schelling* (Stuttgart: Frommann-Holzboog, 2001), 84.

15. Mayr, *Growth of Biological Thought*, 108–9. We now know that the term "biology" was invented well before 1800, though without any systematic consequence. See Peter McLaughlin, "Naming Biology," *Journal of the History of Biology* 35 (2002): 1–4; and more extensively, Kai Torsten Kanz, "Zur Frühgeschichte des Begriffs 'Biologie': Die botanische Biologie (1771) von Johann Jakob Planer (1743–1789)," in *Verhandlungen zur Geschichte und Theorie der Biologie*, ed. German Society for History and Theory of Biology (Berlin: Verlag für Wissenschaft und Bildung, 2000), 269–82; Kai Torsten Kanz, "Von der BIOLOGIA zur Biologie: Zur Begriffsentwicklung und Disziplingenese vom 17. bis zum 20. Jahrhundert," in *Die Entstehung biologischer Disci-*

plinen: Beiträge zur 10. Jahrestagung der DGGTB in Berlin 2001 (Berlin: Verlag für Wissenschaft und Bildung, 2002), 9–30; Ilse Jahn, "Die Ordungswissenschaften und die Begründung biologischer Disziplinen im 18. und zu Beginn des 19. Jahrhunderts," in *Geschichte der Biologie: Theorien, Methoden, Institutionen, Kurzbiographien*, ed. Ilse Jahn, Rolf Löther, and Konrad Senglaub, 2nd ed. (Jena: Fischer, 1985), 264–323; Ilse Jahn, "Untersuchungen zum Phasenunterschied in der Herausbildung der Botanik und Zoologie zur Entstehungszeit der 'Biologie,'" *Rostocker wissenschaftshistorische Manuskripte* 2 (1978): 59–68.

16. Trevor DeJager, "G. R. Treviranus (1776–1837) and the Biology of a World in Transition" (PhD diss., University of Toronto, 1991), 6.

17. Bach, *Biologie und Philosophie bei Kielmeyer und Schelling*, 85.

18. "Natural history developed . . . ultimately into phylogenetic taxonomy, which reconstructs the natural order of its field via common origin [*die natürliche Ordnung ihres Gegenstandbereichs über die gemeinsame Abstammung rekonstruiert*]. Comparative anatomy, on the other hand, demonstrates that not only relative to order but also relative to the organization of organisms, there is a unity of field [*nicht nur bezüglich der Ordnung, sondern auch bezüglich der Organisation der Organismen eine Einheitlichkeit des Gegenstandbereichs vorliegt*]" (ibid., 86–87).

19. Ziche makes a very fruitful distinction between a *Fachgebiet* as a shared field of research and a *Disziplin* in the sense of an organized and institutionalized research community. Paul Ziche, "Von der Naturgeschichte zur Naturwissenschaft: Die Naturwissenschaften als eigenes Fachgebiet an der Universität Jena," *Berichte zur Wissenschaftsgeschichte* 21 (1998): 251–63. I will adopt this distinction for my own reconstruction.

20. Olaf Breidbach, "Transformation statt Reihung—Naturdetail und Naturganzes in Goethes Metamorphosenlehre," in *Naturwissenschaften um 1800: Wissenschaftskultur in Weimar-Jena*, ed. Olaf Breidbach and Paul Ziche (Weimar: Böhlaus, 2001), 46–64, citing 60, 64.

21. Engelhardt has struggled mightily to establish the idea that the German morphological turn was strictly "ideal," not "real"—a typological, not a phylogenetic theory of *Bauplan*. See Dietrich von Engelhardt, "Die Naturwissenschaft der Aufklärung und die romantisch-idealistische Naturphilosophie," in *Idealismus und Aufklärung*, ed. Christoph Jamme and Gerhard Kunz (Stuttgart: Klett-Cotta, 1988), 80–96; Dietrich von Engelhardt, "Zur Naturwissenschaft und Naturphilosophie um 1780 und 1830," in *Hegel und die Chemie: Studie zur Philosophie und Wissenschaft der Natur um 1800* (Wiesbaden: G. Pressler, 1975), 5–30; Dietrich von Engelhardt, "Historical Consciousness in the German Romantic *Naturforschung*," in *Romanticism and the Sciences*, ed. Andrew Cunningham and Nicholas Jardine (Cambridge: Cambridge University Press, 1990), 55–68; Dietrich von Engelhardt, "Natur und Geist, Evolution und Geschichte: Goethe in seiner Beziehung zur romantischen Naturforschung und metaphysischen Naturphilosophie," in *Goethe und die Verzeitlichung der Natur*, ed. Peter Matussek (Munich: Beck, 1998). But even he admits, crucially, that this distinction was not consistently maintained. See, for a similar formulation, Camilla Warnke, "Schellings Idee und Theorie des Organismus und der Paradigmawechsel der Biologie um die Wende zum 19. Jahrhundert," *Jahrbuch für Geschichte und Theorie der Biologie* 5 (1998): 187–234, esp. 188.

22. Bach, *Kielmeyer und Schelling*, 86–87.

23. For a more specific sense of the transnational in German-French interaction, see Michel Espagne and Michael Werner, eds., *Transfert: Les relations interculturelles dans l'espace Franco-Allemand (XVIIIe et XIXe siècle)* (Paris: Editions Recherche sur les Civilisations, 1988); Uwe Steiner, Brunhilde Wehinger, and Barbara Schmidt-Haberkamp, eds., *Europäischer Kulturtransfer im 18. Jahrhundert: Literaturen in Europa—europäische Literatur?* (Berlin: Berliner Wissenschafts-Verlag, 2003).

24. The idea of a "national context," allowing for the more nuanced differentiation of such phenomena as Enlightenment or Romanticism, was the guiding historiographical principle on which my generation of intellectual historians was trained, and it is still a valuable point of departure for the more recent efforts at the integration of a "transnational" cultural history. See Roy

Porter and Mikulas Teich, eds., *The Enlightenment in National Context* (Cambridge: Cambridge University Press, 1981). Less distinguished but equally representative is Roy Porter and Mikulas Teich, eds., *The Scientific Revolution in National Context* (Cambridge: Cambridge University Press, 1992).

25. Two works in the history of life science that I admire most share this attitude. Peter Hanns Reill's *Vitalizing Nature in the Enlightenment*, while it memorably details the formulation of a vital-materialist paradigm shift in the sciences of the mid-eighteenth century, drastically dismisses *Naturphilosophie* as an aberration in natural inquiry. Ron Amundson's *The Changing Role of the Embryo in Evolutionary Thought: Roots of Evo-Devo* (Cambridge: Cambridge University Press, 2005), while it carries out a brilliant revisionism concerning the "modern synthesis" in the history of biology, cannot take seriously the ostensibly thick metaphysics that attached to German contributions to that history.

26. See my essay "Reill's *Vitalizing Nature in the Enlightenment* and German *Naturphilosophie*," in *Life Forms in the Thinking of the Long Eighteenth Century*, ed. Jenna Gibbs and Keith Baker (Toronto: University of Toronto Press, 2016), 70–91; and my review of *Reproduction, Race, and Gender in Philosophy and the Early Life Sciences*, ed. Susanne Lettow, *Critical Philosophy of Race* 3, no. 1 (January 2015): 158–66.

27. Frederick Beiser, *German Idealism: The Struggle against Subjectivism, 1781–1801* (Cambridge, MA: Harvard University Press, 2002), 507. Beiser affirms that "the legacy of positivism remains, and the old image of *Naturphilosophie* persists to this day" (507).

28. According to Richards, "Historians of nineteenth-century science, the really serious historians, usually dismiss anything sounding like Romantic science as an aberration." He elaborates: "The Romantic movement, to be sure, has been dismissed as a source of genuine scientific accomplishment." Robert Richards, *The Romantic Conception of Life: Science and Philosophy in the Age of Goethe* (Chicago: University of Chicago Press, 2002), 3, 512.

29. See my "Reconstructing German Idealism and Romanticism."

30. Daniel Steuer, "Goethe's Natural Investigations and Scientific Culture," in *The Cambridge Companion to Goethe*, ed. Lesley Sharpe (Cambridge: Cambridge University Press, 2002), 160–78, citing 175. For the substantial revival of interest in and appreciation for *Naturphilosophie* and the science of the Romantic epoch, see Cunningham and Jardine, *Romanticism and the Sciences*; Karen Gloy and Paul Burger, eds., *Die Naturphilosophie im deutschen Idealismus* (Stuttgart: Frommann/Holzboog, 1993); Stefano Poggi and Maurizio Bossi, eds., *Romanticism in Science: Science in Europe, 1790–1840* (Dordrecht: Kluwer, 1994); Frederick Amrine, Francis Zucker, and Harvey Wheeler, eds., *Goethe and the Sciences: A Reappraisal* (Dordrecht: Reidel, 1987); Robert Cohen and Marx Wartofsky, eds., *Hegel and the Sciences* (Dordrecht: Reidel, 1984).

31. See my *A Nice Derangement of Epistemes: Post-positivism in the Study of Science from Quine to Latour* (Chicago: University of Chicago Press, 2004). On naturalism, see Philip Kitcher, "The Naturalists Return," *Philosophical Review* 101 (1992): 53–114; Werner Callebaut, ed., *Taking the Naturalistic Turn; or, How Real Philosophy of Science Is Done* (Chicago: University of Chicago Press, 1993); Peter French, Theodore Uehling, and Howard Wettstein, eds., *Philosophical Naturalism*, Midwest Studies in Philosophy 19 (Notre Dame, IN: University of Notre Dame Press, 1994); Joseph Rouse, *Engaging Science: How to Understand Its Practices Philosophically* (Ithaca, NY: Cornell University Press, 1996); Joseph Rouse, *How Scientific Practices Matter: Reclaiming Philosophical Naturalism* (Chicago: University of Chicago Press, 2002); Mario de Caro and David Macarthur, eds., *Naturalism in Question* (Cambridge: Cambridge University Press, 2004).

32. See "Part C: Studies in History and Philosophy of Biological and Biomedical Sciences," special issue, *Studies in History and Philosophy of Science* 43 (2012), on the question of teleology in contemporary biology and philosophy of biology. See also Nicholas Rescher, ed., *Current Issues in Teleology* (Lanham, MD: University Press of America, 1986); Mark Bedau, "Can Biological Teleology Be Naturalized?," *Journal of Philosophy* 88 (1991): 647–55; W. Christensen, "A Complex Systems Theory of Teleology," *Biology and Philosophy* 11 (1996): 301–20; Paul Griffiths and Ronald Gray, "Developmental Systems and Evolutionary Explanation," *Journal of Philosophy* 91 (1994): 277–304; Tim Lewens, "No End to Function Talk in Biology," *Studies in History and Philosophy of Biology*

and the Biomedical Sciences 32 (2001): 179–90; D. M. Walsh, "Organisms as Natural Purposes: The Contemporary Evolutionary Perspective," *Studies in History and Philosophy of the Biological and Biomedical Sciences* 37 (2006): 771–91; Peter McLaughlin, *What Functions Explain: Functional Explanation and Self-Reproducing Systems* (Cambridge: Cambridge University Press, 2001).

33. "What is certain is the positive relation to Protestant religiosity," "a specific element of the German as against the west-European Enlightenment." Hans Poser, "Pietismus und Aufklärung—Glaubensgewißheit und Vernunfterkenntnis im Widerstreit," in *Aufklärung und Erneuerung: Beiträge zur Geschichte der Universität Halle im ersten Jahrhundert ihres Bestehens (1694–1806)*, ed. Günter Jerouschek and Arno Sames (Hanau/Halle: Dausien, 1994), 170–82, citing 173. See also Pangiotes Kondylis, *Die Aufklärung im Rahmen des neuzeitlichen Rationalismus* (Stuttgart: Klett-Cotta, 1981), 538.

34. See Poser, "Pietismus und Aufklärung"; Udo Stäter, "Aufklärung und Pietismus—das Beispiel Halle," in *Universitäten und Aufklärung*, ed. Notker Hammerstein (Göttingen: Wallstein, 1995), 49–61.

35. In this, the German Enlightenment showed strong parallels to the Scottish Enlightenment, and for similar reasons. For some rich explorations of this parallel, see Fania Oz-Salzberger, *Translating the Enlightenment: Scottish Civic Discourse in Eighteenth-Century Germany* (Oxford: Clarendon, 1995).

36. Lester King, *The Medical World in the Eighteenth Century* (Chicago: University of Chicago Press, 1958); Andrew Cunningham and Roger French, eds., *The Medical Enlightenment of the Eighteenth Century* (Cambridge: Cambridge University Press, 1990); Richard Toellner, "Medizin in der Mitte des 18. Jahrhunderts," in *Wissenschaft im Zeitalter der Aufklärung*, ed. Rudolf Vierhaus (Göttingen: Vandenhoeck und Ruprecht, 1985), 194–221.

37. See Jens Häseler and Albert Meier, with the collaboration of Olaf Koch, eds., *Gallophobie im 18. Jahrhundert: Akten der Fachtagung vom 2. / 3. Mai 2002 am Forschungszentrum Europäische Aufklärung* (Berlin: Berliner Wissenschafts-Verlag, 2005); Raymond Heitz, York-Gothard Mix, Jean Mondot, and Nina Birkner, eds., *Gallophilie und Gallophobie in der Literatur und den Medien in Deutschland und in Italien im 18. Jahrhundert / Gallophilie et gallophobie dans la littérature et les médias en Allemagne et en Italie au XVIIIe siècle* (Heidelberg: Winter, 2011). For the traditional stereotyping, see Ruth Florack, "Nationalcharakter als ästhetisches Argument," in Häseler and Meier, *Gallophobie im 18. Jahrhundert*, 33–48, a good résumé of her monograph, *Tiefsinnige Deutsche, frivole Franzosen—Nationale Stereotype in deutscher und französischer Literatur* (Stuttgart: Metzler, 2001).

38. See Avi Lifschitz, *Language and Enlightenment: The Berlin Debates of the Eighteenth Century* (Oxford: Oxford University Press, 2012).

39. The "learned and rational Doctor," in French's terms, had asserted hierarchical preeminence over all other health-care practitioners by an appeal to *learning*, the mastery of a substantial body of texts, and to *discourse*, the ability to formulate persuasive arguments about the nature of human health and illness. Roger French, *Medicine before Science: The Rational and Learned Doctor from the Middle Ages to the Enlightenment* (Cambridge: Cambridge University Press, 2003), 63, 67.

40. See Gerrit Lindeboom, *Descartes and Medicine* (Amsterdam: Rodopi, 1979); Richard B. Carter, *Descartes' Medical Philosophy: The Organic Solution to the Mind-Body Problem* (Baltimore: Johns Hopkins University Press, 1983).

41. Harold Cook, "Physick and Natural History in Seventeenth-Century England," in *Revolution and Continuity: Essays in the History and Philosophy of Early Modern Science*, ed. Peter Barker and Roger Ariew (Washington, DC: Catholic University of America, 1991), 63–82, citing 79. See also Harold Cook, "The New Philosophy and Medicine in Seventeenth-Century England," in *Reappraisals of the Scientific Revolution*, ed. David Lindberg and Robert Westman (Cambridge: Cambridge University Press, 1990), 397–436; Harold Cook, "Physicians and Natural History," in *Cultures of Natural History*, ed. N. Jardine, J. E. Secord, and E. C. Spary (Cambridge: Cambridge University Press, 1996), 91–105.

42. Descartes's appropriation, for this mechanist insurgency, of the greatest empirical discov-

ery in seventeenth-century medicine, William Harvey's work on circulation, proved "a major factor in the loss of traditional theory" (French, *Medicine before Science*, 176). "Descartes read Harvey's book in about 1630 and decided that Harvey's doctrine of the circulation was the ideal vehicle for his own mechanism" (182). On this, see Thomas Fuchs, *The Mechanization of the Heart: Harvey and Descartes*, trans. Marjorie Grene (Rochester, NY: University of Rochester Press, 2001).

43. His formulations are worth citing: "I suppose the body to be nothing but a statue or machine made of earth. . . . We see clocks, artificial fountains, mills, and other such machines which, although only man-made, have the power to move of their own accord in many different ways. But I am supposing this machine to be made by the hands of God, and so I think . . . it capable of a greater variety of movements . . . and of exhibiting more artistry." René Descartes, *Treatise on Man*, in *Philosophical Writings*, 3 vols. (Cambridge: Cambridge University Press, 1984–92), 1:99. "Indeed, one may compare the nerves of the machine I am describing with the pipes in the works of these fountains, its muscles and tendons with the various devices and springs which serve to set them in motion, its animal spirits with the water which drives them, the heart with the source of the water, and the cavities of the brain with the storage tanks. Moreover, breathing and other such activities which are normal and natural to this machine, and which depend on the flow of the spirits, are like the movements of a clock or mill" (100–101). "Think of our machine's heart and arteries, which push the animal spirits into the cavities of the brain, as being like the bellows of an organ, which push air into the wind-chests" (104). "I should like you to consider that these functions follow from the mere arrangement of the machine's organs every bit as naturally as the movements of a clock or other automaton follow from the arrangement of its counter-weights and wheels. In order to explain these functions, then, it is not necessary to conceive of this machine as having any vegetative or sensitive soul or other principle of movement and life, apart from its blood and its spirits, which are agitated by the heat of the fire burning continuously in its heart—a fire which has the same nature as all the fires that occur in inanimate bodies" (108).

44. Thomas Hobbes, *Leviathan* (1651; Oxford: Clarendon, 2012), 1.

45. The key text, of course, is Julien Offray de La Mettrie, *L'homme machine* (1747). Trans. Gertrude C. Bussey, rev. M. W. Calkins, as *Man a Machine* (LaSalle, IL: Open Court, 1912). See Julien Offray de La Mettrie, *Dedication to Haller*, in *Man a Machine and Man a Plant*, trans. Richard A. Watson and Maya Rybalka (Indianapolis: Hackett, 1994).

46. Aram Vartanian, *La Mettrie's "L'homme machine": A Study in the Origins of an Idea* (Princeton, NJ: Princeton University Press, 1960), 59.

47. Robert Young, "Animal Soul," in *Encyclopedia of Philosophy*, 1:122–27, citing 122. According to Rosenfield, "The most unmistakable evolution undergone by the movement of animal automatism was its increasingly religious tone in France as elsewhere. After Descartes it assumed, as time went on, a growing theological motivation." Leonora Cohen Rosenfield, *From Beast-Machine to Man-Machine: Animal Soul in French Letters from Descartes to La Mettrie* (New York: Octagon, 1968), 67. The key figure was the great Oratorian interpreter of Descartes: Nicolas Malebranche. "Before Malebranche, the writers on the subject were preponderantly doctors of medicine. After him, professional churchmen were in the majority" (Rosenfield, *From Beast-Machine to Man-Machine*, 68). Indeed, Malebranche's "phrasing was followed more closely than Descartes'" (67).

48. "What leads the Cartesians to say that beasts are machines is that according to them all matter is incapable of thinking. . . . According to this thesis every man can be convinced of the immortality of his soul. . . . Here is a great advantage for religion." The alternative view, he went on, was clearly pernicious: "by ascribing a soul to beasts that is capable of knowledge, all the natural proofs of our soul's immortality are destroyed." This was the aim of "impious thinkers and Epicureans." Pierre Bayle, "Rorarius," in *Historical and Critical Dictionary: Selections*, trans. Richard Popkin (Indianapolis: Hackett, 1991), 213–34, citing 216–17, 220. Scholastic opponents of Cartesian mechanism put themselves at great risk in seeking to conjecture some lesser soul for animals, for this opened them to the argument that the human soul was not different in kind, a slippery slope to Epicureanism and materialism. "We cannot think without horror of the consequences of this

doctrine: 'Man's soul and that of beasts do not differ substantially....' It follows from this that if their souls are material and mortal, the souls of men are so also, and if the soul of man is an immaterial and spiritual substance, the soul of beasts is so also. These are horrible consequences no matter which way one looks at them" (224–25).

49. See, on physicotheology, Wolfgang Philipp, "Physicotheology in the Age of Enlightenment: Appearance and History," *Studies on Voltaire and the Eighteenth Century* 57 (1967): 1233–67; Richard Toellner, "Die Bedeutung des physico-theologischen Gottesbeweises für die nachcartesianische Physiologie im 18. Jahrhundert," *Berichte zur Wissenschaftsgeschichte* 5 (1982): 75–82; François Russo, "Théologie naturelle et sécularisation de la science au XVIII siècle," *Recherches de science religieuse* 66 (1978): 27–62; Udo Krolzik, "Das physikotheologische Naturverständnis und sein Einfluß auf das naturwissenschaftliche Denken im 18. Jahrhundert," *Medizinhistorisches Journal* 15 (1980): 90–102; Paul Michel, *Physikotheologie: Ursprünge, Leistung und Niedergang einer Denkform* (Zurich: Beer, 2008); Maria Teerea Monti, "Thélogie physique et mécanisme das la physioogie de Haller," in *Science and Relgion / Wissenschaft und Religion*, ed. A. Bäumer and M. Büttner (Bochum: Universitätsverlag N. Brockmeyer, 1989), 68–79. On Epicureanism, see Catherine Wilson, *Epicureanism at the Origins of Modernity* (Oxford: Oxford University Press, 2008); W. R. Johnson, *Lucretius and the Modern World* (London: Duckworth, 2000); Stuart Gillespie and Philip Hardie, eds., *The Cambridge Companion to Lucretius* (Cambridge: Cambridge University Press, 2007); Neven Leddy and Avi Lifschitz, eds., *Epicurus in the Enlightenment* (Oxford: Voltaire Foundation, 2009); Margaret Osler, ed., *Atoms, Pneuma, and Tranquility: Epicurean and Stoic Themes in European Thought* (Cambridge: Cambridge University Press, 1991).

50. French, *Medicine before Science*, 207.

51. Johanna Geyer-Kordesch, "Passions and the Ghost in the Machine: Or What Not to Ask about Science in Seventeenth- and Eighteenth-Century Germany," in *The Medical Revolution of the Seventeenth Century*, ed. Roger French and Andrew Wear (Cambridge: Cambridge University Press, 1989), 145–63, citing 154. As King has aptly noted: "in reality the 17th century physicians applied the laws of mechanics and hydraulics to medicine not by experimental techniques but only by *analogy* and *armchair calculations*." Lester King, *The Background of Herman Boerhaave's Doctrine: Boerhaave Lecture Held on September 17, 1964* (Leiden: University of Leiden, 1965), 17.

52. Allen Debus, *The Chemical Philosophy: Paracelsian Science and Medicine in the Sixteenth and Seventeenth Centuries* (New York: Science History Publications, 1977); Allen Debus, *The French Paracelsians: The Chemical Challenge to Medical and Scientific Tradition in Early Modern France* (Cambridge: Cambridge University Press, 1991); Allen Debus, *Chemistry, Alchemy and the New Philosophy, 1550–1700: Studies in the History of Science and Medicine* (London: Variorum, 1987); Allen Debus, *Chemistry and Medical Debate: Van Helmont to Boerhaave* (Canton, MA: Science History Publications, 2001).

53. Steinke writes of "a current of non-mechanist thought running along or underneath the essentially mechanistic outlook of physiology." Hubert Steinke, *Irritating Experiments: Haller's Concept and the European Controversy on Irritability and Sensibility, 1750–90* (Amsterdam: Rodopi, 2005), 22. To "philosophize mechanistically [*mechanisch philosophieren*]" in medicine was to embrace certain paramount allegiances, within which there had to be plenty of space to accommodate other impulses—whether "iatrochemical" or Galenic or, indeed, "animist."

54. "In distancing themselves from the new empirics, the new rationalists created a problem for themselves.... They could not afford to emphasize observation and experience, which looked rather empirical, at the expense of theory. Worse, the great Hippocrates, widely revered as the Father of Medicine, was—it was generally admitted—without the arts and sciences" (French, *Medicine before Science*, 109). Thus, "the doctors were compelled to consider what they had indignantly rejected for centuries, namely that medicine was an empirical art, not a rational *scientia*" (187).

55. Peter Anstey, "The Creation of the English Hippocrates," *Medical History* 55 (2011): 457–78. See also D. E. Wolfe, "Sydenham and Locke on the Limits of Anatomy," *Bulletin of the History of*

Medicine 35 (1961): 193–200; K. Dewhurst, "Locke and Sydenham on the Teaching of Anatomy," *Medical History* 2 (1958): 1–12; Andrew Cunningham, "Thomas Sydenham: Epidemics, Experiment and the 'Good Old Cause,'" in French and Wear, *Medical Revolution of the Seventeenth Century*, 164–90.

56. "Both animists and mechanists often tried to find common ground in a new, confident and enlightened system of medicine," French tells us (*Medicine before Science*, 207).

57. Thus "Boerhaave welded theory and practice into a system which found recognition and acceptance" in an early eighteenth-century medical community eager for some coherence (King, *Background of Herman Boerhaave's Doctrine*, 19).

58. "It was the great service of Boerhaave to have combined the chemical and the physical, the pathological-anatomical and the microscopic with the Hippocratic conception of medicine." Paul Diepgen, "Hermann Boerhaave und die Medizin seiner Zeit mit besonderer Berücksichtigung seiner Wirkung nach Deutschland," *Hippokrates: Zeitschrift für praktische Heilkunde* 10 (1939): 298–306 and 345–51, citing 302–4. See also F. L. S. Sassen, "The Intellectual Climate in Leiden in Boerhaave's Time," in *Boerhaave and His Time*, ed. G. A. Lindeboom (Leiden: Brill, 1970), 1–16, esp. 15. Diepgen ("Hermann Boerhaave," 302) makes clear the influence of Sydenham's skeptical empiricism on Boerhaave.

59. Haller called him *Communis Europa Praeceptor. Bibliotheca Medicinae Practicae*, 4 vols. (Bern: E. Haller, 1776), 4:142, cited in Gerrit Lindeboom, *Herman Boerhaave: The Man and His Work* (London: Methuen, 1968), 355. One-third of Leiden MDs came from German lands through 1738. Wolfram Kaiser, "Theorie und Praxis in der Boerhaave-Ära und in nachboerhaavianischen Ausbildungssystemen an deutschen Hochschulen des 18. Jahrhunderts," *Clio Medica: Acta Academia Internalialis Historiae Medicinae* 21 (1987): 71–94, citing 72.

60. Some now hold this to be undeserved. Cook, drawing on a study by W. Frithoff, claims that "the period of Boerhaave's professorship shows a decline both in the absolute number of foreigners who took their medical degrees at Leiden and in the percentage of foreigners who studied at Leiden rather than at other Dutch universities," and he supports Frithoff's conclusion that "Boerhaave's considerable reputation may have owed more to the changes in Dutch academic medicine advanced in the seventeenth century than to his personal genius." Harold Cook, "Boerhaave and the Flight from Reason in Medicine," *Bulletin of the History of Medicine* 74 (2000): 221–40, citing 224. To be sure, of the 1,919 students who matriculated at Leiden during Boerhaave's tenure, only 178 completed a dissertation under him (Lindeboom, *Herman Boerhaave*, 356–57). But that does not translate into a lack of influence. See Edgar Ashworth, *Boerhaave's Men at Leyden and After* (Edinburgh: Edinburgh University Press, 1977), for a more nuanced consideration of the behavior of British students at Leiden.

61. In his biography of Haller, Zimmermann goes on for many pages, on the basis of Haller's own attestations, concerning the personality of Boerhaave as the foundation of his pedagogical grandeur. Johann Georg Zimmermann, *Das Leben des Hrn Albrecht von Hallers* (Zurich: n.p., 1755).

62. Urs Boschung et al., eds., *Repertorium zu Albrecht von Hallers Korrespondenz, 1724–1777*, 2 vols. (Basel: Schwabe, 2002), 1:112.

63. Albrecht von Haller, *Bibliotheca Anatomica*, vol. 1 (1774; repr., Hildesheim: G. Olms, 1969), 757, cited in Shirley Roe, "*Anatomia Animata*: The Newtonian Physiology of Albrecht von Haller," in *Transformations and Tradition in the Sciences*, ed. Everett Mendelsohn (Cambridge: Cambridge University Press, 1984), 273–300, citing 287.

64. Adolf Haller, *Albrecht von Hallers Leben* (Basel: Reinhardt, 1954), 24.

65. Ashworth, *Boerhaave's Men*. For the wider context, see Simon Schaffer, "The Glorious Revolution and Medicine in Britain and the Netherlands," *Notes and Records of the Royal Society* 42 (1989): 167–90.

66. See Kaiser, "Theorie und Praxis in der Boerhaave-Ära"; Diepgen, "Hermann Boerhaave und die Medizin seiner Zeit"; Guenther Risse, "Clinical Instruction in Hospitals: The Boerhaavian Tradition in Leyden, Edinburgh, Vienna and Pavia," *Clio Medica* 21 (1987/88): 1–19; Andrew Cun-

ningham, "Medicine to Calm the Mind: Boerhaave's Medical System, and Why It Was Adopted in Edinburgh," in Cunningham and French, *Medical Enlightenment of the Eighteenth Century*, 40–66; and G. E. Lindeboom, *Boerhaave and Great Britain* (Leiden: Brill, 1974).

67. On the Montpellier medical school, see Elizabeth Willliams, *The Physical and the Moral: Anthropology, Physiology, and Philosophical Medicine in France, 1750–1850* (Cambridge: Cambridge University Press, 1994); Elizabeth Williams, *A Cultural History of Medical Vitalism in Enlightenment Montpellier* (Aldershot, UK: Ashgate, 2003).

68. Herman Boerhaave, "Oration on the Usefulness of the Mechanical Method in Medicine," in *Boerhaave's Orations*, trans. E. Kegel-Brinkgreve and A. M. Luyendijk-Elshout (Leiden: Brill, 1983), 85–120.

69. Harm Beukers, "Boerhaavianism in the Netherlands," *Journal of the Japan-Netherlands Institute* 1 (1989): 116–29, citing 117. Beukers points out that Boerhaave himself "indicated that he became less active in medical practice after 1723" and "that he did not spend much time visiting patients." Harm Beukers, "Clinical Teaching in Leiden from Its Beginning until the End of the Eighteenth Century," *Clio Medica* 21 (1987/88): 139–52, citing 147.

70. Irmtraut Scheele, "Grundzüge der institutionellen Entwicklung der biologischen Disziplinen an den deutschen Hochschulen seit dem 18. Jahrhundert," in *"Einsamkeit und Freiheit" neu besichtigt: Universitätsreformen und Disziplinenbildung in Preußen als Modell für Wissenschaftspolitik im Europa des 19. Jahrhunderts* (Stuttgart: F. Steiner, 1991), 144–54, citing 150.

71. The *Fachgebiet* began "to evolve in a direction separate from other branches of medicine." Within the university culture of the eighteenth century, a few "non-clinical members of medical faculties enjoyed the freedom to pursue their interests in a setting largely unencumbered by considerations of how their research might inform or threaten practical doctrines." Thomas Broman, *The Transformation of German Academic Medicine* (Cambridge: Cambridge University Press, 1996), 74, 159.

72. Already in the seventeenth century, William Harvey did poorly as a practicing "learned physician" while pursuing his innovative physiological research. Jan Swammerdam (1637–80), perhaps his most eminent Dutch research peer, had sufficient wealth never to need to practice medicine. Marcello Malpighi (1628–94), the most important Italian research-oriented physician of the seventeenth century, found himself constantly upbraided in the Italian medical community because his inquiries were not "really medical."

73. The relation of *anatomy*, with its strongly interventionist (experimental, not simply observational) orientation, to the "theoretical" enterprise of *physiology* proved very complex. Cunningham has elaborated on this issue, contrasting "the pen and the sword." Andrew Cunningham, "The Pen and the Sword: Recovering the Disciplinary Identity of Physiology and Anatomy before 1800, Part I, Old Physiology: The Pen," *Studies in History and Philosophy of Science* 33 (2002): 631–65; Andrew Cunningham, "The Pen and the Sword: Recovering the Disciplinary Identity of Physiology and Anatomy before 1800, Part II, Old Anatomy: The Sword," *Studies in History and Philosophy of Science* 34 (2003): 51–76. On anatomy in the period, see Andrew Cunningham, *The Anatomist Anatomis'd* (Farham, Surry, UK: Ashgate, 2010); Matthew Landers and Brian Muñoz, eds., *Anatomy and the Organization of Knowledge, 1500–1850* (London: Pickering and Chatto, 2012); Rüdiger Schultka, Josef Neumann, and Susanne Weidemann, eds., *Anatomie und anatomische Sammlungen im 18. Jahrhundert* (Berlin: LIT Verlag Dr. W. Hopf, 2007).

74. Findlen, "Natural History," 462–63.

75. C. E. Raven, *English Naturalists from Neckham to Ray: A Study in the Making of the Modern World* (Cambridge: Cambridge University Press, 1947), 235. See also Neil Gillespie, "Natural History, Natural Theology, and Social Order: John Ray and the 'Newtonian Ideology,'" *Journal of the History of Biology* 20 (1987): 1–49; Phillip Sloan, "John Locke, John Ray, and the Problem of the Natural System," *Journal of the History of Biology* 5 (1972): 1–53. For Réaumur, Dawson puts it clearly: "Réaumur was the unquestioned European authority in natural history prior to 1749 when Georges-Louis Leclerc de Buffon published the first volumes of his more encyclopedic, if

less precise, works of natural history." Virginia Dawson, "Regeneration, Parthenogenesis, and the Immutable Order of Nature," *Archives of Natural History* 18 (1991): 309–21, citing 310.

76. Virginia Dawson, *Nature's Enigma: The Problem of the Polyp in the Letters of Bonnet, Trembley, and Réaumur* (Philadelphia: American Philosophical Society, 1987); P. Speziali, "Réaumur et les savants genevois," *Revue d'histoire des sciences* 11 (1958): 68–80.

77. Jacques Roger, *Buffon: A Life in Natural History* (Ithaca, NY: Cornell University Press, 1997).

78. For the specific persona of the *médecin philosophe*, see my essay "*Médecin-Philosophe*: Persona for Radical Enlightenment," *Intellectual History Review* 18, no. 3 (November 2008): 427–40.

79. Julien Offray de La Mettrie, *Machine Man and Other Writings* (Cambridge: Cambridge University Press, 1996), 5.

80. Théophile Bordeu, *Traité de médecine théorique & pratique* (Paris: Ruault, 1774). See Williams, *The Physical and the Moral*; Elizabeth Haigh, "Vitalism, the Soul, and Sensibility: The Physiology of Théophile Bordeu," *Journal of the History of Medicine* 31 (1976): 30–41.

81. G. Hartung, "Über den Selbstmord: Eine Grenzbestimmung des anthropologischen Diskurses im 18. Jahrhundert," in *Der ganze Mensch*, ed. H.-J. Schings (Stuttgart: Metzler, 1994), 33–51, citing 41.

82. Ibid. As Hatfield puts it, "Ontological questions were bracketed in order to concentrate on the study of mental faculties through their empirical manifestations in mental phenomena and external behavior." Gary Hatfield, "Remaking the Science of Mind: Psychology as Natural Science," in *Inventing Human Science*, ed. C. Fox, R. Porter, and R. Wokler (Berkeley: University of California Press, 1995), 184–231, citing 188.

83. Kathleen Wellman, *La Mettrie: Medicine, Philosophy, and Enlightenment* (Durham, NC: Duke University Press, 1992), 130.

84. John Yolton, *Locke and French Materialism* (Oxford: Clarendon, 1991); Eric Watkins, "Development of Physical Influx in Early Eighteenth-Century Germany: Gottsched, Knutzen, and Crusius," *Review of Metaphysics* 49 (1995): 295–339.

85. H.-J. Schings, "Der philosophische Arzt," in *Melancholie und Aufklärung* (Stuttgart: Metzler, 1977), 11–40; W. Riedel, "Influxus physicus und Seelenstärke," in *Anthropologie und Literatur um 1800*, ed. J. Barkoff and E. Sagarra (Munich: Iudicium, 1992), 24–52.

86. The two greatest names of the German Frühaufklärung, Leibniz and Thomasius, both wrestled energetically with this challenge. See esp. Albert Heinekamp, "Leibniz als Vermittler zwischen Frankreich und Deutschland," in *Aufklärung als Mission / La mission des Lumières*, ed. Werner Schneiders, Das achtzehnte Jahrhundert, Supplementa 1 (Marburg: Hitzeroth, 1993), 93–102; Catherine Julliard, "Christian Thomasius (1655–1728) et son *Discours de l'imitation des Français* (1687): Un plaidoyer gallophile dans un context gallophobe," in Heitz et al., *Gallophilie und Gallophobie*, 1–24. For later developments, see Eric Blackall, *The Emergence of German as a Literary Language, 1700–1775* (Ithaca, NY: Cornell University Press, 1978).

87. Friedrich II, *De la littérature allemande* (1780).

88. See Wolfgang Förster, ed., *Aufklärung in Berlin* (Berlin: Akademie Verlag, 1989); Olav Krämer, "'Welche Gestalt man denen Frantzosen . . . nachahmen solle': Stationen einer Jahrhundertdebatte (Thomasius, Prémontval, Herder, Friedrich II., Möser)," in Häseler and Meier, *Gallophobie im 18. Jahrhundert*, 61–88.

89. Rudolf Vierhaus, "Montesquieu in Deutschland: Zur Geschichte seiner Wirkung als politischer Schriftsteller im 18. Jahrhundert," in *Deutschland im 18. Jahrhundert* (Göttingen: Vandenhoeck und Ruprecht, 1987), 9–32.

90. Herbert Jaumann, "Rousseau in Deutschland: Forschungsgeschichte und Perspektiven," in *Rousseau in Deutschland: Neue Beiträge zur Erforschung seiner Rezeption* (Berlin: De Gruyter, 1995), 1–22; and see the extensive bibliography in that volume, 291–309.

91. Raymond Heitz, "Gallophile et gallophobe Diskurse in der Voltaire-Rezeption der deutschen Publizistik des 18. Jahrhunderts: Identitätssuche und Kampf gegen Entfremdung," in Heitz et al., *Gallophilie und Gallophobie*, 225–47. And see Martin Fontius, *Voltaire in Berlin* (Berlin: Rütten und Loening, 1966).

92. The classic account of the Bodmer/Breitinger challenge to Gottsched is Ernst Cassirer, *Philosophy of the Enlightenment* (Princeton, NJ: Princeton University Press, 1951). For a nuanced view, see Albert Meier, "Plus ultra! Johann Christoph Gottscheds gallophobe Gallophilie," in Heitz et al., *Gallophilie und Gallophobie*, 195–206.

93. See Gerhard Sauder, *Empfindsamkeit*, vol 1, *Voraussetzungen und Elemente* (Stuttgart: Metzler, 1974).

94. See the classic Erich Kästner, *Friedrich der Große und die deutsche Literatur: Die Erwiderungen auf seine Schrift "De la littérature allemande"* (1925; Stuttgart: Kohlhammer, 1972). Frederick's foil was Justus Möser, *Über deutsche Sprache und Literatur: Schreiben an einen Freund nebst einer Nachschrift, die National-Erziehung der alten Deutschen betreffend* (1781). See Winfried Woesler, "'. . . ob unsere Art der Kultur der fremden vorzuziehen sei?': Justus Möser antwortet Friedrich II," *Möser-Forum* 1 (1989): 192–207. Already Möser figured as a forefather in the great manifesto of the "German movement," *Von deutscher Art und Kunst* (1773), edited by Herder and Goethe and launching the Sturm und Drang.

95. Roy Pascal, *The German Sturm und Drang* (Manchester: Manchester University Press, 1953); David Hill, ed., *Literature of the Sturm und Drang* (Rochester: Camden House, 2003); esp. Gerhard Sauder, "The Sturm und Drang and the Periodization of the Eighteenth Century," in Hill, *Literature of the Sturm und Drang*, 309–32.

96. Wilhelm Dilthey, "Die dichterische und philosophische Bewegung in Deutschland 1770–1800," in *Gesammelte Schriften*, vol. 5 (Stuttgart: Vandenhoeck und Ruprecht, 1957), 11–27.

97. Rolf Grimminger, "Aufklärung, Absolutismus und bürgerlichen Individuen: Über den notwendigen Zsusammenhang von Literatur, Gesellschaft und Staat in der Geschichte des 18. Jahrhunderts," in *Hanser Sozialgeschichte der deutschen Literatur vom 16. Jahrhunder bis zur Gegenwart*, vol. 3, *Deutsche Aufklärung bis zur Französsischen Revolution*, ed. Rolf Grimminger (Munich: Hanser, 1980), 15–99, esp. 48ff. In the same volume, see also Reiner Wild, "Städtekultur, Bildungswesen und Aufklärungsgesellschaften," 103–32. Helen Liebel has given us a vivid image of Hamburg in this era; Horst Möller has done the same for Berlin. Helen Liebel, "Laissez-faire vs. Mercantilism: The Rise of Hamburg and the Hamburg Bourgeoisie vs. Frederick the Great in the Crisis of 1763," *Vierteljahrsschrift für Sozial und Wirtschaftsgeschichte* 52 (1965): 207–38; Horst Möller, *Aufklärung in Preußen: Der Verleger, Publizist und Geschichtsschreiber Friedrich Nicolai*, Einzelveröffentlichungen der Historischen Kommission zu Berlin, vol. 15 (Berlin: Colloquium, 1974). See also Helga Schultz, *Berlin, 1650–1800: Sozialgeschichte einer Residenz* (Berlin: Akademie, 1987), 163–320.

98. Paul Raabe, "Die Zeitschriften als Medium der Aufklärung," *Wolfenbütteler Studien zur Aufklärung* 1 (1974): 99ff.

99. Wolfgang Martens, *Botschaft der Tugend: Die Aufklärung im Spiegel der deutschen moralischen Wochenschriften* (Stuttgart: Metzler, 1968).

100. Benjamin Redekop, *Enlightenment and Community: Lessing, Abbt, Herder and the Quest for a German Public* (Montreal, QC: McGill-Queens University Press, 2000); James Van Horn Melton, *The Rise of the Public Sphere in Enlightenment Europe* (Cambridge: Cambridge University Press, 2001); Thomas Broman, "The Habermasian Public Sphere and 'Science *in* the Enlightenment,'" *History of Science* 36 (1998): 123–49; John H. Zammito, "The Second Life of the 'Public Sphere': On Charisma and Routinization in the History of a Concept," in *Changing Perceptions of the Public Sphere*, ed. Christian Emden and David Midgley (New York: Berghahn Books, 2012), 90–119.

101. It is no coincidence that these are Kantian categories. See Norbert Hinske, ed., *Eklektik, Selbstdenken, Mündigkeit* (Hamburg: Meiner, 1986).

102. Lewis White Beck, *Early German Philosophy: Kant and His Predecessors* (1969; repr., Bristol: Thoemmes, 1996), 324ff.

103. See Förster, *Aufklärung in Berlin*; and esp. Lifschitz, *Language and Enlightenment*.

104. Yolton, *Locke and French Materialism*.

105. Lifschitz, *Language and Enlightenment*.

106. See Conrad Grau, *Die Preußische Akademie der Wissenschaften zu Berlin: Eine deutsche Ge-*

lehrtengesellschaft in drei Jahrhunderten (Heidelberg: Spektrum, 1993), 87–114; Dominique Bourel, "Philosophes et politiques français à Berlin au XVIIIe siècle," in Schneiders, *Aufklärung als Mission*, 195–205.

107. Edgar Mass, "Französische Materialisten und deutsche 'Freigeisterei' (1746–1753)," in Schneiders, *Aufklärung als Mission*, 129–56. See also Martin Fontius, "Littérature clandestine et pensée allemande," in *Le matérialisme du XVIIIe siècle et la littérature clandestine*, ed. Olivier Bloch (Paris: Vrin, 1982), 251–62; Arseni Gulyga, *Der deutsche Materialismus am Ausgang des 18. Jahrhunderts* (Berlin: Akademie, 1966); Ann Thomson, *Materialism and Society in the Mid-eighteenth Century: La Mettrie's "Discours préliminaire"* (Geneva: Droz, 1981).

108. "His fondest wish [*Wunschvorstellung*] conformed to the concept of a freelance writer who was forced to make concessions on no sides. He believed this plan could be realized best in a big city in which an urbane air circulated. . . . Thus, he renounced abruptly all his grants and the support of his parents and gave himself over to the daredevil adventure of living as a freelance writer in Berlin." Gustav Stichelschmidt, *Lessing: Der Mann und sein Werk* (Dusseldorf: Droste, 1989), 56. On the notion of the *freier Schriftsteller*, see Hans J. Haferkorn, "Zur Entstehung der bürgerlich-literarischen Intelligenz und des Schriftstellers in Deutschland zwischen 1750 und 1800," in *Deutsches Bürgertum und literarische Intelligenz, 1750–1800*, ed. Bernd Lutz (Stuttgart: Metzler, 1974), 113–276.

109. See Wilhelm Dilthey, "Friedrich der Große und die deutsche Aufklärung," *Gesammelte Schriften*, vol. 3 (Leipzig: Teubner, 1927), 83–209. And see, more recently, Martin Fontius, *Friedrich II. und die europaische Aufklarung* (Berlin: Duncker und Humblot, 1999).

110. Möller, *Aufklärung in Preußen*.

111. See, e.g., Lessing, "Gedanken über die Herrnhuter" (1750), in *Sämtliche Schriften*, ed. Karl Lachmann and Franz Muncker (Stuttgart: Göschen, 1886–1924), 14:154–63.

112. Franklin Kopitzsch, "Gotthold Ephraim Lessing und seine Zeitgenossen im Spannungsfeld von Toleranz und Intoleranz," in *Deutsche Aufklärung und Judenemanzipation*, ed. Walter Grab, Jahrbuch des Instituts für deutsche Geschichte, Beiheft 3 (Tel Aviv: Nateev-Print, 1980), 29–85.

113. Wilfried Barner, "Lessing und sein Publikum in der frühen kritischen Schriften," in *Lessing in heutiger Sicht* (Bremen: Jacobi, 1976), 331. As Martens puts it, "The drive toward truth, seeking, thinking for oneself [*das Selbstdenken*] that can actualize itself in the form of criticism, that is Lessing's thing." Wolgang Martens, "Lessing als Aufklärer: Zu Lessings Kritik an den Moralischen Wochenschriften," in *Lessing in heutiger Sicht*, 244.

114. E. Schmidt, *Lessing: Geschichte seines Lebens und seiner Schriften* (Berlin: Weidmann, 1909), 1:143.

115. Mass, "Französische Materialisten und deutsche 'Freygeisterei' (1746–1753)."

116. Reiner Wild, "Freidenker in Deutschland," *Zeitschrift für historische Forschung* 6 (1979): 253–85, citing 258.

117. Ibid.

118. Johann Anton Trinius, *Freydenker-Lexikon, oder Einleitung in die Geschichte der neuern Freygeister: Ihre Schriften, und deren Widerlegungen* (Leipzig, 1759).

119. Mass, "Französische Materialisten und deutsche 'Freygeisterei' (1746–1753)."

120. See Roland Mortier, *Diderot en l'Allemagne* (Paris: Presses Universitaires de France, 1954); Anne Saada, *Inventer Diderot: Les constructions d'un auteur dans l'Allemagne des Lumières* (Paris: CNRS, 2003), esp. 91–133. On the specific reception of the *Encyclopédie*, see, in addition to these, esp. Jürgen Voss, "Verbreitung, Rezeption und Nachwirkung der Encyclopédie in Deutschland," in *Aufklärungen: Frankreich und Deutschland im 18. Jahrhundert* (Heidelberg: Winter, 1985), 183–91.

CHAPTER ONE

1. Immanuel Kant, *Träume eines Geistersehers*, in *Kants Gesammelte Schriften*, ed. Deutsche Akademie der Wissenschaften zu Berlin (Berlin: Georg Reimer [later Walter de Gruyter], 1900–), 2:331.

Hereafter, this edition of Kant's works will be abbreviated AA (Akademie Ausgabe), followed by volume and page numbers.

2. "In Halle were to be trained above all civil servants for the western territories of the unified state of Brandenburg-Prussia." Anton Schindling, "Kurbrandenburg im System des Reiches während der zweiten Hälfte des 17. Jahrhunderts," in *Preußen, Europa und das Reich*, ed. Oswald Hauser (Cologne: Böhlau, 1987), 33–46, citing 45.

3. The confessional gap between sovereign and subjects was mirrored by a sharp bifurcation in political order between a central-state bureaucracy, ministry, and court that was overwhelmingly composed of Calvinist "outsiders" and a regional political elite that was Lutheran, indigenous, and intent on defending traditional privileges. Günter Mühlpfordt, "Halle-Leipziger Aufklärung—Kern der Mitteldeutschen Aufklärung, Führungskraft der Deutschen Aufklärung, Paradigma in Europa (zur ersten und zweiten Blüte der Universität Halle)," in *Aufklärung und Erneuerung: Beiträge zur Geschichte der Universität Halle im ersten Jahrhundert ihres Bestehens (1694–1806)*, ed. Günter Jerouschek and Arno Sames (Hanau: Dausien, 1994), 46–55, citing 47. Thus, "the primary, central leadership of this state and the secondary, regional and local leadership stood by the eighteenth century in estranged, indeed hostile relations." Klaus Deppermann, "Die politische Voraussetzungen für die Etablierung des Pietismus in Brandenburg-Preußen," *Pietismus und Neuzeit* 12 (1986): 38–53, citing 39–41.

4. Wilhelm Schrader, *Geschichte der Friedrichs-Universität zu Halle* (Berlin: Dümmler, 1894), 1:6. "With the founding of Halle, where Pietism and the early Enlightenment at the outset went hand in hand, Berlin began its competitive struggle for the cultural leadership of Protestant Germany, which was directed above all against the Universities of Leipzig and Wittenberg in Kursaxony, against the strongholds of Lutheran orthodoxy" (Deppermann, "Die politische Voraussetzungen," 50).

5. Arising particularly out of the horrors of the Thirty Years' War, Pietism turned against what it took to be ossifications in orthodox Lutheran teaching and practice, which in some measure seemed responsible for the calamity. A "second Reformation" *within* the Lutheran confession seemed called for, one that restored authentic Christian life values. Carl Hinrichs, *Preußentum und Pietismus: Der Pietismus in Brandburg-Preußen als religiös-soziale Reformbewegung* (Göttingen: Vandenhoeck und Ruprecht, 1971.

6. Richard Gawthrop, *Pietism and the Making of Eighteenth-Century Prussia* (Cambridge: Cambridge University Press, 1993), 120.

7. "An academic city of God was the Pietist concept, while the administration of Brandenburg-Prussia wanted to build a training institute for its population." Johanna Geyer-Kordesch, *Pietismus, Medizin und Aufklärung in Preußen im 18. Jahrhundert: Das Leben und Werk Georg Ernst Stahls* (Tübingen: Niemeyer, 2000), 20. Spiritual indoctrination in Pietism and its service ethos created a substantial cadre of teachers, preachers, and healers who set the tone of local social welfare throughout Prussia in the first half of the eighteenth century. Hartmut Lehmann, "Pietismus und soziale Reform in Brandenburg-Preußen," in Hauser, *Preußen, Europa und das Reich*, 103–20, citing 110. Thus, Pietism became "a fundamental and essential element of the old Prussian state," creating a "*Beamtenreligion*" of dedicated service. Günter Birtsch, "Friedrich Wilhelm I. und die Anfänge der Aufklärung in Brandenburg-Preußen," in Hauser, *Preußen, Europa und das Reich*, 87–102, citing 87. See also Reinhold Dowart, *The Prussian Welfare State before 1740* (Cambridge, MA: Harvard University Press, 1971).

8. Schrader, *Geschichte der Friedrichs-Universität zu Halle*.

9. Martin Brecht, "August Hermann Francke und der Hallische Pietismus," in *Geschichte des Pietismus*, vol. 1, *Der Pietismus von 17. bis zum frühen 18. Jahrhundert* (Göttingen: Vandenhoeck und Ruprecht, 1993), 439–539.

10. Peter Menck, "Unwissenheit und Bosheit: Über die Grundlagen der Erziehung bei August Hermann Francke," in Jerouschek and Sames, *Aufklärung und Erneuerung*, 182–91, citing 182.

11. Jürgen Helm, "Hallesche Medizin zwischen Pietismus und Frühaufklärung," in *Universitäten und Aufklärung*, ed. Notker Hammerstein (Göttingen: Wallstein, 1995), 63–95, citing 64.

12. Wolfram Kaiser and Heinz Krosch, "Zur Geschichte der Medizinischen Fakultät der Universität Halle im 18. Jahrhundert, VII: Die Fakultätsinstitutionen im 18. Jahrhundert," *Wissenschaftliche Zeitschrift der Martin-Luther-Universität Halle-Wittenberg, Mathematisch-naturwissenschaftliche Reihe* 14 (1965): 1–48, citing 2.

13. Johanna Geyer-Kordesch, "German Medical Education in the Eighteenth Century: The Prussian Context and Its Influence," in *William Hunter and the Eighteenth-Century Medical World*, ed. W. F. Bynum and Roy Porter (Cambridge: Cambridge University Press, 1995), 177–205, citing 183. Thus, Schrader makes the crucial point: "The development of the *Anstalt* is clearly visible in its impact on the university in the theological and medical faculties" (*Geschichte der Friedrichs-Universität zu Halle*, 1:122).

14. Francke's *Anstalten* provided massive student aid when no other university in Germany did anything similar; thus, many poor students (and also spiritually motivated ones) came to Halle, swelling enrollments not only in the theology faculty but also eventually in the medical faculty. Geyer-Kordesch notes a report from 1704 that observed "more students received aid at Halle than the entire student population at some other institutions" ("German Medical Education," 192). For a wider conspectus on poverty, education, and careers in Germany in the eighteenth century, see Anthony La Vopa, *Grace, Talent and Merit: Poor Students, Clerical Careers, and Professional Ideology in Eighteenth-Century Germany* (Cambridge: Cambridge University Press, 1988).

15. Thomas Howard, *Protestant Theology and the Making of the Modern German University* (Oxford: Oxford University Press, 2006), 95.

16. Geyer-Kordesch, *Pietismus, Medizin und Aufklärung in Preußen*, 103. See also Friedrich Hoffmann, *Fundamenta Medicinae*, trans. Lester King (London: Macdonald History of Science Library, 1971); Schrader, *Geschichte der Friedrichs-Universität zu Halle*, 1:57; Mühlpfordt, "Halle-Leipziger Aufklärung," 49; Ingo Müller, "Mechanismus und Seele—Grundzüge der frühen halleschen Medizinschulen," in Jerouschek and Sames, *Aufklärung und Erneuerung*, 245–61.

17. Geyer-Kordesch, "German Medical Education," 191; Johanna Geyer-Kordesch, "Die Medizin im Spannungsfeld zwischen Aufklärung und Pietismus: Das unbequeme Werk Georg Ernst Stahls und dessen kulturellen Bedeutung," in *Zentren der Aufklärung*, vol. 1, *Halle Aufklärung und Pietismus*, ed. Norbert Hinske (Heidelberg: Lambert Schneider, 1989), 255–74, citing 257; Wolfram Keiser and Heinz Krosh, "Leiden und Halle als medizinische Zentren des frühen 18. Jahrhunderts: Zum 300. Geburtstag von Hermann Boerhaave (1668–1738)," *Zeitschrift für die gesamte Innere Medizin und ihre Grenzgebiete* 23 (1968): 330–38, citing 330; Helm, "Hallesche Medizin," 63.

18. J. Karcher, "Die animistische Theorie Georg Ernst Stahls im Aspekt der pietistischen Bewegung an der Universität zu Halle an der Saale im zu Ende gehenden 17. und beginnenden 18. Jahrhundert," *Gesnerus* 15 (1958): 1–16, citing 7.

19. Geyer-Kordesch, "German Medical Education," 196.

20. Kaiser and Krosch, "Zur Geschichte," 39.

21. Schrader, *Geschichte der Friedrichs-Universität zu Halle*, 1:97.

22. Kaiser and Krosch, "Zur Geschichte," 27.

23. Schrader, *Geschichte der Friedrichs-Universität zu Halle*, 1:99.

24. Kaiser and Krosch, "Zur Geschichte," 18.

25. "The number of dissections was very small, and even for the few medical students it was completely insufficient. Hoffmann was able to do only twenty dissections in twenty-four years" (Schrader, *Geschichte der Friedrichs-Universität zu Halle*, 1:339).

26. Helm, "Hallesche Medizin," 63. Stäter is of the same view: "In Francke's perspective the theological faculty—perhaps also the medical—assumed the position of an 'annex' to the *Anstalten* in Glaucha" ("Aufklärung und Pietismus—das Beispiel Halle," 54).

27. Johanna Geyer-Kordesch, "Georg Ernst Stahl's Radical Pietist Medicine and Its Influence on the German Enlightenment," in *The Medical Enlightenment of the Eighteenth Century*, ed. Andrew Cunningham and Roger French (Cambridge: Cambridge University Press 1990), 67–87, citing 72. For another view, see Müller, "Mechanismus und Seele," 261.

28. Geyer-Kordesch, *Pietismus, Medizin und Aufklärung in Preußen*, 103.

29. Roger French, "Ethics in the Eighteenth Century: Hoffmann in Halle," in *Doctors and Ethics: The Earlier Historical Setting of Professional Ethics*, ed. Andrew Wear, Johanna Geyer-Kordesch, and Roger French (Amsterdam: Rodopi, 1993), 153–79.

30. Müller, "Mechanismus und Seele," 261.

31. Geyer-Kordesch, *Pietismus, Medizin und Aufklärung in Preußen*, 29–30, 132–34; Helm, "Hallesche Medizin," 81, 83–84.

32. Geyer-Kordesch, *Pietismus, Medizin und Aufklärung in Preußen*, 132.

33. Ibid., 28.

34. Ibid., 32.

35. Ibid., 77, 82, 127–28.

36. See Hinrichs, *Preußentum und Pietismus*, 393–418.

37. Geyer-Kordesch, *Pietismus, Medizin und Aufklärung in Preußen*, 75.

38. Wolfram Kaiser, "Theorie und Praxis in der Boerhaave-Ära und in nachboerhaavianischen Ausbildungssystemen an deutschen Hochschulen des 18. Jahrhunderts," *Clio Medica: Acta Academia Internalialis Historiae Medicinae* 21 (1987): 71–94, citing 73.

39. Helm, "Hallesche Medizin," 89.

40. "Hoffmann was particularly clear about the important role of the applied natural sciences in the progress of medicine; a sign of this that may suffice is that the draft of the founding statute of his own faculty [of medicine] envisioned the combination of a professoriate in the primary field [medicine] with a professoriate in physics." Wolfram Kaiser, "Johann Gottlob Krüger (1715–1759) und Christian Gottlieb Kratzenstein (1723–1795) als Begründer der modernen Elektrotherapie," *Zahn-, Mund-, und Kierferheilkunde mit Zentralblatt* 65 (1977): 539–54, citing 539.

41. Ibid., 540.

42. Schrader, *Geschichte der Friedrichs-Universität zu Halle*, 1:316.

43. Kaiser, "Krüger und Kratzenstein," 540.

44. There are grounds to think the Pietist triumph over Wolff in Halle a Pyrrhic victory. Schrader's section heading in his history of the university puts it precisely: "The Victory and Decay of Pietism" (Sieg und Erstarrung des Pietismus) (*Geschichte der Friedrichs-Universität zu Halle*, 1:219). Indeed, the third book of Schrader's history is entitled "Diminution of Original Potency" (Rückgang der urpsrünglichen Kraft), dating this development already to 1730 (276).

45. Michael Alberti, *Medicinische Betrachtung von den Kräften der Seelen nach den Unterscheid des Leibes und dessen Natürlichen Gesundheit oder Krankheit* (Halle, 1740).

46. Geyer-Kordesch, *Pietismus, Medizin und Aufklärung in Preußen*, 182–83.

47. Friedrich Hoffmann, *Commentarius de Differentia inter Ejus Doctrinam Medico Mechanicam, et Georgii Ernsti Stahlii Medico Organicam*, ed. Ernst Eugen Cohausen (Frankfurt am Main, 1746).

48. Kaiser and Krosch note the prominence of a new generation after the death of Hoffmann in 1742 ("Leiden und Halle als medizinische Zentren," 335).

49. See above all Carsten Zelle, ed., *"Vernünftige Ärzte"—Hallesche Psychomedizin und die Anfänge der Anthropologie in der deutschsprachigen Frühaufklärung* (Tübingen: Neimeyer, 2001).

50. Others include Ernst Anton Nicolai (1722–1802) and Johann Christian Bolten (1727–57).

51. This was a key feature of the third generation. Both Krüger and Unzer obtained double degrees. This was in all likelihood the case for Nicolai and Bolten, as well. Carsten Zelle, "'Zwischen Weltweisheit und Arzeiwissenschaft'—Zur Vordatierung der anthropologischen Wende in die Frühaufklärung nach Halle (eine Skizze)," in *Formen der Aufklärung und ihrer Rezeption / Expressions des Lumières et de leur reception; Festschrift zum 70. Geburtstag von Ulrich Ricken*, ed. Reinhard Bach, Roland Desné, and Gerda Haßler (Tübingen: Stauffenberg, 1999), 35–44, esp. 38.

52. Helm, "Hallesche Medizin," 72.

53. See Carsten Zelle, "Erfahrung, Ästhetik und mittleres Maß: Die Stellung von Unzer, Krüger und E. A. Nicolai in der anthropolgischen Wende um 1750 (mit einem Exkurs über ein Lehrgedichtfragment Moses Mendelssohns)," in *Reiz, Imagination, Aufmerksamkeit: Erregung und Steu-*

rung von Einbildungskraft im klassischen Zeitalter (1680–1830), ed. Jörn Steigerwald and Daniela Wetzke (Würzburg: Königshausen und Neumann, 2003), 203–23, citing 219.

54. Johann Gottlob Krüger, *Grundriss eines neuen Lehrgebäudes der Arzneygelarhtheit* (Halle, 1745).

55. Zelle, *"Vernünftige Ärzte,"* 56.

56. See Hans-Peter Nowicki, *Der wohltempereriete Mensch: Aufklärungsanthropologie im Widerstreit* (Berlin: De Gruyter, 2003), 33–34. Unzer obtained two degrees—one in philosophy and one in medicine—and before he completed either dissertation, he published a number of important works affiliating himself with his prominent mentors at Halle: the rector of the university and leader of philosophical rationalism in the German Enlightenment, Christian Wolff; his immediate supervisor, Krüger; but also Johann Juncker, the ardent Stahlian; and the philosopher Georg Friedrich Meier, with whom he developed a particularly close relation. His *Gedancken vom Einflüsse der Seele in ihren Körper* (1745) drew a devastating review (by Haller) in *Göttingische Gelehrten Anzeigen*, September 1746. Unzer engaged in a disputation on the same material, *De Nexu Metaphysices cum Medicina Generatim* (1749), discussed in Yvonne Wüppen, "Limitierte Anthropologie: Grenzen des medizinisch-philosophischen Wissenstransfers am Beispiel von Johann August Unzer," in *Physis und Norm: Neue Perspektiven der Anthropologie im 18. Jahrhundert*, ed. Manfred Beetz, Jörn Garber, and Heinz Thoma (Göttingen: Wallstein, 2007), 49–68, citing 62. Unzer's culminating work of that period, *Philosophische Beobachtungen des menschlichen Körpers überhaupt* (1750), maintained that there was a complete duplication or correspondence between every mental and every physical event in the human organism.

57. For Thomasius, the core of philosophy was the study of human feelings, *Affektenlehre*, "combining medical therapy, moral philosophy and rhetoric." Ian Hunter, *Rival Enlightenments: Civil and Metaphysical Philosophy in Early Modern Germany* (Cambridge: Cambridge University Press, 2001), 225. He believed that individuals not only could but should improve their conduct, and he advocated a new style of learning, a *Klugheitslehre*, that would accommodate this new social conduct, especially for the middling classes. That entailed taking up the ideas of Baltasar Gracián on "gallant" learning—what Thomasius would propagate as *Philosophia aulica* (courtly philosophy). *Klugheitslehre* (philosophy of prudence/cleverness) had origins in *sophrosyne/phronesis* in Aristotle, then *prudentia* in the Latin tradition.

58. This ambition to constitute a new, special science *between* the existing disciplines was a striking feature of the "rational physicians" in Halle in the 1740s. Unzer called for this explicitly in his *Philosophische Beobachtungen des menschlichen Körpers überhaupt* of 1750, as Zelle notes: "Baumgarten's aesthetics, Krüger's hygiene (*Diät*), and Bolten's soul cure converge" (*"Vernünftige Ärzte,"* 54–55). It is slightly misleading to term Bolten's view a "soul cure" (which resonates all too strongly with *Waisenhausmedizin*) rather than to use his actual term: "psychological cure."

59. Schrader, *Geschichte der Friedrichs-Universität zu Halle*, 1:351–52.

60. Ibid., 255.

61. Ibid., 372.

62. Geyer-Kordesch, "German Medical Education," 194.

63. Helm, "Hallesche Medizin," citing Karl Rothschuh, *Konzepte der Medizin* (Stuttgart: Hippokrates-Verlag, 1978), 232. See Wolfgang Maier and Thomas Zoglauer, eds., *Technomorphe Organismuskonzepte: Modellübertragungen zwischen Biologie und Technik* (Stuttgart: Frommann-Holzboog, 1994).

64. Elizabeth Williams, *The Physical and the Moral: Anthropology, Physiology, and Philosophical Medicine in France, 1750–1850* (Cambridge: Cambridge University Press, 1994); Elizabeth Williams, *A Cultural History of Medical Vitalism in Enlightenment Montpellier* (Aldershot, UK: Ashgate, 2003).

65. Pierre Cabanis, *Sketches of the Revolutions in Medical Science* (1806), cited in L. Rather, "G. E. Stahl's Psychological Physiology," *Bulletin of the History of Medicine* 35 (1961): 37–49, citing 48.

66. T. Blondin, ed. and trans., *Oeuvres medico-philosophiques et pratiques de G.-E. Stahl*, 6 vols. (Paris, 1859–64).

67. Joseph Tissot, "L'animisme et ses adversaires," editorial commentary in Blondin, *Oeuvres medico-philosophiques et pratiques de G.-E. Stahl*, 6:365–444; Albert Lemoine, *Le vitalisme et l'animisme de Stahl* (Paris, 1864). Lemoine was a key instigator of this claim; see his earlier work, *Stahl et l'animisme* (Paris, 1858).

68. J. B. Chancerel, *Recherches sur la pensée biologique de Stahl* (Paris: Dubois et Bauer, 1934).

69. Elizabeth Haigh, "The Roots of the Vitalism of Xavier Bichat," *Bulletin of the History of Medicine* 49 (1975): 72–86; Roselyn Rey, *Naissance et développement du vitalisme en France de la deuxième moitié du 18e siècle à la fin du Premier Empire* (Oxford: Voltaire Foundation, 2000).

70. Bernward J. Gottlieb, "Bedeutung und Auswirkungen des halleschen Professors und kgl. preußischen Leibarztes Georg Ernst Stahl auf den Vitalismus des XVIII. Jahrhunderts, insbesondere auf die Schule von Montpellier," *Nova Acta Leopoldina: Abhandlungen der Kaiserlich Leopoldinisch-Carolinisch Deutschen Akademie der Naturforscher* 12 (1943): 425–502.

71. Karl Rothschuh, *Physiologie: Der Wandel ihrer Konzepte, Probleme und Methoden vom 16. bis 19. Jahrhundert* (Freiburg: Alber, 1968); Karl Rothschuh, *Physiologie im Werden* (Stuttgart: Fischer, 1969); Karl Rothschuh, *History of Physiology*, trans. Guenter Risse (Huntington, NY: Krieger, 1973).

72. "Albrecht von Haller undermined Stahl's reputation in the history of medicine with his ice-cold appraisal," Richard Koch aptly observed. Richard Koch, "War Georg Ernst Stahl ein selbständiger Denker?," *Archiv für Geschichte der Medizin* 18 (1926): 20–50, citing 27.

73. See Albrecht von Haller, *Bibliotheca Anatomica*, vol. 1 (1774; repr., Hildesheim: G. Olms, 1969), 697; Albrecht von Haller, *Bibliotheca Medicinae Practicae*, 4 vols. (Bern: E. Haller, 1776), 3:575, cited in Gottlieb, "Bedeutung," 430n, and in Koch, "War Georg Ernst Stahl ein selbständiger Denker?," 27.

74. Robert Whytt, *An Essay on the Vital and Other Involuntary Motions of Animals* (Edinburgh: Hamiton, Balfour and Neill, 1751).

75. See Roger French, "The Controversy with Haller: Sense and Sensibility," in *Robert Whytt, the Soul, and Medicine* (London: Wellcome Institute of the History of Medicine, 1969), 63–76.

76. Gottlieb comments: "In this sense, Haller has a key role, since it is well known in what measure praise or condemnation from his mouth influenced the history of medicine after him" ("Bedeutung," 429). See, e.g., Paul Diepgen, *Albrecht Haller und die Geschichte der Medizin* (Berlin: J. Springer, 1930).

77. See Jürgen Helm's discussion of the appraisal of Stahl in the wider current of "Romantic medicine" in his "Das Medizinkonzept Georg Ernst Stahls und seine Rezeption im Halleschen Pietismus und in der Zeit der Romantik," *Berichte zur Wissenschaftsgeschichte* 23 (2000): 167–90, esp. 175ff. Ideler's translation of *Theoria Medica Vera* into German is an early nineteenth-century token of this revision: Georg Ernst Stahl, *Theorie der Heilkunde*, trans. Karl Wilhelm Ideler, 3 vols. (Berlin: Enslin, 1831–32).

78. Kant, *Träume eines Geistersehers*, AA 2:331.

79. Immanuel Kant, *De Medicina Corporis, quae Philosophorum est*, AA 15:939–53; translated by Mary Gregor, "On the Philosophers' Medicine of the Body," in Kant, *Anthropology, History, and Education* (Cambridge: Cambridge University Press, 2007), 182–91.

80. Cited in Gottlieb, "Bedeutung," 428.

81. See François Duchesneau, *Leibniz: Le vivant et l'organisme* (Paris: Vrin, 2010), 21–22; and Duchesneau's earlier monograph, *Les modèles du vivant de Descartes à Leibniz* (Paris: Vrin, 1998). On this whole constellation, see esp. T. Cheung, "From the Organism of a Body to the Body of an Organism: Occurrence and Meaning of the Word 'Organism' from the Seventeenth to the Nineteenth Centuries," *British Journal for the History of Science* 39 (2006): 319–39.

82. Geyer-Kordesch is altogether right to observe that "university medicine had not yet disengaged itself from the broader spectrum of debate concerning the nature of man or the theological discussion of his ultimate purpose" ("German Medical Education," 180).

83. Modern commentators make the same complaint, and most have done as I have, working with translations in modern languages, which—at this level of treatment—more than suffices.

84. Stahl, *Theorie der Heilkunde,* 1:xxx.

85. Ibid., v–vi.

86. Helm puts it succinctly: "To Stahl's program for a renewal of medicine belonged precisely the turn away from medicine as book learning to—what he took to be—an observation of the healthy and the sick without presuppositions." Jürgen Helm, "'*Quod naturae ipsae sint morborum medicatrices*': Der Hippocratismus Georg Ernst Stahls," *Medizin-Historisches Journal* 35 (2000): 251–62, citing 252.

87. Georg Ernst Stahl, *De synergeia naturae in medendo* (1695), translated in Bernward J. Gottlieb, "Georg Ernst Stahls *De Synergeia Naturae in Medendo,*" *Sudhoffs Archiv* 43 (1959): 172–82. "One should not believe I have gone on at such length just to uphold my position, to leave others out of consideration, or indeed to make them ridiculous and contemptible. . . . I am far more concerned to move this part of medical science and art forward finally and definitively. Something of this has been passed down over past centuries. Unfortunately, in times closer to our own it has almost all been forgotten" (180). See Helm, "'*Quod naturae,*'" 254–55; Koch, "War Georg Ernst Stahl ein selbständiger Denker?," 39.

88. Georg Ernst Stahl, *De Philosophia Hippocratis* (1704). See Helm, "'*Quod naturae,*'" passim.

89. Stahl, *Theorie der Heilkunde,* 1:xlvii.

90. Ibid., 35.

91. Ibid., xxxiii. "He was interested in medicine, not principally in science. . . . By theory he understood therefore not the intellectual or scientific foundation of medicine but a fundamental fact of great fruitfulness and reliability" (Koch, "War Georg Ernst Stahl ein selbständiger Denker?," 45).

92. Johanna Geyer-Kordesch, "Die 'Theoria Medica Vera' und Georg Ernst Stahls Verhältnis zur Aufklärung," in *Georg Ernst Stahl (1659–1734): Hallesches Symposium 1984,* ed. Wolfram Kaiser and Arina Völker (Halle: Wissenschaftliche Beiträge der Martin-Luther-Universität Halle-Wittenberg, 1985), 89–98, citing 89. See also Johanna Geyer-Kordesch, "Stahl—Leben und seine medizinische Theorie," in *Georg Ernst Stahl (1659–1734) in Wissenschaftshistorischer Sicht,* ed. Dietrich von Engelhardt and Alfred Gierer, Acta Historica Leopoldina 30 (Halle: Deutsche Akademie der Naturforscher Leopoldina, Halle [Salle], 2000), 33–48.

93. Stahl, *Theorie der Heilkunde,* 1:xxiv.

94. Stahl, cited in Geyer-Kordesch, *Pietismus, Medizin und Aufklärung in Preußen,* 144.

95. Haller, *Bibliotheca Anatomica,* 1:697.

96. Stahl, *Theorie der Heilkunde,* 1:64.

97. On the mechanical philosophy of the late seventeenth century, see Roy Laird and Sophie Roux, eds., *Mechanics and Natural Philosophy before the Scientific Revolution* (Dordrecht: Springer, 2008); Daniel Garber, "Remarks on the Pre-history of the Mechanical Philosophy," in *The Mechanization of Natural Philosophy,* ed. Daniel Garber and Sophie Roux (Dordrecht: Springer, 2013), 3–26; Daniel Garber, "Physics and Foundations," in *Cambridge History of Science,* vol. 3, *Early Modern Science,* ed. Katherine Park and Lorraine Daston (Cambridge: Cambridge University Press, 2001), 21–69.

98. Stahl, *Theorie der Heilkunde,* 1:19.

99. Stahl rarely mentioned any other thinkers in his works; he disdained this appeal to "authorities," and so the absence of mention of Newton by name in Stahl's texts cannot of itself prove anything; but the theoretical articulation of inertia in Stahl is patently Cartesian, not Newtonian.

100. Kevin Chang, "Fermentation, Phlogiston and Matter Theory: Chemistry and Natural Philosophy in Georg Ernst Stahl's *Zymotechnia Fundamentalis,*" *Early Science and Medicine* 7 (2002): 31–64.

101. Stahl, *Theorie der Heilkunde,* 1:50. See Martin Carrier, "Zum korpuskularen Aufbau der Materie bei Stahl und Newton," *Sudhoffs Archiv* 70 (1986): 1–17. On the chemistry of "mixtures," see the classic study by Helene Metzger: *Newton, Stahl, Boerhaave et le doctrine chimique* (Paris: Alcan, 1930). On the iatrochemical and alchemical background, see William Newman, *Atoms and*

Alchemy: Chymistry and the Experimental Origins of the Scientific Revolution (Chicago: University of Chicago Press, 2006); William Newman, *Alchemy Tried in the Fire: Starkey, Boyle, and the Fate of Helmontian Chymistry* (Chicago: University of Chicago Press, 2002).

102. This was the theme of his key essay, *De Vera Diversitate Corporis Mixti et Vivi* (1707), reprinted in *Theoria Medica Vera.*

103. Stahl, *Theorie der Heilkunde,* 1:53.

104. Georg Ernst Stahl, "Über den Unterschied zwischen Organismus und Mechanismus," (1714), in *Georg Ernst Stahl,* ed. B. Gottlieb, Sudhoffs Klassiker der Medizin, vol. 36 (Leipzig: J. A. Barth, 1961), 48–53, citing 52.

105. Stahl, *Theorie der Heilkunde,* 1:34.

106. Ibid., 37.

107. Ibid., 73.

108. "Thus the physician will never be able to attain to true therapies at all by even the most meticulous contemplation of the structure of [body] parts" (ibid., 39). The analysis of muscle tissue to its finest fiber would never explain its motions, he insisted. From Glisson through Malpighi to Haller, the best figures in experimental medicine disputed this conviction and changed the course of life science. See Owsei Temkin, "The Classical Roots of Glisson's Doctrine of Irritation," *Bulletin of the History of Medicine* 38 (1964): 297–328.

109. Stahl "regarded anatomical and physiological knowledge and research as altogether [*ohnehin*] superfluous" (Müller, "Mechanismus und Seele," 259). Stahl's *Theoria Medica Vera* "cut medicine away from the foundational sciences of physics and mathematics, in that it formulated a categorical difference of living things from inorganic matter and ascribed to living things an autonomous principle of organization, the soul" (Geyer-Kordesch, *Pietismus, Medizin und Aufklärung in Preußen,* 255). He "considered chemistry useless in the quest for explanations of the life process and even dangerous in the quest for therapeutic measures." Henricus A. M. Snelders, "Iatrochemie und Iatrophysik in den Niederlanden im 17. und 18. Jahrhundert," in *Hallesche Physiologie im Werden: Hallesche Symposium 1981,* ed. Wolfram Kaiser and Hans Hübner (Halle: Wissenschaftliche Beiträge der Martin-Luther-Universität Halle-Wittenberg, 39:41, 1981), 44–54, citing 52.

110. Stahl, *Theorie der Heilkunde,* 1:95. See Lester King, "Stahl and Hoffmann: A Study in Eighteenth Century Animism," *Journal of the History of Medicine* 19 (1964): 118–30, citing 121.

111. Stahl, *Theorie der Heilkunde,* 1:96. See Helm, "Das Medizinkonzept," 168.

112. Stahl, *Theorie der Heilkunde,* 1:52.

113. Huneman and Rey point out the resonances with the later thought of Xavier Bichat in this conception of life and death as informing the theory of organism: Philippe Huneman and Anne-Lise Rey, "La controverse Leibniz-Stahl dite *Negotium otiosum,*" *Bulletin d'histoire, épistémologie et sciences de la vie* 14 (2007): 213–38, esp. 217. See also Philippe Huneman, *Bichat, la vie et la mort* (Paris: Presses Universitaires de France, 1998).

114. Stahl, *Theorie der Heilkunde,* 1:85.

115. On "fermentation," see Chang, "Fermentation, Phlogiston and Matter Theory." See also Walter Pagel, "Helmont, Leibniz, Stahl," *Sudhoffs Archiv* 24 (1931): 19–51.

116. See K. Chang, "*Motus Tonicus*: Georg Ernst Stahl's Formulation of Tonic Motion and Early Modern Medical Thought," *Bulletin of the History of Medicine* 78, no. 4 (Winter 2004): 767–803.

117. Stahl, *Theorie der Heilkunde,* 1:50.

118. "Life seems to exhibit goal and purpose, adaptation, process, function, coordination of parts, integration of units, complex ordered behavior. Temporal relationships and *organization* are of the essence," King aptly summarizes ("Stahl and Hoffmann," 122–23).

119. Koch observed: "Stahl is not the only one who recognized at that time that a purely mechanical biology could not be carried out. . . . But he was certainly by far the most visible, imposing, and influential of them" ("War Georg Ernst Stahl ein selbständiger Denker?," 35).

120. Stahl, *Theorie der Heilkunde,* 1:26. Gottlieb developed a table of all Stahl's synonymic usages. See Gottlieb, "Bedeutung," 470.

121. Stahl, *Theorie der Heilkunde*, 1:xlix.

122. Helm, "'*Quod naturae*,'" 257; Koch, "War Georg Ernst Stahl ein selbständiger Denker?," 44.

123. Stahl, *Theorie der Heilkunde*, 1:20, 43.

124. Gottlieb, "Stahls *De Synergeia Naturae in Medendo*," 177.

125. Stahl, *Theorie der Heilkunde*, 1:xliii; for the program formulation, see 1:xlviii. "The recognition of this self-healing power of nature represented the essential foundation of rational medical practice for Stahl" (Helm, "Das Medizinkonzept," 170).

126. One crucial feature of the development at that time was the shift in the primary sense of the concept of soul from the Aristotelian-medical notion of varying features of organismic order—vegetative, animal, and rational—to the notion of soul as preponderantly the rational faculty itself. See Francisco Vidal, *The Sciences of the Soul: The Early Modern Origins of Psychology* (Chicago: University of Chicago Press, 2011).

127. Gottlieb, "Bedeutung," 454–55.

128. Stahl, *Theorie der Heilkunde*, 1:90–93. See Koch, "War Georg Ernst Stahl ein selbständiger Denker?," 42.

129. Rather notes Stahl's "attempt to fuse the many hierarchical levels of previous writers into the simple conception of the *anima*." L. J. Rather, "G. E. Stahl's Psychological Physiology," *Bulletin of the History of Medicine* 35 (1961): 37–49, citing 46.

130. Georg Ernst Stahl, *Propempticon Inaugurale de Differentia Rationis et Ratiocinationis et Actionum, Quae per et Secundum utrumque Horum Actuum Fiund in Negotio Vitali et Animali* (1701), translated as "Über den Unterschied zwischen *ratio* . . . und *ratiocinatio*. . . ," in Engelhardt and Gierer, *Georg Ernst Stahl (1659–1734) in wissenschaftlicher Sicht*, 275–80.

131. Huneman and Rey offer a clear formulation of this in their "La controverse Leibniz-Stahl," 219.

132. Duchesneau offers the apt formulation of an "instinct of reason[,] . . . a blind spiritual direction." François Duchesneau, "Leibniz et Stahl: Divergences sur le concept d'organisme," *Studia Leibnitiana* 27 (1995): 185–212, citing 194.

133. Geyer-Kordesch finds "an authentic problem in Stahl's teaching: he does not tell us what the soul is. Never does he provide a definition" ("Die 'Theoria Medica Vera,'" 95). That is not quite right; Duchesneau gets the essential point: "the ontological status of the soul does not concern Stahl at all [*ne préoccupe guère Stahl*]. . . . He is satisfied in general to pose the soul as a modern analog of the nature (*physis*) of the ancients . . . [, that is,] an active immaterial substance, both the efficient and the final cause of the vital and organic movements" ("Leibniz et Stahl," 191). As Duchesneau stresses, the soul "does not exist except in and for the body" (191).

134. A starting point is Lester King, "Basic Concepts of Early 18th-Century Animism," *American Journal of Psychiatry* 124 (1967): 797–802; and King, "Stahl and Hoffmann," 118–30.

135. Geyer-Kordesch claims that "Stahl sets out from a religious conception of the soul" ("Die Medizin im Spannungsfeld zwischen Aufklärung und Pietismus," 264). She elaborates: "Stahl appears as a thinker who could bring into physiology and pathology the fundamental faith in the effectiveness of spirit and the place of the soul" ("Die 'Theoria Medica Vera,'" 96).

136. Helm has made very thoughtful criticisms of this misidentification. See Helm, "Das Medizinkonzept," 173–75, 183; Helm, "'*Quod naturae*,'" 260. For King it was quite clear: "The anima was a biological and not a religious concept. This anima inhered in the body, did not exist independently of the body, exerted its activities through bodily members through the laws of motion" ("Basic Concepts of Early 18th-Century Animism," 799).

137. "It was not out of religion that this vitalistic system derived its impetus . . . [but] far more from the reaction of a medical psychologist to the dead end of the research trend in iatromechanism" (Gottlieb, "Bedeutung," 469).

138. It is fruitful to contrast Haller's (protopositivistic) conception of science with Leibniz's notion of the interdependence of science and metaphysics. Haller was no afficionado of Leibniz, as we shall see later.

139. "The frequently raised question concerning the essence of *anima* or *natura* in Stahl's medical theory misses Stahl's intent. In Stahl's eyes this question is irrelevant for physicians and it is therefore also not the topic of his medical writings" (Helm, "Das Medizinkonzept," 171).

140. Johanna Geyer-Kordesch, "Passions and the Ghost in the Machine: Or What Not to Ask about Science in Seventeenth- and Eighteenth-Century Germany," in *The Medical Revolution of the Seventeenth Century*, ed. Roger French and Andrew Wear (Cambridge: Cambridge University Press, 1989), 145–63, citing 159. See Koch, "War Georg Ernst Stahl ein selbständiger Denker?," 26.

141. Stahl, *Theoria Medica Vera*, cited in Geyer-Kordesch, "Stahl—Leben und seine medizinische Theorie," 46.

142. Stahl, *Theorie der Heilkunde*, 1:115.

143. Geyer-Kordesch, "German Medical Education," 195.

144. One of the first dissertations Stahl supervised, defended at Halle in 1695 by Johann Jacob Reich, was entitled *Disputatio Inauguralis de Passionibus Animi Corpus Humanum Varie Alterantibus*. Interestingly, the German translation of this text, *Über den mannigfaltigen Einfluß der Gemüthsbewegungen auf den menschlichen Körper* (in Gottlieb's collection in the Sudhoff Klassiker der Medizin series), rendered the Latin term *passio* by the German *Gemüthsbewegung*, a portentous extrapolation. See the discussion in Gottlieb, "Stahls *De Synergeia Naturae in Medendo*," 174–75.

145. Stahl made repeated reference to this widely shared but misguided judgment: e.g., *Theorie der Heilkunde*, 1:91.

146. Ibid., 1:63, 68. See Paul Hoffmann, "L'âme et les passions dans la philosophie médicale de Georg-Ernst Stahl," *Dix-huitième siècle* 23 (1991): 31–43.

147. Stahl, *Theorie der Heilkunde*, 1:44, 58–59. See also ibid., lii, where Stahl refers to his *Dissertatio de Frequentia morborum in homine prae brutis*. See Helm, "Das Medizinkonzept," 169, for a good discussion.

148. A full English translation has just appeared: *The Leibniz-Stahl Controversy*, ed. François Duchesneau and Justin Smith (New Haven, CT: Yale University Press, 2016], cited hereafter as *LSC*. Modern scholarship seems to have begun with the two-part publication of L. J. Rather and J. B. Frerichs, "The Leibniz-Stahl Controversy—I. Leibniz's Opening Objections to the *Theoria Medica Vera*," *Clio Medica* 3 (1968): 21–40; and "The Leibniz-Stahl Controversy—II. Stahl's Survey of the Principal Points of Doubt," *Clio Medica* 5 (1970): 53–67. These were more in the way of translations with brief introductions than serious interpretive essays. See also Paul Hoffmann, "La controverse entre Leibniz et Stahl sur la nature de l'âme," *Studies on Voltaire and the Eighteenth Century* 199 (1981): 237–49; Sarah Carvallo, *La controverse entre Stahl et Leibniz sur la vie, l'organisme et le mixte* (Paris: Vrin, 2004).

149. Duchesneau offers crucial insights into this influence in his *Leibniz: Le vivant et l'organisme*, esp. chaps. 5–6, and he promises a whole new monograph in this vein soon.

150. Huneman and Rey, "La controverse Leibniz-Stahl," 220.

151. Geyer-Kordesch, *Pietismus, Medizin und Aufklärung in Preußen*, 207.

152. What provoked Leibniz in Stahl's formulations has been variously characterized. For one formulation: "What essentially motivated Leibniz was both Stahl's implicit denial of uniform laws of nature, and Stahl's misunderstanding of the metaphysics of substance and causality that Leibniz was in general elaborating in his own conceptions" (Huneman and Rey, "La controverse Lebiniz-Stahl," abstract [in English], 213).

153. Leibniz, "Preamble to His 'Animadversions,'" *LSC*, 21.

154. See the exchange of letters between Leibniz and Lady Masham in Gottfried Leibniz, *Philosophische Schriften*, ed. Carl Immanuel Gerhard (Hildelsheim: Olm, 1965), 3:333–74; Gottfried Leibniz, "Considerations on Vital Principles and Plastic Natures" (1705), in *Philosophical Papers and Letters*, ed. Leroy Loemker (Chicago: University of Chicago Press, 1956), 2:953–61; Gottfried Leibniz, "Clarification of the Difficulties Which Mr. Bayle Has Found in the New System" (1698), in Loemker, *Philosophical Papers and Letters*, 2:799–807. These ideas found ultimate expression in Leibniz's later publications "Principles of Nature and Grace, Based on Reaon" and *Monadology*,

both 1714. See G. W. Leibniz, *Philosophical Essays,* ed. Roger Ariew and Daniel Garber (Indianapolis: Hackett, 1989), 206–25.

155. Leibniz, "Preamble to His 'Animadversions,'" *LSC,* 25.

156. Ibid., 23.

157. Leibniz, "Animadversion I," *LSC,* 31.

158. Leibniz, "Exception I," *LSC,* 247.

159. Leibniz, "Animadversion II," *LSC,* 31. In his "Exception II" (*LSC,* 255) Leibniz elaborated: organic machines are nothing other than machines in which divine invention and intention are expressed to a greater extent.

160. Leibniz, "Exception I," *LSC,* 249.

161. Leibniz, "Animadversion IV," *LSC,* 33.

162. Leibniz, "Animadversion XX," *LSC,* 43.

163. Leibniz, "Animadversion XXI," *LSC,* 45.

164. Leibniz, "Animadversion XXVII," *LSC,* 49.

165. Leibniz, "Animadversion XXIX," *LSC,* 49.

166. Ibid.

167. Leibniz, "Animadversion III," *LSC,* 31.

168. Leibniz, "Exception XVI," *LSC,* 303.

169. Leibniz, "Exception XXI," *LSC,* 321.

170. Leibniz, "Animadversion X," *LSC,* 35.

171. Leibniz, "Animadversion XI," *LSC,* 37.

172. Leibniz, "Preamble to His 'Animadversions,'" *LSC,* 25.

173. Leibniz, "Exception X," *LSC,* 285.

174. Leibniz, "Exception XI," *LSC,* 289.

175. Ibid., 289–91.

176. Leibniz, "Exception XII," *LSC,* 295.

177. This was, of course, the essential project of the first two essays in *Theoria Medica Vera,* as I have already discussed. These were also the only parts of that work that Leibniz had bothered to read, for they contained all the philosophical problems he wanted to expose.

178. Stahl, "Enodation concerning Animadversion I," *LSC,* 69.

179. Ibid., 67. He elaborated in his response to "Exception IX": "The difficulty that perpetually impedes this new (mechanical) philosophy . . . [is] ultimately [it] ends up ascribing each and everything to matter" (*LSC,* 281–83).

180. Stahl, "Summary of the Principal Points of Doubt," *LSC,* 229.

181. Stahl, "Enodation concerning Animadversion XXI," *LSC,* 175.

182. Ibid., 177.

183. Conversely, Leibniz was, of course, preeminently concerned with this concept.

184. Stahl, "Enodation concerning Animadversion III," *LSC,* 79.

185. Stahl, "Summary of the Principal Points of Doubt," *LSC,* 239–41.

186. Stahl carried this down into the fundamentals-of-matter theory, making an important discrimination between particles and parts in chemical and organic "mixtures." It did not matter which of many identical particles came together to form the parts, though these had determinate form at this higher level. There are parallels, here, with Locke's formulations concerning organism versus physical elements.

187. "The entire force and center of this dispute consists in establishing the seat of the active motor principle" (Stahl, "Response to Exception II," *LSC,* 257).

188. Stahl, "Enodation concerning Animadversion XIX," *LSC,* 145. He added, in response to "Exception IV": "I am concerned with concepts concerning physical things" (*LSC,* 263).

189. Stahl, "Enodation concerning Animadversion IV," *LSC,* 87; "Enodation concerning Animadversion XVI," *LSC,* 123.

190. Stahl, "Enodation concerning Animadversion XVII," *LSC,* 135.

191. Stahl, "Summary of the Principal Points of Doubt," *LSC*, 223.
192. Stahl, "Enodation concerning Animadversion XXI," *LSC*, 183.
193. Stahl, "Enodation concerning Animadversion IV," *LSC*, 83.
194. Stahl, "Summary of the Principal Points of Doubt," *LSC*, 231.
195. Stahl, "Enodation concerning Animadversion XVI," *LSC*, 125.
196. Leibniz,"Animadversion IV," *LSC*, 33.
197. Ibid.
198. Stahl, "Enodation concerning Animadversion IV," *LSC*, 83-85.
199. Ibid., 85-87.
200. Leibniz, "Animadversions X-XI" (*LSC*, 35-37) and "Exceptions X-XI" (*LSC*, 285-91); Stahl, "Enodation concerning Animadversion X," *LSC*, 101.
201. Stahl, "Enodation concerning Animadversion XI," *LSC*, 107.
202. Stahl, "Enodation concerning Animadversion X," *LSC*, 103.
203. Ibid. 103-5.
204. Ibid., 103.
205. Stahl, "Enodation concerning Animadversion XXVII," *LSC*, 207.
206. Stahl, "Enodation concerning Animadversion XII," *LSC*, 111.
207. Stahl, "Enodation concerning Animadversion XI," *LSC*, 109.
208. Duchesneau, *Leibniz*, 25-46.
209. Leibniz, "Exception XI," *LSC*, 289.

CHAPTER TWO

1. Richard Toellner, "Haller und Leibniz: Zwei Universalgelehrte der Aufklärung," in *Akten des II. Internationalen Leibniz-Kongresses Hannover, 17.-22. Juli 1972* (Wiesbaden: F. Steiner, 1973), 1:249-60, citing 255.

2. Haller to Samuel-Auguste Tissot, January 11, 1762, in *Albrecht von Hallers Briefe an Auguste Tissot, 1754-1777*, ed. Erich Hintzsche (Bern: Huber, 1977), 125.

3. Heinrich Buess, "Die Anfänge der pathologischen Psychologie auf dem Gebiet der Kreislaufforschung nach Albrect Hallers *Elementa Physiologiae* (1756/1760)," *Gesnerus* 11 (1954): 121-51, citing 124.

4. Erich Hintzsche, "Albrecht Hallers anatomische Arbeit in Basel und Bern, 1728-1730," *Zeitschrift für Anatomie und Entwicklungsgeschichte* 111 (1941): 452-60.

5. Götz von Selle, *Die Georg-August-Universität zu Göttingen, 1737-1937* (Göttingen: Vandenhoeck und Ruprecht, 1937), 71.

6. Hubert Steinke, *Irritating Experiments: Haller's Concept and the European Controversy on Irritability and Sensiblity, 1750-90* (Amsterdam: Rodopi, 2005), 250.

7. There are grounds for Dijksterhuis to caution that "in the historiography of eighteenth-century science terms like 'Newtonian' and 'Newtonianism' are rather common and often used in an uncritical way," but that should *not* lead us to accept his judgment that "'Newtonianism' does not seem a very fruitful category for doing history of science." Fokko Jan Dijksterhuis, "Low Country Optics: The Optical Pursuits of Lambert ten Kate and Daniel Fahrenheit in Early Dutch 'Newtonianism,'" in *Newton and the Netherlands: How Isaac Newton Was Fashioned in the Dutch Republic*, ed. Eric Jorink and Ad Maas (Amsterdam: Leiden University Press, 2012), 159-83, citing 172-73. Instead, a good point of departure is the formulation of his colleague, Rienk Vermij: "Eighteenth-century 'Newtonianism' was in large part an attempt to create a viable philosophy of nature that on the one hand would account for all the scientific discoveries of the previous century, but on the other would avoid the problems of the mechanical philosophy." Rienk Vermij, "Defining the Supernatural: The Dutch Newtonians, the Bible, and the Laws of Nature," in Jorink and Maas, *Newton and the Netherlands*, 185-206, citing 190. The problems with the mechanical philosophy

were reductionism and materialist atheism, personified in Spinoza but ascribed to Descartes and to Pierre Gassendi's revival of Epicurus.

8. Bernard Cohen, *Franklin and Newton* (Philadelphia: American Philosophical Association and Harvard University Press, 1966). See Henri Guerlac, "Newton's Changing Reputation in the Eighteenth Century," in *Carl Becker's Heavenly City Revisited*, ed. R. Rockwood (Hamden, CT: Archon, 1968), 3–26; Yehuda Elkana, "Newtonianism in the Eighteenth Century," *British Journal for the Philosophy of Science* 22 (1971): 297–306; Robert Schofield, "An Evolutionary Taxonomy of Eighteenth-Century Newtonianism," *Studies in Eighteenth Century Scientific Culture* 7 (1978): 175–92; R. W. Home, "Out of a Newtonian Straightjacket: Alternative Approaches to Eighteenth-Century Physical Science," *Studies in the Eighteenth Century* 4 (1979): 235–49; Steven Shapin, "The Social Uses of Knowledge," in *The Ferment of Knowledge: Studies in the Historiography of Eighteenth-Century Science*, ed. George Rousseau and Roy Porter (Cambridge: Cambridge University Press, 1980), 93–139; Paolo Casini, "Le 'Newtonisme' au siècle des Lumières: Recherches et perspectives," *Dix-huitième siècle* 1 (1969): 139–59; and see "Newton and Newtonianism," special issue, *Studies in History and Philosophy of Science* 35 (2004), with essays by Mandelbrote, Iliffe, Ahnert, van der Wall, and Hutton.

9. Thierry Hoquet, "History without Time: Buffon's Natural History as a Nonmathematical Physique," *Isis* 101, no. 1 (2010): 30–61.

10. I take the term "experimental philosophy" from Peter Anstey, "Experimental versus Speculative Natural Philosophy," in *The Science of Nature in the Seventeenth Century: Patterns of Change in Early Modern Philosophy*, ed. P. Anstey and J. Schuster (Dordrecht: Netherlands: Springer, 2005), 215–42; Peter Anstey and Alberto Vanzo, "Origins of Early Modern Experimental Philosophy," *Intellectual History Review* 22, no. 4 (2012): 499–581.

11. Larry Stewart, *The Rise of Public Science: Rhetoric, Technology, and Natural Philosophy in Newtonian Britain, 1660–1750* (Cambridge: Cambridge University Press, 1992); Margaret Rowbottom, "The Teaching of Experimental Philosophy in England, 1700–1730," in *Actes du XIe Congrès international d'histoire des sciences* (Wrocklaw: Ossilineum, 1968), 46–53.

12. Roger Cotes, preface to second edition of *Principia*.

13. "'Newtonianism' as used by early Dutch Newtonians was the physico-theological program that mobilized the pious response to the allegedly atheist implications of Descartes's and Spinoza's mechanistic philosophies" (Dijksterhuis, "Low Country Optics," 174). Newton "entered the stage at a moment when the discontent with Cartesian physics and Spinozist rationalism was mounting." Eric Jorink and Huis Zuidervaart, "'The Miracle of Our Time': How Isaac Newton Was Fashioned in the Netherlands," in Jorink and Maas, *Newton and the Netherlands*, 13–65, citing 26. The main focus of the reception was "Cotes's foreword to the *Principia* and the remarks in the [*General*] *Scholium*" ("'Miracle of Our Time,'" 26).

14. See Neil Gillespie, "Natural History, Natural Theology, and Social Order: John Ray and the 'Newtonian Ideology,'" *Journal of the History of Biology* 20 (1987): 1–49.

15. Jorink and Zuidervaart, "'Miracle of Our Time,'" 24.

16. "In the Netherlands, Bernard Niewentuijt was the most important representative" of the whole new genre called "physico-theology" (Vermij, "Defining the Supernatural," 189). See Rienk Vermij, "Nature in Defense of Scripture: Physico-theology and Experimental Philosophy in the Work of Bernard Nieuwentijt," in *The Book of Nature in Early Modern and Modern History*, ed. K. van Berkel and A. Vanderjagt (Leuven: Peeters, 2006), 83–96.

17. Jorink and Zuidervaart, "'Miracle of Our Time,'" 33.

18. Vermij, "Defining the Supernatural," 186. See also Eric Jorink, "Honoring Sir Isaac; or, Exorcising the Ghost of Spinoza: Some Remarks on the Success of Newton in the Dutch Republic," in *Future Perspectives on Newton Scholarship and the Newtonian Legacy in Eighteenth-Century Science and Philosophy* (Brussels: Koninklijke Vlaamse Academie van Belgie voor Wetenschappen en Kunsten, 2009), 22–34. Notably, Ad Maas has written that by 1717 the "'Amsterdam mathematicians' like Nieuwentijt and Ten Kate had come to realize the fruitlessness of their attempts to support religion by mathematical argument. They now directed their attention to physico-theological rea-

soning." Ad Maas, "The Man Who Erased Himself: Willem Jacob 'sGravesande and the Enlightenment," in Jorink and Maas, *Newton and the Netherlands*, 113–37, citing 136n50.

19. Jorink and Zuidervaart, "'Miracle of Our Time,'" 26. It should be noted that the second edition of the *Principia* was also crucial for articulating his empiricism.

20. Ibid.

21. Ibid., 25. "The 1714 Amsterdam reprint coincided with a Newtonian offensive not only by Le Clerc, but also by Boerhaave, Nieuwentijt, 'sGravesande and the versatile scholar Lambert ten Kate" (ibid., 28).

22. Ibid., 25.

23. In the Netherlands the émigré British intellectual John Toland invoked Newtonianism in support of his "pantheist" Spinozism. See John Toland, *Letters to Serena* (1704; repr., Dublin: Four Courts Press, 2013). See also Margaret Jacob, *The Radical Enlightenment: Pantheists, Freemasons and Republicans* (London: Allen and Unwin, 1981).

24. "Newton was not the prime mover of eighteenth-century experimental philosophy. . . . The experimental philosophy that is labelled 'Newtonian' had been taking shape well before Newton entered the scene. Wiesenfeld has shown how at Leiden University an experimental physics was established in the 1670s . . . [with] De Volder[, whose] later 'Cartesianism' was quite empirical" (Dijksterhuis, "Low Country Optics," 173).

25. Edward Ruestow, *Physics at Seventeenth and Eighteenth Century Leiden: Philosophy and the New Science in the University* (The Hague: Nijhoff, 1973).

26. Jorink and Zuidervaart, "'Miracle of Our Time,'" 20–21.

27. Catherine Wilson, *The Invisible World: Early Modern Philosophy and the Invention of the Microscope* (Princeton, NJ: Princeton University Press, 1995).

28. Harold Cook, 'The New Philosophy in the Low Countries," in *The Scientific Revolution in National Context*, ed. Roy Porter and Mikulas Teich (Cambridge: Cambridge University Press, 1992), 115–49.

29. Jorink and Zuidervaart, "'Miracle of Our Time,'" 21.

30. Ibid., 25. "In the history of Dutch 'Newtonianism,' the *Opticks* is relatively overshadowed by the *Principia*" (Dijksterhuis, "Low Country Optics," 176). That may have been less true by the time Haller came to Leiden in 1725 than it was earlier in the century, however.

31. Ruestow, *Physics*, 115.

32. Rina Knoeff, "How Newtonian Was Herman Boerhaave?," in Jorink and Maas, *Newton and the Netherlands*, 93–111.

33. "When De Volder retired from his chair in 1705, Boerhaave remained there as the only defender of Newtonian principles and the only partisan of the experimental method, until in 1717 he was joined by 'sGravesande." Gerrit Lindeboom, *Herman Boerhaave: The Man and His Work* (London: Methuen, 1968), 269.

34. Cited in Richard Toellner, "Medizin in der Mitte des 18. Jahrhunderts," in *Wissenschaft im Zeitalter der Aufklärung*, ed. Rudolf Vierhaus (Göttingen: Vandenhoeck und Ruprecht, 1985), 194–221, citing 198.

35. Herman Boerhaave, "Discourse on the Achievement of Certainty in Physics" (1715), in *Boerhaave's Orations*, trans. E. Kegel-Brinkgreve and A. M. Luyendijk-Elshout (Leiden: Brill, 1983), 145–79; Jorink and Zuidervaart, "'Miracle of Our Time,'" 32.

36. "We do not know when Boerhaave first made acquaintance with Newton's views. But it seems probable that he read Newton's *Principia* soon after its appearance. . . . In any case, on the Continent of Europe Boerhaave was the first ardent supporter of Newtonian principles" (Lindeboom, *Herman Boerhaave*, 268). See also Steinke, *Irritating Experiments*, 26.

37. Lindeboom, *Herman Boerhaave*, 22.

38. Ruestow, *Physics*, 110.

39. F. L. S. Sassen, "The Intellectual Climate in Leiden in Boerhaave's Time," in *Boerhaave and His Time*, ed. G. A. Lindeboom (Leiden: Brill, 1970), 1–16, citing 7.

40. Ruestow, *Physics*, 110.

41. Cited in ibid., 111.

42. Ibid.

43. James Axtell, "Locke's Review of the *Principia*," *Notes and Records of the Royal Society of London* 20 (1965): 152–61.

44. Lindeboom, *Herman Boerhaave*, 25.

45. On David Gregory, see C. M. Eagles, "David Gregory and Newtonian Science," *British Journal for the History of Science* 10 (1977): 216–25.

46. A. Guerrini, "Archibald Pitcairne and Newtonian Medicine," *Medical History* 31 (1987): 70–83.

47. Anita Guerrini, "The Tory Newtonians: Gregory, Pitcairne, and Their Circle," *Journal of British Studies* 25 (1986): 288–311, citing 297. See also G. Lindeboom, "Pitcairne's Leyden Interlude Described from the Documents," *Annals of Science* 19 (1963): 273–84. For more on the notion of "Tory" Newtonianism, see John Friesen, "Archibald Pitcairne, David Gregory and the Scottish Origins of English Tory Newtonianism, 1688–1715," *History of Science* 41 (2003): 162–81.

48. Guerrini, "Archibald Pitcairne and Newtonian Medicine," 70.

49. A. Guerrini, "James Keill, George Cheyne, and Newtonian Physiology, 1690–1740," *Journal of the History of Biology* 18 (1985): 247–66, citing 250.

50. Ibid., 247.

51. Ibid., 249.

52. Guerrini, "Archibald Pitcairne and Newtonian Medicine," 70.

53. Ibid., 75.

54. Ibid., 82–83.

55. Steinke, *Irritating Experiments*, 27.

56. On Pitcairn's first "dissertation," see Guerrini, "Archibald Pitcairne and Newtonian Medicine," 75; Guerrini, "Keill, Cheyne," 251.

57. Guerrini, "Keill, Cheyne," 248.

58. J. R. R. Martin, "Explaining John Freind's History of Physick," *Studies in History and Philosophy of Science* 19 (1988): 399–418.

59. Lindeboom, *Herman Boerhaave*, 269.

60. Helene Metzger, *Newton, Stahl, Boerhaave et le doctrine chimique* (Paris: Alcan, 1930).

61. Gerrit Lindeboom, "Boerhaave's Impact on the Relation between Chemistry and Medicine," *Clio Medica* 7 (1972): 271–78, citing 274.

62. Ibid., 273.

63. Herman Boerhaave, "On Chemistry Correcting Its Own Errors," in *Boerhaave's Orations*, 180–213; Lindeboom, "Boerhaave's Impact," 275.

64. Cohen, *Franklin and Newton*, 214–33.

65. T. L. Davis, "The Vicissitudes of Boerhaave's Text-book of Chemistry," *Isis* 10 (1928): 33–46; F. W. Gibbs, "Boerhaave's Chemical Writings," *Ambix* 3 (1959): 117–35; M. Kerker, "Herman Boerhaave and the Development of Pneumatic Chemistry," *Isis* 46 (1955): 36–49; and, most recently, Rina Knoeff, *Herman Boerhaave (1668–1738): Calvinist Chemist and Physician* (Amsterdam: Royal Academy of Sciences, 2002).

66. Jorink and Zuidervaart, "'Miracle of Our Time,'" 32.

67. Knoeff, *Herman Boerhaave*; Knoeff, "How Newtonian Was Herman Boerhaave?," 93.

68. Knoeff, "How Newtonian Was Herman Boerhaave?," 94. The exception is La Mettrie.

69. Ibid.

70. Ibid., 98.

71. Ibid., 100.

72. Ibid., 108.

73. Ibid. See Cohen, *Franklin and Newton*, 214–33.

74. Theodore Brown, "From Mechanism to Vitalism in Eighteenth-Century English Physiology," *Journal of the History of Biology* 7 (1974): 179–216.

75. Knoeff, "How Newtonian Was Herman Boerhaave?," 108.

76. James Keill, *Account of Animal Secretion* (1708), cited in Guerrini, "Keill, Cheyne," 256. On Keill (1673–1719), see F. M. Valadez and C. D. O'Malley, "James Keill of Northampton: Physician, Anatomist, Physiologist," *Medical History* 15 (1971): 317–35.

77. Knoeff, "How Newtonian Was Herman Boerhaave?," 105.

78. "Newton's ideas were appropriated by different people in different ways and for different purposes, and not all concepts which came to be sold under Newton's name actually stemmed from him" (Vermij, "Defining the Supernatural," 195).

79. Knoeff, "How Newtonian Was Herman Boerhaave?," 106. See Robert Schofield, *Mechanism and Materialism: British Natural Philosophy in an Age of Reason* (Princeton, NJ: Princeton University Press, 1970).

80. Henry Pemberton, *A View of Sir Isaac Newton's Philosophy* (repr., Bristol: Thoemmes, 2004). Haller wrote to his friend Johannes Geßner of Pemberton's work that it was "a splendid book in which the mysteries of the Newtonian philosophy are revealed in an easy style and without algebra." Henry E. Sigerist, ed., *Haller-Geßner Correspondence*, Göttingen Academy Papers 11, no. 2 (Berlin: Weidmann, 1923), 22–23.

81. Steinke believes that 'sGravesande "transformed Newtonianism in so far as he admitted not only a mathematical but also an empirical understanding of nature" (*Irritating Experiments*, 90n99). But that is, I believe, to misunderstand Newtonianism and even Newton himself, for this case was explicitly made by him in the *Opticks* and it was the basis of his chemical work (e.g., *De Natura Acidorum* from the early 1690s). Newton was *never* exclusively mathematical in his orientation to nature or science. See Cohen, *Franklin and Newton*, for a classic exposition of all this.

82. Sassen, "Intellectual Climate," 12. "In the Dutch Republic itself, 'sGravesande's influence secured the growing prestige of Newtonianism and experimental science throughout the provinces, and the most eminent and effective of 'sGravesande's followers, Musschenbroek, was ultimately to teach at Leiden with no less acclaim and distinction" (Ruestow, *Physics*, 118–19).

83. Vermij, "Defining the Supernatural," 189.

84. Maas, "The Man Who Erased Himself," 130.

85. Ruestow, *Physics*, 119.

86. Maas, "The Man Who Erased Himself," 126. On Musschenbroek's workshop, see P. De Clercq, *At the Sign of the Oriental Lamp: The Musschenbroek Workshop in Leiden, 1660–1750* (Rotterdam: Erasmus, 1997).

87. Willem 'sGravesande, *Physices Elementa Mathematica, Experimentis Confirmata, sive Introductio ad Philosophiam Newtonianam* (Leiden: Van der Aa, 1720–21). A second edition appeared in 1725.

88. Willem 'sGravesande, *Mathematical elements of physicks, prov'd by experiments: Being an introduction to Sir Isaac Newton's philosophy*, trans. J. Keill (London: Strahan, 1720).

89. Willem 'sGravesande, *Mathematical elements of natural philosophy, confirm'd by experiments; or, An introduction to Sir Isaac Newton's philosophy*, trans. J. T. Desaguliers, 6th ed. (London: Innys, 1747).

90. Ruestow, *Physics*, 118.

91. Ibid.

92. J. B. Shank, *The Newton Wars and the Beginning of the French Enlightenment* (Chicago: University of Chicago Press, 2008).

93. Vermij, "Defining the Supernatural," 188.

94. Kees de Pater, "'The Wisest Man to Whom This Earth Has as Yet Given Birth': Petrus van Musschenbroek and the Limits of Newtonian Natural Philosophy," in Jorink and Maas, *Newton and the Netherlands*, 139–57, citing 140.

95. Pierre Brunet, *L'introduction des théories de Newton en France au XVIIIe siècle avant 1738* (Paris: Blanchard, 1931).

96. De Pater, "'Wisest Man," 144.

97. Ibid., 150.

98. Erich Hintzsche, "Wertung des Tagebuches," in *Albrecht Hallers Tagebuch seiner Studienreise nach London, Paris, Straßburg und Basel, 1727–1728,* 2nd, rev. ed., ed. Erich Hintzsche (Bern: Huber, 1971), 68. (Originally published, St. Gallen: Hausmann, 1948.) Haller gave his disciple/biographer Johann Zimmermann information about his childhood, which Zimmermann reverently reproduced, and scholars have all too willingly followed in his train, until Hintzsche interjected some critical judgment. See Erich Hintzsche, "Einige kritische Bemerkungen zur Bio- und Ergographie Albrecht von Hallers," *Gesnerus* 16 (1959): 1–15. As Hintzsche's successor in directing the Haller archives in Bern, Urs Boschung summarizes this new sense of things: "Haller's comments on his work and his own person need to be taken with particular caution, 'with benign mistrust.'" Urs Boschung, "Albrecht Hallers Aufenthalt in Paris im Lichte eines unbekannten Tagebuch Johannes Geßners," *Medizinhistorisches Journal* 11 (1976): 220–45, citing 229. The internally cited phrase is taken from another great Haller scholar and skeptic, Karl Guthke ("Haller und Pope," *Euphorion* 69 [1975]: 110).

99. Adolf Haller, *Albrecht von Hallers Leben* (Basel: Reinhardt, 1954), 13–14; G. Rudolph, "Albrecht von Haller (1708–1777)," *Annales Universitatis Saraviensis: Medizin* 7 (1959): 273–89, citing 276.

100. Albrecht von Haller reminiscence, cited in Adolf Haller, *Leben,* 14.

101. Rüdiger Robert Beer, *Der grosse Haller* (Säckingen: Stratz, 1947), 17.

102. Haller in *Von und über Albrecht von Haller,* ed. Eduard Bodemann (Hanover: Maeyer, 1885), 89.

103. Rudolph, "Albrecht von Haller," 276.

104. Erich Hintzsche, editorial note to Haller, *Tagebuch seiner Studienreise, 1727–1728,* 102n26.

105. H. M. Jones, "Albrecht von Haller and English Philosophy," *PMLA* 40 (1925): 103–27, citing 104.

106. See Georg Bernhard Bilfinger, *Dilucidationes philosophicae, de deo, anima humana, mundo* (Tübingen, 1725, 1746, 1768).

107. Johann Georg Zimmermann, *Das Leben des Hrn Albrecht von Hallers* (Zurich: n.p., 1755), 24.

108. Haller, *Tagebuch seiner Studienreise, 1727–1728,* 32, 36.

109. Ibid., 37, 38.

110. T. H. Lunsingh Scheurleer and G. H. M. Posthumus Meyjes, eds., *Leiden University in the Seventeenth Century: An Exchange of Learning* (Leiden: Brill, 1975), esp. contributions by Scheurleer, "Un amphithéatre d'anatomie moralisée," 217–78; G. A. Lindeboom, "Dog and Frog—Physiological Experiments," 279–94; C. De Pater, "Experimental Physics," 309–28; and J. W. van Spronsen, "The Beginning of Chemistry," 329–44. See also Harold Cook, "The New Philosophy in the Low Countries," in Porter and Teich, *Scientific Revolution in National Context,* 115–49, esp. 123, on the relation to Padua.

111. Jorink and Zuidervaart, "'Miracle of Our Time,'" 15–16.

112. Haller, *Tagebuch seiner Studienreise, 1727–1728,* 37.

113. Lindeboom, *Herman Boerhaave,* 362.

114. Steinke, *Irritating Experiments,* 52. For more detail on this, see Ulrich Tröhler, "250 Jahre Göttingen Medizin: Begründung—Folgen—Folgerungen," in *Naturwissenschaften in Göttingen: Ein Vortragsreihe,* ed. Han-Heinrich Voigt (Göttingen: Vandenhoeck und Ruprecht, 1988), 9–36.

115. Lindeboom, *Herman Boerhaave,* 356. See most recently Knoeff, *Herman Boerhaave.*

116. See Martin Stuber, Stefan Hächler, and Luc Lienhard, eds., *Hallers Netz: Ein europäischer Gelehrtenbriefwechsel zur Zeit der Aufklärung,* Studia Halleriana 9 (Basel: Schwabe, 2005), esp. 127ff., on the medical network; 67–70, on Haller's early correspondence network.

117. Haller, *Tagebuch seiner Studienreise, 1727–1728,* 37, 80.

118. Lindeboom, *Herman Boerhaave,* 268, citing La Mettrie, *Éloge de Boerhaave,* in *Oeuvres* (London: Chez Jean Nourse, 1751), 1:89–91.

119. Haller, *Tagebuch seiner Studienreise, 1727–1728,* 37.

120. Ibid., 80.

121. Ibid., 82.

122. Ibid., 47.

123. Ibid., 79.

124. H. Punt, *Bernard Siegfried Albinus (1697–1770): On "Human Nature"; Anatomical and Physiological Ideas in Eighteenth-Century Leiden* (Amsterdam: Israël, 1983).

125. "Haller not only attended Albinus's anatomical dissections but, in his last few months of study in spring 1727, he also frequented his private physiological lectures. Haller's lecture notes have survived" (Steinke, *Irritating Experiments*, 97)

126. Albrecht von Haller, *Elementa Physiologiae Corporis Humani*, 8 vols. (Lausanne: Bousquet, 1757–66). Trans. Johann Samuel Halle as *Herrn Albrecht von Hallers Anfangsgründe der Phisiologie des menschlichen Körpers*, 8 vols. (Berlin: Voss, 1762–76), vols. 3–7.

127. Haller, *Tagebuch seiner Studienreise, 1727–1728*, 45, 47.

128. "Haller completed his studies at Leiden already in May 1727, after he expanded the thesis he had composed earlier in Tübingen into a dissertation." Erich Hintzsche, "Albrecht von Haller als Anatom und seine Schule," *Ciba Zeitschrift* 10 (1948): 4068–73, citing 4068.

129. Haller, *Tagebuch seiner Studienreise, 1727–1728*, 59.

130. Ibid., 59–60. See Beer, *Der grosse Haller*, 30.

131. For a good discussion, see Urs Boschung, "Albrecht von Haller und der praktische Arzt seiner Zeit," *Gesnerus* 42 (1985): 252–64, citing 256. See now Ole Peter Grell, Andrew Cunningham, and Jon Arrizabalaga, eds., *Centres of Medical Excellence? Medical Travel and Education in Europe, 1500–1789* (Farnham, UK: Ashgate, 2010).

132. Rudolph, "Albrecht von Haller," 276.

133. See the entry on John Pringle in Urs Boschung et al., eds., *Repertorium zu Albrecht von Hallers Korrespondenz, 1724–1777*, 2 vols. (Basel: Schwabe, 2002), 1:409. Pringle and Haller carried on a lifetime correspondence, which has been edited and published by Otto Sonntag: *John Pringle's Correspondence with Albrecht von Haller* (Basel: Schwabe, 1999).

134. Looking back on the experience and having in the interval learned English, Haller observed: "Here I saw how terrible it is to travel in a country where one does not know the language" (*Tagebuch seiner Studienreise, 1727–1728*, 87).

135. Haller observed: "There is no way of dissection that he has not tried. . . . The man is certainly an anatomist of the highest order" (ibid., 24). See also ibid., 25.

136. Ibid., 18–19, 22–23. See Adolf Haller, *Leben*, 31.

137. Haller, *Tagebuch seiner Studienreise, 1727–1728*, 18. Johann Caspar Scheuchzer was the son of Johann Jakob Scheuchzer (1672–1733), whom Haller met later, after his famous Alps tour of 1728. The elder's major work, *Physic; oder, Natur-wissenschaft* (1729) would prove important for Haller.

138. Haller, *Tagebuch seiner Studienreise, 1727–1728*, 21, 23, 24.

139. March 10, 1730, in Boschung et al., *Repertorium*, 1:478.

140. Erich Hintzsche, ed., *Albrecht Hallers Tagebücher seiner Reisen nach Deutschland, Holland und England, 1723–1727*, 2nd, rev. ed. (Bern: Huber, 1971). (Originally published, St. Gallen: Hausmann, 1948.)

141. Erich Hintzsche, ed., *Albrecht Hallers Tagebuch seiner Studienreise nach London, Paris, Straßburg und Basel, 1727–1728*, 2nd, rev. ed. (Bern: Huber, 1968). (Originally published, Bern: Haupt, 1942.) The difference is striking, but it may be an artifact of Haller's revision of the original diaries, which breaks off precisely in the middle of his account of his visit to England (August 8, 1727). It may reflect as well the interests of the scholar who discovered these diaries, Ludwig Hirzel, who published a selected edition in 1883 that stopped at that same date (Hintzsche, "Einleitung," in *Tagebuch seiner Studienreise*, 12). Hirzel's highly esteemed biography of Haller, the prologue to his critical edition of Haller's poetry, made no bones about leaving all science quite aside in its consideration. Ludwig Hirzel, "Hallers Leben und Dichtungen," in *Albrecht von Hallers*

Gesdichte, ed. Ludwig Hirzel (Frauenfeld: Huber, 1882), i–dxxxvi. Thus, Hirzel was far more taken with Haller's retrospective observations on England, which had direct bearing on his poetry, than with Haller's medical comings and goings, since they were *ex hypothesi* peripheral. The modern editor of these materials, Erich Hintzsche, whose training and interest were utterly medical, took quite the opposite point of view. In 1971 Hintzsche published a revised version of the original Hirzel edition of 1883. But his own editions of 1968 and 1971 are the ones primarily used in this study. They contain all the missing materials with all the details about Haller's medical-scientific experiences in England, together with Haller's accounts of further travels.

142. Haller, *Tagebücher seiner Reisen, 1723–1727*, 93–94.

143. Ibid., 98.

144. Richard Toellner, "Decora Merenti: Glory, Merit and Science: Haller Spellbound by the Newtonian Star," *Janus* 67 (1980): 171–85.

145. Haller, *Tagebücher seiner Reisen, 1723–1727*, 93.

146. Boschung has established that Haller had bought the book in Leiden in 1726, shortly after it was published. See Urs Boschung, "'Mein Vergnügen . . . bey den Büchern': Albrecht von Hallers Bibliothek—von den Anfängen bis 1736," *Librarium* 38 (1995): 154–74.

147. Voltaire, *Letters concerning the English nation* (London: Davis and Lyon, 1733); Voltaire, *Lettres philosophiques: Texte intégral* (1734; Paris: Larousse, 1972).

148. Haller, *Tagebücher seiner Reisen, 1723–1727*, 87.

149. Interestingly, Haller both times included Newton's name as a matter of course, though Newton's exact role in the correspondence has long been a subject of some contention.

150. Haller, *Tagebücher seiner Reisen, 1723–1727*, 17.

151. On Desmaizeaux, see Joseph Almagor, *Pierre Des Maizeaux (1673–1745), Journalist and English Correspondent for Franco-Dutch Periodicals, 1700–1720* (Amsterdam: AP-Holland University Press, 1989).

152. Haller, *Tagebücher seiner Reisen, 1723–1727*, 91.

153. Toby Gelfand, *Professionalizing Modern Medicine: Paris Surgeons and Medical Science and Institutions in the Eighteenth Century* (Westport, CT: Greenwood, 1980), 9. See the latest discussion in Grell, Cunningham, and Arrizabalaga, *Centres of Medical Excellence?*

154. Toby Gelfand, "The 'Paris Manner' of Dissection: Student Anatomical Dissection in Early Eighteenth-Century Paris," *Bulletin of the History of Medicine* 46 (1971): 99–130.

155. Ibid.

156. Boschung, "Albrecht Hallers Aufenthalt in Paris." See also Urs Boschung, "Ein Brief Albrecht Hallers über seinen Aufenthalt in Paris und Strassburg (1728)," *Physis* 19 (1977): 185–96. The new monograph by Boschung summarizing his research on Haller's Paris years—*Albrecht von Haller in Paris, 1727–1728* (Bern: Huber, 2009)—actually offers nothing new on this score.

157. The Geßners wrote to Boerhaave on September 11, 1727: "Everything that there is to learn in Paris about medicine is available only for substantial money. . . . We are particularly concerned not just to watch but to get our hands into it" (Boschung, "Albrecht Hallers Aufenthalt in Paris," 231).

158. Heinrich Ernst Jenny, "Haller als Philosoph: Ein Versuch" (diss., University of Bern, 1902), 8.

159. Johann Geßner wrote to Benedikt Stähelin in Basel, on September 12, 1727: "upon the particular advice of Mr. Boerhaave we have traveled to Paris to get some practical experience in anatomy and surgery and to unify theory with our own experience. . . . M. Winslow, the omnicompetent anatomist, the unquestioned master of his field, can be persuaded by nothing to give a private course in anatomy, even if all the others do" (Boschung, "Albrecht Hallers Aufenthalt in Paris," 233). That Geßner already knew Stähelin (MD Basel, 1716, and professor of physics from 1747) suggests that Geßner had contacts in Basel before going to Leiden, where he met Haller. This would also explain Geßner's desire to move on to Basel from Paris, not only to study mathematics with Bernoulli but to attain his own MD there in 1730.

160. Urs Boschung, *Albrecht Haller in Paris, 1727–1728* (Bern: Huber, 2009), 71.

161. The phrase is from Hintzsche's "Wertung des Tagebuches," in Haller, *Tagebuch seiner Studienreise, 1727–1728*, 68–70, citing 69.

162. Erich Hintzsche, "Albrecht Hallers 'Manuscripta Winslowiana': Ein wieder aufgefundenes Tagebuch aus seiner Pariser Studienzeit," *Centaurus* 4 (1955): 97–121.

163. Rudolph, "Albrecht von Haller," 278.

164. Hintzsche, "Einige kritische Bemerkungen," 7.

165. Hintzsche, "Albrecht Hallers 'Manuscripta Winslowiana,'" 97.

166. Hintzsche cites Zimmermann to this effect in ibid., 118. Hintzsche brings to the fore Haller's critical distance from Winslow in the articles Haller wrote for the Yverdon *Encylopédie*, in which Haller now more properly assigned to Winslow's teacher, Joseph Guichard Duverney (1648–1730), and to Duverney's Danish model, Niels Stensen (1638–86), much that had been ascribed to Winslow. Haller went so far as to assert: "Duverney . . . is the real author of the anatomy that was presented and taught in Paris by Winslow." Albrecht von Haller, "Histoire de la physiologie," in *Encyclopédie d'Yverdon*, ed. Fortunato Bartolomeo De Felice, cited in Erich Hintzsche, "Niels Stensen, Winslow und Haller," in *International Sumposium on Steno and Brain Research in the Seventeenth Century: Proceedings*, ed. Gustav Scherz (Oxford: Symposium, 1968), 207–17, citing 212.

167. Boschung, "Albrecht Hallers Aufenthalt in Paris," 228n41.

168. Adolf Haller, *Leben*; Hintzsche, "Einige kritische Bemerkungen."

169. Beer, *Der grosse Haller*, 34.

170. Geßner notes this in his Paris diary for December 12, 1727 (Boschung, "Albrecht Hallers Aufenthalt in Paris," 243).

171. Geßner Paris diary entry, December 19, 1727. It is not entirely clear which Duverney this is, but it is more plausible that Geßner was referring to Joseph Guichard Duverney's son than to the octogenarian giant of French medicine himself.

172. Boschung, "Albrecht Hallers Aufenthalt in Paris," 229.

173. Hintzsche, "Albrecht Hallers 'Manuscripta Winslowiana,'" 119.

174. Boschung, "Albrecht Hallers Aufenthalt in Paris," 226.

175. Only a letter from Albinus, responding to Haller's inquiries, suggests that Haller was concerned with questions of physiology even in January 1728 (ibid.).

176. In his *Bibliotheca Anatomica* Haller writes about needing to get out of Paris: "To escape severe punishment, even perhaps the gallows, I had to make myself scarce and give up dissecting." Albrecht von Haller, *Bibliotheca Anatomica*, vol. 2 (Tiguri, 1777), 196, cited in Boschung, "Ein Brief Albrecht Hallers," 187).

177. Hintzsche, "Einige kritische Bemerkungen," 8, correcting Rudolph, "Albrecht von Haller," 284n18.

178. There is even a work of (relative) fiction about this: Paul De Gendre, "Le séjour du Docteur Albert Haller à Paris en 1728: Botanique, poésie, flirt et dissection," *Le concours médical* 13 (1936): 1429–32.

179. Haller to Johann Philipp Burggrave (1700–1775), March 7, 1728: "In the beginning of February [1728] I began to prepare for a trip to Italy since the dissections at the Charité had been for various reasons interrupted" (cited in Boschung, "Ein Brief Albrecht Hallers," 194).

180. Curiously, Haller reports this to his friend Burggrave in the following terms: "Circumstances compelled me, then, to travel to Basel and to study mathematics under the direction of the famous Bernoulli" (ibid.). How could circumstances compel that specific action?

181. Haller did compose a poem in Leiden about his homesickness for Switzerland, but that came years earlier, and homesickness does not suffice as an explanation here. See Urs Boschung, "Hallers Lebens-stationen," in *Albrecht von Haller im Göttingen der Aufklärung*, ed. Norbert Elsner and Nicolaas Rupke (Göttingen: Wallstein, 2009), 21–46, citing 30. Neither does Zimmermann's account of an illness that kept Haller from continuing to Italy explain it, even if this was Haller's retrospective account of his decision.

182. On the other hand, Boschung notes that in all Haller's notebooks of the Paris period, "he attaches no particular significance to the friendship with the Geßner brothers" ("Albrecht Hallers Aufenthalt in Paris," 228).

183. Rudolph, "Albrecht von Haller," 278.

184. Hintzsche, preface to Haller, *Tagebücher seiner Reisen, 1723–1727*, 8.

185. Beer, *Der grosse Haller*, 35–36.

186. Bernard de Fontenelle, *The Elogium of Sir Isaac Newton* (London: Tonson and Osborn-Longman, 1728); reprinted in *Isaac Newton's Papers and Letters on Natural Philosophy and Related Documents*, 2nd ed., ed. Bernard Cohen (Cambridge, MA: Harvard University Press, 1978), 445–74.

187. Haller, *Tagebücher seiner Reisen, 1723–1727*.

188. Shank, *Newton Wars*; Pierre Brunet, *L'introduction des théories de Newton en France au XVIIIe siècle avant 1738* (Paris: Blanchard, 1931).

189. Shank, *Newton Wars*, 50–51.

190. Ibid., 51.

191. Ibid., 52.

192. On occult qualities in the seventeenth century, see John Henry, "Occult Qualities and the Experimental Philosophy: Active Principles in Pre-Newtonian Matter Theory," *History of Science* 24 (1986): 335–81.

193. Mary Hesse, "Action at a Distance," in *Forces and Fields: The Concept of Action at a Distance in the History of Physics* (New York: Philosophical Library, 1962), 157–88; Koffi Maglo, "The Reception of Newton's Gravitational Theory by Huygens, Varignon, and Maupertuis: How Normal Science May Be Revolutionary," *Perspectives on Science* 11 (2003): 135–69.

194. Shank, *Newton Wars*, 147.

195. Ibid., 47.

196. In *Newton Wars* Shank gets to "English experimental Newtonianism" at page 147 and thereafter. He notes the key role of Desaguliers and Pemberton and the parallel endeavors of 'sGravesande in Holland, but he see these as merely stubborn, ideological opposition to the prevailing style of natural philosophy and mathematics, as exemplified, for Shank, by Jean-Pierre de Crousaz (1663–1750) and Pierre Rémond de Montmort (1678–1719).

197. Ibid., 187.

198. Ibid., 112.

199. Ibid., 124.

200. Stewart, *Rise of Public Science*.

201. Newton established complete institutional domination of the Royal Society and of British science, especially after the death of Robert Hooke and his own ascendancy to permanent president of the society. That Newton was aggressively building a coterie and becoming the "autocrat of science," as Frank Manuel aptly dubbed him, is perhaps clearest in his privileging of David Gregory, his devoted follower, over Edmond Halley, one of his most important advocates and a man who had invested fortune with reputation in sponsoring the original version of the *Principia*. This "cult of personality" dimension to Newtonianism starting in the late 1690s is a factor that Shank (*Newton Wars*) does not dwell upon. See Mordechai Feingold, "Mathematicians and Naturalists: Sir Isaac Newton and the Royal Society," in *Isaac Newton's Natural Philosophy*, ed. Jed Buchwald and Bernard Cohen (Cambridge, MA: MIT Press, 2001), 77–102; Peter Anstey, "Experimental Pedagogy and the Eclipse of Robert Boyle in England," *Intellectual History Review* 25 (2015): 115–31.

202. Shank, *Newton Wars*, 73.

203. Ibid., 78–82, on Réaumur; 94–102, on Dortous de Mairan.

204. Ibid., 248.

205. John Greenberg, *The Problem of the Earth's Shape from Newton to Clairaut* (Cambridge: Cambridge University Press, 1995), 246.

206. Shank, *Newton Wars*, 250.

207. Ibid., 251.

208. Ibid., 220.

209. Ibid., 202.

210. Greenberg, *Problem of the Earth's Shape*, 246.

211. Shank, *Newton Wars*, 252.

212. Ibid.

213. Ibid., 263.

214. Ibid., 290–91.

215. Haller, *Tagebücher seiner Studienreise, 1727–1728*, 48.

216. Haller–Geßner correspondence, 1729, in Boschung et al., *Repertorium*, 1:171.

217. "However I had become more acquainted with the English poets and had adopted from them the love of thought and the preference for a more severe poetic form. The philosophical poets, whose greatness I admired, swiftly suppressed in me the dreary [*geblähte*] and puffed-up character" of German baroque poetry. Albrecht Haller, preface to the fourth edition of his *Versuch schweizerischer Gedichte* (1748), cited by Jones, "Albrecht von Haller and English Philosophy," 107.

218. Jones, "Albrecht von Haller and English Philosophy," 108. See also the dedication to King George in his *Historia Stirpium Indigenarum Helvetiae Inchoata* (Bern: sumptibus Societatis Typographicae, 1768): "Nata est et floret in Britannia paulo seculo junior illa societas cujus ideam meditatus erat parens melioris philosophiae, Geometria, Algebra, Mechanica, Chemia experimenta difficillima et impediossima conspiraverunt in restaurationem Physices. Dedit orbi Newtonum Providentia, qui doceret, quantum humanum ingenium posset in inveniendo, et limites figeret ultra quos nihil posteris sperendum esset." Cited in Jones, "Albrecht von Haller and English Philosophy," 108n.

219. Jones, "Albrecht von Haller and English Philosophy," 107.

220. *Versuch schweizerischer Gedichte* was first published anonymously (Bern: E. Haller, 1732); the second edition (Bern: Haller, 1734) acknowledged his authorship.

221. Marjorie Nicolson, *Mountain Gloom and Mountain Glory: The Development of the Aesthetics of the Infinite* (Ithaca, NY: Cornell University Press, 1959); Ernest Tuveson, "Space, Deity, and the 'Natural Sublime,'" *Modern Language Quarterly* 12 (1951): 20–38; Frederick Staver, "'Sublime' as Applied to Nature," *Modern Language Notes* 70 (1955): 485–87; Christopher Thacker, *The Wildness Pleases: The Origins of Romanticism* (New York: St. Martin's Press, 1983).

222. Uwe Hentschel, "Albrecht von Hallers Alpen-Dichtung und ihre zeitgenössische Rezeption," *Wirkendes Wort* 48 (1998): 183–91; Urs Boschung, "Haller botaniste et poète—à la découverte des Alps," in *Une cordée originale*, ed. J.-C. Pont et al. (Chêne-Bourg: Georg, 2000), 96–119.

223. Beat Ludwig von Muralt, *Lettres sur les Anglois et les François* (1725).

224. Hentschel, "Albrecht von Hallers Alpen-Dichtung," 183–84.

225. Haller's last publication in Bern would be his account of his Alpine tours, of which he made a second extended one in 1731 and several more in the local Bern highlands thereafter. Albrecht Haller, *Iter Alpinum* (Tiguri: Heidegger, 1736).

226. Shteir notes that "from the very first edition of Haller's poem, the text was accompanied by footnotes containing botanical nomenclatures and other botanical particulars about the plants characterized in the verses." She notes further that in the preface to his *Enumeratio Methodica Stirpium Helvetiae Indigenarum qua Omnium Brevis Descriptio et Synonyma Compendium Virium Medicarum Dubiarum Declaration Novarum et Rariorum Uberior Historia et Icones Continentur* (Göttingen: Vandenhoeck, 1742), Haller traced the "plan to compile, describe, and catalogue indigenous Swiss plants from the 1728 Alpine excursion and the preceding months of botanical enthusiasm with friends in Basel." Ann Shteir, "Albrecht von Hallers' Botany and 'Die Alpen,'" *Eighteenth-Century Studies* 10 (1976/77): 169–84, citing 173, 176.

227. Haller, *Enumeratio Methodica Stirpium Helvetiae Indigenarum*. There is reason to believe that Geßner made contact with Linnaeus or the Linnaeans (they were a very tightly knit organization) at least shortly after his first Alpine expedition with Haller and gave the Linnaean community advance notice of their botanical discoveries.

228. Letter dated August 28, 1728, in Boschung et al., *Repertorium*, 1:347.

229. Hintzsche, "Albrecht Hallers anatomischc Arbeit in Basel und Bern."

230. As Terrall has noted, Maupertuis appeared at the optimal time to serve as Bernoulli's advocate in Paris. Mary Terrall, *The Man Who Flattened the Earth* (Chicago: University of Chicago Press, 2002), 45–49.

231. Cited in Latin by Jones, "Albrecht von Haller and English Philosophy," 106.

232. Ibid., 107.

233. Rudolph, "Albrecht von Haller," 278.

234. Boschung et al., *Repertorium*, 1:433.

235. Ibid.

236. Sonntag claims that "Haller did not, to be sure, agree with the low estimation of mathematics expressed around the middle of the century by Buffon and Diderot. . . . [He was] in principle sympathetic to attempts at employing quantitative and mathematical analysis in the biological sciences." Otto Sonntag, "The Mental and Temperamental Qualities of Haller's Scientist," *Physis* 19 (1977): 173–84, citing 181. But in practice, Haller was quite critical of efforts at mathematization among his contemporaries and in his own work. As he explicitly advised readers of his masterpiece, the *Elementa*, there was little beyond arithmetic and simple geometry. For Haller, calculus was not a tool; it was a credential.

237. Toellner, "Decora Merenti," 178.

238. Beer, *Der grosse Haller*, 36–37.

239. "Six years after Haller, Pope brought these experiences together in his famous couplet." (Toellner, "Medizin in der Mitte des 18. Jahrhunderts," 209).

240. Jones, "Albrecht von Haller and English Philosophy," 119; Shirley Roe, "*Anatomia Animata*: The Newtonian Physiology of Albrecht von Haller," in *Transformations and Tradition in the Sciences*, ed. Everett Mendelsohn (Cambridge: Cambridge University Press, 1984), 273–300, citing 288.

241. Albrecht Haller, *Gedanken über Vernunft, Aberglauben, und Unglauben*, cited in Roe, "*Anatomia Animata*," 288.

242. Albrecht Haller, *Falschheit menschlicher Tugenden*, cited in Roe, "*Anatomia Animata*," 288. De Angelis is correct to observe that, notwithstanding all this gushing enthusiasm for Newton, "[h]itherto, even the literary Haller scholarship has not been able to provide precise results concerning Haller's engagement with the writings of Isaac Newton." Simone De Angelis, *Von Newton zu Haller: Studien zum Naturbegriff zwischen Empirismus und deduktiver Methode in der Schweizer Frühaufklärung* (Tübingen: Niemeyer, 2003), 178. To be sure, the German didactic poem of the day entailed engaging with broad contours of Newton's cosmology, but that did not attest to any detailed or specific grasp of his arguments or methods (De Angelis, *Von Newton zu Haller*, 111).

243. Toellner, "Decora Merenti," 178.

244. Aiton takes a stronger view of Bernoulli's vorticism: "Although admitting that, as a way of describing phenomena, the attraction was a fertile concept, Bernoulli still found the attraction unacceptable as a cause, and a physicist, he believed, should seek causes." E. J. Aiton, *The Vortex Theory of Planetary Motions* (London: Macdonald, 1972), 147.

245. Toellner, "Decora Merenti," 178.

246. Ibid.

247. Ibid., 182.

248. Ibid., 179.

249. François Duchesneau, "Physiological Mechanism from Boerhaave to Haller," in *Man and Nature / L'homme et la nature: Proceedings of the Canadian Society for Eighteenth-Century Studies* (London: Faculty of Education, University of Western Ontario, 1982), 1:209–81, citing 217.

250. Albrecht Haller, review of *Enumeratio Methodica Stirpium Helvetiae Indigenarum*, *Bibliothèque raisonnée* 29 (1742): 270, cited in Sonntag, "Mental and Temperamental Qualities of Haller's Scientist," 175.

251. Albrecht von Haller, *Briefe über die wichtigsten Wahrheiten der Offenbarung* (Bern: Neue Buchhandlung, 1772), 6.

252. Steinke, *Irritating Experiments*, 139.

253. Roe, "*Anatomia Animata*," 287.

254. Ibid.

255. Karl Guthke, "Zur Religionsphilosophie des jungen Albrecht von Haller," *Colloquia Germanica* 1 (1967): 142–55, comment on 146, and full Haller text on 150.

256. Ibid., 150.

257. See Haller's journal comments on Pemberton as reproduced in ibid., 150.

258. In a brilliant study, De Angelis argues that Haller was profoundly shaped by 'sGravesande's Newtonianism, which in turn contained an important legacy from the natural-law theory of Cumberland and Pufendorf: "in Leiden the young Haller could pursue 'sGravesande's grounding of Newton's methodological rules for natural-scientific research upon the natural-law juridical principles [of Cumberland and Pufendorf] on the basis of study of the *Physices elementa*, which the future physiologist read in its second edition of 1725" (*Von Newton zu Haller*, 382). While there is much to suggest that the mature Haller of circa 1750 conforms to De Angelis's model, important quibbles need to be registered about how carefully Haller had assimilated 'sGravesande's ideas already in 1725 and when and with what impact Haller read the balance of 'sGravesande's publications. It is, as De Angelis notes from Maria Monti (*Congettura ed esperienza nella fisiologia di Haller* [Florence: Olschki, 1990], 105), all about "what actually were the ways and the stages of Haller's comprehension of the works of Newton [*quale siano effettivamente stati per Haller modi e tempi nella conoscenza dell'opera di Newton*]," and De Angelis is correct to contend that "the central meaning of 'sGravesande for the development of a scientific epistemology on the basis of Newtonian methodological principles and his influence on the young Haller has not yet been taken sufficiently into consideration by the history-of-science-oriented Haller scholarship." We should therefore welcome her endeavors to fill the gap by providing "an exact analysis of the importance of . . . 'sGravesande's writings for Haller's concept of science." The danger that De Angelis does not quite avoid is believing that their views are immediately correlated where, at best, inferences and imputations to Haller are all the evidence really permits (9). Thus, after 1728 De Angelis's argument about the centrality of 'sGravesande makes more sense, but it still needs to be connected with Pemberton, and we need more direct evidence of Haller's careful study of 'sGravesande's subsequent writings on the epistemology of science.

259. Roe, "*Anatomia Animata*," 289.

260. Steinke, *Irritating Experiments*, 114–16.

261. Ibid., 121n66, referring to P. M. Heimann and J. E. McGuire, "Newtonian Forces and Lockean Powers: Concepts of Matter in Eighteenth-Century Thought," *Historical Studies in the Physical Sciences* 3 (1971): 233–306.

262. Steinke, *Irritating Experiments*, 115.

263. Ibid., 95. A line from one of Haller's poems phrased this memorably: "Ins innre der Natur dringt kein erschaffner Geist," from *Falschheit menschlicher Tugenden* (1730), cited in Richard Toellner, *Albrecht von Haller: Über die Einheit im Denken des letzten Universalgelehrten*, Sudhoffs Archiv, Beiheft 10 (Wiesbaden: Steiner, 1971), 65.

264. Steinke, *Irritating Experiments*, 95.

265. Ibid., 158.

266. Ibid.

267. Lester King, introduction to Albrecht von Haller, *First Lines of Physiology*, trans. William Cullen (1786; repr., New York: Johnson, 1966), ix–lxxii, citing liv. Originally published as *Anfangsgründe der Physiologie* (1747).

268. Ibid., lxvi.

269. De Angelis, *Von Newton zu Haller*, 205.

270. Ibid., 207.

271. Ibid., 277.

272. Ibid., 213.

273. Ibid., 292.

274. Ibid., 355.

275. Ibid., 300. "In particular, around 1750, Haller's physiology of the senses, Bonnet's psychophysiology, and Euler's optics provided the sciences of empirical inquiry with a natural-scientific, experimentally grounded basis of knowledge of the neurosensory substrate of cognitive processes in the human brain and created therewith the presuppositions for the conceptualization of an empirical logic of science" (ibid., 310). See Hans Adler, "Fundus Animae—der Grund der Seele: Zur Gnoseologie des Dunklen in der Aufklärung," *Deutsche Vierteljahrsschrift für Literaturwissenschaft und Geistesgeschichte* 62 (1988): 197–220; Hans Adler, *Die Prägnanz des Dunklen: Gnoseologie, Ästhetik, Geschichtsphilosophie bei J. G. Herder*, Studien zum 18. Jahrhundert, vol. 13 (Hamburg: Meiner, 1990).

276. De Angelis, *Von Newton zu Haller*, 387ff., esp. 398–99. De Angelis associates this particularly with Calvinism, going back to Bacon (392).

277. Ibid., 338–39.

278. Ibid., 382.

279. Ibid., 428.

280. Ibid., 420.

281. On Haller as an arbiter of German Enlightenment, see esp. Otto Sonntag and Hubert Steinke, "Der Forscher und Gelehrte" (317–46); Hubert Steinke and Martin Stuber, "Haller und die Gelehrtenrepublik" (381–414); and Wolfgang Proß, "Haller und die Aufklärung" (415–58); all in *Albrecht von Haller: Leben—Werk—Epoche*, ed. Hubert Steinke, Urs Boschung, and Wolfgang Proß (Göttingen: Wallstein, 2008). For Haller as a "universal scholar," see the title of Toellner's major monograph: *Albrecht von Haller: Über die Einheit im Denken des letzten Universalgelehrten*.

282. Herder on Haller, cited in Sonntag and Steinke, "Der Forscher und Gelehrte," 324. See also Herder's tribute in *Ideen*, cited by Gerhard Rudolph, "Hallers Lehre von der Irritabilität und Sensibilität," in *Von Boerhaave bis Berger*, ed. K. E. Rothschuh (Stuttgart: Fischer, 1964), 30n.

283. Johann Friedrich Blumenbach, *Medicinische Bibliothek* (Göttingen, 1785), 2:177, cited in Toellner, "Decora merenti," 175.

284. Stephen D'Irsay, *Albrecht von Haller: Eine Studie zur Geistesgeschichte der Aufklärung* (Leipzig, 1930), reprinted in Shirley Roe, *The Natural Philosophy of Albrecht von Haller*, (New York: Arno Press, 1981), with original pagination.

285. P. Wernle, *Der schweizerische Protestantismus im XVIII. Jahrhundert*, 3 vols. (Tübingen: n.p., 1923–25).

286. Toellner, "Haller und Leibniz," 250, 255.

287. Ibid., 255.

288. Guthke, "Zur Religionsphilosophie des jungen Albrecht von Haller."

289. Albrecht Haller, review of *Commercium Philosophicum & Mathematicum*, by Gottfried Leibniz and Johann Bernoulli, *Bibliothéque raisonnée* 37 (1746): 179–80.

290. This is evident in the passage where Haller celebrates Bernoulli. It is largely at Leibniz's expense.

291. Haller, review of *Commercium philosophicum & mathematicum*. See Otto Sonntag, "Albrecht von Haller on Academies and the Advancement of Science: The Case of Göttingen," *Annals of Science* 32 (1975): 379–91, esp. 382.

292. "In Haller's opinion, Leibniz had worked on a wide range of scientific fields at the expense of his accomplishments in its different parts. He would have contributed more to our knowledge had he cultivated only one science" (Steinke, *Irritating Experiments*, 256).

293. Roe, "*Anatomia Animata*," 276.

294. Albrecht von Haller, "Anzeige der Holmannischen Logik und Metaphysik," in *Sammlung kleiner Hallerischer Schriften* (Bern: Emanuel Haller, 1772), 3:9, 305–33.

295. On Hollmann, see Konrad Cramer, "Die Stunde der Philosophie: Über Göttingens ersten Philosophen und die philosophische Theorielage der Gründungszeit," in *Zur geistigen Situation der Zeit der Göttingen Unversitätsgründung 1737*, ed. Jürgen von Stackelberg, Göttingen Universitätsschriften, vol. A/12 (Göttingen: Göttingen Universitätsschriften, 1988), 101–43.

296. Haller on Hollmann, cited in Roe, "*Anatomia Animata*," 275.

297. "What Descartes understands as the conceptual products of mind, with which he could deduce the laws of physical nature, are for Haller 'inventions of imagination,' hence products of fantasy" (De Angelis, *Von Newton zu Haller*, 245).

298. Albrecht von Haller, preface to Buffon's *Natural History*, trans. Phillip Sloan in *From Natural History to the History of Nature: Readings from Buffon and His Critics*, ed. John Lyon and Phillip Sloan (Notre Dame, IN: University of Notre Dame Press, 1981), 297.

299. Albrecht von Haller, *Rede an dem Geburtage Georg des Zweyten, Die Köngliche Gesellschaft der Wissenchaften sich zum erstenmale öffentlich versamlete, den 10 November 1751*, in *Kleine Schriften* 2:173–206.

300. De Angelis, *Von Newton zu Haller*, 195–96.

301. Ibid., 201.

302. See Albrecht von Haller, "Vorrede zur Prüfung der Seckte die an allem zweifelt" (1751), in *Sammlung kleiner Hallerischen Schriften*, 2nd ed., 3 vols. (Bern: Haller, 1772), 1:1ff.; the preface to Jean Henri Samuel Formey's abridged translation of Crousaz, *Examen du pyrrhonisme ancien & moderne* (The Hague: De Hondt, 1733). Haller expressed himself "critically toward the Wolffian system" in alignment with Crousaz (Boschung et al., *Repertorium*, 1:219).

CHAPTER THREE

1. Rüdiger Robert Beer, *Der grosse Haller* (Säckingen: Stratz, 1947), 61–64.

2. Lester King, introduction to Albrecht von Haller, *First Lines of Physiology*, trans. William Cullen (1786; repr., New York: Johnson, 1966), ix–lxxii, citing xiv. Originally published as *Anfangsgründe der Physiologie* (1747).

3. Erich Hintzsche, "Albrecht Hallers anatomische Arbeit in Basel und Bern 1728–1736," *Zeitschrift für Anatomie und Entwicklungsgeschichte* 111, no. 3 (1941): 452–60, citing 457.

4. Urs Boschung et al., eds., *Repertorium zu Albrecht von Hallers Korrespondenz, 1724–1777*, 2 vols. (Basel: Schwabe, 2002), 1:504.

5. Hubert Steinke and Claudia Profos, eds., with the assistance of Pia Burkhalter, *Bibliographia Halleriana: Verzeichnis der Schriften von und über Albrecht von Haller*, Studia Halleriana 8 (Basel: Schwabe, 2004), 187.

6. Boschung et al., *Repertorium*, 1:540.

7. Hintzsche, "Albrecht Hallers anatomische Arbeit in Basel und Bern 1728–1736," 458.

8. Frans Stafleu, *Linnaeus and the Linnaeans: The Spreading of Their Ideas in Systematic Biology, 1735–1789* (Utrecht: Oosthoek, 1971).

9. Hintzsche, "Albrecht Hallers anatomische Arbeit in Basel und Bern 1728–1736," 459.

10. Gerlach Adolf von Münchhausen, "Nachträgliches Votum Münchhausesn über die Einrichtung der Universität in der Sitzung des geheimen Rats-Collegium. 14. April 1733," cited in R. Vierhaus, "1737—Europa zur Zeit der Universitätsgründung," in *Stationen der Göttinger Universitätsgeschichte, 1737—1787—1837—1887—1937*, ed. Bernd Moeller (Göttingen: Vandenhoeck und Ruprecht, 1988), 9–26, citing 22.

11. Vierhaus, "1737," 25; Thomas Howard, *Protestant Theology and the Making of the Modern German University* (Oxford: Oxford University Press, 2006), 107.

12. Vierhaus, "1737," 20–21. See also Peter Hanns Reill, "'Pflanzgarten der Aufklärung': Haller und die Gründung der Göttingen Universität," in *Albrecht von Haller im Göttingen der Aufklärung*, ed. Norbert Elsner and Nicolaas Rupke (Göttingen: Wallstein, 2009), 47–70; Ulrich Hunger, "Die

Universitätstadt Göttingen," in *Albrecht von Haller: Leben—Werk—Epoche*, ed. Hubert Steinke, Urs Boschung, and Wolfgang Proß (Göttingen: Wallstein, 2008), 99–118.

13. Götz von Selle, *Die Georg-August-Universität zu Göttingen, 1737–1937* (Göttingen: Vandenhoeck und Ruprecht, 1937), 23.

14. Ibid., 29.

15. Howard, *Protestant Theology*, 106.

16. Selle, *Die Georg-August-Universität zu Göttingen*, 24.

17. Bernd Moeller, "Johann Lorenz Mosheim und die Gründung der Göttinger Universität," in *Theologie in Göttingen: Eine Vorlesungsreihe*, ed. Bernd Moeller (Göttingen: Vandenhoeck und Ruprecht, 1987), 9–40.

18. Howard, *Protestant Theology*, 107–10.

19. Werlhof and Haller would have a massive correspondence; Werlhof sent 1,587 letters to Haller between 1736 and Werlhof's death in 1767. He was Haller's most important confidant in the entire period of Haller's Göttingen career, as he was instrumental in envisioning the kind of medical school at Göttingen in which Haller would thrive. See Boschung et al., *Repertorium*, 1:568.

20. Ulrich Tröhler, "250 Jahre Göttingen Medizin: Begründung—Folgen—Folgerungen," in *Naturwissenschaften in Göttingen: Ein Vortrragsreihe*, ed. Han-Heinrich Voigt (Göttingen: Vandenhoeck und Ruprecht, 1988), 9–36, citing 13.

21. Werlhof, cited in W. Brednow, *Jena und Göttingen: Medizinische Beziehungen im 18. und 19. Jahrhundert* (Jena: Fischer, 1949), 2.

22. Selle, *Die Georg-August-Universität zu Göttingen*, 24.

23. This started with the semihagiographic authorized biography of Haller by his student Johann Georg Zimmermann: *Das Leben des Hrn Albrecht von Hallers* (Zurich: n.p., 1755).

24. Erich Hintzsche, "Einige kritische Bemerkungen zur Bio- und Ergographic Albrecht von Hallers," *Gesnerus* 16 (1959): 1–15, citing 9.

25. Julius Pagel, *Über die Geschichte der Göttinger medizinischen Schule im XVIII. Jahrhundert* (Berlin: Itzkowski, 1875), 17; Selle, *Die Georg-August-Universität zu Göttingen*, 59; Georg Gruber, *Naturwissenschaftliche und medizinische Einrichtungen der jungen Georg-August-Universität in Göttingen* (Göttingen: Musterschmidt, 1955), 10.

26. Pagel, *Über die Geschichte der Göttinger medizinischen Schule*, 21–23.

27. Brednow, *Jena und Göttingen*, 3. Hamberger would later enter into a fierce controversy with Haller, as I will note.

28. Pagel, *Über die Geschichte der Göttinger medizinischen Schule*, 18–19.

29. Ibid., 18.

30. Urs Boschung, "Göttingen, Hanover, and Europe: Haller's Correspondence," in *Göttingen and the Development of the Natural Sciences*, ed. Nicolaas Rupke (Göttingen: Wallstein, 2002), 33–49, citing 37.

31. Ibid., 38. See Boschung et al., *Repertorium*, 1:256.

32. Gruber, *Naturwissenschaftliche und medizinische Einrichtungen.*

33. Richard Toellner, "Medizin in der Mitte des 18. Jahrhunderts," in *Wissenschaft im Zeitalter der Aufklärung*, ed. Rudolf Vierhaus (Göttingen: Vandenhoeck und Ruprecht, 1985), 194–221, citing 196.

34. Boschung et al., *Repertorium*, 1:361.

35. Ibid., 362.

36. Ulrich Joost, "'Trübselige kleine Stadt in einem trübseligen Land'? Hallers Göttingen," in Elsner and Rupke, *Albrecht von Haller im Göttingen der Aufklärung*, 71–106.

37. Hubert Steinke, *Irritating Experiments: Haller's Concept and the European Controversy on Irritability and Sensibility, 1750–90* (Amsterdam: Rodopi, 2005), 53; Hubert Steinke, "Divergierende Resultate eines Forschungslabors des 18. Jahrhunderts," *Cardanus: Jahrbuch für Wissenschaftsgeschichte* 1 (2000): 93–115, citing 100.

38. Johann Georg Zimmermann, notes on Albrecht Haller in *Von und über Albrecht von Haller:*

Ungedruckte Briefe und Gedichte Hallers sowie ungedruckte Briefe und Notizen über denselben, ed. Eduard Bodemann (Hanover: Meyer, 1885), 216n11.

39. Zimmermann, *Das Leben des Hrn Hallers.*

40. Albrecht von Haller, *Tagebücher seiner Beobachtungen über Schriftsteller und über sich selbst*, ed. Johann Georg Heinzmann (1787; repr., Frankfurt: Athenäum, 1971).

41. G. Rudolph, "Albrecht von Haller (1708–1777)," *Annales Universitatis Saraviensis: Medizin* 7 (1959): 273–89, citing 278.

42. Pagel, *Über die Geschichte der Göttinger medizinischen Schule*, 8–9.

43. Tröhler, "250 Jahre Göttingen Medizin," 17.

44. Pagel, *Über die Geschichte der Göttinger medizinischen Schule*, 14.

45. Ibid., 12.

46. Tröhler, "250 Jahre Göttingen Medizin."

47. Haller's relation with Carl Linnaeus (1707–78) was brittle but sustained. Heinz Goerke, "Die Beziehung Hallers zu Linné," *Sudhoffs Archiv* 38 (1954): 367–77. Their correspondence began in 1737, concentrating, understandably, on topics botanical. They had become somewhat estranged by 1745 because of criticisms of Haller in Linnaeus's *Flora Suecica*. In 1747 Haller announced to Linnaeus that he was putting botany aside to pursue more intensely his interests in anatomy and physiology. See Boschung et al., *Repertorium*, 1:309.

48. Albrecht Haller, *De Methodico Studio Botanices absque Praeceptore cum Botanices Anatomiae et Chirurgiae Professionem Gottingae Capesseret* (Göttingen, October 1736).

49. Albrecht Haller, *Historia Stirpium Indigenarum Helvetiae Inchoata* (Bern: sumptibus Societatis Typographicae, 1768). See G. R. De Beer, "Haller's *Historia Stirpium*," *Annals of Science* 9 (1953): 1–46.

50. Haller to Linnaeus, 1747 (Boschung et al., *Repertorium*, 1:309). See Heinrich Zoller, "Albrecht von Hallers Pflanzensammlung in Göttingen, sein botalisches Werk, und sein Verhältnis zu Carl von Linné," *Nachrichten der Akademie der Wissenschaften in Göttingen, II. Mathematsich-Physikalische Klasse* 10 (1958): 238–49.

51. Richard Toellner, *Albrecht von Haller: Über die Einheit im Denken des letzten Universalgelehrten*, Sudhoffs Archiv, Beiheft 10 (Wiesbaden: Steiner, 1971).

52. Steinke, *Irritating Experiments*, 50–51.

53. Richard Toellner, "Die Verbindung von Lehren und Forschung an den jungen Georgia Augusta zu Göttingen," *Hippokrates: Zeitschrift für praktisxche Heilskjunde* 39 (1968): 859–63, citing 863n12. He refers to Eduard Bodemann, ed., *Von und über Albrecht von Haller* (Hanover: Maeyer, 1885), 190, 200–201.

54. Toellner, "Die Verbindung von Lehren und Forschung," 862.

55. Albrecht von Haller, *Bibliotheca Anatomica*, cited in Rudolph, "Albrecht von Haller," 286n30.

56. Renato Mazzolini, "Le dissertazioni degli allievi di Albrecht von Haller a Göttingen (1736–1753): Una indagine bio-bibliografica," *Nuncius* 2 (1987): 125–94. One of Haller's students was Johann A. H. Reimarus (1729–1814), who worked with Haller in 1752; in 1760 he sent Haller a copy of the book by his eminent father, Hermann Reimarus, on animal instinct, seeking his evaluation, presumably in *GGA*. See Boschung et al., *Repertorium*, 1:422.

57. Steinke, "Divergierende Resultate," 99.

58. Boschung et al., *Repertorium*, 1:332. It was Meckel's view that Haller rejected the invitation to the Berlin Academy because of the presence of La Mettrie in Berlin.

59. Ibid., 613–14.

60. Ibid., 606.

61. On Haller's reviews in *GGA*, see Claudia Profos, "Literaturkritik," in Steinke, Boschung, and Proß, *Albrecht von Haller: Leben—Werk—Epoche*, 182–98, which is a reprise of her *Gelehrte Kritik: Albrecht von Hallers literarisch-wissenschaftliche Rezensionen in den Göttingischen Gelehrten Anzeigen* (Basel: Schwabe Basel, 2008), which revises Karl Guthke's pioneering *Haller und die Literatur* (Göttingen: Vandenhoeck und Ruprecht 1962). See Walter Zimmerli, "Der Rezensent und Präses

perpetuus der 'Königlichen Societät der Wissenschaften zu Göttingen,'" in *Albrecht von Haller: Zum 200. Todestag* (Göttingen: Vandenhoeck und Ruprecht, 1977), 12–19. See also Karl Guthke, *Literarisches Leben im achtzehnten Jahrhundert in Deutschland und in der Schweiz* (Bern: Francke, 1975).

62. Peter-Eckhard Knabe, *Die Rezeption der Französischen Aufklärung in den "Göttingischen Gelehrten Anzeigen" (1739–1779)* (Frankfurt: Klostermann, 1978).

63. Richard Toellner, "Entstehung und Programm der Göttinger Gelehrten Gesellschaft unter besonderer Berücksichtigung des Hallerschen Wissenschaftsbegrifes," in *Der Akademiegedanke im 17. und 18. Jahrhundert*, ed. Fritz Hartmann (Bremen: Jacobi, 1977), 97–115.

64. Haller, memorandum to Münchhausen, January 11, 1751, cited in Otto Sonntag "Albrecht von Haller on Academies and the Advancement of Science: The Case of Göttingen," *Annals of Science* 32 (1975): 379–91, citing 381.

65. Albrecht von Haller, *Rede an dem Geburtage Georg des Zweyten, Die Königliche Gesellschaft der Wissenschaften sich zum erstenmale öffentlich versamlete, den 10 November 1751*, in *Sammlung kleiner Hallerischer Schriften: Zweyter Teil* (Bern: Verlag Emanuel Haller, 1772) 2:173–206, cited in Sonntag, "Albrecht von Haller on Academies," 383.

66. Sonntag, "Albrecht von Haller on Academies," 390.

67. Ibid., 387.

68. Cited in Toellner, "Die Verbindung von Lehren und Forschung," 860.

69. Boschung et al., *Repertorium*, 1:288.

70. Johann Gottlob Krüger, *Versuch einer Experimental-Seelenlehre* (1756). See Carsten Zelle, "Experimentalseelenlehre und Efrahrungsseelenkunde: Zur Unterscheidung von Erfahrung, Beobachtung und Experiment bei Johann Gottlob Krüger und Karl Philipp Moritz," in *"Vernünftige Ärzte"—Hallesche Psychomedizin und die Anfänge der Anthropologie in der deutschsprachigen Frühaufklärung* (Tübingen: Neimeyer, 2001), 173–85.

71. Boschung et al., *Repertorium*, 1:296.

72. Kant wished to dedicate his *Kritik der reinen Vernunft* to Lambert, but the latter died before Kant could do this.

73. Shirley Roe, "*Anatomia Animata*: The Newtonian Physiology of Albrecht von Haller," in *Transformations and Tradition in the Sciences*, ed. Everett Mendelsohn (Cambridge: Cambridge University Press, 1984), 273–300, citing 295n.

74. Erna Lesky, "Albrecht von Haller, Gerard von Swieten und Boerhaavens Erbe," *Gesnerus* 15 (1958): 120–40, citing 122–24.

75. Steinke, *Irritating Experiments*, 98.

76. Ibid., 259.

77. H. Punt, *Bernard Siegfried Albinus (1697–1770): On "Human Nature": Anatomical and Physiological Ideas in Eighteenth-Century Leiden* (Amsterdam: Israël, 1983).

78. As King notes, Haller was "scarcely a pleasant character[,] . . . quarrelsome and vindictive." Thus, he had "an uncanny knack of making professional enemies. He was constantly involved in controversies" (King, introduction to Haller, *First Lines of Physiology*, xxiii, xxi).

79. Haller, preface to *Elementa Physiologiae*, vol. 5 (1763), cited in Lesky, "Albrecht von Haller, Gerard von Swieten und Boerhaavens Erbe," 129.

80. Lesky, "Albrecht von Haller, Gerard von Swieten und Boerhaavens Erbe," 137. Steinke makes the same point: Haller "considered controversies as a necessary and productive element of scientific research" (*Irritating Experiments*, 259).

81. Haller, preface to Buffon's *Natural History* (1750), trans. Phillip Sloan in *From Natural History to the History of Nature: Readings from Buffon and His Critics*, ed. John Lyon and Phillip Sloan (Notre Dame, IN: University of Notre Dame Press, 1981).

82. Ibid.

83. For twenty years Van Swieten (MD, Leiden, 1725) attended and transcribed in shorthand Boerhaave's lectures. In 1745 he moved to Vienna, since as a Roman Catholic he had no career

prospects in the Netherlands. Haller knew Van Swieten and his word-for-word note-taking in Boerhaave's lectures from their days together in Leiden. See Haller, *Tagebücher seiner Beobachtungen über Schriftsteller und über sich selbst*, 81.

84. This comparison with Van Swieten was made by Haller in his review of (his own) *Praelectiones* in *Bibliothèque raisonnée* 33 (1744): 33–50.

85. Boschung et al., *Repertorium*, 1:511.

86. Kathleen Wellman, *La Mettrie: Medicine, Philosophy, and Enlightenment* (Durham, NC: Duke University Press, 1992), 107.

87. Ibid., 126.

88. For Haller's reviews, see Haller, review of *Institutions de medicine*, by Herman Boerhaave, trans. Julien Offray de La Mettrie, vol. 1 (Paris, 1740), *GGA*, 1745, 377–78; and Haller, review of *Institutions de medicine*, by Herman Boerhaave, trans. Julien Offray de La Mettrie, vol. 2 (Paris, 1740), *GGA*, 1748, 950–52.

89. Haller, review of *Histoire naturelle de l'âme*, by Julien Offray de La Mettrie, *GGA*, June 1747, 413–15.

90. Aram Vartanian, *La Mettrie's "L'homme machine": A Study in the Origins of an Idea* (Princeton, NJ: Princeton University Press, 1960), 200.

91. Julien Offray de La Mettrie, *Man a Machine and Man a Plant*, trans. Richard A. Watson and Maya Rybalka (Indianapolis: Hackett, 1994), 20n.

92. Haller, *GGA*, December 1747, 907.

93. Vartanian, *La Mettrie's "L'homme machine,"* 104.

94. Haller, *Journal des savants*, March 1749, cited in Roe, "*Anatomia Animata*," 282.

95. Vartanian, *La Mettrie's "L'homme machine,"* 200.

96. Haller, review in *GGA*, December 1747, 907.

97. Vartanian, *La Mettrie's "L'homme machine,"* 87.

98. King, introduction to Haller, *First Lines of Physiology*, lxviii. On Hamberger's approach, see ibid., lxvi–vii.

99. In his authorized biography, Zimmermann documents this dispute blow-by-blow.

100. Haller, preface to Buffon's *Natural History* (1750), cited in Steinke, *Irritating Experiments*, 259.

101. "Replicable experiment triumphed after the midcentury over mathematical speculation." Wolfram Kaiser, "Hallesche Physiologie im Werden," in *Hallesche Physiologie im Werden: Hallesches Symposium 1981*, ed. Wolfram Kaiser and Hans Hübner (Halle: Wissenschaftliche Beiträge der Martin-Luther Universität Halle-Wittenberg, 1981), 10–24, citing 18.

102. Steinke, *Irritating Experiments*, 131.

103. Ibid., 80.

104. Rudolph, "Albrecht von Haller," 280.

105. All citations here will be from the English translation, *Dissertation on the Sensible and Irritable Parts of Animals*, trans. Samuel Auguste David Tissot (1755), ed. Owsei Temkin (Baltimore: Johns Hopkins University Press, 1936). Note the discrepancy in the titles. The lecture to the academy explicitly refers to the human body, while the English translation, emulating its French source, refers to the parts of animals.

106. Nikolaus Mani, "Physiologische Konzepte von Galen bis Haller," *Gesnerus* 45 (1988): 165–90; G. E. Lindeboom, "Boerhaave's Concept of the Basic Structure of the Body," *Clio Medica* 5 (1970): 203–8; Lester King, *The Growth of Medical Thought* (Chicago: University of Chicago Press, 1962), 177–85.

107. Steinke, *Irritating Experiments*, 103.

108. Haller, *Dissertation on the Sensible and Irritable Parts of Animals*, 8.

109. Ibid., 21.

110. Richard Toellner, "Decora merenti: Glory, Merit and Science: Haller Spellbound by the Newtonian Star," *Janus* 67 (1980): 171–85, citing 181.

111. Haller, *Dissertation on the Sensible and Irritable Parts of Animals*, 22, 7. In correspondence with his friend the physician Paul Werlhof in late 1750, it was clear that Haller's whole concentration was on specific allocation of the distinct properties of irritability and sensibility to distinct tissues and parts of the body, with the only "theoretical" or "metaphysical" issue being that of the relation of the "soul" to bodily motions (Steinke, "Divergierende Resultate," 105).

112. Haller, *Dissertation on the Sensible and Irritable Parts of Animals*, 8, 9.

113. Ibid., 40.

114. Ibid., 24.

115. Ibid., 8.

116. Haller, *Dissertation on the Sensible and Irritable Parts of Animals*, cited in Roe, "*Anatomia Animata*," 279.

117. Haller, *Dissertation on the Sensible and Irritable Parts of Animals*, 42.

118. Steinke, *Irritating Experiments*, 106.

119. Haller, *Dissertation on the Sensible and Irritable Parts of Animals*, 42.

120. Alexander Berg, *Die Lehre von der Faser als Form- und Funktionselement des Organismus: Die Geschichte des biologisch-medizinischen Grundproblems vom kleinsten Bauelement des Körpers bis zur Begründung der Zellenlehre*, Virchows Archiv, vol. 309 (Berlin: Reimer, 1942); M. D. Grmek, "La notion de fibre vivante chez les médecins de l'école iatrophysique," *Clio Medica* 5 (1970): 297–318; L. J. Rather, "Some Relations between Eighteenth-Century Fiber Theory and Nineteenth-Century Cell Theory," *Clio Medica* 4 (1969): 191–202.

121. Steinke, "Divergierende Resultate," 106.

122. Haller, *Dissertation on the Sensible and Irritable Parts of Animals*, 9, 47.

123. Seeking the real behind the distortion of the particular has been grasped by Lorraine Daston and Peter Galison as the specific manifestation of "objectivity" that motivated eighteenth-century science. See their *Objectivity* (New York: Zone, 2007), 64–75.

124. Haller, *Dissertation on the Sensible and Irritable Parts of Animals*, 26. This became one of the most explicit points he criticized in the dissertation of his student Zimmermann.

125. Steinke has tried to infer an experimental regimen from Haller's text. He comes up with five components: (1) the contrast of experiment to observation; (2) no prejudgment; (3) care in execution; (4) replication; (5) restriction of inferential hypothesizing. Hubert Steinke, "'Die Ehre des Rechthabens': Experiment und Theorie im Streit um de Lehre der Irritabilität," *Sudhoffs Archiv* 82 (1998): 141–69, esp. 148–52. The very generality of these rules suggests the primitiveness of the definition of experimental procedure upon which Haller's audience had to base their assessment and, even more important, their efforts to test his results by replication.

126. Steinke, *Irritating Experiments*, 94.

127. Ibid., 101. Steinke ("Divergierende Resultate," 94) reports that all the experimental protocols of their dissections have been preserved.

128. Steinke, "Divergierende Resultate."

129. "Boerhaave's pupils in the Netherlands . . . actually made the step that their teacher only contemplated, viz. to openly declare innate bodily faculties responsible for movement that are independent of the soul" (Steinke, *Irritating Experiments*, 34).

130. Ibid., 102.

131. Ibid., 20.

132. Zimmermann's dissertation, *On Irritability*, appeared in 1751.

133. Steinke, "Divergierende Resultate," 105.

134. Ibid., 111.

135. Ibid., 162.

136. Haller, *Dissertation on the Sensible and Irritable Parts of Animals*, 12.

137. Ibid., 13.

138. Lesky, "Albrecht von Haller, Gerard von Swieten und Boerhaavens Erbe."

139. Haller, *Dissertation on the Sensible and Irritable Parts of Animals*, 17.

140. Ibid., 23.

141. Ibid., 28, 41.

142. On Whytt, see R. G. French, *Robert Whytt, the Soul, and Medicine* (London: Wellcome Institute of the History of Medicine, 1969).

143. Haller, *Dissertation on the Sensible and Irritable Parts of Animals*, 28.

144. Ibid.

145. Boschung et al., *Repertorium*, 1:301.

146. Haller on La Metrie in his *Dissertation on the Sensible and Irritable Parts of Animals*, 66–67.

147. Heinrich Ernst Jenny, "Haller als Philosoph: Ein Versuch" (diss., University of Bern, 1902).

148. "In anatomical circles, Haller already based on the Göttingen years, assumed a dominant standing." Erich Hintzsche, "Albrecht von Haller als Anatom und seine Schule," *Ciba Zeitschrift* 10 (1948): 4068–73, citing 4071.

149. Steinke, *Irritating Experiments*, 127.

150. That monuments can also be roadblocks is an idea that Bruno Latour made central to his construal of Immanuel Kant's paradoxical relation to "modernity." See Bruno Latour, *We Have Never Been Modern* (Cambridge, MA: Harvard University Press, 1993).

151. Richard Toellner, "Mechanismus—Vitalismus: Ein Paradigmawechsel? Testfall Haller," *Studien zur Wissenschaftstheorie* 10 (1977): 61–72.

152. Ibid. See also Richard Toellner, "Anima et Irritabilitas: Hallers Abwehr von Animismus und Materialismus," *Sudhoffs Archiv* 51 (1967): 130–44, esp. 143: "thus, Haller had opened the way for the investigation of living nature with resources and theoretical models that were more appropriate to the highly organized objects of inquiry than the old mechanical conceptions and methods." And yet he refused to abandon mechanism; he proved, as it were, "a secret vitalist," Steinke observes. "Haller is not—as Toellner writes—the creator of the vitalist paradigm; he is merely its initiator" (Steinke, "'Die Ehre des Rechthabens," 161n). But, first, that is exactly what Toellner was arguing, and second, even "initiator" is too strong. A better term would be "catalyst," precisely because in the chemical reaction the catalyst does not itself change.

153. Steinke, *Irritating Experiments*, 128.

154. Haller to Tissot, September 17, 1764, cited in Steinke, *Irritating Experiments*, 217.

155. Steinke, *Irritating Experiments*, 159.

156. Tissot, introduction to Haller, *Dissertation on the Sensible and Irritable Parts of Animals*.

157. Hubert Steinke, "Haller's Concept of Irritability and Sensibility and Its Reception in France," in *Méchanisme et vitalisme, la belle et la bête: Naissance et métamorphose d'un conte* (Oxford: Maison Française, 2001), 37–69.

158. Steinke, *Irritating Experiments*, 279.

159. Andrew Cunningham, "The Pen and the Sword: Recovering the Disciplinary Identity of Physiology and Anatomy before 1800, Part I, Old Physiology: The Pen," *Studies in History and Philosophy of Science* 33 (2002): 631–65; Andrew Cunningham, "The Pen and the Sword: Recovering the Disciplinary Identity of Physiology and Anatomy before 1800, Part II, Old Anatomy: The Sword," *Studies in History and Philosophy of Science* 34 (2003): 51–76.

160. Steinke, *Irritating Experiments*, 162.

161. Ibid., 139.

162. Ibid., 162.

163. Ibid., 279.

164. Ibid., 212.

165. Ibid., 144.

166. Ibid., 145.

167. Ibid.

168. Ibid., 77.

169. Ibid., 138.

170. Ibid., 163.

171. Cited in ibid., 137.

172. Ibid.

173. Ibid., 77. On Blumenbach et al., see ibid., 162.

174. Roselyn Rey, *Naissance et développement du vitalisme en France de la deuxième moitié du 18e siècle à la fin du Premier Empire* (Oxford: Voltaire Foundation, 2000).

175. Steinke, "'Die Ehre des Rechthabens,'" 168.

176. Steinke recognizes this shift (*Irritating Experiments*, 20) and points to Sergio Moravia's important essay "From *Homme Machine* to *Homme Sensible*: Changing Eighteenth-Century Models of Man's Image," *Journal of the History of Ideas* 39 (1978): 45–60. See also Sergio Moravia, "'Moral'—'Physique': Genesis and Evolution of a 'Rapport,'" in *Enlightenment Studies in Honor of Lester G. Crocker*, ed. A. Bingham and V. Topazio (Oxford: Voltaire Foundation, 1979), 163–74; Sergio Moravia, "Philosophie et médecine en France à la fin du XVIIIe siècle," *Studies on Voltaire and the Eighteenth Century* 89 (1972): 1089–1151.

177. Boschung et al., *Repertorium*, 1:144.

178. Cited in Rudolph, "Albrecht von Haller," 285n23.

179. See Albrecht von Haller, "Schreiben des Herrn von Haller, an den Herrn von Maupertuis wegen einer Schrift des Herrn Ofrai de la Mettrie, nebst einer Antwort des Herrn von Maupertuis" (1751), in *Sammlung kleiner Hallerischen Schriften*, 2nd ed., 3 vols. (Bern: Haller, 1772), 1:317–41.

180. Boschung et al., *Repertorium*, 1:328.

181. King, introduction to Haller, *First Lines of Physiology*, xx.

182. François de Capitani, "Hallers Bern," in Steinke, Boschung, and Proß, *Albrecht von Haller: Leben—Werk—Epoche*, 83–98; Urs Boschung, "Ein Berner Patriot: Hallers Lebensstationen," in Elsner and Rupke, *Albrecht von Haller im Göttingen der Aufklärung*, 21–46.

183. King, introduction to Haller, *First Lines of Physiology*, xii.

184. "After his return to Switzerland in 1753, Haller had no laboratory at his disposal and thus only rarely had the opportunity to experiment with larger animals" (Steinke, *Irritating Experiments*, 81).

185. Albrecht von Haller, *Sur la formation du cœur dans le poulet: sur l'œil, sur la structure du jaune, etc.*, 2 vols. (Lausanne: Bousquet, 1758). See Amor Cherni, *Épistémologie de la transparence: Sur l'embryologie de A. Von Haller* (Paris: Vrin, 1998); Maria Monti, *Cogetture ed esperienza nella fisilogia de Haller: La riforma dell'anatomia animata e il sistema della generazione* (Florence: Olschki, 1989); Maria Monti, "Difficultés et arguments de l'embryologie d'Albrecht von Haller: La reconversion des catégories de l'*anatome animata*," *Revue des sciences philosophiques et théologiques* 72 (1988): 301–12; Maria Monti, "Théologie physique et mécanisme dans la physiologie de Haller," in *Science and Religion / Wissenschaft und Religion*, Eighteenth International Congress of the History of Science, Hamburg and Munich, vol. 1 (Bochum: Brockmeyer, 1989), 68–79.

186. Lazzaro Spallanzani, *Nouvelles recherches sur les découvertes microscopiques, [microform] et la génération des corps organisés. Ouvrage traduit de l'Italien de M. l'Abbé Spalanzani, . . . Avec des notes, des recherches physiques & métaphysiques . . . & une nouvelle théorie de la terre. Par M. de Needham* (Paris: Lacombe, 1769). See Elena Agazzi, "Haller, Spallanzani e Bonnet: Il dibattito sul preformismuo tra poesia e scienza," in *Il pimsa di Goethe: Letteratura di viaggio e scienza nell'etá classicoromantica* (Naples: Guida, 1996), 19–38; Massino Aloisi, "Biology in the 18th Century as Seen through the Letters of L. Spallanzani to C. Bonnet," *Acta Medica Historiae Patavina* 21 (1974/75): 9–26; Carlo Castellani, "I rapporti tra Lazzaro Spallanzani e John T. Needham," *Physis* 15 (1973): 73–106.

187. Cunningham, "The Pen and the Sword."

188. Erich Hintzsche, "Albrecht von Haller als Anatom und seine Schule," *Ciba Zeitschrift* 10 (1948): 4068–73, citing 4072.

189. Martin Stuber, Stefan Hächler, and Luc Lienhard, eds., *Hallers Netz: Ein europäischer Gelehrtenbriefwechsel zur Zeit der Aufklärung*, Studia Halleriana 9 (Basel: Schwabe, 2005).

190. Boschung et al., *Repertorium*, 1:354.

191. Ibid., 479. See Erich Hintzsche, ed., *Briefwechsel 1745–1768 [von] Albrecht von Haller [und] Giambattista Morgagni* (Bern: Huber, 1964).

192. Boschung et al., *Repertorium*, 1:10.

193. Renato Mazzolini, "Die Entdeckung der Reizbarkeit: Haller als Anatom und Physiologe," in Elsner and Rupke, *Albrecht von Haller im Göttingen der Aufklärung*, 283–306.

194. Boschung et al., *Repertorium*, 1:84.

195. Ibid., 152.

196. Ibid., 484.

197. Marita Hübner, *Jean-André Deluc (1727–1817): Protestantische Kultur und moderne Naturforschung* (Göttingen: Vandenhoek und Ruprecht, 2010), 211.

198. Jacques Marx, *Charles Bonnet contre les Lumières, 1738–1850*, Studies on Voltaire and the Eighteenth Century 156–57 (Oxford: Voltaire Foundation at the Taylor Institution, 1976).

199. Renato Mazzolini and Shirley Roe, eds., *Science against the Unbelievers: The Correspondence of Bonnet and Needham, 1760–1780* (Oxford: Voltaire Foundation, 1986).

200. Margarete Hochdorfer, *The Conflict between the Religious and the Scientific View of Albrecht von Haller (1708–1777)* (Lincoln: University of Nebraska Press, 1932), reprinted in Shirley Roe, ed., *The Natural Philosophy of Albrecht von Haller* (New York: Arno, 1981), original pagination.

201. Toellner, *Albrecht von Haller*.

202. Karl Guthke, review of *Albrecht von Haller*, by Richard Toellner, *Monatshefte* 64 (1972): 386–88.

203. See the newer work on Haller and religion, especially the most comprehensive and thoughtful discussion of these matters: Thomas Kaufmann, "Über Hallers Religion: Ein Versuch," in Elsner and Rupke, *Albrecht von Haller im Göttingen der Aufklärung*, 307–80; Cornelia Rémi, "Religion und Theologie," in Steinke, Boschung, and Proß, *Albrecht von Haller: Leben—Werk—Epoche*, 199–225.

204. Toellner, *Albrecht von Haller*, 1–32.

205. This is the implication of the title of Hochdorfer's work, though her text is more ambiguous. It is unequivocally the upshot of Guthke's interpretation of Haller's literary output.

206. While this was already tacit in his monograph, it is most clearly argued in Toellner's later essay, "Haller als Christ: Zur Deutung der 'Fragmente religiöser Empfindungen,'" in Steinke, Boschung, and Proß, *Albrecht von Haller: Leben—Werk—Epoche*, 485–94.

207. Haller was reported to have said "that he not only never believed but that he found it impossible to believe anything, no matter how much he would have liked to," in the words of one of those in attendance, conveyed two days after Haller's death (December 12, 1777), in a letter to Christian Gottlob Heyne (Toellner, "Haller als Christ," 485).

208. His former student and then biographer, Zimmermann, who found Haller's religiosity increasingly distasteful, doubted that the report could be valid, because he took Haller for an increasingly doctrinaire "hyperorthodox" Calvinist.

209. "Where a Hobbes doubted, a Newton believed; where an Offray [de La Mettrie] scoffed, a Boerhaave worshiped." Albrecht von Haller, *Briefe über die wichtigsten Wahrheiten der Offenbarung* (Bern: Neue Buchhandlung, 1772), 6, cited in Otto Sonntag, "The Motivations of the Scientist: The Self-Image of Albrecht von Haller," *Isis* 65 (1974): 336–51, citing 339.

210. Albrecht von Haller, "Fragmente religiöser Empfindungen," in *Tagebücher seiner Beobachtungen über Schriftsteller und über sich selbst*, 2:219ff. See Toellner, "Haller als Christ," 486.

211. Rémi ("Religion und Theologie," 199) observes that Toellner's view, starting with his monograph of 1971, has gradually won over scholarship to this judgment.

212. Kaufmann, "Über Hallers Religion," 332–33.

213. Haller, "Fragmente religiöser Empfindungen," 226, 286, 316, cited in Toellner, "Haller als Christ," 488, 491.

214. Noted in Toellner, "Haller als Christ," 485.

215. Zimmermann's views on Haller's religious excesses are discussed in ibid., 485; 492–3n. See also Kaufmann, "Über Hallers Religion," 311.

216. Goethe on Haller, cited in Kaufmann, "Über Hallers Religion," 376.

217. Ibid., 319.

218. Rémi, "Religion und Theologie," 205.

219. Kaufmann, "Über Hallers Religion," 335, 337n.

220. Albrecht von Haller, "Vorrede zur Prüfung der Seckte die an allem zweifelt" (1751), in *Sammlung kleiner Hallerischen Schriften*, 1:1ff. See also Rémi, "Religion und Theologie," 206.

221. Otto Sonntag, ed., *The Correspondence between Albrecht von Haller and Charles Bonnet* (Bern: Hans Huber, 1983). The correspondence began in 1754.

222. Bartolomeo De Felice, *De Newtoniana Attractione Unica Cohaerentia Naturalis Causa Dissertatio* (1757). De Angelis uses De Felice's book effectively to uncover Haller's own positions. See Simone De Angelis, *Von Newton zu Haller: Studien zum Naturbegriff zwischen Empirismus und deduktiver Methode in der Schweizer Frühaufklärung* (Tübingen: Niemeyer, 2003), 406, 436, etc.

223. Hübner, *Jean-André Deluc (1727–1817)*, 84. That project may have been particularly triggered by the materialism rampant in the Baron d'Holbach's *Système de la nature* (1770), which had recently occasioned Europe-wide horror. Thus, Goethe, full of Sturm und Drang sensibility for the liveliness of nature, found d'Holbach's rendering lifeless and uninspiring. Indeed, the horror was felt not just among the orthodox enemies of Enlightenment; it evoked consternation among Enlightenment moderates and even posed difficulties for the radical Enlightenment, in the measure that *reductive* physical mechanism was not in fact the agenda that vital materialism wished to promote, especially not in life science.

224. Eugène Maccabez, *F. B. de Felice, 1723–1789, et son encyclopédie: Yverdon, 1770–1780* (Basel: Imprimerie Emile Birkhaeuser, 1903); Clorinda Donato, "L'*Encyclopédie* d'Yverdon et l'*Encyclopédie* de Diderot et de d'Alembert: Éléments pour une comparaison," *Annales Benjamin Constant* 14 (1993): 75–83; Kathleen Hardesty Doig, "The Yverdon *Encyclopédie*," in *Notable Encyclopedias of the Late Eighteenth Century: Eleven Successors of the "Encyclopédie,"* ed. Frank Kafker (Oxford: Voltaire Foundation, 1994), 85–116; Christian de Félice, *L'Encyclopédie d'Yverdon: Une encyclopédie suisse au siècle des lumières* (Yverdon: Fondation de Felice, 1999); Henri Cornaz, "Fortunato Bartolomeo De Felice and the *Encyclopédie d'Yverdon*," in *The Encyclopédie and the Age of Revolution*, ed. Clorinda Donato and Robert Maniquis (Los Angeles: University of California; Yverdon: Société du Musée et Vieil Yverdon, Switzerland, 1992), 39–48.

225. Erich Hintzsche, "Albrecht von Hallers Tätigket als Enzyklopädist," *Clio Medica* 1 (1966): 235–54.

226. Rémi, "Religion und Theologie," 215.

227. Albrecht von Haller, *Briefe über einige noch lebenden Freigeister: Einwürfe wider der Offenbarung*, rev. ed., 3 vols (Bern, 1778).

228. Roe, "*Anatomia Animata*," 284.

CHAPTER FOUR

1. Denis Diderot, *De l'interprétation de la nature*, in *Oeuvres philosophiques*, ed. Paul Vernière and Jean-Christophe Abramovici (Paris: Classiques Garnier, 1998), 177–244, citing 188.

2. On French materialism in the eighteenth century, see the classic anthology edited by Roland Desné, *Les matérialistes français de 1750 à 1800* (Paris: Buchet-Chastel, 1965); and Béatrice Fink and Gerhardt Stenger, eds., *Être matérialiste à l'âge des Lumières: Hommage offert à Roland Desné* (Paris: Presses Universitaires de France, 1999). For an older approach, see Aram Vartanian, *Diderot and Descartes: A Study of Scientific Materialism in the Enlightenment* (Princeton, NJ: Princeton University Press, 1953). For more recent treatments, see Ann Thomson, *Bodies of Thought: Science, Religion, and the Soul in the Early Enlightenment* (Oxford: Oxford University Press, 2008); Charles T. Wolfe, "Forms of Materialist Embodiment," in *Anatomy and the Organization of Knowledge, 1500–*

1850, ed. Matthew Landers and Brian Muñoz (London: Pickering and Chatto, 2012), 129–44; Charles T. Wolfe, *Materialism: A Historico-philosophical Introduction* (Heidelberg: Springer, 2016).

3. Lenoir was the first to use the concept of "vital materialism" to denote German thought about life science. Timothy Lenoir, "Kant, Blumenbach, and Vital Materialism in German Biology," *Isis* 71 (1980): 77–108. Reill has made it the crucial concept for a major shift in scientific thinking across Europe in the eighteenth century. See Peter Hanns Reill, *Vitalizing Nature in the Enlightenment* (Berkeley: University of California Press, 2005).

4. Among Peter Hanns Reill's essays, see esp. "*Bildung, Urtyp* and Polarity: Goethe and Eighteenth-Century Physiology," *Goethe Yearbook* 3 (1986): 139–48; "Anti-mechanism, Vitalism and Their Political Implications in Late Enlightened Scientific Thought," *Francia* 16, no. 2 (1989): 195–212; "Buffon and Historical Thought in Germany and Great Britain," in *Buffon 88: Actes du Colloque international pour le bicentenaire de la mort de Buffon*, ed. Jean Gayon et al. (Paris: Vrin, 1992), 667–79; "Die Historisierung von Natur und Mensch: Der Zusammenhang von Naturwissenschaften und historischen Denken im Entstehungsprozeß der modernen Naturwissenschaften," in *Geschichtsdiskurs*, vol. 2, *Anfänge modernen historischen Denkens* (Hamburg: Fischer, 1993), 48–61; "Analogy, Comparison, and Active Living Forces: Late Enlightenment Responses to the Skeptical Critique of Causal Analysis," in *The Skeptical Tradition around 1800*, ed. Johann van der Zande and Richard Popkin (Dordrecht: Kluwer, 1998), 203–11.

5. Peter Hanns Reill, "Between Mechanism and Hermeticism: Nature and Science in the Late Enlightenment," in *Frühe Neuzeit—Frühe Moderne?*, ed. Rudolf Vierhaus (Göttingen: Vandenhoeck und Ruprecht, 1992), 393–421, citing 401, 400, 402, 405–6, 416.

6. Pierre-Louis Moreau de Maupertuis, *Vénus physique*, in *Vénus physique suivi de la Lettre sur le progrès des sciences*, ed. Patrick Tort (Paris: Aubier Montaigne, 1980). Trans. Simone Boas as *The Earthy Venus*, Sources of Science, no. 29 (New York: Johnson Reprint, 1966).

7. Georges-Louis Leclerc de Buffon, "Preliminary Discourse: On the Manner of Studying and Expounding Natural History," trans. John Lyon, in *From Natural History to the History of Nature: Readings from Buffon and His Critics*, ed. John Lyon and Phillip Sloan (Notre Dame, IN: University of Notre Dame Press, 1981), 97–128.

8. Aram Vartanian, *La Mettrie's "L'homme machine": A Study in the Origins of an Idea* (Princeton, NJ: Princeton University Press, 1960); Ann Thomson, *Materialism and Society in the Mid-Eighteenth Century: La Mettrie's "Discours préliminaire"* (Geneva: Droz, 1981); and Kathleen Wellman, *La Mettrie: Medicine, Philosophy, and Enlightenment* (Durham, NC: Duke University Press, 1992).

9. Georges LeClerc de Buffon, *Histoire naturelle, générale et particulière* (Paris: Imprimerie royale, 1812).

10. Elizabeth Williams, *A Cultural History of Medical Vitalism in Enlightenment Montpellier* (Aldershot, UK: Ashgate, 2003).

11. Thomas Hankins, *Science and the Enlightenment* (Cambridge: Cambridge University Press, 1985), 125.

12. Roselyne Rey, *Naissance et développement du vitalisme en France de la deuxième moitié du 18e siècle à la fin du Premier Empire* (Oxford: Voltaire Foundation, 2000); Roselyne Rey, "Naissance de biologie et redistribution des savoirs," *Revue de synthèse* 4, nos. 1–2 (1994): 167–97.

13. See Lester Crocker, "Diderot and Eighteenth-Century French Transformationism," in *Forerunners of Darwin: 1745–1859*, ed. Bentley Glass, Owsei Temkin, and William Straus (Baltimore: Johns Hopkins University Press, 1959), 114–43.

14. Robert Young, "Animal Soul," in *Encyclopedia of Philosophy*, 1:122–27.

15. Étienne Bonnot de Condillac, *Traité des animaux*, ed. Michel Malherbe (Paris: Vrin, 2004).

16. Robert Whytt, *An Essay on the Vital and Other Involuntary Motions of Animals* (Edinburgh: Hamiton, Balfour and Neill, 1751). See Roger K. French, *Robert Whytt, the Soul, and Medicine* (London: Wellcome Institute of the History of Medicine, 1969).

17. Paul Wood, "The Natural History of Man in the Scottish Enlightenment," *History of Science* 27 (1989): 89–123, citing 93.

18. On the Hunter brothers, see W. F. Bynum and Roy Porter, eds., *William Hunter and the*

Eighteenth-Century Medical World (Cambridge: Cambridge University Press, 1995); Stephen Cross, "John Hunter, the Animal Oeconomy, and Late-Eighteenth-Century Physiological Discourse," *Studies in History of Biology* 5 (1981): 1–110.

19. Michael Hoffheimer, "Maupertuis and the Eighteenth-Century Critique of Preexistence," *Journal of the History of Biology* 15 (1982): 119–44.

20. The eroticism of the title was carried through the text in its address to a female reader and its almost pornographic descriptions of sexual relations. Mary Terrall has provided a thorough and balanced consideration of these aspects, first in her essay "Salon, Academy, and Boudoir: Generation and Desire in Maupertuis's Science of Life," *Isis* 87 (1996): 217–29, and then in her book *The Man Who Flattened the Earth: Maupertuis and the Sciences in the Enlightenment* (Chicago: University of Chicago Press, 2002). In that moment, Maupertuis himself was engaged in a riot of sexual escapades in Paris high society. See Elisabeth Badinter, *Les passions intellectuelles*, vol. 1, *Désirs de gloire, 1735–1751* (Paris: Fayard, 1999).

21. J. B. Shank, *The Newton Wars and the Beginning of the French Enlightenment* (Chicago: University of Chicago Press, 2008).

22. Maupertuis, *Vénus physique*, 172n, 176; see Hoffheimer, "Maupertuis and the Eighteenth-Century Critique of Preexistence," 128; Terrall, *The Man Who Flattened the Earth*, 209n.

23. Hoffheimer, "Maupertuis and the Eighteenth-Century Critique of Preexistence," 124.

24. Maupertuis, *Earthy Venus*, 51.

25. C. U. M. Smith, *The Problem of Life: An Essay in the Origins of Biological Thought* (New York: Wiley, 1976), 264.

26. Maupertuis, *Vénus physique*, 120; *Earthy Venus*, 55.

27. Maupertuis, *Vénus physique*, 120; *Earthy Venus*, 55–56.

28. Maupertuis, *Vénus physique*, 121; *Earthy Venus*, 55.

29. Maupertuis, *Vénus physique*, 121; *Earthy Venus*, 56.

30. Maupertuis, *Vénus physique*, 140; *Earthy Venus*, 79.

31. Maupertuis, *Vénus physique*, 141; *Earthy Venus*, 80. "The final outcome of any mating is never completely determinate, since factors such as climate, diet, and even chance combinations of elements, contribute to individual variations" (Terrall, *The Man Who Flattened the Earth*, 217).

32. Maupertuis, *Vénus physique*, 140; *Earthy Venus*, 79; and again, Maupertuis, *Vénus physique*, 141; *Earthy Venus*, 80.

33. Maupertuis, *Vénus physique*, 123; *Earthy Venus*, 58. As Hoffheimer writes, the "theoretical limitations of preexistence were most trenchantly fixed by familiar macroscopic phenomena: hereditary resemblances, hybrids, and monsters" ("Maupertuis and the Eighteenth-Century Critique of Preexistence," 123). For the central role of heredity in eighteenth-century thought, see Staffan Müller-Wille and Hans-Jörg Rheinberger, *A Cultural History of Heredity* (Chicago: University of Chicago Press, 2012); Staffan Müller-Wille and Hans-Jörg Rheinberger, eds., *Heredity Produced: At the Crossroads of Biology, Politics, and Culture, 1500–1870* (Cambrdige, MA: MIT Press, 2007).

34. Maupertuis, *Vénus physique*, 134; *Earthy Venus*, 71.

35. Maupertuis, *Vénus physique*, 134; *Earthy Venus*, 72.

36. Buffon, *Histoire naturelle*, vol. 2 (1749), cited in Hoffheimer, "Maupertuis and the Eighteenth-Century Critique of Preexistence," 129.

37. Maupertuis, "Letter on the Generation of Animals," trans. in Hoffheimer, "Maupertuis and the Eighteenth-Century Critique of Preexistence," 140.

38. Diderot, *De l'interprétation de la nature*, 180.

39. Wilda Anderson, *Diderot's Dream* (Baltimore: Johns Hopkins University Press, 1990), 13.

40. "Buffon's notion of molecule . . . means the minimal particle that carries identifying characteristics and can act and react. Therefore it need not necessarily be thought of as a solid fixed particle" (ibid., 30n).

41. Aram Vartanian, "Trembley's Polyp, La Mettrie and Eighteenth-Century French Materialism," *Journal of the History of Ideas* 11 (1950): 259–86.

42. Jacques Roger, *The Life Sciences in Eighteenth-Century French Thought*, ed. Keith R. Benson, trans. Robert Ellrich (Stanford: Stanford University Press, 1997), 666–67. The Benson edition is an abridged translation of Jacques Roger, *Les sciences de la vie dans la pensée française du XVIIIe siècle* (Paris: Armin Colin, 1963). On spontaneous generation, see John Farley, *The Spontaneous Generation Controversy from Descartes to Oparin* (Baltimore: Johns Hopkins University Press, 1977).

43. Herbert Dieckmann, "The Influence of Francis Bacon on Diderot's *Interprétation de la Nature*," in *Studien zur europäischen Aufklärung* (Munich: Fink, 1974), 34–57, citing 37.

44. Roland Mortier, *Diderot en Allemagne* (Paris: Presses Universitaires de France, 1954), 334, 339.

45. La Mettrie *himself* made the same error, writing of the "physician Diderot" in *L'homme machine*.

46. R. James, *Dictionnaire universel de médecine, chirugie, anatomie*, ed. Denis Diderot, Marc-Antoine Eidous, and François-Vincent Toussaint, 6 vols. (Paris, 1746–48).

47. Cited in Arthur Wilson, *Diderot* (New York: Oxford University Press, 1957), 93. For this medical fascination, see Kathleen Wellman, "Medicine as a Key to Defining Enlightenment Issues: The Case of Julien Offray de La Mettrie," *Studies in Eighteenth-Century Culture* 17 (1987): 75–89, esp. 89n43.

48. See my "*Médecin Philosophe*: Persona for Radical Enlightenment," *Intellectual History Review* 18, no. 3 (November 2008): 427–40.

49. On the notion of "radical" Enlightenment, see Margaret Jacob, *Radical Enlightenment: Pantheists, Freemasons and Republicans* (London: Allen and Unwin, 1981); Jonathan Israel, *Radical Enlightenment: Philosophy and the Making of Modernity, 1650–1750* (Oxford: Oxford University Press, 2001).

50. Aram Vartanian, "From Deist to Atheist: Diderot's Philosophical Orientation, 1746–1749," *Diderot Studies* 1 (1949): 46–63.

51. He had already met and had numerous conversations with Buffon. Jacques Roger, "Buffon et Diderot en 1749," *Diderot Studies* 4 (1963): 221–36.

52. Denis Diderot, "Animal," in *Encyclopédie*, reprinted in *Oeuvres philosophiques*, 5:381–401.

53. Denis Diderot, *Lettre sur les aveugles à l'usage de ceux qui voient* (1749; repr., Paris: Droz, 1951).

54. Roger, *Les sciences de la vie*, 601, 617. See also Émile Callot, *La philosophie de la vie au XVIIIe siècle* (Paris: Rivière, 1965), 246.

55. See Anderson, *Diderot's Dream*; Suzanne Pucci, *Diderot and a Poetics of Science* (New York: Peter Lang, 1986); Angelica Goodden, *Diderot and the Body* (Oxford: Legenda, for European Humanities Research Center, University of Oxford, 2001); Paolo Quintili, *La pensée critique de Diderot: Matérialisme, science et poésie à l'âge de l'Encyclopédie, 1742–1782* (Paris: Champion, 2001); Paolo Quintili, *La matière et l'homme dans l'Encyclopédie: Actes du colloque de Joinville (10–12 juillet 1995)*, ed. Sylvaine Coppola and Anne-Marie Chouillet (Paris: Klincksieck, 1998).

56. Jacques Proust, *Diderot et l'Encyclopédie* (Paris: Martel, 1995).

57. Thomas Hankins, *Jean d'Alembert: Science and the Enlightenment* (Oxford: Clarendon Press, 1970).

58. Jean Le Rond d'Alembert, *Traité de dynamique: dans lequel les loix de l'equilibre & du mouvement des corps sont réduits au plus petit nombre possible, & démontrées d'une maniére nouvelle: & où l'on donne un principe général pour trouver le mouvement de plusieurs corps qui agissent les uns sur les autres, d'une maniére quelconque* (1758; New York: Readex Microprint, 1968).

59. Dieckmann, "Influence of Francis Bacon on Diderot's *Interprétation de la nature*," citing 55.

60. Diderot, *De l'interprétation de la nature*, 178.

61. Thus, "with both Bacon and Diderot we find the same thorough comprehension of the unlimited wealth and variety of nature which the mathematical scheme excluded or ignored" (Dieckmann, "Influence of Francis Bacon on Diderot's *Interprétation de la nature*," 39).

62. Diderot, *De l'interprétation de la nature*, 178.

63. See the discussion of eighteenth-century objectivity in Lorraine Daston and Peter Galison, *Objectivity* (New York: Zone, 2007), 64–75, for a sympathetic reconstruction.

64. Dieckmann, "Influence of Francis Bacon on Diderot's *Interprétation de la nature*," 49.

65. See esp. Pucci, *Diderot and a Poetics of Science*.

66. Sergio Moravia, "The Enlightenment and the Sciences of Man," *History of Science* 18 (1980): 247–68, citing 247.

67. Dieckmann, "Influence of Francis Bacon on Diderot's *Interprétation de la nature*," 53.

68. Daston and Galison, *Objectivity*, 67–74.

69. Diderot, *De l'interprétation de la nature*, 188.

70. Diderot cites both Buffon and Maupertuis but omits any mention of La Mettrie, except in the strange "postscript" to the advertisement he inserted into the second edition: "Always keep in mind that *nature* is not *God*; that a *man* is not a *machine*; that a *hypothesis* is not a *fact*" (Diderot, *De l'interprétation de la nature*, 173). As Wellman observes, Diderot expressed disdain for La Mettrie in other publications: "He said of La Mettrie: 'Dissolute, impudent, a buffoon, a flatterer; made for life at court and the favor of nobles. He dies as he should have, a victim of his own intemperance and his folly. He killed himself by ignorance of the art he professed'" ("Medicine as a Key to Defining Enlightenment Issues, 86n6).

71. Diderot, *De l'interprétation de la nature*, 190.

72. Ibid., 186–87.

73. Ibid., 187.

74. Ibid.

75. Ibid., 188.

76. Hankins, *Science and the Enlightenment*, 127.

77. Diderot to Duclos, October 10, 1765, cited in ibid., 130. Later, in a sustained monograph of his own, Diderot attempted a synthesis of the whole view: Denis Diderot, *Éléments de physiologie* (1778), ed. Paolo Quintili (Paris: Champion, 2004). See G. Rudolph, "Diderots *Elemente der Physiologie*," *Gesnerus* 24 (1967): 24–45.

78. "We shall see, after experimental physics has become more advanced, that all the phenomena, whether of gravity or of elasticity or of attraction or of magnetism or of electricity, are nothing more than different aspects of the same impulse [*affection*]" (Diderot, *De l'interprétation de la nature*, 220).

79. Ibid., 229.

80. "Spinozism is for them [the neo-Spinozists and their opponents] a hylozoism in which God gets assimilated into matter and becomes a useless appendage." Paul Vernière, *Spinoza et la pensée française avant la Révolution* (Paris: Presses Universitaires de France, 1954), 554.

81. Pierre-Louis Moreau de Maupertuis, *Réponse aux objections de M. Diderot*, in *Essai de cosmologie, Système de la nature, Réponse aux objections de M. Diderot*, ed. François Azouvi (Paris: Vrin, 1984).

82. Editor's note in Diderot, *De l'interprétation de la nature*, 229.

83. Mortier, *Diderot en Allemagne*, 334, 339.

84. Israel ends his *Radical Enlightenment* with a treatment of La Mettrie and Diderot, claiming that their "materialism" represented a decisive moment in the reception of Spinoza in the eighteenth century (704–71). Yet he minimizes the degree to which this reception, as is often the case, involved a substantial degree of creative misunderstanding.

85. Vernière, *Spinoza et la pensée française*, 544–45.

86. See my "'The Most Hidden Conditions of Men of the First Rank'—The Pantheist Current in Eighteenth-Century Germany 'Uncovered' by the Spinoza Controversy," *Eighteenth-Century Thought* 1 (2003): 335–68.

87. Vernière, *Spinoza et la pensée française*, 529.

88. Ibid., 553.

89. Denis Diderot, "Spinosiste," in *Encyclopédie*, reprinted in Diderot, *Oeuvres philosophiques*, ed. J. Lough and J. Proust (Paris: Gallimard, 2010), 8:328–29.

90. See Condillac's dissertation on the monad, submitted to the Berlin Academy for the contest of 1747 and published together with the winning entry for that prize contest shortly thereafter: Étienne Bonnot de Condillac, *Les monades*, ed. L. Bongie (Oxford: Voltaire Foundation, 1980). See also Denis Diderot, "Leibnizianisme," in *Encyclopédie*, vol. 5 (1765); W. H. Barber, *Leibniz in France, from Arnauld to Voltaire: A Study in French Reactions to Leibnizianism, 1670–1760* (Oxford: Clarendon Press, 1955); Claire Fauvergue, *Diderot, lecteur et interprète de Leibniz* (Paris: Champion, 2006).

91. Vernière, *Spinoza et la pensée française*, 530, 553.

92. Denis Diderot, *D'Alembert's Dream*, in *Rameau's Nephew / D'Alembert's Dream*, trans. Leonard Tancock (Harmondsworth: Penguin, 1966).

93. The phrase cited in the text is from Bordeu's very important *Recherches de la médecine* (1768), his clearest statement of vitalist materialism as the "speculation of the most modern philosophical physicians" (Roger, *Les sciences de la vie*, 629).

94. Roger, *Les sciences de la vie*, 641.

95. See Herbert Dieckmann, "Théophile Bordeu und Diderots *Rêve de d'Alembert*," *Romanische Forschungen* 52 (1938): 55–122.

96. Roger, *Les sciences de la vie*, 657.

97. Diderot, *D'Alembert's Dream*, 154.

98. Ibid., 176–77.

99. See Crocker, "Diderot and Eighteenth-Century French Transformationism."

100. Roger, *Les sciences de la vie*, 663.

101. Ibid., 681. See Baron d'Holbach, *Système de la nature* ([Amsterdam], 1770). On Diderot's place in d'Holbach's circle, see Alan Kors, *D'Holbach's Coterie: An Enlightenment in Paris* (Princeton, NJ: Princeton University Press, 1976).

102. Phillip Sloan, "Natural History, 1670–1802," in *Companion to the History of Modern Science*, ed. R. C. Olby, G. N. Cantor, J. R. R. Christie, and M. J. S. Hodge (London: Routledge, 1990), 295–313, esp. 304–6; Phillip Sloan, "Natural History," in *Cambridge History of Eighteenth Century Philosophy*, ed. Knud Haakonssen (Cambridge: Cambridge University Press, 2006), 903–37, esp. 913–25 on Buffon.

103. Lesley Hanks, *Buffon avant l'Histoire naturelle* (Paris: Presses Universitaires de France, 1966).

104. Phillip Sloan, "From Logical Universals to Historical Individuals: Buffon's Idea of Biological Species," in *Histoire du concept d'espèce dans la science de la vie*, ed. J.-L. Fischer and J. Roger (Paris: Singer-Polignac, 1985), 101–40.

105. Henri Guerlac, "The Continental Reputation of Stephen Hales," *Archives internationales de l'histoire des sciences* 4 (1951): 393–404.

106. Georges-Louis Leclerc de Buffon, preface to translation of Stephen Hales, *Vegetable Statics*, in Lyon and Sloan, *From Natural History to the History of Nature*, 37.

107. Shank, *Newton Wars*, 408.

108. Thierry Hoquet, "History without Time: Buffon's Natural History as a Nonmathematical Physique," *Isis* 101, no. 1 (2010): 30–61.

109. Ibid., 30.

110. Ibid., 34.

111. Ibid., 37.

112. On eighteenth-century Newtonianism, see discussion in chapter 3.

113. Bernard Cohen, *Franklin and Newton* (Philadelphia: American Philosophical Association and Harvard University Press, 1966).

114. Shirley Roe, "*Anatomia Animata*: The Newtonian Physiology of Albrecht von Haller," in *Transformations and Tradition in the Sciences*, ed. Everett Mendelsohn (Cambridge: Cambridge University Press, 1984), 273–300.

115. Thierry Hoquet, *Buffon: Histoire naturelle et philosophie* (Paris: Champion, 2005), 143.

116. Ibid., 60. On Buffon's connection with these views, see Gordon Wattles, "Buffon,

d'Alembert and Materialist Atheism," *Studies on Voltaire and the Eighteenth Century* 266 (1989): 285–341.

117. Hoquet, "History without Time," 42.

118. Ibid., 30, 36.

119. Ibid., 36. As Hoquet rightly argues, "*Histoire naturelle* claims to be the real *physique*" (ibid., 30).

120. Diderot, *De l'interprétation de la nature*.

121. Buffon, "Preliminary Discourse," 121.

122. Ibid., 126.

123. Ibid., 123.

124. Ibid. Compare Hume's distinction between relations of ideas and matters of fact.

125. Kant would reassert the mathematical approach emphatically near the end of the century: "There is only so much science in a discipline as there is mathematics in it." Kant, *Metaphysical Foundations of Natural Science* (1786), AA 4:470.

126. Hoquet, "History without Time," 37. Thus D'Argenson observed: "Really, Buffon was not called upon to do more than give a description of the royal cabinet, and he goes off from that to deduce a system of general and speculative physics, a system both new and impossible, because he himself denounces general systems" (cited in Hoquet, *Buffon*, 61).

127. Hoquet, "History without Time," 33.

128. Ibid., 37.

129. "I have just read the three volumes of Monsieur Buffon's Natural History that appeared some time ago. They contain some curious facts, but not enough in proportion to their size. There is a great number of conjectures, several of which are very rash. It often happens to Monsieur de Buffon to give them at first as conjectures, and afterwards to make use of them as demonstrated principles. . . . Monsieur de Buffon aspires to explain almost everything about reproduction, but I confess that I can only consider his system as a hazardous hypothesis. He makes the facts on which he builds it, prove too much. It seems to me that he allows himself to be carried away by his imagination. If his work is highly esteemed, I fear that he may injure natural history by restoring the taste for hypotheses." Abraham Trembley to his patron, Count Bentinck, cited in John Baker, *Abraham Trembley of Geneva: Scientist and Philosopher, 1710–1784* (London: Arnold, 1952), 182; Hoquet (*Buffon*, 61) dates this letter to January 20, 1750. "If nature did not make him an observer, instead she endowed him with the most brilliant of gifts and made him the most eloquent man of his century. If he is not a *Malpighi*, a *Réaumur*, he is a *Plato*, a *Milton*, and his writings, full of fire and life, will say to posterity that the painter of nature was not always its definer." Charles Bonnet, *Contemplation de la nature* (orig. 1764), new, corrected and expanded ed., with an "Avertissement" by Bonnet, dated September 1781 (vol. 1, v–xi), explaining the development of the text, plus notation of changes and revisions throughout, 2 vols. (Hamburg: Virchaux, 1782), 2:204. Not only was this the choral strain of the Réaumur coterie, from the master himself to his key epigones like Bonnet and Trembley and, of course, de Lignac; it was also the view of the taxonomists in the Linnean vein. C.-G. de Lamoignon de Malesherbes (1721–94), for instance, wrote as much in his *Observations sur l'Histoire naturelle* in 1751: "We do not deny that the author is a man of great intelligence. . . . We simply deny that he is a naturalist, and consequently that he is competent to speak pertinently about natural history." Cited in Jacques Roger, *Life Science in Eighteenth-Century French Thought* (Stanford: Stanford University Press, 1997), 481–82.

130. Hoquet illustrates this with two salient referents: d'Alembert and Condorcet; see "History without Time," 32. Elsewhere, he elaborates: "Starting in 1750, Raynal construed Buffon as an author rejected by the specialists but over whom the ladies went wild and of whom even the valets took note in terms of worldly success" (*Buffon*, 15).

131. Hoquet, *Buffon*, 18–19. As Daniel Mornet has established, Buffon's *Histoire naturelle* ranked third in frequency in library inventories of the epoch (Mornet, cited in ibid., 11n). That suggests that not just ladies or valets but *everyone*—and especially natural scientists—read Buffon, even if many did not find what they read entirely congenial.

132. Lyon and Sloan, introduction to *From Natural History to the History of Nature*, 1–27, citing 3.

133. Buffon, "Preliminary Discourse," 122, 127.

134. Ibid., 123.

135. Ibid., 101.

136. Ibid., 125.

137. Ibid.

138. Ibid., 102.

139. Ibid.

140. Ibid., 108.

141. See Phillip Sloan, "The Buffon-Linnaeus Controversy," *Isis* 67 (1976): 356–75; James Larson, "Linnaeus and the Natural Method," *Isis* 58 (1967): 305–20; G. Barsanti, "Linné et Buffon: Deux visions différentes de la nature et de l'histoire naturelle," *Revue de synthèse*, ser. 3, 113–14 (1984): 83–111.

142. Buffon, "Preliminary Discourse," 107.

143. Ibid., 108.

144. Ibid., 111.

145. The contrast of *esprit systematique* with *esprit de système* was articulated virtually simultaneously by the key epistemologist of the French Enlightenment, the abbé de Condillac, and then popularized in d'Alembert's *Preliminary Discourse to the Encyclopédie*. Buffon enacted the distinction in his practices as natural historian.

146. Phillip Sloan, "The Idea of Racial Degeneracy in Buffon's *Histoire naturelle*," in *Racism in the Eighteenth Century*, ed. H. Pagliaro, Studies in Eighteenth-Century Culture, vol. 3 (Cleveland: Case Western Reserve University Press, 1973), 293–321.

147. Phillip Sloan, "The Gaze of Natural History," in *Inventing Human Science: Eighteenth-Century Domains*, ed. Christopher Fox, Roy Porter, and Robert Wokler (Berkeley: University of California Press, 1995), 112–51, citing 122.

148. Ibid., 133.

149. Ibid., 120.

150. Buffon, "Preliminary Discourse," 102.

151. Sloan, "From Logical Universals to Historical Individuals," 123.

152. "By reinterpreting the issue of generation in epigenetic terms, Buffon provided a means by which the contingencies of geography and climate, acting upon the *molécules*, could affect the actual reproductive lineage of the species" (Sloan, "Gaze of Natural History," 133).

153. A. O. Lovejoy, *The Great Chain of Being* (New York: Harper, 1960); Giorgio Tonelli, "The Law of Continuity in the Eighteenth Century," *Studies on Voltaire and the Eighteenth Century* 27 (1963): 1619–38.

154. Lyon and Sloan, introduction to *From Natural History to the History of Nature*, 2.

155. Jean Torlais, "Un rivalité célèbre: Réaumur et Buffon," *Presse medicale* 66 (1958): 1057–58.

156. The alliance between the physicotheology of the abbé Pluche and the school of Réaumur is most transparent in the writings of Charles Bonnet. Jacques Marx, *Charles Bonnet contre les Lumières, 1738–1850*, Studies on Voltaire and the Eighteenth Century 156–57 (Oxford: Voltaire Foundation at the Taylor Institution, 1976); Virginia Dawson, "La théologie des insects dans las pensée de Charles Bonnet," in *Charles Bonnet: Savant et philosophe (1720–1793)*, ed. Marino Buscaglia, René Sigrist, Jacques Trembley, and Jean Wüest (Geneva: Editions Passé Présent, 1993), 79–90. That Buffon and his party set about deliberately to debunk physicotheology is most palpable in the writing of Buffon's ally Maupertuis, such as the latter's *Essai de cosmologie*. See my "Kant's Early Views on Epigenesis: The Role of Maupertuis," in *The Problem of Animal Generation in Early Modern Philosophy*, ed. Justin E. Smith (Cambridge: Cambridge University Press, 2006), 317–54.

157. On "insect theology," see Friedrich Christian Lesser, *Théologie des insectes, ou, Demonstration des perfections de Dieu dans tout ce qui concerne les insectes. Avec des remarques de Pierre Lyonnet* (1742).

158. Joseph Lelarge de Lignac, *Lettres à unAmériquain sur l'Histoire naturelle, générale et par-*

ticuliere de Monsieur de Buffon (À Hambourg, 1751). "De Lignac's *Lettres à un Amériquain* . . . were at the very least instigated if not actually in part composed by Réaumur." Jean Torlais, "Réaumur philosophe," *Revue d'histoire des sciences* 11 (1958): 13–33, citing 27.

159. As Hoquet explains, Epicureanism signified an immanent, naturalist account of the world, as against a natural-theological one (*Buffon*, 47, 125, 147ff.). Buffon clearly embraced that sense of the project of natural science, and the age therefore assigned him to the Epicurean camp.

160. Robert Shackelton, "When Did the French 'Philosophes' Become a Party?," *Bulletin of the John Rylands University Library of Manchester* 60 (1977): 181–99; Franck Bourdier and Yves François, "Buffon et les Encyclopédistes," *Revue d'histoire des sciences* 4 (1951): 228–32; David Goodman, "Buffon's *Histoire naturelle* as a Work of the Enlightenment," in *The Light of Nature: Essays in the History and Philosophy of Science Presented to A. C. Crombie*, ed. J. D. North and J. J. Roche (Dordrecht: M. Nijhoff, 1985), 57–65; Otakar Matousek, "Buffon and the Philosophy of His Natural History," *Archives internationales d'histoire des sciences* 3 (1950): 312–19; Robert Wohl, "Buffon and His Project for a New Science," *Isis* 51 (1960): 186–99.

161. On Réaumur, see Jean Torlais, *Un esprit encyclopédique en dehors de "l'Encyclopédie": Réaumur d'après des documents inédits* (Paris: Blanchard, 1961); and Jean Torlais, "Réaumur philosophe," *Revue d'histoire des sciences* 11 (1958): 13–33.

162. "Réaumur stood opposed to Buffon. The spiritualist philosopher against the materialist philosopher" (Torlais, "Réaumur philosophe," 24).

163. See Virginia Dawson, "Trembley, Bonnet and Réaumur and the Issue of Biological Continuity," *Studies in Eighteenth Century Culture* 12 (1984): 43–63. Jacques Marx has emphasized this "party" formation of the era in *Charles Bonnet*, 140, and in "L'art d'observer au XVIIIe siècle: Jean Senebier et Charles Bonnet," *Janus* 61 (1974): 201–20, citing 208.

164. Raymond de Saussure, "Haller and La Mettrie," *Journal of the History of Medicine* (Autumn 1949): 431–49; Karl Guthke, "Haller, La Mettrie und die anonyme Schrift *l'homme plue que machine*," *Études germaniqes* 17 (1962): 137–43; Erich Hintzsche, "Neue Funde zum Thema L'homme machine und Albrecht Haller," *Gesnerus* 25 (1968): 135–66.

165. The importance of Trembley is well captured in the title and substance of Sylvia Lenhoff and Howard Lenhoff, *Hydra and the Birth of Experimental Biology—1744: Abraham Trembley's Memoirs concerning the Natural History of a Type of Freshwater Polyp with Arms Shaped Like Horns* (Pacific Grove, CA: Boxwood, 1986). See also Virginia Dawson, *Nature's Enigma: The Problem of the Polyp in the Letters of Bonnet, Trembley, and Réaumur* (Philadelphia: American Philosophical Society, 1987).

166. This was a central feature of the comments of Bonnet, Haller, and Reimarus, as I will discuss below.

167. Charles Bodemer, "Regeneration and the Decline of Preformationism in Eighteenth Century Embrology," *Bulletin for the History of Medicine* 38 (1964): 20–31.

168. "The story of the Phoenix which is reborn from its ashes, fabulous as it is, offers nothing more marvellous than the discovery of which we are going to speak. The chimerical ideas of Palingenesis or regeneration of Plants & Animals, which some Alchemists believed possible by the bringing together and the reunion of their essential parts, only leads to restoring a Plant or an Animal after its destruction; the serpent cut in half, & which is said to be rejoined, gives but one & the same serpent; but here is Nature which goes farther than our chimeras. From one piece of the same animal cut in 2, 3, 4, 10, 20, 30, 40 pieces, & so to speak, chopped up there are reborn as many complete animals similar to the first." Report of the *Académie royale*, 1741, cited in Virginia Dawson, "Regeneration, Parthenogenesis, and the Immutable Order of Nature," *Archives of Natural History* 18 (1991): 309–21, citing 317–18.

169. "Yet was the polyp really an animal? . . . [W]hat criterion could be used to mark the distinction between plants and animals? Were polyps zoophytes or plant-animals—the point of passage in a continuous chain of being from plants to animals?" (Dawson, "Regeneration, Parthenogenesis," 316).

170. Trembley wrote: "It was the shape of these polyps, their green colour, and their immobility that gave birth to my idea that they were plants" (Abraham Trembley, *Mémoires*, cited in Baker, *Abraham Trembley of Geneva*, 28).

171. Dawson, "Regeneration, Parthenogenesis," 317. Trembley was unaware that Antonie van Leeuwenhoek had already reached this conclusion in his research in the seventeenth century.

172. That Trembley's observations *were* epochal, two great biologists of the early nineteenth century, Karl Ernst von Baer (1792–1876) and Georges Cuvier (1769–1832), clearly believed (Baker, *Abraham Trembley of Geneva*, 47).

173. Georges Cuvier, *Histoire des sciences naturelles*, 5 vols. (Paris: Fortin, Masson, 1841–45), 3:256. In our times, several historians of science have made similarly emphatic claims. Schiller identified "the birth of experimental biology" with 1740 and Trembley's discovery, and the Lenhoffs have made a similar characterization in their book title, using the year 1744, when Trembley published his monograph. See Joseph Schiller, "Queries, Answers and Unsolved Problems in Eighteenth Century Biology," *History of Science* 12 (1974): 184–99, citing 185; Lenhoff and Lenhoff, *Hydra and the Birth of Experimental Biology—1744*. Jacques Roger (*Life Sciences*, 475ff.) uses 1745 as his point of inflection between two epochs of research in life science, again pivoting on Trembley's discovery. Charles Bodemer ("Regeneration") has made a strong case that it triggered, albeit unintentionally, the shift away from preformation in embryology, and Aram Vartanian ("Trembley's Polyp") has argued that it similarly stimulated, again against Trembley's own inclinations, the rise of materialism. Dawson concurs: "materialism and vitalism, sometimes associated with pantheistic or atheistic notions, took on a new intellectual respectability" (*Nature's Enigma*, 8). The discovery "seemed to introduce an unwelcome element of chance into the regular structure of living beings and implied that regeneration might be an attribute of matter independent of rational plan . . . [and thus] brought into focus the problem of animal soul" (8). Dawson goes so far as to claim that Trembley's *Mémoires* on the polyp "ushered in a new era of philosophical speculation" (Dawson, "Regeneration, Parthenogenesis," 310). More subtly, Ritterbush, Mornet, and Hazard all see this as an important moment in the shift away from mathematical natural science toward experimental Newtonianism or empiricism in scientific practice more generally. See Philip Ritterbush, *Overtures to Biology: The Speculation of Eighteenth-Century Naturalists* (New Haven, CT: Yale University Press, 1964); Daniel Mornet, *Les sciences de la nature en France au XVIIIe siècle, un capitre de l'histoire des idées* (Paris: Colin, 1911); Paul Hazard, *La pensée européenne au XVIIIe siécle, de Montesquieu à Lessing* (Paris: Furne, 1946).

174. Buffon made the famous disparagement in his *Discours de la nature des animaux* (1754) that a fly should occupy no more important a place in the mind of man than it did in the physical universe, a savage swipe at Réaumur's life's work. See René-Antoine Ferchault de Réaumur, *The Natural History of Ants*, trans. William Wheeler (New York: Knopf, 1926).

175. Hoquet, *Buffon*, 21.

176. Buffon, "Preliminary Discourse," 110ff., esp. 120–24.

177. Ibid., 116.

178. Hoquet makes the point that even Claude Perrault would not go so far: "Perrault condemns the Baconian strategy: it is vain to restrict oneself to historical collection with the hope that some day the time of a true general system will arise. It is important on the contrary that one not hesitate to go beyond history and propose some systems, all the while not pretending to provide the veritable and definitive system but restricting oneself always to particular hypotheses" (*Buffon*, 118).

179. "From the beginnings of his career as a naturalist, there gathered around [Bonnet] a group hostile to Buffon, a gathering of spirits deeply antipathetic to the famous *systems*" (ibid., 156).

180. Ibid., 39. Hoquet argues for an eighteenth-century "epicurization of philosophy," in which Buffon served as the great spokesman of this tendency whereby physics displaced metaphysics (ibid., 147ff.).

181. On Palissot and the conception of *philosophes* as freethinking coterie, see ibid., 139–41.

The term *cacouac* was invented in 1757 by Jacob Nicolas Moreau (1717–1803) and popularized in a pamphlet entitled *Nouveau mémoire pour servir à l'histoire des Cacouacs,* the thrust of which was to assert that these cacouacs substituted nature for God and denied revealed religion. Diderot, for one, embraced the label: "I am a cacouac to the devil[;] . . . there is no one of good mind and integrity [*de bon esprit et d'honnête homme*] who would not be more or less of this clique." Diderot, "Paradox sur le comédien" (1761), in *Oeuvres esthétiques* (Paris: Garnier, 1968), 355. See Gerhardt Stenger, *L'affair des cacouacs: Trois pamphlets contre les philosophes des lumières* (Saint Étienne: Publications de l'Université de Saint Étienne, 2004). By the same token, religious conservatives like Haller and Bonnet used the term to disparage figures like Buffon and Diderot and to express their distaste generally for the *philosophes* and freethinking. See Charly Guyot, "Albert de Haller et Charles Bonnet, juges de l'*Encyclopédie,*" *Literature and Science: Proceedings of the 6th Triennial Congress* (Oxford: Blackwell, 1955), 205–12.

182. Jean Senebier, *Essai sur l'art d'observer et de faire des expériences* (Geneva: Paschoud, 1802). The *Essai* won a prize in 1768 and was first published in 1775; it appeared in revised and expanded form in 1802.

183. Marx, "L'art d'observer," 203.

184. Senebier, *Essai sur l'art d'observer,* preface, cited in Marx, "L'art d'observer," 204.

185. Senebier to Bonnet, August 1782, cited in Marx, "L'art d'observer," 211.

186. Marx, "L'art d'observer," 205.

187. Senebier, *Essai sur l'art d'observer,* 1:4.

188. Charles Bonnet, *Traité d'insectologie* (Paris: Durand, 1745), preface, ii–iii.

189. *Oeuvres d'histoire naturelle et de philosophie de Charles Bonnet,* vols. 5–6, *Considérations sur les corps organisés* (orig. 1762), rev. ed., with an "Avertissement" by Bonnet, dated October 1778 (vol. 1, 1–4), explaining the changes (Neuchâtel: Fauche, 1779), 1:363.

190. Ibid., 379.

191. Bonnet, *Contemplation de la nature,* 2:289.

192. Ibid., vi–viii.

193. Charles Bonnet, *Mémoires autobiographiques* (Paris: Vrin, 1948), 23 (from a letter to Haller), cited in Pierre Speziali, "Réaumur et les savants genevois," *Revue d'histoire des sciences* 11 (1958): 68–80, citing 68n. In a letter from Réaumur to Bonnet, July 22, 1738, at the very outset of their contact, the French academician lavished praise on Bonnet: "you are a disciple it will always bring me glory to acknowledge." Cited in H. Erhard, "Die Entdeckung der Parthenogenesis durch Charles Bonnet," *Gesnerus* 3 (1946): 15–27, citing 16. See Raymond Savioz, "Un maitre et un disciple au XVIIIe siècle (Charles Bonnet et Réaumur)," *Thalès* 4 (1940): 100–112.

194. Bonnet, *Contemplation de la nature,* 2:287.

195. Ibid., 512.

196. Ibid., 182.

197. Ibid., 187.

198. Ibid., 189; and see 131: "Natural history is a better logic, because it is one that teaches us better to suspend our judgment."

199. Ibid., 188.

200. Ibid., 189.

201. Ibid., 1:93.

202. Bonnet, *Considérations sur les corps organisés,* 1:266.

203. Bonnet, *Contemplation de la nature,* 2:68.

204. Ibid., 267.

205. Cited in Marc Ratcliff, "Une métaphysique de la méthode chez Charles Bonnet," in Buscaglia et al., *Charles Bonnet: Savant et philosophe,* 51–60, citing 52, referencing a letter from Bonnet to Haller, November 8, 1757, in *The Correspondence between Albrecht von Haller and Charles Bonnet,* ed. Otto Sonntag (Bern: Huber, 1983), 124.

206. Bonnet, *Considérations sur les corps organisés,* 1:261.

207. Bonnet to Haller, September 7, 1757, in Sonntag, *Correspondence*, 110.

208. Peter Bowler, "Bonnet and Buffon: Theories of Generation and the Problem of Species," *Journal of the History of Biology* 6 (1973): 259–81; Carlo Casellani, "The Problem of Generation in Bonnet and in Buffon: Critical Comparison," in *Science, Medicine and Society in the Renaissance: Essays to Honor Walter Pagel*, ed. Allen G. Debus (New York: Science History Publications, 1972), 265–88.

209. Bonnet, *Contemplation de la nature*, 2:201.

210. Bonnet, *Considérations sur les corps organisés*, 1:308.

211. Ibid., 313.

212. Ibid., 181.

213. Bonnet, *Contemplation de la nature*, preface (June 1764), 1:xxii.

214. Ibid., II:204.

215. Bonnet, *Considérations sur les corps organisés*, preface (March 1762), 1:71.

216. Ibid., 174ff.

217. Ibid., 312n.

218. Ibid., 202n.

219. Ibid., 2:398n.

220. Ibid., 220n.

221. Ibid., 1:224n.

222. Ibid., 2:185.

223. Ibid., 190.

224. Ibid., 1:266–75.

225. Ibid., 1:307.

226. Ibid., 1:311.

227. Ibid., 2:398n.

228. Haller to Bonnet, January 5, 1759, cited in ibid., 1:78.

229. Dawson, *Nature's Enigma*, 23.

230. On Cramer see P. Speziali, "Une correspondence inédite entre Clairault et Cramer," *Revue d'histoire des sciences* 8 (1955): 193–97.

231. René Sigrist, "Science et société à Genève au XVIIIe siècle: L'example de Charles Bonnet," in Buscaglia et al., *Charles Bonnet: Savant et philosophe*, 21–22.

232. See chapter 2.

233. With others, Dawson stresses the Protestant—indeed, Reform—character of this scientific trend, though a liberalized, rationalized Reform Protestantism (*Nature's Enigma*, 23).

234. Shank is particularly concerned with Buffon's correspondence with Cramer and its evidence for Buffon's shift from mathematics to life science. See Shank, *Newton Wars*, 404–7.

CHAPTER FIVE

1. J. G. Herder, *Treatise on the Origin of Language*, in *Philosophical Writings*, trans. and ed. Michael Forster (Cambridge: Cambridge University Press, 2002), 87.

2. Hans Aarsleff, "The Berlin Academy and Frederick the Great," *History of the Human Sciences* 2 (1989): 193–206; Harcourt Brown, "Maupertuis *Philosophe*: Enlightenment and the Berlin Academy," *Studies on Voltaire* 24 (1963): 255–69; Ronald Calinger, "Frederick the Great and the Berlin Academy of Sciences (1740–1766)," *Annals of Science* 24 (1968): 239–49.

3. Ann Thomson, *Materialism and Society in the Mid-eighteenth Century: La Mettrie's "Discours préliminaire"* (Geneva: Droz, 1981).

4. See the classic essay by Phillip Sloan: "Buffon, German Biology, and the Historical Interpretation of Biological Species," *British Journal for the History of Science* 12 (1979): 109–53.

5. Amor Cherni, "Haller et Buffon: À propos des *Réflexions* / Haller and Buffon: Concerning the

Reflexions," in "Les sciences de la vie au siècle des Lumières: In memoriam Roselyn Rey," special issue, *Revue d'histoire des sciences* 48 (1995): 267–306.

6. [Albrecht von Haller], review of *Die vornehmsten Wahrheiten der natürlichen Religion*, by Hermann Samuel Reimarus, *Göttingische Gelehrten Anzeigen* (*GGA*) 11 (1754): 1–35; 12 (1754): 267– 308.

7. For a pioneering assessment, see Shirley Roe, *Matter, Life and Generation: Eighteenth-Century Embryology and the Haller-Wolff Debate* (Cambridge: Cambridge University Press, 1981). Tracing Wolff's intellectual development and his conflict with Haller in the 1760s proves decisive for the reconstruction of German life science.

8. Johann August Unzer, *Erste Gründe einer Physiologie der eigentlichen thierischen Natur thierischer Körper* (Leipzig: Weidmann, 1771); abridged translation by Thomas Laycock, "*The Principles of Physiology*" *by John Augustus Unzer and "A Dissertation on the Functions of the Nervous System" by George Prochaska* (London: Sydenham Society, 1851).

9. That was certainly Wolff's suspicion. In a letter of 1748 he wrote of the Berlin Academy as the combination of "so-called Newtonian philosophy with the French world of flattery." Cited in Mary Terrall, *The Man Who Flattened the Earth: Maupertuis and the Sciences in the Enlightenment* (Chicago: University of Chicago Press, 2002), 251n. Terrall elaborates: "The Wolffians looked at Berlin and saw foreign mathematicians overreaching the bounds of their competence to dictate to German metaphysicians" (265). Avi Lifschitz makes the same observation: "Wolff and his followers interpreted the renewal of the Academy as a concerted assault on German philosophy." Avi Lifschitz, *Language and Enlightenment: The Berlin Debates of the Eighteenth Century* (Oxford: Oxford University Press, 2012), 71.

10. Much of Euler's work of the 1750s involved metaphysical sallies against German school metaphysics—some of them singularly uninspired. See Rüdiger Thiele, "Euler und Maupertuis vor dem Horizont des teleologischen Denkens: Über die Begründung des Prinzips der kleinsten Aktion," in *Schweizer im Berlin des 18. Jahrhunderts*, ed. Martin Fontius and Helmut Holzhey (Berlin: Akademie, 1996), 2:373–90.

11. Terrall, *The Man Who Flattened the Earth*, 257ff. Compare the earlier studies cited in n. 2. I believe Terrall (257) has a better case than David Beeson, *Maupertuis: An Intellectual Biography* (Oxford: Voltaire Foundation, 1992), 109.

12. Hartmut Hecht, "Pierre Louis Moreau de Maupertuis oder die Schwierigkeit der Unterscheidung von Newtonianern und Leibnitianern," in *Leibniz und Europa: Vorträge des VI. Internationalen Leibniz-Kongresses, Hannover, 18. bis 23 Juli 1994* (Hanover: Leibniz Gesellschaft, 1994), 1:331–38. Terrall cautions: "It is not clear whether Maupertuis appreciated how close his position was to that of Leibniz" (*The Man Who Flattened the Earth*, 284n).

13. Ernst Cassirer, *Philosophy of the Enlightenment* (Princeton, NJ: Princeton University Press, 1951).

14. See Thomas Broman, "Metaphysics for an Enlightened Public: The Controversy over Monads in Germany, 1746–1748," *Isis* 103 (2012): 1–23.

15. Ibid., 8.

16. Ibid., 10.

17. Even Hartmut Hecht, ed., *Pierre Louis Moreau de Maupertuis: Eine Bilanz nach 300 Jahren* (Berlin: Arno Spitz, 1999), still devotes much attention to this controversy. Goldenbaum has made the controversy a crucial event in the emergence of a public sphere in Germany, demonstrating that it highlighted the abuse of executive authority in suppressing dissent and thus spoke directly to the idea of freedom of opinion. Ursula Goldenbaum, *Appell an das Publikum: Die öffentliche Debatte in der deutschen Aufklärung, 1687–1796* (Berlin: Akademie, 2004).

18. Claire Salomon-Bayet, "Maupertuis et l'institution," in *Actes de la journée Maupertuis (Créteil, 1er décembre 1973)* (Paris: Vrin, 1975), 183–202.

19. Brown, "Maupertuis, *Philosophe*," 256.

20. Ilse Jahn, "Maupertuis zwischen Präformations- und Epigenesis-theorie: Sein Beitrag zu biologischen Fragen des 18. Jahrhunderts," in Hecht, *Maupertuis: Bilanz nach 300 Jahren*, 89–101.

21. Pierre-Louis Moreau de Maupertuis, “Devoirs de l’académicien” (1750), in *Oeuvres* (Hildesheim: Olms, 1965), 3:283–302.

22. Beeson (*Maupertuis,* 191, 206–14, 241) gives an extended account of the relationship between Maupertuis and La Mettrie, ascribing great influence to the latter in the development of Maupertuis’s thought in life science.

23. Ibid., 214.

24. Gordon Wattles, “Buffon, d’Alembert and Materialist Atheism,” *Studies on Voltaire and the Eighteenth Century* 266 (1989): 285–341.

25. Terrall, *The Man Who Flattened the Earth*, 122. I thank Jennifer Mensch for providing me with a copy of this rare text.

26. See Aram Vartanian, “Diderot and Maupertuis,” *Revue internationale de philosophie* 38 (1984): 46–66.

27. Terrall, *The Man Who Flattened the Earth*, 224. It is possible that this review was by Haller.

28. Ibid., 323.

29. Indeed, his “response” to Diderot was only a half-hearted repudiation, and as Mary Terrall concludes, “Maupertuis’s defense amounted to showing how close his position was to that of his interlocutor” (ibid., 346).

30. Buffon, *Histoire naturelle*, vol. 2 (1749), cited in Michael Hoffheimer, “Maupertuis and the Eighteenth-Century Critique of Preexistence,” *Journal of the History of Biology* 15 (1982): 119–44, citing 129.

31. Pierre-Louis Moreau de Maupertuis, “Lettre sur la génération des animaux,” in *Lettres* (Dresden: Walther, 1752). Trans. as “Letter on the Generation of Animals” in an appendix to Hoffheimer, “Maupertuis and the Eighteenth-Century Critique of Preexistence,” 138–44.

32. Hoffheimer, “Maupertuis and the Eighteenth-Century Critique of Preexistence,” 137.

33. Beeson, *Maupertuis*, 211.

34. Terrall, *The Man Who Flattened the Earth*, 310.

35. Ibid., 155–56. Thus, Glass is wrong to suggest that Maupertuis found vitalism “abhorrent” and that he was “a consistent mechanist.” Bentley Glass, “Maupertuis, Pioneer of Genetics and Evolution,” in *Forerunners of Darwin: 1745–1859*, ed. Glass et al. (Baltimore: Johns Hopkins University Press, 1959), 51–83, citing 62.

36. See Hoffheimer, “Maupertuis and the Eighteenth-Century Critique of Preexistence,” 135–37.

37. Terrall, *The Man Who Flattened the Earth*, 202.

38. Mary Terrall, “Salon, Academy, and Boudoir: Generation and Desire in Maupertuis’s Science of Life,” *Isis* 87 (1996): 217–29, citing 225.

39. Pierre-Louis Moreau de Maupertuis, *Système de la nature*, in *Essai de cosmologie, Système de la nature, Réponse aux objections de M. Diderot*, ed. François Azouvi (Paris: Vrin, 1984), 158.

40. Ibid., 147.

41. Ibid., 146–47.

42. Ibid., 158–59.

43. Ibid., 179.

44. Ibid., 180.

45. Ibid., 158.

46. Pierre-Louis Moreau de Maupertuis, *Essai de cosmologie*, in Azouvi, *Essai de cosmologie, Système de la nature, Réponse aux objections de M. Diderot*, 11–12.

47. Ibid., 72–73.

48. Ibid., 73.

49. Ibid., 74.

50. Ibid., 170.

51. Ibid., 164–65.

52. Terrall, *The Man Who Flattened the Earth*, 339.

53. Terrall, “Salon, Academy, and Boudoir,” 224; Terrall, *The Man Who Flattened the Earth*, 340.

54. Maupertuis, *Système de la nature*, 166.

55. Ibid., 167

56. Ibid., 169.

57. Ibid., 166–67.

58. Annie Ibrahim, "Matière inerte et matière vivante: La théorie de la perception chez Maupertuis," *Dix-huitième siècle* 24 (1992): 95–103.

59. "These active properties allow for many possible outcomes, within certain parameters, and they locate the capability to produce order in matter itself" (Terrall, "Salon, Academy, and Boudoir," 223).

60. That is, it embraced "the idea that matter contained a plastic, vital, even divine principle continuously at work," in the words of C. U. M. Smith, *The Problem of Life* (New York: Wiley, 1976), 268.

61. Hoffheimer, "Maupertuis and the Eighteenth-Century Critique of Preexistence," 126. Hoffheimer observes, very strikingly, that this "hylozoism converges with various forms of the Leibniz-Wolffian philosophy" then current in Germany (136). Terrall makes the important point that Maupertuis endeavored to make this theologically and scientifically palatable (*The Man Who Flattened the Earth*, 329).

62. As Glass characterizes the context: "By 1740 there remained very few epigenesists indeed, and the encasement theory of preformation prevailed almost universally" ("Maupertuis, Pioneer," 62).

63. Albrecht von Haller, "Buffon on Hypotheses: The Haller Preface to the German Translation of the *Histoire Naturelle* (1750)," trans. Phillip Sloan, in *From Natural History to the History of Nature: Readings in Buffon and His Critics*, ed. John Lyon and Phillip Sloan (Notre Dame, IN: University of Notre Dame Press, 1981), 295–310; Albrecht von Haller, *Réflexions sur le système de la génération de M. de Buffon* (Geneva: Barillot et fils, 1751; German version, March 1752), trans. Phillip Sloan as "Haller on Buffon's Theory of Generation," in Lyon and Sloan, *From Natural History to History of Nature*, 311–28.

64. Georges-Louis Leclerc de Buffon, *Allgemeine Historie der Natur nach allen ihren besondern Theilen behandelt . . . mit einer Vorrede [von] Herrn Doktor Albrecht von Haller*, vol. 1 (Hamburg: Grund und Holle, 1750). See the acknowledgments in this volume.

65. Kästner, a prominent mathematician but also a creative writer of distinction, who was in the process of winning a Berlin Academy prize in the belles-lettres division, parlayed the translation enterprise and his new association with Haller into a position at the latter's university, Göttingen. The move from Leipzig enabled him to assume a prominent role at his new university, as I will have occasion to note in later chapters.

66. Haller's promotion of the German translation in his *GGA* review particularly praised the editorial annotations by Kästner: "The notes on some of the ideas of the author restrict these ideas [*schränken dieselben ein*], warn the reader, soften the harsher judgments of M. de B., and increase the value of the work" (*GGA*, January 1751, 3–4). In his *Bibliothèque raisonnée* review, pt. 2, Haller makes similar observations: "This work has been translated into German. . . . This edition has been produced with great precision. . . . Mr. Kästner, a sharp-witted mathematician of Leipzig, has accompanied the translation with comments which are not always of the same mind as the author. Mr. Zink directed the translation" (Lyon and Sloan, *From Natural History to History of Nature*, 281).

67. Haller, *Réflexions sur le système de la génération de M. de Buffon*. See Urs Boschung et al., eds., *Repertorium zu Albrecht von Hallers Korrespondenz, 1724–1777*, 2 vols. (Basel: Schwabe, 2002), 1:612.

68. See Cherni, "Haller et Buffon," for the most perspicacious account of the scholarly debate and a persuasive reconstruction of the whole matter.

69. He reviewed each succeeding volume as it appeared both in his own *GGA* and in the Francophone *Bibliothèque raisonnée*, published in Holland. The *Bibliothèque raisonnée* reviews are translated in Lyon and Sloan, *From Natural History to History of Nature*: pt. 1 (1750), 255–68; pt. 2 (1751), 269–82.

70. Indeed, Haller saw it mainly as a defense of the beleaguered position on the utility of hypothesis in science taken by Dortous de Mairan in the preface to his *Dissertation sur la glace*. Haller, review of Buffon in *Bibliothèque raisonnée*, pt. 2, in Lyon and Sloan, *From Natural History to History of Nature*, 281.

71. In the Sloan translation, Haller begins to deal with Buffon at p. 306, after some ten pages (out of twelve) on the general problem of hypothesis in natural science.

72. Haller, "Buffon on Hypotheses," 307.

73. Haller's review of Bonnet's *Insectology* of 1746 (*GGA*, 1746) is reproduced in *Sammlung kleiner Hallerisiche Schriften* (Bern: Haller, 1772), 3:281–303.

74. On Haller's short-lived epigenetic orientation, see Cherni, "Haller et Buffon"; Amor Cherni, *Épistémologie de la transparence: Sur l'embryologie de A. von Haller* (Paris: Vrin, 1998).

75. We have already seen how carefully Bonnet considered these experiments, and how vigorously Réaumur's associate de Lignac scrutinized and condemned them. We know from Bonnet's later writings how prominently Spallanzani's replication experiments figured in the debate about epigenesis. And we will see a similar concern with these experiments and their interpretive significance for Reimarus below.

76. Haller, review of Buffon in *Bibliothèque raisonnée*, pt. 2, in Lyon and Sloan, *From Natural History to History of Nature*, 275. See Renato Mazzolini, "Two Letters on Epigenesis from John Turberville Needham to Albrecht von Haller," *Journal of the History of Medicine and Allied Sciences* 31 (1976): 68–77; Renato Mazzolini and Shirley Roe, eds., *Science against the Unbelievers: The Correspondence of Bonnet and Needham, 1760–1780* (Oxford: Voltaire Foundation, 1986).

77. Haller, review of Buffon in *Bibliothèque raisonnée*, pt. 2, in Lyon and Sloan, *From Natural History to History of Nature*, 269, 270. But Haller was not persuaded that Buffon's theory actually "claimed to reestablish the opinions of Lucretius," as those of La Mettrie clearly did (280).

78. Ibid., 280.

79. Ibid., 272.

80. Haller, *Réflexions*, trans. Sloan as "Haller on Buffon's Theory of Generation," 320.

81. Haller had become somewhat estranged from Linnaeus by 1745 because of criticisms of Haller in Linnaeus's *Flora Suecica*. In 1747 Haller announced to Linnaeus that he was putting botany aside to pursue more intensely his interests in anatomy and physiology. See Urs Boschung et al., *Repertorium zu Albrecht von Hallers Korrespondenz, 1724–1777*, 1:309.

82. Haller defended Linnaean taxonomy in botany against Buffon's criticisms in virtually all the texts under consideration.

83. Sloan discerned all this in his pioneering essay "Buffon, German Biology."

84. Haller, review of Buffon, in *Bibliothèque raisonnée*, pt. 2, in Lyon and Sloan, *From Natural History to History of Nature*, 280, 279.

85. Ibid., 276.

86. Haller, *Réflexions*, trans. Sloan as "Haller on Buffon's Theory of Generation," 322.

87. "As for my own view, it seems obvious to me that the mountains, rivers, and the sea are to be found on the earth just as they came from the Creator's hand. . . . [It is] probable that the earth has been just about the same since creation." Haller adds: "Nothing gives us a warrant to posit causes different from those which exist in our day." Review of Buffon in *Bibliothèque raisonnée*, pt. 1, in Lyon and Sloan, *From Natural History to History of Nature*, 263, 264. Haller accepted the idea of a universal deluge along scriptural lines as sufficient to explain all geological anomalies: "one can get an idea of the changes which may have taken place over 4,000 years on the earth by judging by the sorts of changes which have taken place in our times" (264). Similarly, Haller evinced skepticism regarding the organic origins of fossils (264).

88. On the widespread recognition of the extent of geological time by 1750, see Martin Rudwick, *Bursting the Limits of Time: The Reconstruction of Geohistory in the Age of Revolution* (Chicago: University of Chicago Press, 2005).

89. Joseph Lelarge de Lignac, *Lettres à un Amériquain sur l'Histoire naturelle, générale et particuliere de Monsieur de Buffon* (1751).

90. Review of Reimarus, *GGA* 11 (1754): 1–35; 12 (1754): 267–308.

91. Hermann Samuel Reimarus, *Die vornehmsten Wahrheiten der natürlichen Religion* (1754), ed. Günter Gawlick, 2 vols. (Göttingen: Vandenhoeck und Ruprecht, 1985).

92. See Kant, *Kritik der Urteilskraft* (1790), AA 5:476. All the while, we know now, Reimarus was composing but concealing his most momentous work, the *Apologie oder Schutzschrift für die vernünftigen Verehrer Gottes*, which would serve as the basis for Gotthold Efraim Lessing's explosive publication (1774–78) known as the *Wolfenbüttel Fragmente*—the most uncompromising work of deism in the German Enlightenment. Indeed, the introduction to the critical edition of *Vornehmsten Wahrheiten* goes so far as to suggest deliberate deception in the 1754 text, pretending to serve—as it was, indeed, received at the time—as a prolegomenon to revealed religion. Michael Emsbach and Wilfried Schröder, "Einleitung," in Reimarus, *Vornehmsten Wahrheiten*, 1:9–50, citing 13.

93. Reimarus had a longer-standing interest in the technical question of animal instinct, and it was the convergence of this issue in life science with the larger questions of natural theology in the epoch that galvanized his work. Already as a student Reimarus had composed an essay entitled "Instinctum Brutorum Existentis Dei Ejusdemque Sapientissimi Indicem" (1725), reprinted and translated into German in Hermann Samuel Reimarus, *Allgemeine Betrachtungen über die Triebe der Thiere, hauptsächlich über ihre Kunsttriebe*, ed. Jürgen von Kempski (Göttingen: Vandenhoeck und Ruprecht, 1982), 757–80. These preoccupations in the German philosophical landscape were reflected in works by the key Wolffian philosopher at Halle, Georg Friedrich Meier: *Beweiss: Daß keine Materie denken könne* (1742), *Rettung der Ehre der Vernunft wider die Freygeister* (Halle, 1747), and *Lehre von der Thierseele* (Halle, 1747). See Friedrich Niewöhner and Jean-Loup Seban, eds., *Die Seele der Tiere* (Berlin: Harrassowitz, 2001), esp. Günter Frank, "Seele oder Maschine? Der Streit um die Tierseele in der deutschen Aufklärung," 249–66; Justin E. Smith, ed., *The Problem of Animal Generation in Early Modern Philosophy* (Cambridge: Cambridge University Press, 2006).

94. This was the punch in La Mettrie's *L'homme machine* (1747): he called Descartes's bluff in a way that mortified all Europe.

95. Hermann Samuel Reimarus, "Vorbericht," in *Vornehmsten Wahrheiten*, 1:56. In an extended review in *GGA*, probably by Haller, the work was praised precisely as an antidote to the French "deluge of freethinking [*freidenkerische Überschwemmung*]." *GGA* 11 (1754): 1–35; 12 (1754): 267–308. That this "deluge" was taking place is attested among other things by the publication, at the end of the decade, of Johann Anton Trinius's *Freydenker-Lexikon, oder Einleitung in die Geschichte der neuern Freygeister: Ihre Schriften, und deren Widerlegungen* (Leipzig, 1759).

96. Julian Jaynes and William Woodward, "In the Shadow of the Enlightenment: I. Reimarus against the Epicureans" and "In the Shadow of the Enlightenment: II. Reimarus and His Theory of Drives," *Journal of the History of the Behavioral Sciences* 10 (1974): 3–15, 144–59. The editors of the critical edition produce all the documentation to establish the matter (Reimarus, *Vornehmsten Wahrheiten*, 1:9).

97. Cited in Reimarus, *Vornehmsten Wahrheiten*, 1:9.

98. "It is injurious to the honor of these three gentlemen, whom I esteem for their merits, and is contrary to what I have expressed in my work, to call them followers of Epicurus, and to arrange them in the same class with La Mettrie" (Reimarus, letter to *Monthly Review* [London], November 18, 1766, reprinted in Reimarus, *Vornehmsten Wahrheiten*, 1:865–69, citing 867).

99. Ibid., 868.

100. Richard Wynne, letter to the *Monthly Review* [London], February 15, 1767, reprinted in Reimarus, *Vornehmsten Wahrheiten*, 1:870–74, citing esp. 872–73. Wynne offered copious citations from the Reimarus text to support his rebuttal.

101. For himself, he added: "I would by no means tax these Gentlemen with the vices of *La Mettrie*, who was an Epicurean in the worst sense of the word, but of adopting some of the opinions, and reviving the objections of Lucretius against Providence, &c." (ibid., 872–73). Wynne went on to characterize what he took to be the main doctrines of Epicurus and Lucretius and alleged that the authors in question subscribed to them.

102. Reimarus, *Vornehmsten Wahrheiten*, 1:877.

103. Ibid., 72. "A dead, unfeeling matter and blind nature knows nothing of itself, and cannot enjoy or sense its own existence, but it is rather a matter of indifference whether it exists or not, or whether it should be something else" (ibid., 202). But everything actual must be determinate: there must be sufficient reason for it to be so and not otherwise (ibid., 89). Since the physical world could not boast an intrinsic reason for its own determinacy, it was an imperfect being, requiring its constitution by some other being. This argument was developed most extensively in the third *Abhandlung*.

104. Ibid., 893.

105. Ibid.

106. Ibid., 150.

107. Ibid., 151n. Reimarus read La Mettrie's *Système d'Epicure*, *Traité de l'âme* (the revised version of his earlier *Histoire naturelle de l'âme* of 1745), and *Discours préliminaire* in the posthumous edition of La Mettrie's *Oeuvres philosophiques* (1753). Though La Mettrie was dead, the rage against his rampant "Epicureanism" persisted here as it did with Haller (though the latter had the motivation of a personal grievance).

108. Reimarus, *Vornehmsten Wahrheiten*, 1:152.

109. Ibid., 153. He meant Francesco Redi (1626–97), *Experiments on the Generation of Insects* (1668; Open Court, 1909). The parallels to Bonnet are striking.

110. Reimarus, *Vornehmsten Wahrheiten*, 1:179–80. He scorned this both in the ancients and in the moderns, especially La Mettrie. He insisted that the complex unity of an organism could simply not be the product of chance (ibid., 183).

111. Ibid., 206. It was no improvement to view the natural world as a world of forces, not masses, for forces had no intrinsic purposiveness either, and their interactions did not of themselves generate harmony but rather a "desolate chaos" (ibid., 267–69, citing 269).

112. Ibid., 205. Such, too, was his explicit condemnation of Spinoza: not only did Reimarus contest the coherence of Spinoza's argument about substance and the nature of God, but he urged that what Spinoza offered in his system was simply "fatal and unlimited necessity" (ibid., 256–59, citing 259).

113. Reimarus not only criticized his usual targets—La Mettrie, Buffon, and Maupertuis—for this but also attacked d'Alembert (ibid., 359n).

114. Ibid., 291.

115. Ibid., 339ff., esp. 341. This is a powerful confirmation of Hoquet's reconstruction of the reception of Buffon. See Thierry Hoquet, *Buffon: Histoire naturelle et philosophie* (Paris: Champion, 2005), 147ff.

116. Reimarus, *Vornehmsten Wahrheiten*, 1:348.

117. On Maupertuis being no different from La Mettrie in his Epicureanism, see ibid., 311.

118. Ibid., 364–65.

119. Ibid., 382–83, 433.

120. "A crowd of natural scientists, after Newton, have found God in the stars, in insects, in plants, in water," Maupertuis pronounced acerbically (*Essai de cosmologie*, 7). He illustrated by describing one text that used the hide of a rhinoceros to argue for God's Providence. Such proofs verged on the "indecent," and Maupertuis urged, "let us leave such trivialities [*bagatelles*] to those who do not sense their frivolity" (12).

121. Reimarus, *Vornehmste Wahrheiten*, 1:155ff., 881ff.

122. Ibid., 887.

123. Ibid., 882.

124. Ibid., 887.

125. The question of microscopic unreliability was a significant issue in that epoch. See Luigi Belloni, "Micrografia illusoria e animalcula," *Physis* 4 (1962): 65–73.

126. Reimarus, *Vornehmsten Wahrheiten*, 1:882. The irony was that Needham made a much stronger case for spontaneous generation than did Buffon, yet he was also a Catholic priest intent

upon affirming intelligent design. See Shirley Roe, "John Turberville Needham and the Generation of Living Organisms," *Isis* 74 (1983): 158–84. And see Mazzolini and Roe, introduction to *Science against the Unbelievers*.

127. Reimarus, *Vornehmsten Wahrheiten*, 1:883.

128. Ibid., 172.

129. The parallels between Reimarus and Bonnet on all these matters are striking.

130. Ibid., 891.

131. Ibid., 221.

132. Ibid., 273.

133. Ibid. This stance was replicated emphatically in his later *Triebe der Thiere*.

134. Reimarus, *Vornehmsten Wahrheiten*, 1:222, 239.

135. Ibid., 444.

136. Étienne Bonnot de Condillac, *Traité des animaux*, ed. Michel Malherbe (Paris: Vrin, 2004).

137. "But the reading of this work will demonstrate that it is impossible that I have derived anything from those [works] of Buffon" (ibid., 112n). And see ibid., 132, where Condillac notes that it had been alleged that his *Treatise on Sensation* was based on Buffon's treatment.

138. Georges-Louis Leclerc de Buffon, *Discours de la nature des animaux*, in *Histoire naturelle, générale et particulière*, vol. 4 (Paris: Imprimerie Royale, 1754).

139. Diderot, "Animal," in *Encyclopédie*, reprinted in Diderot, *Oeuvres philosophiques*, ed. J Lough and J. Proust (Paris: Gallimard, 2010), 5:381–401.

140. Condillac, *Traité des animaux*, 117. He evoked Haller's Academy Address on irritability as a model of proper empirical-scientific modesty, especially in its restraint regarding any naturalistic explanation of sensibility, and he noted with satisfaction Haller's "refutation" of Buffon's theory of generation (ibid., 128n).

141. Ibid., 112.

142. Ibid., 129; title of pt. 1, chap. 5.

143. Ibid., 150ff.

144. Ibid., 151.

145. Ibid., 153.

146. Ibid., 123, 146, proposition xi. See Reimarus, *Triebe der Thiere*, preface, 65, where he makes clear how central this problem is to his own articulation.

147. Condillac, *Traité des animaux*, 123.

148. Ibid., 168.

149. See Étienne Bonnot de Condillac, *Condillac's Treatise on the Sensations*, trans. Margaret Geraldine Spooner Carr (London: Favil Press, 1930); and Étienne Bonnot de Condillac, *Essay on the Origin of Human Knowledge* (Cambridge: Cambridge University Press, 2001).

150. Condillac, *Traité des animaux*, 165.

151. Reimarus, *Vornehmsten Wahrheiten*, 1:379.

152. Ibid., 387.

153. As Jaynes and Woodward ("In the Shadow of the Enlightenment") point out, the whole of the *Triebe der Thiere* sought to conceptualize the notion of "drive." Disappointingly, the authors miss the connection with Johann Friedrich Blumenbach's crucial concept *Bildungstrieb*. On the other hand, they do point out the shift in treatment of Reimarus and his notion of instinct from the original to the supplementary editions of the Paris *Encyclopédie*, for the former took no note of Reimarus in the article on instinct (1765), but the latter (1777) in fact consisted in a summary of his interpretation (and may have been the work of Haller).

154. Reimarus, *Triebe der Thiere*, 65. See chap. 7 of *Triebe der Thiere* for this elaborate taxonomy. See Jürgen von Kempski, "Hermann Samuel Reimarus as Ethologe," in Reimarus, *Triebe der Thiere*, 21–56, citing 5.

155. Reimarus, *Triebe der Thiere*, 130.

156. Reimarus was prepared to allow a degree of innovation to animal creative drives: "I will

later show that the animals even in their creative drives do not behave in so completely uniform and mechanical a manner that one cannot concede to them the capacity to make different determinations here and there, according to circumstances" (ibid., 110).

157. Ibid., 66.

158. It was taken up by Schiller and by Schelling, and as I will note, it figured importantly in Blumenbach's notion of the *Bildungstrieb.*

159. Reimarus, *Triebe der Thiere,* 65, 78.

160. Ibid., 147.

161. Ibid., 66.

162. Ibid.

163. "Buffon assumes with Descartes that animals are merely bodily machines that have no soul, no concepts, no imagination and memory, to say nothing of understanding, judgment, and reason" (ibid., 302).

164. In ibid., §117, 319–21, Reimarus offers a clear summary of Condillac's position. In his "Opening Remarks," Mayr insists that Condillac was not the original provocation of Reimarus's project, which is certainly true, since Reimarus had been concerned with these matters since 1725, and Condillac's work appeared only in 1755, but Mayr himself makes the crucial point: "Condillac's book notably receives more pages of analysis than the work of any other author" (Ernst Mayr, "Geleitwort," in *Triebe der Thiere,* 9–18, citing 14).

165. This was already his project in the *Vernunftlehre* of 1756.

166. Reimarus, *Triebe der Thiere,* 66–67.

167. Mayr drew a clear contrast between the philosophical and the observational approaches to animal behavior and recognized Reimarus as bridging these two (Mayr, "Geleitwort," 13).

168. Reimarus, *Triebe der Thiere,* 325.

169. Ibid., 266.

170. Ibid., 395.

171. Ibid., 406.

172. Ibid., 434.

173. Ibid., 176.

174. Hence, Reimarus takes up in detail the work of the Halle philosopher Georg Friedrich Meier, *Versuche eines neuen Lehrgebäudes von den Seelen der Thiere* (1750), concentrating on the key school-philosophical distinction between the "higher" and "lower" faculties of mind, the key move in the rational psychology of the Wolffian school. See *Triebe der Thiere,* 336ff.

175. Condillac, *Traité des animaux,* 111.

176. Reimarus, *Triebe der Thiere,* 381.

177. Ibid., 107.

178. Reimarus, *Triebe der Thiere,* 110.

179. Ibid., 111. The continuities with the Kantian theory of concept construction are patent here.

180. Ibid., 113.

181. Ibid., 121.

182. Ibid., 158.

183. Ibid., 123.

184. Ibid., 126.

185. Reimarus insisted that the proportionality between efficacy and restriction was essential: "All the drives of animals are enclosed within the restrictions of their sensual awareness and desires" (ibid., 222). "No species of animal has unnecessary or superfluous creative drives" (224). "No single animal has by nature alien, false, or perverse creative drives" (225). Moreover, "there arise just as few new arts among the animals as are lost or become weaker" (234). "Every animal expresses the creative drives of its species, from the very first trial, with a complete, regular competence, without previous instruction, exercise, or mulling over" (235). Still, it is possible for animals

to make discriminating use of their creative drives, under given circumstances (248). They can also err in the application of their instincts (256).

186. Ibid., 170.

187. It is noteworthy that Reimarus ascribed "opinions" to the ancients but "hypotheses" to the moderns. It is not altogether clear that the latter term suggested any superiority.

188. Reimarus, *Triebe der Thiere*, 363.

189. Ibid., 364.

190. Ibid., 364–65.

191. Moses Mendelssohn, review of *Briefe, die neueste Litteratur betreffend*, letter 130 (1760), by Hermann Reimarus, reprinted in *Triebe der Thiere*, 781–827, citing 795; referring to Reimarus, *Triebe der Thiere*, 364–65.

192. Mendelssohn, review of *Briefe*, 799–802.

193. Ibid., 812–13.

194. Reimarus reacted with outrage, composing a hundred-page "Appendix" for the second edition of his *Triebe der Thiere* (1762), in which he accused Mendelssohn of deliberate distortion and ad hominem abuse. Hermann Samuel Reimarus, *Anhang von der Verschiedenen Determination der Naturkräfter, und ihren mancherley Stufen, zur Erläuterung des zehnten Capitels* (1762), reprinted in *Triebe der Thiere*, 477–580. Mendelssohn admired Reimarus, and he hardly meant to offend him: "I really admire Mr. Reimarus very highly, and sincerely wish not to be suspected of having wished to harm this admirable old man" (letter to Thomas Abbt, November 20, 1763, reprinted in *Triebe der Thiere*, 860). He found Reimarus's vehement response all-too-pedantically academic, as he complained to his colleague Abbt (letter to Abbt, November 2, 1762, reprinted in *Triebe der Thiere*, 857–58). He responded only to defend himself from these personal accusations rather than to renew the original dispute (Mendelssohn, *Briefe, die neueste Litteratur betreffend*, July 1–8, 1762, reprinted in *Triebe der Thiere*, 828ff.).

195. J. G. Herder, *Abhandlung über den Ursprung der Sprache* (1772). Trans. as *Treatise on the Origin of Language* in *Philosophical Writings*, trans. and ed. Michael Forster (Cambridge: Cambridge University Press, 2002), 65–164. For this section I draw on my essays "Physiological Psychology: Herder's Engagement with Haller in the 1770s," in *Der frühe und der späte Herder: Kontinuität und/oder Korrektur / Early and Late Herder: Continuity and/or Correction; Beiträge sur Konferenz der Internationalen Herder-Gesellschaft Saarbrücken 2004* (Heidelberg: Synchron, 2007), 433–48; and "Herder between Reimarus and Tetens," in *Herder: Philosophy and Anthropology*, ed. Anik Waldow and Nigel DeSouza (Oxford: Oxford University Press, 2017), 127–46.

196. Herder, *Treatise on the Origin of Language*, 77–78.

197. Ibid., 78.

198. Ibid., 79.

199. "[T]he smaller the sphere of animals is, the less they need language" (ibid., 79).

200. Ibid., 74 (my emphasis).

201. This notion that humans were defined by the ubiquity of their settlement on the planet was one of the crucial ideas of emergent anthropology and of comparative zoology. It was decisive for Kant, for Herder, and for the German natural-historical community led by Blumenbach, the Forsters, and E. A. W. Zimmermann.

202. *Treatise on the Origin of Language*, 128.

203. Ibid., 82.

204. Ibid.

205. Ibid., 84.

206. Ibid.

207. Ibid., 81, 78.

208. Thus, he put "special emphasis upon mankind's physiological constitution. . . . [He] stressed the importance of understanding man's place in nature and the affinity of our species with others in the animal world. . . . [M]an's freedom from the constraints of animal instincts . . . makes

possible our acquisition of speech and reason . . . [and] transforms man into the 'first freeborn of creation.'" Wolfgang Pross, "Naturalism, Anthropology, and Culture," in *The Cambridge History of Eighteenth-Century Political Thought*, ed. Mark Goldie and Robert Wokler (Cambridge: Cambridge University Press), 218–47, citing 244. Thus, Pross summarizes, Herder was committed to the following view: "Nature brings forth life on this earth, which follows a pattern of development until it reaches the most complicated form of organization, in man; human life, therefore, cannot be regarded as set apart from other forms of animate beings" (234).

209. Herder, *Treatise on the Origin of Language*, 85.

210. J. G. Herder, *Vom Erkennen und Empfinden den zwo Hauptkräften der Menschlichen Seele* (1775), and *Vom Erkennen und Empfinden der menschlichen Seele: Bemerkungen und Träume* (1778), in *Herders Sämmtliche Werke*, ed. Bernhard Suphan, vol. 8 (Berlin: Weidmann, 1892), 263–333, 165–262, here 180. Herder's 1778 version, plus the preface to the 1775 version, can be found in English translation in Michael Forster, ed., *Herder's Philosophical Writings* (Cambridge: Cambridge University Press, 2002), 178–243. Hereafter, references to this source will have the form Herder, *Vom Erkennen und Empfinden* (year), page number, trans. page number.

211. Herder, *Vom Erkennen und Empfinden* (1775), 265, trans. 181.

212. J. G. Herder, *Journal meiner Reise im Jahre 1769*, in *Werke in zehn Bänden* (Frankfurt: Deutsche Klassiker Verlag, 1985–2000), 9:119.

213. "Thus, from the reworking of the impressions of sight arise the ideas of distance, space, and substance, from the reworking of those of hearing arise the representations of succession and thus of time, and finally from the feeling of touch, the representation of unity and multiplicity and also of cause and effect." Wolfgang Pross, "Commentary on *Viertes Kritischen Wäldchen*," in J. G. Herder, *Werke*, vol. 2, *Herder und die Anthropologie der Aufklärung*, ed. Wolfgang Pross (Munich: Hanser, 1987), 864–84, citing 876.

214. He modeled his approach explicitly on Diderot: "Diderot can be the model for making experiments but not simply to build on his experiments and to systematize them! A work of that sort can become the first psychology, and since from this all the [other] sciences follow, simultaneously a philosophy or encyclopedia for all that!" (Herder, *Journal meiner Reise im Jahre 1769*, 110).

215. Wolfgang Pross, "Herder und die Anthropologie seiner Zeit," in Herder, *Werke*, 2:1128–1216, citing 1182, 1187; Raymond Immerwahr, "Diderot, Herder, and the Dichotomy of Touch and Sight," *Seminar* 14 (1978): 84–96; Jean Chabbert, "Le jeune Herder et Diderot: Une relation paradoxale?," *Beiträge zur romanischen Philologie* 24 (1985): 281–87.

216. *Condillac's Treatise on the Sensations* (1754), trans. Carr.

217. J. G. Herder, *Viertes kritischen Wäldchen*, in *Werke in zehn Bänden*, 2:325.

218. Ibid.

219. Hence, the representation of Herder as a notorious anti-French cultural nationalist needs considerable nuancing. That he was committed to a German cultural revival entailing liberation from overriding French influence is unquestionably true. But at the same time, he felt that the creation of this new German cultural efflorescence would require discriminating adaptation of foreign—and notably French—cultural resources.

220. Regine Otto, "Herder auf dem Weg zu Spinoza," *Weimarer Beiträge* 10 (1978): 165–77.

221. Pross, "Naturalism, Anthropology, and Culture," 225. Pross aligns Herder's views with those of Spinoza in the *Theological-Political Treatise*: "Neither man's reason nor God's providence . . . regulated the course of history, which is determined rather in a totally arbitrary way by mankind's instincts and passions of self-preservation" (228).

222. For Pross, Herder's new approach, "notwithstanding all lip service to a faith in providence[, encompasses] a vision of the world and history that runs diametrically counter to the established dogmatic faith in a divine guidance of the universe extending into the life of the individual, who profits from this divinity." J. G. Herder, *Werke*, vol. 3.1, *Ideen zur Philosophie der Geschichte der Menschheit*, ed. Wolfgang Pross (Munich: Hanser, 2002), 881. See my commentary "Herder and Naturalism: The Views of Wolfgang Pross Examined," in *Herder and Religion*, ed. Staffan Bengtsson,

Heinrich Clairmont, Robert E. Norton, Johannes Schmidt, and Ulrike Wagner (Stuttgart: Metzler, 2015), 13–24.

223. As Adler explains, "Psychology should not be structured according to presupposed faculties, in order to subordinate affect and sensations under them, but rather, all that in actuality manifests itself in affects and sensations needs in the first place to be observed [*beobachtet*] and grasped as unique, individual. 'No speculating, only gathering' is Herder's demand for a 'human science.'" Hans Adler, *Die Prägnanz des Dunklen: Gnoseologie, Ästhetik, Geschichtsphilosophie bei J. G. Herder*, Studien zum 18. Jahrhundert, vol. 13 (Hamburg: Meiner, 1990), 71.

224. "The sense of touch is according to Herder the sense for the inner forces of the soul, through which it constructed its body, hence the sense for the force of attraction of one's own soul and the force of repulsion of the world, the consciousness of the strivings of the self and at the same time the consciousness of being limited by external forces. On the basis of the universal claim of this statement, the sense of touch is the metaphysical sense." Reiner Wisbert, "Commentary on 'Gesetze der Welt, Gesetze der Körper,'" in Herder, *Werke in zehn Bänden*, 9:1077.

225. Herder, *Zum Sinn des Gefühls*, in *Werke in zehn Bänden*, 1:111.

226. Ibid.

227. Ibid., 106.

228. As Hagner has observed aptly, "Haller's physiology and its success expressed the existing need to find a middle ground on the question of the relevance of nature for the determination of the place of man between materialism and animism of the Stahlian variety." Michael Hagner, "Aufklärung über das Menschenhirn: Neue Wege der Neuroanatomie im späten 18. Jahrhundert," in *Der ganze Mensch: Anthropologie und Literatur im 18. Jahrhundert; DFG Symposion 1992*, ed. Hans-Jürgen Schings (Stuttgart: Metzler, 1994), 147. He goes on to claim, "Herder appears among the earliest to grasp this process of transformation . . . taking account of a metaphysical as well as a natural-historical determination of man" (148). See also Johann Marbach, "Beiträge Albrecht von Hallers zu Herders Anthropologie unter besonderer Berücksichtigung sprachlich-literarischer Aspekte," *Annali* 7 (1964): 41–60.

229. Herder, *Vom Erkennen und Empfinden* (1775), 265, trans. 181.

230. Ibid., 266, trans. 182.

231. Ibid., 277 (not translated).

232. Ibid., 27, 3 (not translated).

233. Ibid., 272. See the slight revision of 1778 in trans. 189.

234. Ibid. (1778), 178, trans. 195.

235. Ibid. (1775), 276 (not translated).

236. Ibid., 277 (not translated).

237. Ibid. (1778), 185, trans. 201.

238. Ibid.

239. Nisbet is outraged that Herder "completely misunderstood Haller, and ignored almost everything of scientific value in his work." H. B. Nisbet, *Herder and the Philosophy and History of Science* (Cambridge: Modern Humanities Research Association, 1970), 265. Herder "misinterpreted Haller's vitalism, linking it to his own metaphysics of '*Kräfte*' rather than to precise physiological functions, as Haller had intended" (257). Could Herder have been so ignorant of what Haller was about? Had he failed to understand the context of Haller's work in the life sciences of the day? Hardly. Nisbet seems insensitive to the theoretical desiderata of the physiological and psychological sciences in extricating themselves from the palpable *limitations* of Haller's intentions.

240. Herder to Johann Heinrich Merck (1741–91), September 1771: "I have thoroughly studied his [Haller's] sections on physiology (senses, powers of the soul, and economy of life)." Herder went on to add that while he had the greatest respect for Haller, he could not bring himself to "enthusiasm" for the man (cited in Nisbet, *Herder and the Philosophy and History of Science*, 263).

241. Simon Richter, "Medizinistischer und ästhetischer Diskurs im 18. Jahrhundert: Herder und Haller über Reiz," *Lessing Yearbook* 25 (1993): 83–95, citing 89.

242. Ibid., 88.

243. Ibid., 83.

244. Kant would therefore seek in his *Kritik der Urteilskraft* systematically to ban *Reiz* from "pure" aesthetic judgments, which should be "disinterested." On the residual fear of feminine beauty and allure that informed this Kantian strategy in the third *Critique*, see Susan Shell, *The Embodiment of Reason: Kant on Spirit, Generation, and Community* (Chicago: University of Chicago Press, 1996), 219–24.

245. Richter, "Medizinistischer und ästhetischer Diskurs," 90.

246. Kurt Danziger, "Generative Metaphor and the History of Psychological Discourse," in *Metaphor in the History of Psychology* (Cambridge: Cambridge University Press, 1990), 331–56. For Herder, analogy, metaphor, observation, and experience, bringing thought back down to earth and concerning oneself with the whole man, were the sum and substance of thinking in an enlightened manner. For this "epistemological liberalization" in Enlightenment thought, see Sergio Moravia, "The Enlightenment and the Science of Man," *History of Science* 18 (1980): 247–68; and for Herder specifically, see my monograph *Kant, Herder, and the Birth of Anthropology* (Chicago: University of Chicago Press, 2002).

247. Herder, *Vom Erkennen und Empfinden* (1775), 277, trans. 197n.

248. Ibid. (1778), 180, trans. 196.

249. Imre Lakatos, "Falsification and the Methodology of Scientific Research Programmes," in *Criticism and the Growth of Knowledge*, ed. Imre Lakatos and Alan Musgrave (Cambridge: Cambridge University Press, 1970), 91–196.

250. See my "'Method' vs. 'Manner'?—Kant's Critique of Herder's *Ideen* in Light of the Epoch of Science, 1790–1820," *Herder Yearbook*, 1998, 1–25.

251. This is further confirmation of Toellner's notion that there is a paradigm shift around Haller—but, ironically, against his own intentions. See Richard Toellner, "Mechanismus—Vitalismus: Ein Paradigmenwechsel? Testfall Haller," *Studien zur Wissenschaftsgeschichte* 10 (1977): 61–72.

252. Ilse Jahn, "Wer regte Caspar Friedrich Wolff (1734–1794) zu seiner Dissertation 'Theoria generatonis' an?," cahier spécial, *Philosophia Scientiae* 2 (1998–99): 35–54; Ilse Jahn, "Maupertuis zwischen Präformations- und Epigenesis-theorie."

253. Shirley Roe, "Rationalism and Embryology: Caspar Friedrich Wolff's Theory of Epigenesis," *Journal of the History of Biology* 12 (1979): 1–43; Roe, *Matter, Life and Generation.*

254. For other important work on Wolff, see Richard Aulie, "Caspar Friedrich Wolff and His 'Theoria Generationis,' 1759," *Journal of the History of Medicine* 16 (1961): 124–44; Julius Schuster, "Der Streit um die Erkennthis des organischen Werdens im Lichte der Briefe C. F. Wolffs an A. von Haller," *Sudhoffs Archiv* 34 (1941): 196–218, an account marred by its German *völkisch* rhetoric but accompanied by German translations of Wolff's crucial correspondence with Haller; Georg Uschmann, *Caspar Friedrich Wolff: Ein Pionier der modernen Embryologie* (Leipzig: Urania, 1955); Robert Herrlinger, "C. F. Wolffs 'Theoria generationis' (1759): Die Geschichte einer epochenmachendedn Dissertation," *Zeitschrift für Anatomie und Entwicklungsgeschichte* 121 (1959): 245–70; Gerhard Müller, "La conception de l'épigénèse chez Caspar Friedrich Wolff (1734–1794)," *Rivista di biologia* 77 (1984): 343–62; Tat'jana Lukina, "Caspar Friedrich Wolff und die Petersburger Akademie der Wissenschafter," *Acta Historica Leopoldina* 9 (1975): 411–25, with a detailed account of Wolff's long years of service at the Saint Petersburg Academy of Sciences and of his research projects during that time; A. E. Gaissinovich, "C. F. Wolff on Variability and Heredity," *History and Philosophy of the Life Sciences* 12 (1990): 179–201, a consideration of these materials from Wolff's Saint Petersburg years; A. E. Gaissinovich, "Le rôle du Newtonianisme dans la renaissance des idées épigénétiques en embryologie du XVIIIe siècle," in *Actes du XIe Congrès international d'histoire des sciences (1965)* (Wrocław: Ossolineum, 1968), 5:105–10, the only accessible presentation of Gaissinovich's Russian monograph on Wolff, which is beyond my linguistic skills; Reinhard Mocek, "Caspar Friedrich Wolffs Epigenesis-Konzept—ein Problem im Wandel der Zeit," *Biologisches Zentralblatt* 114 (1995): 179–90.

255. Gaissinovich, "Le rôle du Newtonianisme."

256. Roe observes that "the biological sciences in the eighteenth century . . . were experiencing fundamental changes . . . not due primarily to new discoveries or an influx of new data. Rather, biology was undergoing an intense philosophical reexamination" ("Rationalism and Embryology," 1). There is no question that there was a major philosophical dimension, but it is not clear that new discoveries and data played so little a role. Here, I find more persuasive Peter Hanns Reill's characterizations of the new paradigm of vital materialism. Peter Hanns Reill, *Vitalizing Nature in the Enlightenment* (Berkeley: University of California Press, 2005). Yet Roe hits the crucial connection: "Could the properties of matter themselves be responsible for the activities of a living creature? If so, what would this imply for the existence of a soul in man, and a Creator of the universe?" ("Rationalism and Embryology," 1). That was exactly the linkage this whole chapter seeks to elucidate between life science and vital materialism: the threat of "atheism" felt by a whole range of figures at the time. Roe is equally apt in explaining preformation's popularity in the epoch 1650–1750 along these ideological lines: if organisms were all original creations of God, preformed at the moment of creation, then all that needed to be accounted for in natural science was simple enlargement: "the preformationist could explain embryological development mechanistically without fear of materialism" ("Rationalism and Embryology," 2).

257. See Shirley Roe, "*Anatomia Animata*: The Newtonian Physiology of Albrecht von Haller," in *Transformations and Tradition in the Sciences*, ed. Everett Mendelsohn (Cambridge: Cambridge University Press, 1984).

258. Biographical information on Wolff is drawn from Uschmann, *Caspar Friedrich Wolff*, and other sources.

259. On the education of persons of this social background in eighteenth-century Prussia, see Anthony La Vopa, *Grace, Talent and Merit: Poor Students, Clerical Careers, and Professional Ideology in Eighteenth-Century Germany* (Cambridge: Cambridge University Press, 1988).

260. As Roe notes, "Wolff's dissertation was unusual in length, originality of content, and the fact that the name of his dissertation director was not printed on the title page" ("Rationalism and Embryology," 5n). *That* it was published, immediately and in Halle, deserves additional attention. This, too, seems strikingly unusual.

261. Caspar Friedrich Wolff, "Notizen von C. F. Wolff über die Bemerkungen der Opponenten zu seiner Dissertation," discussed first by A. E. Gaissinovich ("Notizen von C. F. Wolff über die Bemerkungen der Opponenten zu seiner Dissertation," *Wissenschaftliche Zeitschrift der Friedrich-Schiller-Universität Jena: Mathematische-Naturwissenschaftliche Reihe* 6, nos. 3/4 [1956/57]: 121–24) and then, more extensively, by Jahn and reproduced in Jahn, "Wer regte Caspar Friedrich Wolff . . . an?"

262. Jahn, "Wer regte Caspar Friedrich Wolff . . . an?," 43.

263. Ibid., 47–50. Jahn notes that Maupertuis's *Lettres sur les sciences* was a text that Voltaire pilloried mercilessly in his *Doktor Akakia* pamphlet (1752), creating a great brouhaha especially in Berlin. That made it not only very likely that a young intellectual like Wolff would have been intrigued to see what Maupertuis had written, but also, as Jahn notes at the end of her essay, not something that he would wish openly to cite or be identified with (51).

264. Ibid., 50–51. Maupertuis also wrote a great deal about "monsters"—birth deformities, etc.—and this became one of Wolff's main areas of research in Saint Petersburg, where a remarkable collection of such anomalies had been collected in the imperial museum.

265. Ibid., 49–50; Jahn, "Maupertuis zwischen Präformations- und Epigenesis-theorie," 97.

266. Crucial here was Haller's introduction to the second volume of the translation of Buffon's *Histoire naturelle*, which had already appeared as a freestanding publication in French: *Réflexions sur le système de la génération de M. de Buffon*.

267. Cited in Jahn, "Maupertuis zwischen Präformations- und Epigenesis-theorie," 98, from Harnack's history of the Prussian Academy.

268. Ibid.

269. See the rationales offered by the various members of the Collegium Medico-Chirugicum

for denying Wolff the opportunity to offer courses in their institution. That he was nonetheless allowed to offer private instruction only more fiercely embittered them. Street fights between their respective student bodies ensued. We have a detailed account of this unfortunate set of developments from Wolff's assistant and admirer of these years, Christian Ludwig Mursinna, which Goethe saw into publication: Christian Ludwig Mursinna, "Caspar Friedrich Wolffs erneuertes Andenken," in *Zur Morphologie*, ed. J. W. von Goethe (Stuttgart: Cotta, 1817–21), 1:252–56, cited in Roe, "Rationalism and Embryology," 4n.

270. The Meckel family remained bound up with the reception of Wolff's work in Germany notwithstanding the bitter hostility that the elder Meckel manifested toward him. See Bernd Fischer, "Die halleschen Anatomen Meckel Vater und Sohn: Genial oder skurril?," in *Bedeutende Gelehrte der Universität Halle seit ihrer Gründung im Jahre 1694*, ed. Hans-Hermann Hartwick and Gunnar Berg (Opladen: Leske and Budrich, 1995), 75–80. Philip Meckel edited (anonymously) a second, annotated edition of Wolff's dissertation in 1774: *Theoria generationis . . . editio nova, aucta et emendata* (Halle, 1774). There is now a detailed consideration of Philip Meckel's contributions in Rüdiger Schultka, Josef Neumann, and Susanne Weidemann, eds., *Anatomie und anatomische Sammlungen im 18. Jahrhundert* (Berlin: LIT Verlag Dr. W. Hopf, 2007). And, of course, Johann Friedrich Meckel the Younger translated and introduced Wolff's great monograph on the development of the intestine in 1812: *Über die Bildung des Darmkanals im bebrüteten Hünchen* (Halle: Renger, 1812), inaugurating the belated but enthusiastic reception of Wolff among German life scientists, including such eminences as Pander, Goethe, and von Baer. See Arthur William Meyer, *Human Generation: Conclusions of Burdach, Döllinger and von Baer* (Stanford: Stanford University Press, 1956).

271. Roe, "Rationalism and Embryology," 4–5.

272. Euler's recommendation of Wolff to the Saint Petersburg Academy characterized his dissertation as "having built its teachings on entirely new grounds and experiments and aroused the wonder of all the experts [*von allen Kennern bewundert*]" (cited in Jahn, "Wer regte Caspar Friedrich Wolff . . . an?," 43n).

273. Ibid., 45–47.

274. Ibid.

275. See the anonymous review of Reimarus's *Vormehmsten Wahrheiten* in the *Bibliothèque raisonnée* and my discussion above.

276. The opening line of his review is widely cited, and it was welcomed with great joy by Wolff. What followed is a different matter. See Julius Schuster, "Der Streit um die Erkennthis des organischen Werdens im Lichte der Briefe C. F. Wolffs an A. von Haller," *Sudhoffs Archiv* 34 (1941): 196–218.

277. The first letter to Haller, which accompanied a copy of the dissertation, gives a clear indication of Wolff's hopes. The dissertation itself has no armature of self-defense against so formidable an opponent, which prudence would have dictated had he anticipated such opposition. The later *Theorie von der Generation* (1764) demonstrates that Wolff was quite capable of developing such a defense—indeed, a counterattack—when he understood its necessity.

278. We now have a very fine analysis of Haller's stance toward Buffon in the early 1750s which suggests that Haller was a very long time in converting, and that one could still find strong evidence of epigenetic leanings in Haller even in his critique of Buffon: Cherni, "Haller et Buffon."

279. Haller's *Sur la formation du cœur dans le poulet: sur l'œil, sur la structure du jaune, etc.*, 2 vols. (Lausanne: Bousquet) appeared in 1758, obviously in French. How swiftly this became accessible to the research community in Berlin and Halle is not altogether clear. There is no mention of Haller's work on chicken embryos in Wolff's dissertation, though this was the dissertation's central concern, and Wolff did cite Malpighi on chicken embryos as well as other works by Haller, so it would seem that had he known of Haller's monograph, he would have referenced it.

280. Caspar Friedrich Wolff, *De Formatione Intestinorum = La formation des intestins: 1768–1769*, ed. and trans. Jean-Claude Dupont (Turnhout: Brepols, 2003).

281. Roe ("Rationalism and Embryology," 12n) cites from Haller's review in *GGA*, 1770 and 1771.

282. Albrecht von Haller, *Elementa Physiologiae Corporis Humani*, 8 vols. (Lausanne: Bousquet, 1757–66), vol. 8 (1766).

283. Wolff's response to Haller's *Elementa* publication is translated and reprinted in Schuster, "Der Streit um die Erkennthis des organischen Werdens im Lichte der Briefe C. F. Wolffs an A. von Haller," 210–13.

284. Wolff to Haller, October 6, 1766, translated and reprinted in ibid., 212.

285. Wolff to Haller, April 17, 1767, translated and reprinted in ibid., 213–14.

286. Otto Sonntag, ed., *The Correspondence between Albrecht von Haller and Charles Bonnet* (Bern: Huber, 1983); Charles Bonnet, *Considérations sur les corps organisées* (1779 ed.); Charles Bonnet, *Contemplation de la nature* (1782 ed.).

287. "Original production, in all systems, is a miracle," Maupertuis averred (*Système de la nature*, 157). The issue for natural philosophy lay elsewhere: "the universe once formed, by what laws is it preserved? What are the means whereby the Creator provided for the reproduction of individuals that perish? Here we have a free field [for inquiry]" (155).

288. Wolff to Haller, April 17, 1767, translated in Roe, "Rationalism and Embryology," 17 (translation amended).

289. Roe, "Rationalism and Embryology," 13.

290. Ibid. Roe refers to Tore Frängsmyr, "Christian Wolff's Mathematical Method and Its Impact on the Eighteenth Century," *Journal of the History of Ideas* 36 (1975): 653–68, among others. That Christian Wolff's views had an impact is unquestionable. That he was disposed to a mathematicized model of general knowledge is also valid. But his actual conduct of thought, most exemplarily in the relation between empirical and rational psychology, which Caspar Friedrich Wolff explicitly invokes, was far more complex and eclectic.

291. Roe, "Rationalism and Embryology," 14.

292. Werner Schneiders, ed., *Christian Wolff, 1679–1754: Interpretationen zu seiner Philosophie und deren Wirkung mit einer Bibliographie der Wolff-Literatur*, Studien zum achtzehnten Jahrhundert, vol. 4 (Hamburg: Meiner, 1983).

293. See Giorgio Tonelli, "Der Streit über die mathematische Methode in der Philosophie in der ersten Hälfte des 18. Jahrhunderts und die Entstehung von Kants Schrift über die 'Deutlichkeit,'" *Archiv für Philosophie* 9 (1959): 37–66. The best evidence came in the famous prize competition query of the Berlin Academy for 1762, the apex moment for the German Hochaufklärung. See Paul Guyer, "Mendelssohn and Kant: One Source of the Critical Philosophy," *Philosophical Topics* 19 (1991): 119–52.

294. Roe, "Rationalism and Embryology," 14.

295. Ibid.

296. Ibid., 15, citing Wolff, *Theoria Generationis* (1759), 8.

297. Willem de Jong and Arianna Betti, "The Classical Model of Science I: A Millennia-Old Model of Scientific Rationality," *Synthese* 174, no. 2 (May 2010): 185–203; Willem de Jong and Arianna Betti, "The Classical Model of Science II: The Axiomatic Method, the Order of Concept and the Hierarchy of Sciences," *Synthese* 183, no. 1 (November 2011): 1–6.

298. Gaissinovich, "Le rôle du Newtonianisme," 108–9.

299. Olaf Breidbach, "Die Geburt des Lebendigen—Embryogenese der Formen order Embryogene der Natur?—Anmerkungen zum Bezug von Embryologie und Organismustheorien vor 1800," *Biologisches Zentralblatt* 114 (1995): 191–99.

300. Roe, "Rationalism and Embryology," 13.

301. Caspar Friedrich Wolff, *Theoria Generationis*, trans. into German by Paul Samassa, Ostwald's Klassiker der exakten Wissenschaften, 84–85 (Leipzig: Engelmann, 1896), §§253–55, 70–76.

302. Roe, "Rationalism and Embryology," 13.

303. See, above, chap. 1.

304. Roe, "Rationalism and Embryology," 20, 20n.

305. Wolff, *Theoria Generationis*, §255, 81; English trans. in Roe, "Rationalism and Embryology," 20.

306. Gaissinovich, "Le rôle du Newtonianisme," 109.

307. See my discussion of Stahl's concept of the soul in chapter 1.

308. See the special issue of *Studies in History and Philosophy of the Medical and Biomedical Sciences* 37, no. 4 (2006).

309. In that light, the procedure of Wolff's argumentation here is doubly salient. That is, first it is constructed empirically, and second, it formulates a model in one domain in order to extend it more widely and inclusively. This is a procedure that we can recognize in contemporary theory of science as well.

310. Not only was Hales a pioneer in plant physiology (*Vegetable Statics* [London, 1723]), but he was also a prominent spokesman for experimental Newtonianism, as we have seen in my discussion of Buffon's French translation and its cultural context. See Henri Guerlac, "The Continental Reputation of Stephen Hales," *Archives internationales de l'histoire des sciences* 4 (1951): 393–404. This immediate reference to Hales makes that experimental Newtonian tradition more relevant to our understanding of Wolff than has hitherto been considered.

311. That was standard scientific judgment then (and even seems straightforward enough now).

312. Wolff, *Theoria Generationis*, §84, 11.

313. Newton's four rules of method in *Mathematical Principles of Natural Philosophy*, in *Newton's Philosophy of Nature: Selections from His Writings*, ed. H. S. Thayer (New York: Haffner, 1953), 3–5.

314. That fluids have no structure was a principle insisted upon by all, and thus an organism could not be completely fluid.

315. Wolff, *Theoria Generationis*, §84, 17.

316. Hume's distinction of inference from deduction was felt strongly in the German context, particularly in the Berlin Academy, by the time Wolff was writing.

317. Wolff, *Theoria Generationis*, §84, 17.

318. Ibid., 19.

319. Ibid., 21.

320. Ibid., 22.

321. There is something breezily self-confident in Wolff's spinning out of definitions of method and substance in these opening propositions. He does not appear to think his procedure here, perhaps in contrast to his results, is controversial.

322. There is a monumental precedent in Newton's *Mathematical Principles*, which presented his newly invented calculus in the guise of traditional geometrical demonstrations.

323. Caspar Friedrich Wolff, *Theorie von der Generation* (1764), in *Theorie von der Generation in zwei Abhandlungen erklärt und bewiesen, Theoria generationis, Mit einer Einführung von Robert Herrlinger* (Hildesheim: Olms, 1966), preface (unpaginated, [4–5]).

324. He quickly revised this judgment, admitting that he was now aware of one such theory—proferred by Descartes, who had a correct understanding of what an explanation required. However, Descartes's theory was all wrong (ibid., 6).

325. Ibid., 8. The distinction of philosophical from historical knowledge was, to be sure, a talisman of Wolffianism, but it was not exclusively his, nor was his construction of the distinction the only one. Kant, for example, used it extensively. See Kant, *Kritik der reinen Vernunft*, A836/B864.

326. Wolff, *Theorie von der Generation*, 3.

327. Ibid., 8.

328. Ibid.

329. Ibid.

330. Ibid., 9.

331. Ibid., 10.

332. That seems to be an uptake of Haller's definition of physiology as *anatomia animata*. See Erich Hintschke, "Anatomia Animata," *Ciba Zeitschrift* 10 (1948): 4042–73; Lloyd Stevenson, "Anatomical Reasoning in Physiological Thought," in *The Historical Development of Physiological Thought: A Symposium Held at the State University of New York, Downstate Medical Center* (New York: Hafner, 1959), 27–38.

333. Wolff, *Theorie von der Generation*, 13.

334. Ibid., 11.

335. On Christian Wolff's empirical versus rational psychology, see Charles Corr, "Wolff's Distinction between Empirical and Rational Psychology," in *Akten des II. Internationalen Leibniz-Kongresses Hannover, 17.–22. Juli 1972*, Studia Leibnitiana, Supplementa 14 (Wiesbaden: Steiner, 1972), 3:195–215.

336. Maria Monti, *Congettura ed esperienza nella fisilogia di Haller* (Florence: Olschki, 1990).

337. Wolff, *Theoria Generationis*, trans. Samassa, pt. 2, §85, 1.

338. Ibid., 5.

339. Ibid.

340. Ibid., 60.

341. Ibid., 6.

342. Roe, "Rationalism and Embryology," 21.

343. Ibid., 7.

344. Wolff, *Theorie von der Generation*, 103.

345. This was the view of Haller and Buffon before, and of Kant and Blumenbach after, Caspar Friedrich Wolff. It goes back to Locke's discrimination between real and nominal essences and the explicit epistemological restriction that distinction placed on the claims of natural science. Gaissinovich has made an excellent case for the prominence of this view of experimental Newtonianism, and he highlights C. F. Wolff's participation in it.

346. The clearest and most insistent formulation is by Bonnet.

347. Wolff, *Theorie von der Generation*, 103.

348. Wolff, *Theoria Generationis*, §85, 6.

349. Ibid.

350. Albrecht von Haller, review of *Theoria Generationis*, by Caspar Friedrich Wolff, *GGA* 143 (November 29, 1760): 1226–31.

351. Wolff, *Theorie von der Generation*, 79.

352. Ibid., 50ff., on Bonnet, *Considérations sur les corps organisés*.

353. Wolff, *Theorie von der Generation*, 47–48.

354. Wolff, *Theoria Generationis*, §85, 48.

355. Ibid., 49.

356. Haller, review of *Theoria Generationis*.

357. Wolff, *Theorie von der Generation*, 28–30.

358. Wolff, *Theoria Generationis*, 49.

359. On the prominence of the nutritive in Wolff's theory and on the prize contest of Saint Petersburg, see Caspar Friedrich Wolff, *Von der eigenthümlichen und wesentlichen Kraft der vegetabilischen, sowohl als auch der animalischen Substanz* (St. Peterburg: Kayserlichen Academie der Wissenschaften, 1789).

360. Wolff, *Theorie von der Generation*, 73. This is the vein that Olaf Breidbach ("Die Geburt des Lebendigen") had mined to grasp Wolff in terms of an immanent teleology with Spinozistic origins and *Naturphilosophisch* repercussions.

361. Wolff, *Theorie von der Generation*, 117.

362. Ibid., 102.

363. This emerges in Wolff's letter to Haller, October 6, 1766 (in Schuster, "Der Streit um die Erkennthis des organischen Werdens im Lichte der Briefe C. F. Wolffs an A. von Haller," 211–12), in which he defends himself from this aspersion.

364. Johann August Unzer, *Erste Gründe einer Physiologie der eigentlichen thierischen Natur thierischer Körper* (Leipzig: Weidmann, 1771). While I have verified against this German original, I will be quoting from the mid-nineteenth-century abridged translation by Thomas Laycock, *"The Principles of Physiology" by John Augustus Unzer and "A Dissertation on the Functions of the Nervous System" by George Prochaska* (London: Sydenham Society, 1851), which proved the vehicle for Unzer's long-term reception in neurophysiology, long after his immediate influence had been spent.

365. Thomas Laycock, "Translator's Introduction," to *"Principles of Physiology" by John Augustus Unzer*, ii.

366. Thomas Broman, *Transformation of German Academic Medicine* (Cambridge: Cambridge University Press, 1996).

367. See Laycock, "Translator's Introduction," ii.

368. Unzer, *Erste Gründe einer Physiologie*, preface, trans. 16.

369. Ibid.

370. Ibid., trans. 12.

371. Ibid., §13, trans. 18. One of the English translator's strategies for abridgement was precisely to excise the extensive Haller citations. Unzer claimed that he had "nothing to add" (preface, trans. 7) to Haller's anatomical account of the nervous system, including his description of the nerves as hollow tubes with a liquid core.

372. Ibid., preface, trans. 4.

373. Ibid., §379, trans. 207.

374. Ibid., §162, trans. 88.

375. Ibid., §25, trans. 25–27, for Unzer's incorporation of Wolffian faculty-psychology from Baumgarten's *Metaphysica*.

376. Ibid., §432, trans. 232.

377. Roger K. French, "The Controversy with Haller: Sense and Sensibility," in *Robert Whytt, the Soul, and Medicine* (London: Wellcome Institute of the History of Medicine, 1969), 63–76.

378. Unzer, *Erste Gründe einer Physiologie*, §10, trans. 17.

379. Ibid., preface, trans. 5; §351, trans. 186. He never mentioned names when referencing the materialists, but we can presume that these would have included La Mettrie, Helvétius, and perhaps d'Holbach, with his newly published European shocker *System of Nature* (1770).

380. Unzer, *Erste Gründe einer Physiology*, §1, trans. 17.

381. We should note this language. Unzer extensively invoked the idea of the animal as *machine*, but he upheld Leibniz's distinction between organic machines and human artifacts in their infinite complexity. The idea of "animal economy" had additional resonances, driving toward a unified conception of the organism as against the sharp dualism advocated by Haller.

382. Ibid., §361, trans. 194. The general animal economy entailed, as well, basic physical and mechanical elements (e.g., hydraulics) and shared irreducibly organic forces (e.g., metabolism) present in other organisms such as plants (ibid., preface, trans. 13; §600, trans. 309). But animals had something more (and humans, something more still).

383. Ibid., preface, trans. 12, 6, translation amended.

384. Haller recognized the physiology of nerves, Unzer noted, "but there he has stopped. I have ventured to extend the outline" (ibid., preface, trans. 9).

385. Alexander Monro II, *Observations on the Structure and Function of the Nervous System* (Edinburgh, 1783); Felice Fontana, *Treatise on the Venom of the Viper* (London: Murray, 1787), originally published in 1781 in French.

386. Charles Bonnet, *Essai de psychologie* (1754); Charles Bonnet, *Essai analytique sur les facultés de l'âme* (1760). See Ed Claparède, *La pscyhologie animale de Charles Bonnet* (Geneva: Georg, 1909); Fernando Vidal, "La psychologie de Charles Bonnet comme 'miniature' de sa metaphysique," in *Charles Bonnet: Savant et philosophe (1720–1793)*, ed. Marino Buscaglia, René Sigrist, Jacques Trembley, and Jean Wüest (Geneva: Editions Passé Présent, 1993), 43–50; Patrick Baud, "L'âme et les sensations selon Charles Bonnet," *Gesnerus* 49 (1991): 323–32. Unzer dismissed as misguided Bon-

net's attempt "to analyze the different mental faculties by means of movements in the brain, of which we are entirely ignorant" (*Erste Gründe einer Physiology*, preface [trans., 8]); one could not *derive* the mental from the physical, as Bonnet attempted (§25 [trans., 26]).

387. Laycock, translator's note, 29.

388. Unzer, *Erste Gründe einer Physiology*, §623, trans. 317.

389. Ibid., §403, trans. 218.

390. Ibid., preface, trans. 4.

391. "Everything is ordered according to laws altogether unknown, and which we can never fathom" (ibid., §40, trans. 37). "No one knows how the structure of the brain is adapted to material ideas, how these ideas are formed in it, or what is their nature, or how the vital spirit is formed in them, or in what the animal-sentient force of the brain differs from its primary vital force" (§719 [trans., 358]). "We know nothing of the nature of the cerebral forces, or of the mode in which the conceptions [*Vorstellungen*] excite the cerebral functions. . . . But how all this happens *psychologically*, is taught by metaphysics" (ibid., §111 [trans., 61]).

392. "The nerves have in themselves no visible movements. . . . We infer their existence in the external impressions from their action on the brain" (ibid., §145 [trans., 77]). This was an application of the general methodological principle of inferring forces from their effects (ibid., §190 [trans., 101]).

393. Ibid., §15, trans. 20.

394. "What in [lower animals] appears to be volitional, only *appears* so, because we draw conclusions as to other animals from the nature and working of our own minds" (ibid., §625, trans. 323).

395. Ibid., §25, trans. 25.

396. Ibid., §15, trans. 19.

397. Unzer recognized the theoretically central issue of the velocity of nerve transmission, even if he had no account for it.

398. Ibid., §126; trans. 66–67.

399. Ibid., §48, trans. 41; §460, trans. 244; §469, trans. 248.

400. Ibid., §400, trans. 216.

401. Ibid., §557, trans. 285.

402. Ibid., §366, trans. 198.

403. Ibid., §642, trans. 331.

404. The idea "that the soul was diffused throughout the entire organism" had been developed by the ancients and "renewed by Whytt" (ibid., §404, trans. 218), but it was utterly erroneous, Unzer insisted: "Their souls must be extended, and be everywhere present in their bodies, since polypes may be cut into pieces, and each piece becomes a new animal. . . . How opposed is all this to sound theory and to common sense!" (ibid. §624 [trans., 319]).

405. Ibid., §621, trans. 317.

406. Hence the classic etymological bond of *anima* to "animal" in the tradition of "psychology" dating back to Aristotle's key texts and promulgated in European university curricula from the medieval period all the way to the eighteenth century. See Francisco Vidal, *The Sciences of the Soul: The Early Modern Origins of Psychology* (Chicago: University of Chicago Press, 2011).

407. Unzer, *Erste Gründe einer Physiology*, §623, trans. 317.

408. Instincts were analogous to external stimulations in their involuntary force and even more deterministic in terms of their behavioral compulsion. Unzer followed Reimarus in denying that these were learned responses or that animals had any rational insight or subjective wisdom directing their behavior (ibid., §90 [trans., 57]; elaborated in §§262ff. [trans., 130–32]).

409. Ibid., §262, trans. 130.

410. Ibid., §262, trans. 132.

411. Ibid., §299, trans. 163.

412. Ibid., §296, trans. 162; §298, trans. 162.

413. Ibid., §648, trans. 332–33.

414. Ibid., §171, trans. 93–94.

415. Ibid., §647, trans. 332.

416. Anonymous, *GGA* 65 (1772). That the review of Unzer's monograph in *GGA* was by Haller was recognized by Laycock already in 1851 (translator's note, 34).

417. Johann August Unzer, *Physiologische Untersuchungen: Auf Veranlassung der Göttingischen, Frankfurter, Leipziger und Hallischen Recensionen seiner Physiologie der thierischen Natur* (Leipzig: Weidmann, 1773), 7–62.

418. Johann Nicolaus Tetens, *Philosophische Versuche über die menschliche Natur und ihre Entwickelung*, 2 vols. (Leipzig: Weidman, 1777).

419. Johannes Speck notes that it deflated the enormous enthusiasm for mechanistic psychology on the model of Bonnet in German philosophical circles, especially at Göttingen (Feder, Meiners, Hißmann). See Johannes Speck, "Bonnets Einwirkung auf die deutsche Psychologie des vorigen Jahrhunderts," *Archiv für Geschichte der Philosophie* 10 (1897): 504–19; 11 (1898), 58–72, 181–211.

420. Puech writes of "a materialist psychology that came from the French Enlightenment." Michel Puech, "Tetens et la crise de la métaphysique allemande en 1775 (Über die allgemeine spekulative Philosophie)," *Revue philosophique de la France et de l'Étranger* 182, no. 1 (1992): 3–29, citing 12. Bonnet was almost schizophrenic in his effort to explain the "soul" as an epiphenomenon of the action of brain fibers while yet embracing the conventional Christian notion of an immortal, rational soul. He was never comfortable with acknowledging his most drastically mechanist-materialist account, published anonymously in 1755 as *Essai de psychologie*. When he dared more publicly to revisit these issues in his *Essai analytique sur les forces de l'âme* (1760), the theory did not change substantially, and Bonnet seemed to the Germans to be propounding a materialist reduction of the soul to the brain. See Speck, "Bonnets Einwirkung auf die deutsche Psychologie des vorigen Jahrhunderts"; Gary Hatfield, "Remaking the Science of Mind: Psychology as Natural Science," in *Inventing Human Science: Eighteenth-Century Domains*, ed. C. Fox, R. Porter, and R. Wokler (Berkeley: University of California Press, 1995), 184–231.

421. Speck makes this point throughout his "Bonnets Einwirkung auf die deutsche Psychologie des vorigen Jahrhunderts." Barnouw draws Tetens interestingly into alignment with Friedrich Schiller's medical dissertations as efforts to develop—against Bonnet—an empirical *conative* psychology. Jeffrey Barnouw, "The Philosophical Achievement and Historical Significance of Johann Nicolas Tetens," *Studies in Eighteenth-Century Culture* 9 (1979): 301–35, citing 306, 328n3. This juxtaposition involved what has come to be called the commitment to the "whole man" in German Enlightenment anthropology. See Schings, *Der ganze Mensch*. In Barnouw's view, Tetens, "like Herder and Schiller, was constantly working to undermine faculty psychology" ("Philosophical Achievement," 309). That is, he resisted rigid segmentation, insisting rather on a continuity from sensation to concepts. Moreover, he insisted that everything began with sensation, a principle that in fact brought the human and the animal experience far closer together than German school philosophy and its traditional view of the animal soul found comfortable (or than would Kant from his full, transcendental-philosophical position). See Kenneth Dewhurst and Nigel Reeves, *Friedrich Schiller: Medicine, Psychology, and Literature* (Berkeley: University of California Press, 1978), esp. 98, where they stress "a significant difference between Hallerian vitalism and Stahlian animism." And see Wolfgang Riedel, *Die Anthropologie des jungen Schiller: Zur Ideengeschichte der medizinischen Schriften und der "Philosophische Briefe"* (Würzburg: Königshausen und Neumann, 1985); Wolfgang Riedel, "Influxus physicus und Seelestärke: Empirische Psychologie und moralische Erzählung in der deutschen Spätaufklärung und bei Jacob Friedrich Abel," in *Anthropologie und Literatur um 1800*, ed. Jürgen Barkhoff and Eda Sagarra (Munich: Iudicium Verlag, 1992), 24–52.

422. Tetens wrote at the very outset of his eleventh essay: "Not hypotheses, but rather observations" (*Philosophische Versuche*, 1:730). "Forces can be recognized by us only through the effects they occasion and can be characterized only in their terms" (ibid., 733). This was the cardinal

methodological maxim of the empirical sciences in late-eighteenth-century Germany: Tetens wished to assimilate empirical psychology to this methodology of the natural sciences.

423. Ibid., 2:302–3.

424. Tetens, *Philosophische Versuche*, preface, iv–viii.

425. Unzer, *Erste Gründe einer Physiologie*, preface, trans. 8.

426. Tetens, *Philosophische Versuche*, 2:350.

427. Ibid., 459–64, esp. 469.

428. Ibid., 488.

429. Ibid., 460.

430. Ibid., 456.

431. Ibid., 497.

432. Ibid., 464.

433. Ibid., 498.

434. Ibid., 500. Tetens elaborated his intermediate notion extensively (ibid., 502ff.).

435. Johann Friedrich Blumenbach, *Über den Bildungstrieb* (Göttingen: Dieterich, 1781).

436. Carl Friedrich Kielmeyer, *Über die Verhältniße der organischen Kräfte unter einander in der Reihe der verschiedenen Organisationen, die Gesetze und Folgen dieser Verhältniße* (1793), facsimile reproduction, ed. Kai Torsten Kanz (Marburg: Basilisken, 1993).

CHAPTER SIX

1. Friedrich Schelling, *First Outline of a System of the Philosophy of Nature*, trans. Keith Peterson (Albany: State University of New York Press, 2004), 53.

2. Reijer Hooykaas, "The Parallel between the History of the Earth and the History of the Animal World," *Archives internationales d'histoire des sciences* 10 (1957): 3–18.

3. Ibid., 3. Rudwick makes the same point: "the reconstruction of a contingent geohistory was historically distinct from, as well as being an indispensable condition for, the slightly later construction of an equally contingent history of life." Martin Rudwick, *Bursting the Limits of Time: The Reconstruction of Geohistory in the Age of Revolution* (Chicago: University of Chicago Press, 2005), 7. Rudwick notes, further, that it is "significant that the words 'geology' and 'biology' were both coined at this time" (9).

4. Lyon and Sloan put this into a pithy title: *From Natural History to the History of Nature: Readings from Buffon and His Critics*, ed. John Lyon and Phillip Sloan (Notre Dame, IN: University of Notre Dame Press, 1981). That very formulation had been employed earlier by Wolf Lepenies, *Das Ende der Naturgeschichte: Wandel kultureller Selbstverständlichkeiten in den Wissenschaften des 18. und 19. Jahrhunderts* (Munich: Hanser, 1976). The question of the historicization of nature has been addressed by a number of other authors as well: e.g., Dietrich von Engelhardt, *Historisches Bewußtsein in der Naturwissenschaft von der Aufklärung bis zum Positivismus* (Freiburg im Breisgau: Alber, 1979); Wolfhart Langer, "Verzeitlichungs- und Historisierungstendenzen in der frühen Geologie und Paläontologie," *Berichte zur Wissenschaftsgeschichte* 8 (1985): 87–97. "The entry of time and history into biological systems of classification is perhaps the single most significant development in the history of biological systematics in the modern era," Sloan wrote in one of the pioneering essays on the history of biology in eighteenth-century Germany. See Phillip Sloan, "Buffon, German Biology, and the Historical Interpretation of Biological Species," *British Journal for the History of Science* 12 (1979): 109–53, citing 109.

5. François Russo, "Théologie naturelle et sécularisation de la science au XVIII siècle," *Recherches de science religieuse* 66 (1978): 27–62, citing 43.

6. Isaac Newton, "General Scholium," in *Mathematical Principles of Natural Philosophy*, 2nd ed. (1713), in *Newton's Philosophy of Nature: Selections from His Writings* (New York: Hafner, 1953), 41–67, 116–34.

7. This represents a dramatic and still not thoroughly acknowledged revision in the fundamental notion of what natural science is and how it can construe the physical world—that is, the idea of a *historical* natural science. This *literal* sense of the phrase "history of nature" raised ontological, epistemological, and empirical questions concerning *historical* science—that is, what exactly "history" betokened, and what could be done with it: not merely how things came to be, but how humans could attain knowledge of this.

8. Michel Foucault, *The Order of Things: An Archaeology of the Human Sciences* (New York: Vintage, 1973), 127–28.

9. Such major figures as Leibniz, Bourguet, Linnaeus, Buffon, and others made contributions to this shift.

10. See Rudwick, *Bursting the Limits of Time*; and the earlier work Francis Haber, *The Age of the World: Moses to Darwin* (Baltimore: Johns Hopkins University Press, 1959).

11. Gottfried Leibniz, *Protogaea* (1749), trans. C. Cohen and A. Wakefield (Chicago: University of Chicago Press, 2008); Benôit de Maillet, *Telliamed; or, Conversations between an Indian Philosopher and a French Missionary on the Diminution of the Sea* (1749), trans. and ed. Albert V. Carozzi (Urbana: University of Illinois Press, 1968).

12. Georges-Louis Leclerc de Buffon, *Les époques de la nature*, critical ed., ed. Jacques Roger (Paris: Éditions du Muséum, 1988).

13. Rudwick, *Bursting the Limits of Time*, 6.

14. Cited in ibid., 199.

15. Ibid., 200.

16. Cited in ibid., 202.

17. For historians of geology who differ with Rudwick on this crucial issue, see Kenneth Taylor, "The Historical Rehabilitation of Theories of the Earth," *Compass: Earth Science Journal of Sigma Gamma Epsilon* 69 (1992): 334–45; Rachel Laudan, "Tensions in the Concept of Geology: Natural History or Natural Philosophy," *Earth Science History: Journal of the History of the Earth Sciences Society* 1 (1982): 7–13.

18. Louis Bourguet, *Lettres philosophiques sur la formation des sels et des crystaux, et sur la generation et le mechanisme organique des plantes et des animaux* (1729; Amsterdam: Rey, 1762), 173.

19. See Daniel Garber, "*De ortu et antiquissisimis fontibus protogaeae Leibnizianae dissertatio*: Observation, Exploration, and Natural Philosophy," in *Leibniz y las ciencias empíricas / Leibniz and the Empirical Sciences* (Granada: Comares, 2011), 165–85.

20. Rhoda Rappaport, *When Geologists Were Historians* (Ithaca, NY: Cornell University Press, 1997), 157–58.

21. Martin Rudwick, *The Meaning of Fossils: Episodes in the History of Paleontology*, 2nd ed. (Chicago: University of Chicago Press, 1976).

22. Haber, *Age of the World*, 36ff.; Ezio Vaccari, "European Views on Terrestrial Chronology from Descartes to the Mid-eighteenth Century," in *The Age of the Earth: From 4004 BC to AD 2002*, ed. C. L. E. Lewis and S. J. Knell (London: Geological Society, 2001), 25–38; Paolo Rossi, *The Dark Abyss of Time* (Chicago: University of Chicago Press, 1984); Stephen Toulmin and June Goodfield, *The Discovery of Time* (New York: Harper and Row, 1965); Rudwick, *Meaning of Fossils*.

23. Roy Porter, *The Making of Geology: Earth Science in Britain, 1660–1815* (Cambridge: Cambridge University Press, 1971), 137. "For naturalists interested in marine fossils, the universality of the Flood offered an instant and obvious explanation of the ubiquity of such forms" (ibid., 140).

24. See Haller's reviews of Buffon in *Bibliothèque raisonnée*. The *Bibliothèque raisonnée* reviews are translated in Lyon and Sloan, *From Natural History to History of Nature*: pt. 1 (1750), 255–68; pt. 2 (1751), 269–82.

25. "*Insensiblement, la durée s'introduit dans les schémas* [Unawares, duration inserted itself into the models]." Gabriel Gohau, *Naissance de la géologie historique: La terre, des 'théories' a l'histoire* (Paris: Vuibert, 2003), 40. This marvelous line from Gohau offers a path into the historical problem. First, note the impersonal construction. Second, note the adverb: "unawares." Finally, note

that there were *already* "models." Duration "inserted itself" into them. By the mid-eighteenth century, "long duration of the geological past turned into a commonplace"; indeed, "*the sudden awareness of the necessity of long geological durations reflects a collective movement of European thought.*" François Ellenberger, *History of Geology*, vol. 2, *The Great Awakening and Its First Fruits—1660–1810* (Rotterdam: Balkema, 1999), 38, 40. Roy Porter notes, similarly, that by the late eighteenth century "all pointed unambiguously to an 'enormous' antiquity for Earth," though most refused to specify *how* long the timescale was (*Making of Geology*, 159–60). In short, historicization *befell* theory in (proto)geology.

26. Jean-André Deluc, *Lettres physiques et morales sur l'histoire de la terre et de l'homme: Addressées à la Reine de la Grande-Bretagne* (1779/80).

27. François Ellenberger and Gabriel Gohau, "A l'aurore de la stratigraphie paleontologique: Jean-André De Luc, son influence sur Cuvier," *Revue d'histoires des sciences* 34 (1981): 217–57.

28. John H. Eddy Jr., "Buffon's *Histoire naturelle*: History? A Critique of Recent Interpretations," *Isis* 85 (1994): 644–61. See also Hans-Jörg Rheinberger, "Buffon: Zeit, Veränderung und Geschichte," *History and Philosophy of the Life Sciences* 12 (1990): 203–23.

29. Jacques Roger, "Buffon et l'introduction de l'histoire dans l'*Histoire naturelle*," in *Buffon 88*, ed. Jean Gayon (Paris: Vrin, 1992), 193–205.

30. Georges-Louis Leclerc de Buffon, *De la dégéneration des animaux* (1766), cited in ibid., 201–2.

31. Ibid., 202.

32. "Buffon encapsulates the history of life within the general history of the terrestrial globe to the point of not utilizing paleontology except to confirm his theory of cooling and even abandoning his earlier research on the modification of life-forms" (ibid., 204). Thus, in the *Époques*, Roger sees Buffon tending "to substitute a deductive scheme for a historical reconstruction in the proper sense" (ibid.).

33. Eddy, "Buffon's *Histoire naturelle*: History?," 645.

34. Phillip Sloan, "From Logical Universals to Historical Individuals: Buffon's Idea of Biological Species," in *Histoire du concept d'espèce dans les sciences de la vie* (Paris: Gauthier-Villars, 1987), 101–40, citing 120.

35. Eddy, "Buffon's *Histoire naturelle*: History?," 648 and note.

36. Ibid., 650.

37. Ibid., 657. Eddy continues: "History, in the sense of unique and irreversible developments in time, did not exist for Buffon," even in the case of degeneration, for he insisted that degeneration (at least in the case of man) could be reversed. Eddy continues: "nowhere in Buffon's theory of degenerative organic alteration is there the idea that the past informs or imbues the future with existential elements it was not already destined to have." Not even *extinction* is historical for Eddy: "[T]he epochal disappearance of great numbers of living things does not presuppose anything like the developmental history of the organic world" (654).

38. Ibid., 654.

39. Ibid., 646n. He claims strong corroboration of his view from essays by M. J. S. Hodge and Kenneth Taylor in the compendium, *Buffon 88*: M. J. S. Hodge, "Two Cosmogonies (Theory of the Earth and Theory of Generation) and the Unity of Buffon's Thought," 241–54; Kenneth Taylor, "The *Époques de la nature* and Geology during Buffon's Later Years," 371–85. Taylor, according to Eddy, argues that "Buffon's geological research in the 'Époques' did not reveal truly historical methods of investigation" ("Buffon's *Histoire naturelle*: History?," 655n). That thrusts upon us the question: what exactly are "*truly* historical methods of investigation"? As a philosopher of history, I find the pronouncements of these scholars exasperatingly uncritical.

40. Rudwick, *Bursting the Limits of Time*, 150.

41. This recognition of Kant's importance in the eighteenth-century revision of natural history is prominent in the writings of Phillip Sloan. See Phillip Sloan, "Natural History," in *Cambridge History of Eighteenth Century Philosophy*, ed. Knud Haakonssen (Cambridge: Cambridge University Press, 2006), 903–37, citing 926–27; Phillip Sloan, "Natural History, 1670–1802," in *Compan-*

ion to the History of Modern Science, ed. R. C. Olby, G. N. Cantor, J. R. R. Christie, and M. J. S. Hodge (London: Routledge, 1990), 295–313, citing 306–9. See also Bernhard Fritscher, "Kant und Werner: Zum Problem einer Geschichte der Natur und zum Verhältnis von Philosophie und Geologie um 1800," *Kant-Studien* 83 (1992): 417–35.

42. Kant coined the term "archaeology of nature" in his *Kritik der Urteilskraft* in 1790, but he was thinking about adequate terminology for a "history of nature" already from the early 1750s.

43. See my epigraph for this chapter from Friedrich Schelling.

44. This is a position I have defended in many previous works. For some important corroborations, see Phillip Sloan, "Kant on the History of Nature: The Ambiguous Heritage of the Critical Philosophy for Natural History," *Studies in History and Philosophy of the Biological and Biomedical Sciences* 37 (2006): 627–48. See also Gian Franco Frigo, "'Der stete und feste Gang der Natur zur Organisation'—Von der Naturgeschichte zur Naturphilosophie um 1800," in *Naturwissenschaft um 1800: Wissenschaftskultur in Jena-Weimar*, ed. Olaf Breidbach and Paul Ziche (Weimar: Böhlaus, 2001), 27–45. For the broader context, see Lepenies, *Das Ende der Naturgeschichte*.

45. Kant, "... ob die Erde in ihrer Umdrehung ...," AA 1:190.

46. Ibid.

47. We have evidence for this from his physical geography materials that date to the mid-1750s.

48. Bourguet, *Lettres philosophiques*.

49. Kant, *Allgemeine Naturgeschichte*, AA 1:215–368.

50. Kant, *Physische Geographie*, AA 9:151–436; and lesser writings on the Lisbon earthquake, etc., of 1756, AA 1:417–72.

51. Kant, "Das Konzept zur Vorlesung über physische Geographie (1757–1759) [aufgrund der Handschrift >Holstein<]," AA 26/1:66–79.

52. Erich Adickes, *Untersuchungen zu Kants physischer Geographie* (Tübingen: Mohr, 1911); Erich Adickes, *Kants Ansichten zu Geschichte und Bau der Erde* (Tübingen: Mohr, 1911); Joseph May, *Kant's Concept of Geography and Its Relation to Recent Geographical Thought* (Toronto: University of Toronto Press, 1970); Stuart Erden and Eduardo Marieta, eds., *Reading Kant's Geography* (Albany: State University of New York Press, 2011).

53. Kant, "Von den verschiedenen Racen der Menschen," AA 2:434n.

54. But Buffon's far more radical *Époques de la nature* (1779) had not yet appeared when Kant composed this essay. It may have had a very different impact on Kant than the earlier work.

55. Kant, "Von den verschiedenen Racen der Menschen," AA 2:434n.

56. Ibid., 429.

57. Sloan, "Kant on the History of Nature," 635.

58. Eddy, "Buffon's *Histoire naturelle*: History?," 646n.

59. Kant, "Von den verschiedenen Racen der Menschen," AA 2:443.

60. See Erich Adickes, *Kant als Naturforscher* (Berlin: Walter de Gruyter, 1924).

61. See my "Kant's Early Views on Epigenesis: The Role of Maupertuis," in *The Problem of Animal Generation in Early Modern Philosophy*, ed. Justin E. Smith (Cambridge: Cambridge University Press, 2006), 317–54.

62. Maupertuis was the early Kant's paradigmatic instance of a modern hylozoist. Kant's *Dreams of a Spirit-Seer* treats him in exactly that context: "*Hylozoism* invests everything with life, while *materialism*, when carefully considered, deprives everything of life. Maupertuis ascribes the lowest degree of life to the organic particles of nourishment consumed by animals; other philosophers regard such particles as nothing but dead masses, merely serving to magnify the power of the levers of animal machines" (AA 2:330).

63. Reiner Wisbert, commentary on *Journal meiner Reise im Jahre 1769*, in Herder, *Werke in zehn Bänden* (Frankfurt: Deutsche Klassiker Verlag, 1985), 9:898.

64. Roland Mortier, *Diderot en Allemagne (1750–1830)* (Geneva: Slatkin Reprints, 1986), 25.

65. Herder to J. F. Hartknoch, December 1769, in *Briefe*, vol. 1, ed. Wilhelm Dobbek and Günter Arnold (Weimar: Böhlaus, 1984), 183.

66. Leonard Tancock, introduction to Denis Diderot, *D'Alembert's Dream*, in *Rameau's Nephew / D'Alembert's Dream*, trans. Leonard Tancock (Harmondsworth: Penguin, 1966), 134.

67. See Peter Hanns Reill, *Vitalizing Nature in the Enlightenment* (Berkeley: University of California Press, 2005).

68. Thus, I undertake revisions of the great study published more than thirty years ago by H. B. Nisbet: *Herder and the Philosophy and History of Science* (Cambridge: Modern Humanities Research Association, 1970).

69. Elias Palti, 'The "Metaphor of Life": Herder's Philosophy of History and Uneven Developments in Late Eighteenth-Century Natural Science," *History and Theory* 38 (1999): 322–47, citing 323n.

70. "We shall see, after experimental physics has become more advanced, that all the phenomena, whether of gravity or of elasticity or of attraction or of magnetism or of electricity, are nothing more than different aspects of the same impulse [*affection*]." Denis Diderot, *De l'interprétation de la nature*, in *Oeuvres philosophiques* (Paris: Classiques Garnier, 1998), 220.

71. See Heinrich Clairmont, "'Die Leute wollen keinen Gott, als in ihrer Uniform, ein menschliches Fabeltier': Herders anthropologisch fundierte Gnoseologie und seine Spinozadeutung in Gott," in *Spinoza im Deutschland des achtzehnten Jahrhunderts: Zur Erinnerung an Hans-Christian Lucas*, ed. Evan Schürmann, Norbert Waszek, and Frank Weinreich (Stuttgart: Frommann-Holzboog, 2002), 329–55.

72. Roger is emphatic about the pervasive primacy of Lucretius in Diderot's philosophy of nature. Jacques Roger, *Les sciences de la vie dans la pensée française du XVIIIe siècle* (Paris: Armin Colin, 1963), 664.

73. On Herder's theological scruples, see Pangiotes Kondylis, *Die Aufklärung im Rahmen des neuzeitlichen Rationalismus* (Stuttgart: Klett-Cotta, 1981), 622.

74. Diderot, *D'Alembert's Dream*, 177.

75. Willi Vollrath, "Die Auseinandersetzung Herders mit Spinoza" (PhD diss., Gießen, 1911); Regine Otto, "Herders Weg zu Spinoza," *Weimarer Beiträge* 10 (1978): 165–77. There is evidence of this engagement in Herder's "Grundsätzen der Philosophie" of 1769: "Spinoza believed that everything existed in God. He . . . assumed only One Center; he called it God and World. One can therefore term him with equal justice an idealist as an atheist; he never was this latter. . . . Thus, God belongs to the world, as the world to God. He is the principle: everything is therefore contingency insofar as it has its ground in God, but also necessary insofar as it belongs necessarily to the Thought of God." See Bernhard Suphan, ed., *Herders Sämmtliche Werke*, 33 vols. (Berlin: Weidmann, 1877–1913), 32:228. Without any knowledge of them, Herder replicated the philosophical line of thought that Lessing's brief texts had worked out only a few years earlier for his correspondence with Mendelssohn. See my essay "'The Most Hidden Conditions of Men of the First Rank'—The Pantheist Current in Eighteenth-Century Germany 'Uncovered' by the Spinoza Controversy," *Eighteenth-Century Thought* 1 (2003): 335–68.

76. Marion Heinz, *Sensualistischer Idealismus* (Hamburg: Meiner, 1994), xxi.

77. As Lindner correctly surmises, Herder's "transformation of Spinozism was initiated by Leibniz's philosophy." Herbert Lindner, *Problem des Spinozismus im Schaffen Goethes und Herders* (Weimar: Arion, 1960), 89. "The fixed substance of Spinoza was transcended by the admixture of the Leibnizian principle of force" (ibid., 92).

78. "By rendering the 'active side' of Leibniz in a materialistic manner . . . to overcome the mechanistic tendencies of Spinozism[,] nature could now be seen in constant movement and development. The events of nature took on a process character" (ibid., 93).

79. Ibid., 90–91.

80. "In the second half of the eighteenth century Leibniz was readily simplified in a monistic manner—and not just by Herder—to be set against the two-substance dualism of Wolff." Helmut Pfotenauer, *Literarische Anthropologie* (Stuttgart: Metzler, 1987), 13. This way of interpreting Leibnizian monadology permeated the philosophical milieu of the precritical Kant. Knutzen and Cru-

sius took this line; Kant himself worked on physical monadology for the bulk of his precritical period. See Alison Laywine, *Kant's Early Metaphysics and the Origins of the Critical Philosophy*, North American Kant Society Studies in Philosophy, vol. 3 (Atascadero, CA: Ridgeview, 1993); Martin Schönfeld, *The Philosophy of the Young Kant: The Precritical Project* (Oxford: Oxford University Press, 2000); Hans-Joachim Waschkies, *Physik und Physikotheologie des jungen Kant* (Amsterdam: B. R. Grüner, 1987). It was just here, Ritzel suggests, that Kant—the *critical* Kant—and Herder parted ways. Kant gave up the "force metaphysics" and Herder did not; "this is when the ways of thought of Kant and Herder part." Wolfgang Ritzel, *Immanuel Kant: Eine Biographie* (Berlin: De Gruyter, 1985), 69. See my *Kant, Herder, and the Birth of Anthropology* (Chicago: University of Chicago Press, 2002).

81. Dreike has noted that the most important features of Herder's engagement with Leibniz were his acceptance of the theory of dynamism and his rejection of the idea that substances could not interact. Beate Dreike, *Herders Naturaffassung in ihrer Beeinflussung durch Leibniz' Philosophie*, Studia Leibnitiana, Supplementa, 10 (Wiesbaden: Steiner, 1973). Pross agrees: "The main point of Herder's criticism is the incommunicability of the monads, which already in Diderot's article [in the *Encyclopédie*] had been identified as the critical point of the entire system." Wolfgang Pross, commentary on "Wolff, Baumgarten und Leibniz," in J. G. Herder, *Werke*, vol. 2, *Herder und die Anthropologie der Aufklärung*, ed. Wolfgang Pross (Munich: Hanser, 1987), 863.

82. Suphan, *Herders Sämmtliche Werke*, 9:536–40.

83. J. G. Herder, *God: Some Conversations*, trans. F. Burkhardt (New York: Veritas, 1943), 105.

84. J. G. Herder, *Ideen zur Philosophie der Geschichte der Menschheit*, in J. G. Herder, *Werke*, ed. Wolfgang Pross (Munich: Hanser, 2002), 3:1.

85. Hans-Dietrich Irmscher, "Grundfragen der Geschichtsphilosophie Herders bis 1774," in *Bückeburger Gespräche über Johann Gottfried Herder, 1783*, ed. Brigitte Porschman (Rinteln: C. Bösendahl, 1984), 10–32, citing 27.

86. See my "Stealing Herder's Thunder: Kant's Debunking of Herder on History in 'Conjectural Beginning of the Human Race,'" in *Immanuel Kant: German Professor and World-Philosopher / Deutscher Professor und Weltphilosoph*, ed. Günter Lottes and Uwe Steiner (Saarbrücken: Wehrhan Verlag, 2007), 43–72.

87. Martin Bollacher, "'Natur' und 'Vernunft' in Herders Entwurf einer Philosophie der Geschichte der Menschheit," in *Johann Gottfried Herder, 1744–1803*, ed. Gerhard Sauder (Hamburg: F. Meiner, 1987), 123. For a thorough consideration of the tension between Kant and Herder over history, see Hans-Dietrich Irmscher, "Die geschichtsphilosophische Kontroverse zwischen Kant und Herder," in *Hamann—Kant—Herder: Acta des 4. Internationalen Hamann-Kolloqiums im Herder-Institut zu Marburg/Lahn, 1985*, ed. Bernhard Gajek (Frankfurt: Peter Lang, 1987), 111–92.

88. Reinhard Brandt, "Kant—Herder—Kuhn," *Allgemeine Zeitschrift für Philosophie* 5 (1980): 27–36.

89. Kant, *Allgemeine Naturgeschichte und Theorie des Himmels* (1755), AA 1:215–368; Georges LeClerc de Buffon, *Histoire naturelle, générale et particulière* (1749–); C. F. Wolff, *Theorie von der Generation* (1764).

90. J. G. Herder, *Vom Erkennen und Empfinden: Bemerkungen und Träume* (1778), in *Herders Sämmtliche Werke*, 165–262. Trans. in Michael Forster, ed., *Herder's Philosophical Writings* (Cambridge: Cambridge University Press, 2002), 195.

91. Herder, *Ideen*, bk. 5, chap. 3, 163.

92. Ibid., 164.

93. Ibid., 78–86. See R. Clarke, "Herder's Concept of '*Kraft*,'" *PMLA* 57 (1942): 737–52.

94. Herder, *Ideen*, bk. 3, chap. 4, 95. "In every living creature the circle of organic powers seems to be whole and complete, only differently modified and distributed in each" (ibid., chap. 2, 85). "A certain uniformity of structure, and as it were a *standard form* . . . convertible into the most abundant variety" (ibid., 66). "One single principle of life seems to prevail throughout all nature: this the *ethereal* or *electric stream* . . . more and more finely elaborates, till it produces all those won-

derful instincts and mental faculties" (ibid., chap. 1, 74). "Thus, it is anatomically and physiologically true that the analogy of *one organization* prevails through the whole animated creation of our globe" (ibid., bk. 2, chap. 4, 69).

95. Ibid., bk. 2, chap. 4, 66.

96. Ibid., bk. 1, chap. 2, 25–26.

97. Ibid., 26.

98. Ibid., 26.

99. Ibid., 154.

100. In "Grundfragen der Geschichtsphilosophie," *Bückeburger Gespräche über Johann Gottfried Herder, 1971*, ed. Johann Maltusch (Bückeburg: Grimme, 1973), 17–57, H. D. Irmscher notes Herder's early and distinctive embrace of the idea of epigenesis. Palti suggests a more ambivalent relationship, offering a number of distinctions and tensions in the biological theories and in Herder's reception of them, which he conceives as "uneven developments." Elias Palti, "The 'Metaphor of Life': Herder's Philosophy of History and Uneven Developments in Late Eighteenth-Century Natural Sciences," *History and Theory* 38 (1999): 322–47. For more on this, see my "Epigenesis: Concept and Metaphor in Herder's *Ideen*," in *Vom Selbstdenken: Aufklärung und Aufklärungskritik in Herders "Ideen zur Philosophie der Geschichte der Menschheit,"* ed. Regine Otto and John H. Zammito (Heidelberg: Synchron 2001), 129–44.

101. Herder, *Ideen*, bk. 7, chap. 4, 245.

102. Ibid., 159–60.

103. Ibid., 178.

104. Kant, "Recension von J. G. Herders *Ideen zur Philosophie der Geschichte der Menschheit*," Theil 2 (1785), AA 8:62–63.

105. Kant wrote: "The reviewer is fully in agreement with him here, but with this reservation: if the cause that organizes *from within* were limited by its nature to only a certain number and degree of differences in the development of the creature that it organizes (so that, once these differences were exhausted, it would no longer be free to work from another archetype [*Typus*] under altered circumstances), one could well describe this natural development of formative nature in terms of germs [*Keime*] or original dispositions [*Anlagen*], without thereby regarding the differences in question as originally implanted and only occasionally activated mechanisms or buds [*Knospen*] (as in the system of evolution); on the contrary, such differences should be regarded simply as limitations imposed on a self-determining power, limitations that are inexplicable, because the power itself is incapable of being explained or rendered comprehensible" (ibid.).

106. Herder, *Ideen*, bk. 7, chap.1, 230.

107. Heinz Stolpe, "Herder und die Ansätze einer naturgeschichtlichen Entwicklungslehre im 18. Jahrhundert," in *Neue Beiträge zur Literatur der Aufklärung*, ed. Werner Krauss (Berlin: Rütten und Loening, 1964), 289–316, 454–68, citing 315.

108. Charlotte von Stein wrote to Karl Ludwig von Knebel (May 1, 1784) about the impact of Herder's *Ideen* on Goethe: "Herder's new text makes it appear that we started out as plants and animals." Cited in Manfred Wenzel, "Die Anthropologie Johann Gottfried Herders und das klassische Humanitätsideal," in *Die Natur des Menschen: Probleme der Physischen Anthropologie und Rassenkunde (1750–1850)*, ed. Gunter Mann and Franz Dumont, Soemmerring Forschungen 6 (Stuttgart: G. Fischer, 1990), 137–67, citing 137.

109. Nisbet, *Herder and the Philosophy and History of Science*, 109.

110. On Kant's attitudes, see my essays "'Method' vs. 'Manner'?—Kant's Critique of Herder's *Ideen* in Light of the Epoch of Science, 1790–1820," *Herder Yearbook*, 1998, 1–25; "'This Inscrutable *Principle* of an Original *Organization*': Epigenesis and 'Looseness of Fit' in Kant's Philosophy of Science," *Studies in History and Philosophy of Science* 34 (2003): 73–109. And see Phillip Sloan, "Preforming the Categories: Eighteenth-Century Generation Theory and the Biological Roots of Kant's A Priori," *Journal of the History of Philosophy* 40 (2002): 229–53.

111. "Man is the first *liberated being* [*Freigelassene*] of creation; he stands erect" (Herder, *Ideen*, bk. 4, chap. 4, 135). See also bk. 3, chap. 6, 69ff.; bk. 4, passim.

112. Ibid., 133–34.

113. "The central issue in anthropological theory of the period concerned the status of man in the growing taxonomic systematization of nature. . . . Linnaeus had triggered the problem by his inclusion of man among the Quadrupeda in the order Anthropomorpha, along with the apes and sloths" (Sloan, "Buffon, German biology, and the Historical Interpretation of Biological Species," 109).

114. Goethe to Knebel, November 17, 1784, cited in Hans-Dietrich Irmscher, "Beobachtungen zur Funktion der Analogie im Denken Herders," *Deutsche Vierteljahrsschrift für Literaturwissenschaft und Geistesgeschichte* 55 (1981): 64–97, citing 70.

CHAPTER SEVEN

1. J. G. Zimmermann to Johann Friedrich Blumenbach, September 15, 1779, in Frank Dougherty, ed., *The Correspondence of Johann Friedrich Blumenbach*, vol. 1, *1773–1782* (Göttingen: Klatt, 2006), 177.

2. Toellner has written that Blumenbach "at the end of the first century of the Georgia Augusta [University of Göttingen], as before him only at its beginning Albrecht von Haller, brought the world to Göttingen and made Göttingen a worldwide idea." Richard Toellner, "Opening Remark" to *Commercium Epostolicum J. F. Blumenbachii: Aus einem Briefwechsel des klassischen Zeitalters der Naturgeschichte; Katalog zur Ausstellung im Foyer der Niedersächsichen Staats- und Universitätsbibliothek Göttingen, 1. Juni–21 Juni 1984*, ed. F. W. P. Dougherty (Göttingen: n.p., 1984), unpaginated [5].

3. As Blumenbach's biographer, Marx, noted, "For more than half a century the most important events of this University [i.e., Göttingen] are bound up with his memory and his name." K. F. H. Marx, "Life of Blumenbach," in *The Anthropological Treatises of John Friedrich Blumenbach. . .*, ed. Thomas Bendysche (London: Longman, Green, etc., for the Anthropological Society of London, 1865), separately paginated, 1. Thus, Frank Dougherty notes: "all of academic Germany from Bonn to Königsberg celebrated in 1825 the golden anniversary of the doctoral promotion of someone who had educated, excited, and supported generations of scholars" (*Commercium Epostolicum J. F. Blumenbachii*, 54).

4. Tanya Van Hoorn, *Dem Leibe abgelesen* (Tübingen: Niemeyer, 2004), 3n, 96n. The most important work on Blumenbach was that of the late Frank Dougherty, and his early demise largely accounts for the lacuna van Hoorn laments. See Frank Dougherty, *Gesammelte Aufsätze zu Themen der klassischen Periode der Naturgeschichte* (Göttingen: Klatt, 1996).

5. There have been other important essays that shed light on Blumenbach. See, e.g., Phillip Sloan, "Buffon, German Biology, and the Historical Interpretation of Biological Species," *British Journal for the History of Science* 12 (1979): 109–53; Timothy Lenoir, "Kant, Blumenbach, and Vital Materialism in German Biology," *Isis* 71 (1980): 77–108; Timothy Lenoir, "The Göttingen School and the Development of Transcendental *Naturphilosophie* in the Romantic Era," *Studies in History of Biology* 5 (1981): 111–205; Peter McLaughlin, "Blumenbach und der Bildungstrieb: Zum Verhältnis von epigenetischer Embryologie und typologischen Artbegriff," *Medizinhistorisches Journal* 17 (1982): 357–72; Georgette Legée, "Johann Friedrich Blumenbach (1752–1840): La naissance de l'anthropologie à l'époque de la révolution française," in *Scientifiques et sociétés pendant la Révolution et l'Empire* (Paris: Editions du CTHS, 1990), 395–420; Nicholas Jardine, "Scenes of Natural History," in *The Scenes of Inquiry: On the Reality of Questions in the Sciences* (Oxford: Clarendon, 2000), 11–55; Robert Richards, "Kant and Blumenbach on the *Bildungstrieb*: A Historical Misunderstanding," *Studies in History and Philosophy of the Biological and Biomedical Sciences* 31 (2000): 11–32.

6. John Gascoigne, "Blumenbach, Banks, and the Beginnings of Anthropology," in *Göttingen and the Development of the Natural Sciences*, ed. N. Rupke (Göttingen: Wallstein, 2002), 86–98, citing 94.

7. Strickland, drawing on Rudolf Stichweh, observes: "Eighteenth-century writers tended to

use the terms *Physik* and *Naturlehre* interchangeably," but there was an emergent tendency for *Naturlehre* to be used for "experimental sciences whose phenomena had thus far eluded mathematical representation." Stuart Strickland, "Galvanic Disciplines: The Boundaries, Objects, and Identities of Experimental Science in the Era of Romanticism," *History of Science* 33 (1995): 449–68, citing 452. Summarizing, he writes: "*Naturlehre* and *Physik* began the eighteenth century as indistinguishable global terms. By mid-century or slightly later, a division had opened between the sciences of the organic and those of the inorganic, a division that seems to have been based on a distinction drawn from natural differences in the objects and reinforced by the methods applied to those objects" (463). See Rudolf Stichweh, *Zur Entstehung des modernen Systems wissenschaftlicher Disziplinen: Physik in Deutschland, 1740–1890* (Frankfurt am Main: Suhrkamp, 1984). See also Gian Franco Frigo, "'Der stete und feste Gang der Natur zur Organisation'—von der Naturgeschichte zur Naturphilosophie um 1800," in *Naturwissenschaft um 1800: Wissenschaftskultur in Jena-Weimar*, ed. Olaf Breidbach and Paul Ziche (Weimar: Böhlaus, 2001), 27–45; Volker Hess, "Das Ende der 'Historia naturalis'? Die naturhistorische Methode und Klassifikation bei Kielmeyer," in *Philosophie des Organischen in der Goethezeit: Studien zu Werk und Wirkung des Naturforschers Carl Friedrich Kielmeyer (1765–1844)*, ed. Kai Torsten Kanz (Stuttgart: Steiner, 1994), 153–73, esp. 154, 157.

8. Erxleben completed his dissertation at Göttingen in 1767, became extraordinary professor of philosophy there in 1771, a member of the Göttingen Academy in 1774, and ordinary professor of philosophy in 1775, a position he held until his premature death in 1777. He wrote his (fourteen-page) dissertation on the classification of mammals. His supervisor was the mathematician (and man of letters) Abraham Gotthelf Kästner (1719–1800), who had joined the Göttingen faculty in 1756. Before that, Kästner had edited and annotated the first two volumes of the German translation of Buffon's *Histoire naturelle*.

9. One year after his dissertation, Erxleben published *Anfangsgründe der Naturgeschichte* (1768). After his death it was reissued twice (in 1782 and 1790), edited by the noted Göttingen chemist J. F. Gmelin (1748–1804). It never achieved the prominence in its field that the physics textbook enjoyed. Instead, Blumenbach's *Handbuch der Naturgeschichte* became the standard text in the field for the next several decades. In any event, Erxleben's 1768 work would be dwarfed in its impact by his monumental textbook in physical science, *Anfangsgründe der Naturlehre*, with six editions from 1772 to 1794, most of the later ones edited and annotated after Erxleben's demise by the eminent Göttingen physicist Georg Lichtenberg (1742–99). See William Clark, "German Physics Textbooks in the *Goethezeit*," pt. 1, *History of Science* 35 (1997): 219–39; pt. 2, *History of Science* 35 (1997): 295–363. Erxleben actually composed *three* textbooks: the last dealing with chemistry—which illuminates the semantic field of natural science at the time still further. Strickland argues that "by the 1780s, . . . especially due to the promotion of the discipline in the textbooks of Erxleben and Lichtenberg, chemistry not only came to hold a central position among experimental sciences, but expanded its domain to encompass the bulk of the *Naturlehre*." Strickland goes so far as to assert: "Little remained for physics beyond the study of electrostatics" (Strickland, "Galvanic Disciplines," 453).

10. Hess, "Das Ende der 'Historia naturalis'?," 172.

11. On Baldinger, see Klaus Mross, "Ernst Gottfried Baldinger (1738–1804), gelehrter Arzt der Aufklärungszeit, und sein Schüler Samuel Thomas Soemmerring," in *Samuel Thomas Soemmerring und die Gelehrten der Goethezeit*, ed. Gunter Mann and Franz Dumont (Stuttgart: Fischer, 1985), 245–61.

12. Ernst Gottfried Baldinger, *Die Grenzen der Naturlehre werden bestimmt: Eine academische Abhandlung* (Jena: Ruedel, 1762).

13. Paul Ziche, "Von der Naturgeschichte zur Naturwissenschaft: Die Naturwissenschaften als eigenes Fachgebiet an der Universität Jena," *Berichte zur Wissenschaftsgeschichte* 21 (1998): 251–63, esp. 253.

14. Olaf Breidbach, "Transformation statt Reihung—Naturdetail und Naturganzes in Goethes Metamorphosenlehre," in Breidbach and Ziche, *Naturwissenschaft um 1800*, 46–64, citing 47, 60.

15. Robert Paul Willem Visser, *The Zoological Work of Petrus Camper (1722–1789)* (Amsterdam: Rodopi, 1995), 74ff., esp. 79.

16. F. J. Cole called him an "incorrigible speculator" in *Early Theories of Sexual Generation* (Oxford: Oxford University Press, 1930), 100. Rey observed: "Charles Bonnet was a philosopher as much as a naturalist, logician as much as experimentalist." Roselyn Rey, "La partie, le tout, et l'individu: Science et philosophie dans l'œuvre de Chalres Bonnet," in *Charles Bonnet: Savant et philosophe (1720–1793)*, ed. Marino Buscaglia, René Sigrist, Jacques Trembley, and Jean Wüest (Geneva: Editions Passé Présent, 1993), 61–75, citing 62. Particularly central for Bonnet was the tradition out of Leibniz; see Olivier Rieppel, "The Reception of Leibniz's Philosophy in the Writings of Charles Bonnet (1720–1793)," *Journal of the History of Biology* 21 (1988): 119–45; François Duchesneau, "Charles Bonnet et le concept Leibnizien d'organisme," *Medicina nei Secoli—arte e scienza* 15 (2003): 351–69; Tobias Cheung, "Die Ordnung des Organischen: Zur Begriffsgeschichte organismischer Einheit bei Charles Bonnet, Spinoza und Leibniz," *Archiv für Begriffsgeschichte* 46 (2004): 87–108.

17. Camper, cited in Visser, *Camper*, 78.

18. On the "tree of Diana," see Pierre-Louis Moreau de Maupertuis, *Système de la nature*, in *Essai de cosmologie, Système de la nature, Réponse aux objections de M. Diderot*, ed. François Azouvi (Paris: Vrin, 1984), 167.

19. Crucial here is the work of Louis Bourguet, *Lettres philosophiques sur la formation des sels et des crystaux, et sur la generation et le mechanisme organique des plantes et des animaux* (1729; Amsterdam: Rey, 1762), and his elaboration of the notion of "intussusception" from Réaumur. See Joseph Schiller, "La notion d'organisation dans l'œuvre de Louis Bourguet (1678–1742)," *Gesnerus* 32 (1975): 87–97; Olivier Rieppel, "'Organization' in the *Lettres philosophiques* of Louis Bourguet Compared to the Writings of Charles Bonnet," *Gesnerus* 44 (1997): 125–32. This idea was embraced by Kant and by Blumenbach.

20. Karen Reeds, *Botany in Medieval and Renaissance Universities* (New York: Garland, 1991).

21. On Linnaeus, see Gunnar Broberg, ed., *Linnaeus: Progress and Prospects in Linnaean Research* (Stockholm: Almqvist and Wiksell, 1980); Frans Stafleu, *Linnaeus and the Linnaeans: The Spreading of Their Ideas in Systematic Biology, 1735–1789* (Utrecht: Oosthoek, 1971); Tore Frängsmyr, ed., *Linnaeus: The Man and His Work* (Canton, MA: Science History Publications, 1994); James Larson, *Reason and Experience: The Representation of Natural Order in the Work of Carl von Linné* (Berkeley: University of California Press, 1971); James Larson, "Linnaeus and the Natural Method," *Isis* 58 (1967): 305–20. At the same time, others, like Hales, made significant advances in plant physiology. See Henri Guerlac, "The Continental Reputation of Stephen Hales," *Archives internationales de l'histoire des sciences* 4 (1951): 393–404.

22. On the new importance of zoology, see Scott Atran, *Cognitive Foundations of Natural History: Towards an Anthropology of Science* (Cambridge: Cambridge University Press, 1990), 190ff., 210.

23. Lamarck, *Histoire naturelle des animaux sans vertèbres . . . précédée d'une introduction offrant la détermination des caractères essentiels de l'animal, sa distinction du végétal et des autres corps naturels, enfin, l'exposition des principes fondamentaux de la zoologie*, 7 vols. (Paris: Verdière, 1815–22). On Lamarck, see R. W. Burkjardt, *The Spirit of System: Lamarck and Evolutionary Biology* (Cambridge, MA: Harvard University Press, 1977); Pietro Corsi, *The Age of Lamarck* (Berkeley: University of California Press, 1988).

24. Londa Schiebinger has written on this new category from the important vantage of gender in science: "Why Mammals Are Called Mammals," in *Nature's Body: Gender in the Making of Modern Science* (New Brunswick, NJ: Rutgers University Press, 2004), 40–74. It was Linnaeus who conceived of the class of mammals, a very important focus for German zoological taxonomists of the second half of the eighteenth century. Most notably, Erxleben's work at Göttingen in the very years Blumenbach studied there concentrated on this topic. The last work of Erxleben's published during his lifetime was *Systema regnis animalis: Classis I: Mammalia* (1777), a return to the topic of his dissertation. The connection between his conception of natural history and that of the early Blu-

menbach is striking. Compare, for example, the structure of his *Anfangsgründe der Naturgeschichte* (1768) with the structure of Blumenbach's *Handbuch der Naturgeschichte.*

25. "The eighteenth century vertebrate paleontologists were the first to apply the comparative method systematically and consistently" (Visser, *Camper,* 121).

26. Peter Simon Pallas, *Reise durch verschiedene Provinzen des Russischen Reichs,* 3 vols. (1771–76); James Larson, *Interpreting Nature: The Science of Living Form from Linnaeus to Kant* (Baltimore: Johns Hopkins University Press, 1994).

27. Conjectures on the Ohio creature preoccupied, among others, William Hunter, Johann Heinrich Merck (1741–91), and Thomas Jefferson, before Cuvier. Meijer notes: "Camper himself became active in paleontology after the fossil remains of the giant land vertebrates were confronted, i.e. recognized for what they were, around 1770." Miriam Meijer, *Race and Aesthetics in the Anthropology of Petrus Camper (1722–1789)* (Amsterdam: Rodopi, 1999), 63.

28. Nikolaas Rupke, "The Study of Fossils in the Romantic Philosophy of History and Nature," *History of Science* 21 (1983): 389–413. Rupke (396–97) points to the central importance, in German research, of Johann Friedrich Esper (1732–81), who published *Ausführliche Nachrichten von neuentdeckten Zoolithen* (1774), facsimile reproduction, ed. Armin Geus (Wiesbaden: Guido Pressler, 1978).

29. The vagueness of the history of science on the establishment of animal extinction in this period is striking. Visser registers the conversion of Camper from resistance to the notion of extinction, largely on religious grounds still in 1770, to full conversion to extinction by the end of the 1770s (Visser, *Camper,* 135–38; see also Meijer, *Race and Aesthetics,* 64).

30. C. D. O'Malley and H. W. Magoun, "Early Concepts of the Anthropomorpha," *Physis: Revista di storia della scienza* 4 (1962): 39–64. An outstanding study of this whole topic and its wider context is Franck Tinland, *L'homme sauvage:* Homo ferus *et* Homo sylvestris*—de l'animal à l'homme* (Paris: Payot, 1968).

31. Meijer, *Race and Aesthetics,* 127. See Miriam Meijer, "The Century of the Orangutan," *New Perspectives in the Eighteenth Century* 1 (2004): 62–78, for the story of the "orangutan war of 1777," which took place in Holland over the disposition of an orangutan cadaver. The keeper of the prince's museum wanted to stuff the animal; Peter Camper had secured permission to dissect it, but the keeper was not informed in time. The result was a damaged specimen for Camper and a fierce round of mutual recriminations.

32. O'Malley and Magoun, "Early Concepts of the Anthropomorpha."

33. Ibid.

34. Jessie Dobson, "John Hunter and the Early Knowledge of the Anthropoid Apes," *Proceedings of the Zoological Society of London* 13 (1953): 1–12, citing 6.

35. Linnaeus, *Systema naturae* (1735). See Gunnar Broberg, "*Homo sapiens*: Linnaeus's Classification of Man," in Frängsmyr, *Linnaeus: The Man and His Work,* 156–94, here 169.

36. Linnaeus, *Systema naturae,* 10th ed. (1758). In 1764 a dissertation associated with his student Christann Emannuel Hippius and entitled "Anthropomorpha" was published by Linnaeus in *Amoenitatae Accademiae* (vol. 6, 63–76), elaborating the typology. See Meijer, *Race and Aesthetics,* 43n.

37. This bafflement over albinism proved a major concern over the eighteenth century. It was triggered most notably by the introduction of an albino African boy in Paris, leading to a number of publications, most notably those of Maupertuis, which I have discussed earlier. Blumenbach would be a decisive figure in confirming that albinism was a disease, not a separate kind, of human beings.

38. This is not to say that there was no credulity or desire to include all sorts of source material, no matter how dubious, for the sake of completeness. These were weaknesses in Linnaeus, and Buffon made the most of them in his harsh castigations. But real problems of indeterminacy of evidence featured quite prominently here, from which Buffon himself suffered.

39. To be sure teratology was a crucial concern of eighteenth-century natural historians and

medical men, and one of Blumenbach's major contributions was the insight that even malformations reveal and embody developmental laws in their deviations.

40. Edward Tyson, *Orang-Outang, sive Homo Sylvestris: or, The Anatomy of a Pygmie compared with that of a Monkey, an Ape, and A Man. T o which is added a Philological Essay concerning the Pygmies, the Cynocephali, the Satyrs, and Spinges of the Ancients, wherein it will appear that they are all either Apes or Monkeys, and not Men, as formerly pretended* (1699), facsilime reproduction (London: Dawsons, 1966) (hereafter, *Anatomy of a Pygmie*). On Tyson, see Ashley Montagu, *Edward Tyson, M.D., F.R.S., 1650–1708, and the Rise of Human and Comparative Anatomy in England: A Study in the History of Science* (Philadelphia: American Philosophical Society, 1943); Robert Wokler, "Tyson, Buffon, and the Orangutan," *Studies on Voltaire and the Eighteenth Century* 155 (1976): 2301–19; and, most recently, Justin E. H. Smith, "Language, Bipedalism, and the Mind-Body Problem in Edward Tyson's *Orang-Outang* (1699)," *Intellectual History Review* 17, no. 3 (2007): 291–304.

41. After the work of Vesalius, it was widely recognized that Galen had conducted dissections of animals—in particular, of apes—to approximate human anatomy. One of the inciting motivations of Camper's comparative anatomy was to ascertain exactly which animals Galen had used, especially which apes. See Meijer, *Race and Aesthetics*, 123.

42. Tyson's philological appendix was extensive and wide-ranging in its exploration of ancient and medieval lore concerning the apes. One can compare his observations at the close of the seventeenth century with a more recent monograph on the same theme: Horst Janson, *Apes and Ape Lore in the Middle Ages and the Renaissance* (London: Warburg Institute, University of London, 1952).

43. One of the important concerns of this study is to establish that comparative anatomy as a project did not spring full formed from the head of Georges Cuvier in the early nineteenth century. It was a central project of the eighteenth century, from Tyson to Camper to Blumenbach, well before Cuvier even took up a scalpel. Here I disagree, regretfully but firmly, with Andrew Cunningham, *The Anatomist Anatomis'd* (Farnham, Surry, UK: Ashgate, 2010).

44. The traditional sense of the Pygmy was of a diminutive race of humans. Tyson wished to dispel that notion. Our own understanding makes room both for his distinction of the apes from such a form and for the empirical existence of such humans.

45. Buffon, "Nomenclature des singes," in *Histoire naturelle*, vol. 14 (1766), 1–42; and "Les orangs-outangs, ou le Pongo et le Jocko," in ibid., 43–71.

46. On Nicolaes Tulp, see O'Malley and Magoun, "Early Concepts of the Anthropomorpha."

47. On the linguistic origins of the term "chimpanzee," see ibid., 59.

48. Thus, the problem of classing humans among the animal forms was not unique to the European mind, and the issues of human degeneracy that seem salient in these cases prove central to a critical engagement with the construal of human difference across cultures.

49. One crucial association, here, is with the satyr of antiquity, with all its associated lusts, and another is, via the tail, with the devil. Conrad Gessner presented a tailed human as "Lucifer" in the sixteenth century.

50. Camper was particularly irate about this, criticizing even Buffon for such damaging indulgences. See Meijer, *Race and Aesthetics*, 188.

51. The question of the relation between the "rational" and the "animal" souls in the Aristotelian tradition and the reception of his text *De anima* became a crucial concern for the eighteenth-century naturalists in the wake of Descartes's *bête machine* hypothesis and the whole controversy over whether animals had any soul whatsoever. Christian accruals to this "rational" soul from Aristotle only accentuated the dangers of this discourse, as Pierre Bayle made painfully clear in his notorious article "Rorarius." See Francisco Vidal, *The Sciences of the Soul: The Early Modern Origins of Psychology* (Chicago: University of Chicago Press, 2011).

52. On negative reaction to Linnaeus, see Broberg, "*Homo sapiens*."

53. The animal/human boundary has emerged as a central concern in this study, as it has become for our own epoch, when the extravagant privileging of the human has come under much more thoroughgoing assault.

54. This proved a major theme in the case of Blumenbach. See, for a start, my essay "Policing Polygeneticism in Germany, 1775: (Kames,) Kant, and Blumenbach" in *The German Invention of Race*, ed. Sara Eigen and Mark Larrimore (Albany: State University Press of New York, 2006), 35–54.

55. Robert Wokler, "The Ape Debates in Enlightenment Anthropology," *Studies on Voltaire and the Eighteenth Century* 192 (1980): 1164–75, citing 1168. See also Robert Wokler, "Apes and Races in the Scottish Enlightenment: Monboddo and Kames on the Nature of Man," in *Philosophy and Science in the Scottish Enlightenment*, ed. Peter Jones (Edinburgh: Donald, 1988), 145–68.

56. Meijer, "Century of the Orangutan." See also Carl Niekerk, "Man and Orangutan in Eighteent-Century Thinking: Retracing the Early History of Dutch and German Anthropology," *Monatshefte* 96 (2004): 477–502.

57. The classic work is, of course, Arthur Lovejoy, *The Great Chain of Being* (New York: Harper, 1960).

58. Tyson, Dedicatory Preface, *Anatomy of a Pygmie* (unpaginated).

59. Smith, "Language, Bipedalism," 291.

60. Tyson, *Anatomy of a Pygmie*, 53–54; Smith, "Language, Bipedalism," 301; Wokler, "Tyson, Buffon," 2308.

61. Wokler, "Tyson, Buffon," 2306. See William F. Bynum, "The Anatomical Method, Natural Theology, and the Functions of the Brain," *Isis* 64 (1973): 445–68, for the classic account of the issue at the close of the seventeenth century. The basic premise was that there could be nothing in an organism (indeed, in nature anywhere) that was unnecessary. This idea was still preeminent in Kant's thinking, as is illustrated in the opening claims of his brief essay on conjectural history: "Idee zu einer allgemeinen Geschichte in weltbürgerlicher Absicht" (1784), AA 8:15–32. The notion of vestigial organs would not gain prominence until the frame of the mutability of species was firmly established in the nineteenth century.

62. As Tyson acknowledged, Charles Perrault and his circle of French naturalists had already made this case in 1671 (Wokler, "Tyson, Buffon," 2307).

63. Linnaeus could find no physiological difference: "I well know what a splendidly great difference there is [between] a man and a *bestia* . . . [yet] as a naturalist . . . I know scarcely one feature by which man can be distinguished from apes" (cited in Broberg, "*Homo sapiens*," 167). Similarly, in the preface to *Fauna Svecica* (1746), he wrote: "as a natural historian I have yet to find any characteristic which enables man to be distinguished on scientific principles from the ape" (cited in Broberg, "*Homo sapiens*," 170).

64. Buffon accentuated the failure of materialism to explain human difference: "Can there be a more certain proof that matter by itself, no matter how perfectly organized, is incapable of producing either thought or utterance, which signifies that there must be animation by a higher principle?" (*Histoire naturelle*, 14:61, cited in Wokler, "Tyson, Buffon," 2313).

65. Not only was this a conjecture European travelers themselves ventured, but it was one that they were provided by native informants as well. See O'Malley and Magoun, "Early Concepts of the Anthropomorpha," with many citations from the early texts that are quite explicit; and esp. Lorna Schiebinger, "The Gendered Ape: Early Representations of Primates in Europe," in *A Question of Identity* (1993), 119–51; reprinted in Schiebinger, *Nature's Body*, 75–114.

66. Johann Christian Fabricius, *Betrachtungen über die allgemeine Einrichtungen in der Natur* (1781), cited in Meijer, *Race and Aesthetics*, 125.

67. Meijer, *Race and Aesthetics*, 124–25; Schiebinger, "Gendered Ape."

68. Buffon made several comments on the ugliness of "Kalmucks" (Meijer, *Race and Aesthetics*, 54, 144).

69. "In effect, from the last two decades of the eighteenth century onwards, the crucial divisions in anthropology came to be those which distinguished the several races *within* the human species, rather than those that separated man from ape" (Wokler, "Ape Debates," 1174). "With the exhaustion of the Enlightenment discussion of the primate limits of humanity, anthropology came

instead to be focused upon the boundaries and distinctions within our species, upon the study of races, in effect, rather than the study of apes and language." Robert Wokler, "Anthropology and Conjectural History in the Enlightenment," in *Inventing Human Science: Eighteenth-Century Domains*, ed. Christopher Fox, Roy Porter, and Robert Wokler (Berkeley: University of California Press, 1995), 31–52, citing 40.

70. First of all, it was, as Buffon expressly stated, a *humiliation* to be presented with such kinship in ancestry; second, it was generally inconceivable that there could be a departure from the fixity of species, so that these distinct creatures of today could never have been related to or derived one from the other over time.

71. See Phillip Sloan, "The Idea of Racial Degeneracy in Buffon's *Histoire naturelle*," in *Racism in the Eighteenth Century*, ed. H. Pagliaro (Cleveland: Case Western Reserve University Press, 1973), 293–332.

72. Cornelius De Pauw, *A general history of the Americans, of their customs, manners, and colours. An history of the Patagonians, of the Blafards, and White Negroes. History of Peru. An history of the manners, customs, &c. of the Chinese and Egyptians* (1771), trans. Daniel Webb (1806; repr., Rochdale, UK: T. Wood, 1999). See Robert Wokler, "Apes and Races in the Scottish Enlightenment: Monboddo and Kames on the Nature of Man," in *Philosophy and Science in the Scottish Enlightenment*, ed. Peter Jones (Edinburgh: Donald, 1988), 145–68, citing 159–60.

73. "Anthropology sprang from a great thought of Buffon. Up to his time, man had never been studied, except as an individual; Buffon was the first who, in man, studied the species." M. Flourens, "Memoir of Blumenbach," in *Anthropological Treatises*, by J. Blumenbach (London: Longman, 1865), 55.

74. Kathleen Wellman, *La Mettrie: Medicine, Philosophy, and Enlightenment* (Durham, NC: Duke University Press, 1992), 211.

75. Phillip Sloan, "The Gaze of Natural History," in Fox, Porter, and Wokler, *Inventing Human Science*, 112–51, citing 126.

76. Hans-Werner Ingensiep, "Der Mensch im Spiegel der Tier- und Pflanzenseele: Zur Anthropomorphie der Naturwahrnehmung im 18. Jahrhundert," in *Der ganze Mensch: Anthropologie und Literatur im 18. Jahrhundert; DFG Symposion 1992*, ed. Hans-Jürgen Schings (Stuttgart: Metzler, 1994), 54–79, citing 61.

77. And that explains the centrality of the controversy over the origins of language, from Condillac to Herder. See Hans Aarsleff, "The Tradition of Condillac: The Problem of the Origin of Language in the Eighteenth Century and the Debate in the Berlin Academy before Herder," in *Studies in the History of Linguistics*, ed. Dell Hymes (Bloomington: Indiana University Press, 1974), 93–156.

78. La Mettrie wrote in *L'homme machine* on the orangutan: "I would take the great ape in preference to any other, until chance leads us to discover another species more similar to ours. . . . This animal bears such a strong resemblance to us that naturalists have called it the 'wild man' or the 'man of the woods.'" "I hardly doubt at all that if this animal were perfectly trained, we would succeed in teaching him to utter sounds and consequently to learn a language. Then he would no longer be a wild man, nor an imperfect man, but a perfect man." "Why . . . should the education of apes be impossible? . . . I do not presume to decide whether the ape's speech organs will never be able to articulate anything whatever we do, but such an absolute impossibility would surprise me, in view of the close analogy between ape and man." "In general, the form and composition of the quadruped's brain is more or less the same as man's. Everywhere we find the same shape and the same arrangement, with one essential difference: man, of all the animals, is the one with the largest and most convoluted brain, in relation to the volume of his body." Julien Offray de La Mettrie, *Machine Man and Other Writings* (Cambridge: Cambridge University Press, 1996), 9–12.

79. J.-J. Rousseau, *Discourse on the Origins of Inequality* (1754); trans. in J.-J. Rousseau, *Basic Political Writings* (Indianapolis: Hackett, 1987). Robert Wokler, "Perfectible Apes in Decadent Cultures: Rousseau's Anthropology Revisited," *Daedalus* 107 (1978): 107–34; Francis Moran,

"Between Primate and Primitive: Natural Man as the Missing Link in Rousseau's Second Discourse," *Journal of the History of Ideas* 54 (1993): 37–58.

80. Wokler, "Perfectible Apes," 113.

81. James Burnet, Lord Monboddo, *Origin and Progress of Language* (1773; 2nd, improved ed., 1774). Arthur Lovejoy, "Monboddo and Rousseau," in *Essays in the History of Ideas* (Baltimore: Johns Hopkins University Press, 1948), 38–61; Wokler, "Apes and Races," passim; Stefann Blancke, "Lord Monboddo's *Ourang Outang* and the Origin and Progress of Language," in *The Evolution of Social Communication in Primates*, ed. Marco Pina and Nathalie Gontier (Heidelberg: Springer International, 2014); Alan Barnard, "Monboddo's Orang Outang and the Definition of Man," in *Ape, Man, Apeman: Changing Views since 1800* (Leiden: Department of Prehistory, Leiden University, 1995), 71–85.

82. Wokler, "Apes and Races," 158.

83. Peter Camper, "Account of the Organs of Speech of the Orang Outang," *Philosophical Transactions* 69 (1779): 139–59.

84. Pietro Moscati's work appeared in Italian in 1771 and was translated immediately into German: *Von dem körperlichen wesentlichen Unterscheide zwischen der Struktur der Thiere und der Menschen: Eine akademische Rede gehalten auf den anatomischen Theater zu Pavia*, trans. Johann Beckmann (Göttingen, 1771).

85. Kant, "Recension von Moscatis Schrift: Von dem körperlichen . . . ," AA 2:421–26.

86. Camper considered his view ridiculous (Visser, *Camper*, 109).

87. Blumenbach noted: "As to those who make out the erect position to be the fomenter of disorders, they must forget both veterinary practice and the diseases which we find afflict both wretched men and fierce quadrupeds." Johann Friedrich Blumenbach, *On the Natural Variety of Mankind* [*De Generis Humani Varietate Nativa*] (diss., Göttingen, 1775); trans. Thomas Bendyshe as *The Anthropological Treatises of Johann Friedrich Blumenbach* (1865; repr., Boston, MA: Elibron Classics, 2005), 88.

88. Louis-Jean-Marie Daubenton, "Mémoires sur les différences de la situation du grand trou occipital dans l'homme et dans les animaux," in *Histoire de l'Académie des sciences avec les mémoires de mathématique et de physique* (Paris, 1764), 568–79. See Meijer, *Race and Aesthetics*, 110; Claude Blanckaert, "Le trou occipital et la 'craniotomie' comparée des races humaines (XVIIIe–XIXe siècle)," in *Le Trou*, ed. Jacquers Hainard and Roland Kaehr (Neuchâtel: Musée d'Enthnographie, 1990), 253–99.

89. Visser, *Camper*, 35.

90. Thomas Soemmerring, "Etwas vernüftiges über den Orangoutang," *Göttinger Taschen-Kalender auf das Jahr 1781*, 40–64.

91. Visser, *Camper*, 110–111.

92. Indeed, Blumenbach would uphold the priority of this assertion against Herder's view that upright posture wasn't decisive for human nature, in light of a review of both claims by his colleague Christoph Meiners: *GGA* 1 (January 15, 1785): 65–68.

93. Hermann Bräuning-Oktavio, *Vom Zwischenkierferknochen zur Idee des Typus: Goethe als Naturforscher in den Jahren 1780–1786*, Nova Acta Leopoldina: Abhandlungen der Deutschen Akademie der Naturforscher (Leopoldina), n.s., vol. 18, no. 126 (Leipzig: Johann Ambrosius Barth, 1956), 1–140.

94. Meijer (*Race and Aesthetics*, 138) minimizes the significance of this. Visser (*Camper*, 115) takes it more seriously.

95. Meijer, *Race and Aesthetics*, 298.

96. Bräuning-Oktavio, *Vom Zwischenkierferknochen zur Idee des Typus.*

97. Visser, *Camper*, 4, 16–17; Meijer, *Race and Aesthetics*, 10, 27–31.

98. Visser, *Camper*, 5.

99. Ibid., 9.

100. Meijer, *Race and Aesthetics*, 18.

101. Ibid., 10.

102. "After he resigned his professoriate, in 1773, Camper entirely devoted himself to comparative anatomy and became one of the best known scientists in this field" (Visser, *Camper*, 1). "Camper devoted particular attention to zoology. . . . Three of his students wrote dissertations on comparative anatomy" (ibid., 11).

103. Peter Camper, *Natural History of the Orang-Outang, the Double Horned Rhinoceros, and the Reindeer* (Dutch, 1782; German, 1792; French, 1803).

104. Visser, *Camper*, 94.

105. Ibid., 15; Meijer, *Race and Aesthetics*, 16.

106. Visser, *Camper*, 39.

107. Blumenbach began teaching comparative anatomy in 1785. Frank Dougherty, "Der Begriff der Naturgeschichte nach J. F. Blumenbach anhand seiner Korrespondenz mit Jean-André DeLuc—ein Beigtrag zur Wissnschaftsgeschichte bei der Entdeckung der Geschichtlichkeit *ihres* Gegenstandes," in Dougherty, *Gesammelte Aufsätze*, 148–59, citing 155. His textbook appeared only in 1805.

108. He was "very swiftly and everyday more convinced of the truth of Rousseau's [corrected to "Leibniz's" in later editions] explanation that such *anatome comparata* is the living soul of the entire natural history of animals." Peter Camper, *Handbuch der vergleichenden Anatomie* (1805), preface, v–vi.

109. J. G. Zimmermann to J. F. Blumenbach, September 15, 1779, in Dougherty, *Correspondence of Blumenbach*, 1:177.

110. On Blumenbach and Camper, see Carlos Gysel, "Les relations du jeune Blumenbach avec Camper viellissant," *Histoire des sciences medicales* 17 (1983): 135–39. The correspondence and documentation is more complete and accurate in Dougherty, *Correspondence of Blumenbach*, vol. 1. See also Meijer, *Race and Aesthetics*, 170.

111. Blumenbach did, however, acknowledge Camper's eminence and influence with some good grace.

112. Camper himself was inspired to adopt the new vision of natural history by his encounters with Buffon in Paris in 1749: "Camper was fascinated by Buffon's ideas about natural science, since they gave biology, which had hitherto only been able to collect phenomenological facts, a scientific foundation." Antonie Luyendijk-Elshout, "'Les beaux esprits se rencontrent': Petrus Camper und Samuel Thomas Soemmerring," in Mann and Dumont, *Samuel Thomas Soemmerring und die Gelehrten der Goethezeit*, 57–72, citing 62. Thus, there is a very strong line of affiliation in the constitution of the field.

113. Gascoigne, "Blumenbach, Banks," 92.

114. "The Göttingen medical faculty is to be regarded without question as the premier and most distinguished one at the end of the eighteenth century in Germany." Julius Leopold Pagel, *Über die Geschichte der Göttinger medizinische Schule im XVIII. Jahrhundert* (diss., Berlin, 1875), 8. See also Max Neuburger, "British Medicine and the Göttingen Medical School in the Eighteenth Century," *Bulletin of the History of Medicine* 14 (1943): 449–66; Ulrich Tröhler, "250 Jahre Göttinger Medizin: Begründung—Folgen—Folgerungen," in *Naturwissenschaften in Göttingen: Eine Vortragsreihe* (Göttingen: Vandenhoeck und Ruprecht, 1988), 9–36; Götz von Selle, *Die Georg-August-Universität zu Göttingen, 1737–1937* (Göttingen: Vandenhoeck und Ruprecht, 1937); Luigi Marino, *Praeceptores Germaniae: Göttingen, 1770–1820* (Göttingen: Vandenhoeck und Ruprecht, 1995).

115. On Heyne's role at the university, see Friedrich Leo, "Heyne," in *Festschrift zur Feier des hundertfünfzigjährigen Bestehens der Königlichen Gesellschaft der Wissenschaften zu Göttigen: Beiträge sur Gelehrtengeschichte Göttingens* (Berlin: Weidmann, 1901), 153–234. On Heyne and Haller, see Frank Dougherty, ed., *Christian Gottlob Heyne's Correspondence with Albrecht and Gottlieb Emanuel von Haller* (Göttingen: Klatt, 1997).

116. J. F. Blumenbach, Autobiographical Statement, reproduced in Marx, "Life of Blumenbach," 4.

117. On the acquisition of the Büttner collection by the university, see the correspondence reproduced in Dougherty, *Correspondence of Blumenbach*, 1:1–4. See also Rudolph Wagner, "On the Anthropological Collection of the Physiological Institute of Göttingen," in Bendysche, *Anthropological Treatises of John Friedrich Blumenbach*, 347–55; Hans Plischke, *Die Ethnographische Sammlung der Universität Göttingen, ihre Geschichte und ihre Bedeutung* (Göttingen: Vandenhoeck und Ruprecht, 1931).

118. Blumenbach, *On the Natural Variety of Mankind*, 69–70; and later testimonials.

119. Blumenbach, Autobiographical Statement, reproduced in Marx, "Life of Blumenbach," 5.

120. Ibid.

121. Pagel, *Über die Geschichte der Göttinger medizinische Schule im XVIII. Jahrhundert*; Rupke, *Göttingen and the Development of the Natural Sciences.*

122. See William Clark, "From Enlightenment to Romanticism: Lichtenberg and Göttingen Physics," in Rupke, *Göttingen and the Development of the Natural Sciences*, 72–85.

123. J. L Heilbron, "Physics and Its History at Göttingen around 1800," in Rupke, *Göttingen and the Development of the Natural Sciences*, 50–71.

124. Clark, "From Enlightenment to Romanticism," 82.

125. Ibid., 80–81. See Karl Hufbauer, *The Formation of the German Chemical Community* (Berkeley: University of California Press, 1982), 82ff.

126. Lichtenberg introduced Deluc to Blumenbach in 1775, when he brought Deluc to Göttingen.

127. On Soemmerring, see F. Dumont and G. Mann, eds., *Gehirn—Nerven—Seele: Anatomie und Psychologie im Umfeld S. Th. Soemmerrings*, Soemmerring-Forschungen, vol. 3 (Stuttgart: G. Fischer, 1988); Franz Dumont, "Einführung in die Soemmerring-Briefedition," in *Briefwechsel, 1761/65–Oktober 1784*, in *Werke*, vol. 18 (Stuttgart: G. Fischer, 1996).

128. Baldinger's comment was printed in Latin on the official version of Soemmerring's dissertation. Soemmerring confirmed this promise with his masterpiece, the multivolume *Vom Bau des menschlichen Körpers*, of the 1790s.

129. Samuel Thomas Soemmerring, curriculum vitae, 15, cited in Marx, *Life of Blumenbach*, 28n.

130. Camper became Forster's lifelong mentor and friend. See Hans Querner, "Samuel Thomas Soemmerring und Johann Georg Forster—eine Freundschaft," in Mann and Dumont, *Samuel Thomas Soemmerring und die Gelehrten der Goethezeit*, 229–44.

131. After Göttingen, chairs in natural history appeared in German universities in the following sequence: Halle, 1769; Erlangen, 1770; Innsbruck, 1774; Kiel, 1775; Freiburg im Breisgau, 1775; Greifswald, 1781; Jena, 1786; Frankfurt am Oder, 1788; Würzburg, 1788; Rostock, 1791; and Dorpat, 1802. Irmtraut Scheele, "Grundzüge der institutionellen Entwicklung der biologischen Disziplinen an den deutschen Hochschulel seit dem 18. Jahrhundert," in *"Einsamkeit und Freiheit" neu besichtigt. Universitätsreformen und Disziplinenbildung in Preußen als Modell für Wissenschaftspolitik im Europa des 19. Jahrhunderts* (Stuttgart: Steiner, 1991), 144–54, citing 149.

132. E. A. W. Zimmermann, *Geographische Geschichte des Menschen, und der vierfüßigen Thiere*, 3 vols. (Leipzig: Weygand, 1778–83). See Petra Feuerstein-Herz, *Eberhard August Wilhelm von Zimmermann (1743–1815) und die Tiergeographie* (Braunschweig: Technische Universität Braunschweig, 2004).

133. Blumenbach's journal notice of meeting G. Forster, November 21, 1777, cited in Marx, "Life of Blumenbach," 31n.

134. See Joseph Gordon, "Reinhold and Georg Forster in England, 1766–1780" (PhD diss., Duke University, 1975), esp. 265–320.

135. On Forster, Soemmerring, and Kassel, see Querner, "Samuel Thomas Soemmerring und Johann Georg Forster"; Franz Dumont, "Naturkenntnis—Weltkenntnis: Das 'Seelenbündnis' zwischen Georg Forster und Samuel Thomas Soemmerring," in Dumont and Mann, *Gehirn—Nerven—Seele*, 381–440.

136. On European race discourse in the eighteenth century, a good place to start is Roxann

Wheeler, *The Complexion of Race: Categories of Difference in Eighteenth-Century British Culture* (Philadelphia: University of Pennsylvania Press, 2000). See also an anthology of key texts from the period: Emmanuel Chukwudi Eze, ed., *Race and the Enlightenment: A Reader* (Oxford: Blackwell, 1997); and another anthology extending farther into the nineteenth century: H. F. Augstein, ed., *Race: The Origins of an Idea, 1760–1850* (Bristol: Thoemmes, 1996).

137. Blumenbach, *On the Natural Variety of Mankind*, 775. Indeed, Blumenbach was among those discussing it. In two anonymous articles in the *Göttingische Taschen-Calender vom Jahr 1776*, he addressed the topics of "anthropology" and the "variety of mankind." (See A. G. Kästner's report in *GGA* 2 [October 24, 1775]: 1089–91, esp. 1090–91.) He continued to address this topic in this journal in an essay for the 1778 volume: "Physiologie des Laufes menschlichen Lebens," *Göttinger Taschen-Calender vom Jahr 1778*, 38. The issue of "race"—especially with regard to Africans—was becoming a very testy issue as criticism of colonialism, slavery, and their attendant inhumanities entered the German public sphere. In this context, the pronouncements of Christoph Meiners at Göttingen deserve more attention. The fierce public dispute between Blumenbach and Meiners on these matters in the 1790s originated, in my view, here in the mid-1770s. Frank Dougherty, "Christoph Meiners und Johann Friedrich Blumenbach im Streit um den Begriff der Menschenrasse," in *Die Natur des Menschen: Probleme der Physischen Anthropologie und Rassenkunde (1750–1850)*, ed. Gunter Mann and Franz Dumont, Soemmerring Forschungen 6 (Stuttgart: G. Fischer, 1990), 89–111; reprinted in Dougherty, *Gesammelte Aufsätze*, 176–90. See my "Policing Polygeneticism in Germany, 1775." Blumenbach would face similar conflicts on these matters with his colleague Soemmerring after 1784.

138. Henry Home, Lord Kames, *Sketches of the History of Man* (Edinburgh: W. Creech; London: Strahan and Cadell, 1774), 6–10; Sloan, "Buffon, German Biology, and the Historical Interpretation of Biological Species," 124.

139. For a grand conspectus of this issue in the eighteenth century, see Henri Daudin, *De Linné à Jussieu: Méthodes de la classification et idée de série en botanique et en zoologie (1740–1790)* (Paris: Alcan, 1926).

140. A major source for this idea came from the work in archaeology of Christian Heyne. See Lenoir, "Göttingen School and the Development of Transcendental *Naturphilosophie*," 127. On Heyne's epochal influence, see Hermann Bräuning-Oktavio, *Christian Gottlob Heynes Vorlesungen über die Kunst der Antike und ihr Einfluß auf Johann Heinrich Merck, Herder und Goethe* (Darmstadt: Liebig, 1971).

141. Blumenbach would be far clearer about all this in later editions of the dissertation and especially in his famous *Handbuch der Naturgeschichte*.

142. Blumenbach, *On the Natural Variety of Mankind*, 69.

143. "In the history of literature, Blumenbach emulated his original and pattern, Albrecht von Haller." Marx also notes that "scarcely anyone was so well acquainted with all the writings of that most famous of Göttingen teachers" ("Life of Blumenbach," 15, 15n).

144. Johann Friedrich Blumenbach, *Institutiones Physiologiae* (Göttingen: Dieterich, 1787); trans. Joseph Eyerel as *Anfangsgründe der Physiologie* (Vienna: Wappler, 1789); trans. John Elliotson as *Institutions of Physiology* (London: Bensley and Son, 1817). Johann Friedrich Blumenbach, *Handbuch der vergleichenden Anatomie* (Göttingen: Dieterich, 1805); trans. William Lawrence, rev. and augmented by William Coulson, as *A Manual of Comparative Anatomy*, 2nd ed. (London: Simpkin and Marshall, 1827).

145. Blumenbach, *On the Natural Variety of Mankind*, 73. Indeed, for the rest of his life Blumenbach actively investigated hybrid fertility and collected documentation in a massive research dossier. See L. V. Karolyi, foreword to *Über den Bildungstrieb und das Zeugungsgeschäfte* (repr., Stuttgart: G. Fischer, 1971), viii.

146. Blumenbach, *On the Natural Variety of Mankind*, 75–76.

147. Ibid., 77–80. See Conrad Zirkle, "The Jumar," *Isis* 33 (1947): 486–506.

148. Blumenbach, *On the Natural Variety of Mankind*, 73.

149. Ibid., 74.

150. Ibid.

151. Ibid., 83.

152. Ibid., 86–88.

153. Blumenbach was provoked by Meiners's review of the first volume of Herder's *Ideen*, where the issue of priority was raised. Christoph Meiners, review of *Ideen*, vol. 1, by J. G. Herder, *GGA* 1 (January 15, 1785): 65–68. See Meiners's letter to Blumenbach in *Correspondence of Johann Friedrich Blumenbach*, vol. 2, *1783–1785*, ed. Frank Dougherty (Göttingen: Klatt, 2007), 233–34.

154. Blumenbach, *On the Natural Variety of Mankind*, 83. Blumenbach took a far less sanguine view of Moscati's achievement than did Kant.

155. Ibid., 91.

156. Ibid., 95.

157. Ibid. Blumenbach believed at this time that there were no anatomical or physiological differences in apes to account for their failure at speech: "I have myself found the uvula in apes, and the other parts of the larynx exactly like those in man" (ibid., 84n).

158. Ibid., 96.

159. Ibid., 97.

160. Ibid., 96.

161. Ibid., 93.

162. Bräuning-Oktavio, *Vom Zwischenkierferknochen zur Idee des Typus*.

163. Blumenbach, *On the Natural Variety of Mankind*, 98.

164. Ibid., 98–99.

165. Ibid., 109.

166. Ibid., 81. Even Buffon, in "Nomenclature des singes" (*Histoire naturelle*, vol. 14 [1766]), had indulged in this fantasy of apes fornicating with women, which allowed for overtones of the simian origins of the black race.

167. Blumenbach, *On the Natural Variety of Mankind*, 100.

168. Ibid., 101.

169. Ibid.

170. Ibid., 103.

171. Ibid.

172. Kant, "Von den verschiedenen Racen der Menschen" (1775/1777), AA 2:427–44.

173. Blumenbach, *On the Natural Variety of Mankind*, 111.

174. Ibid., 107.

175. This was a decisive difference in Blumenbach's theory of human variety in 1775 relative to that developed by Kant. Indeed, irreversibility was central to Kant's whole theory of generation, but it was not *initially* accepted by Blumenbach (or *ever* by Buffon). Blumenbach's conversion to irreversibility in 1795 is the clearest evidence of Kant's influence on his later theories of race. See Robert Bernasconi, "Who Invented the Concept of Race?," in *Race*, ed. Robert Bernasconi (Oxford: Blackwell, 2001), 11–36.

176. Blumenbach, *On the Natural Variety of Mankind*, 113.

177. Ibid., 115.

178. Visser, *Camper*, chap. 4: "Man's Place in Nature," makes clear that while Camper had taken a clear stance against claims of the racial inferiority of blacks in an inaugural oration of 1764, he was more famous for his lectures at Groningen in the 1770s, in which he developed his idea of the "facial angle." For the most recent and complete consideration of this matter, see Meijer, *Race and Aesthetics*.

179. See Stephen Jay Gould, "The Geometer of Race" (1994); reprinted in *The Concept of "Race" in Natural and Social Science*, ed. Nathaniel Gates (New York: Garland, 1997), 1–5; and, more extensively, *The Mismeasure of Man* (New York: W. W. Norton, 1981). See also Thomas Junker, "Blumenbach's Racial Geometry," *Isis* 89 (1998): 490–501, for a good corrective of Gould. And see Carlos

Gysel, "L'anthropologie crânio-faciale de Blumenbach (1752–1840)," *L'orthodontie française: Extraits du volume 52*** *1981*, 707–24.

180. *The Works of the late Professor Camper, on the Connexion between the Science of Anatomy and the Arts of Drawing, Painting, Statuary, etc, in Two Books containing a Treatise on the Natural Difference of Features in Persons of Different Countries and Periods of Life; and on Beauty, as exhibited in ancient sculpture; with a new method of sketching heads, national features, and portraits of individuals, with accuracy, etc.*, trans. Thomas Cogan (London: Dilly, 1794). Originally published in Dutch in 1791.

181. Blumenbach, *On the Natural Variety of Mankind*, 93, 97.

182. Ibid., 116–17.

183. Ibid., 117.

184. Ibid., 121.

185. Blumenbach, "Über Künsteleyen oder zufällige Verstümmelungen am thierischen Körper, die mit der Zeit zum erblichen Schlag ausgeartet," *Magazin für das Neueste aus der Physik und Naturgeschichte* 6 (1789): 13–23.

186. Therefore, we need to recognize *change* in Blumenbach's view over the course of his writings, perhaps signaled by his actual adoption in later writings of the term "race" and perhaps also under the direct influence of Immanuel Kant's views, as Robert Bernasconi has argued. See Bernasconi, "Who Invented the Concept of Race?"

187. Blumenbach, *On the Natural Variety of Mankind*, 99.

188. Ibid. See the original Latin version, reprinted in Robert Bernasconi, ed., *The Concept of Race in the Eighteenth Century* (Bristol: Sterling, 2001), 4:41. This perplexity in the text has not been noted. I believe he meant North *Africa*, as his later accounts clearly state. Perhaps this anomaly in the text went unobserved because by the time the English translation was made (1865), most of the North American population (i.e., immigrants from Europe) were naturally identified with the white race, especially by an Anglo-Saxon writer.

189. Blumenbach, *On the Natural Variety of Mankind*, 99.

190. Ibid.

191. Blumenbach introduced the "Malay" variety in the first part of his original edition of the *Handbuch der Naturgeschichte* (1779) and incorporated it into the revised edition of his dissertation in 1781 (which Bendysche and a tradition following his authority have falsely taken as the first instance of Blumenbach's famous five-variety theory). On Blumenbach and the "Malay" variety, see Hans Plischke, "Die malayische Varietät Blumenbachs," *Zeitschrift für Rassenkunde* 8 (1938): 225–31.

192. "[T]he Pacific . . . , to European eyes, was a new world offering a virtual laboratory to test theories about nature in general and the development of humankind in particular" (Gascogine, "Blumenbach, Banks," 90). See Bernard Smith, *European Vision and the South Pacific, 1768–1850*, 2nd. ed. (New Haven, CT: Yale University Press, 1985); Bernard Smith, *Imagining the Pacific in the Wake of the Cook Voyages* (New Haven, CT: Yale University Press, 1992).

193. "Blumenbach was able to add to his museum in 1782 more than 350 objects from the second and third voyages of James Cook (1728–1779) thanks to a gift by George III (1738–1820)" (Gascoigne, "Blumenbach, Banks," 88).

194. For details of this, see the correspondence assembled in Dougherty, *Correspondence of Blumenbach*, 1:70–71.

195. The candid criticisms of Wrisberg in Blumenbach's correspondence with Georg Brandes and C. G. Heyne make all this clear. See ibid., 33–34 (and see notes to letter from privy council to Heyne, July 11, 1775, in ibid., 25–27, esp. n. 9).

196. Johann Friedrich Blumenbach, *Prolusio Anatomica de Sinibus Frontalibus*; Blumenbach, announcement in *GGA* 2 (September 16, 1779): 961–65.

197. J. G. Zimmermann to Blumenbach, September 15, 1779, in Dougherty, *Correspondence of Blumenbach*, 1:177.

198. Particularly interesting in this regard is the correspondence Blumenbach undertook with his classmate Simon Ludwig Eberhard de Marées in 1775–76. We have only de Marées's letters, but they suggest amply that he and Blumenbach wrangled through details of Haller's physiology together, and not uncritically. See esp. de Marées to Blumenbach, January 6, 1776, in Dougherty, *Correspondence of Blumenbach*, 1:58–63.

199. Johann Friedrich Blumenbach, "Physiologie des Laufes menschlichen Lebens," *Göttinger Taschen-Calender vom Jahr 1778*, 38, cited in Daugherty, *Correspondenceof Blumenbach*, 1:239.

200. See the opening section of the book version: Johann Friedrich Blumenbach, *Über den Bildungstrieb und das Zeugungsgeschäfte* (Göttingen: Dieterich, 1781; repr., Stuttgart: G. Fischer, 1971).

201. Note the strikingly different treatment that the established Blumenbach gave the respective views of Haller and Bonnet in such later works as his third edition of *On the Natural Variety of Mankind* (Göttingen: Vandenhoeck und Ruprecht, 1795); trans. in Bendysche, *Anthropological Treatises*.

202. Georg Forster, *A Voyage Round the World*, 2 vols. (London: B. White et al., 1777).

203. J. R. Forster, *Observations Made on a Voyage Round the World*, reprint of 1778 text with commentaries by N. Thomas, H. Guest, and M. Dettelbach (Honolulu: University of Hawai'i Press, 1996).

204. Querner, "Samuel Thomas Soemmerring und Johann Georg Forster," 232–33.

205. Johann Caspar Lavater, *Essays on physiognomy: For the promotion of the knowledge and the love of mankind* (Boston: Spotswood and West, 1794).

206. On Lichtenberg on Lavater, see, e.g., Kevin Joel Berland, "'The Air of a Porter': Lichtenberg and Lavater Test Physiognomy by Looking at Dr. Johnson," *Age of Johnson* 10 (1999): 219–30.

207. Kant, "Von den verschiedenen Racen der Menschen."

208. See J. J. Engel, ed., *Philosophie für die Welt*, Klassiker des In- und Auslandes, 13 (1777; repr., Berlin: Hofmann, 1860). I have discussed the incongruity of this publication context for Kant and its significance in "Policing Polygeneticism in Germany, 1775"; and in my monograph *Kant, Herder, and the Birth of Anthropology* (Chicago: University of Chicago Press, 2002).

209. Bernhard Christoph Breitkopf to Kant, March 21, 1778, and response from Kant, April 1, 1778, in *Kant Briefwechsel*, AA 10:227–28, 229–30.

210. Blumenbach refers to Kant's 1777 essay in the second edition of his book on human variety, published in 1781, but, as van Hoorn (*Dem Leibe abgelesen*, 98n) notes, without any indication of intellectual influence. Lovejoy was wrong to think either Kant or Blumenbach knew of or drew upon the other's work in composing their respective 1775 essays on race. See Arthur Lovejoy, "Kant and Evolution," in *Forerunners of Darwin: 1745–1859*, ed. Bentley Glass et al. (Baltimore: Johns Hopkins University Press, 1959), 173–206.

211. Zimmermann, *Geographische Geschichte des Menschen*.

212. See van Hoorn, *Dem Leibe abgelesen*, 105–7, for an excellent discussion of Zimmermann's prominence. She establishes quite convincingly that his work "must be considered the synthesis of the contemporary state of science and debate" (107).

213. J. R. Forster, *Observations Made on a Voyage Round the World*.

214. Blumenbach, *Handbuch der Naturgeschichte* (Göttingen: Dieterich, 1779); Blumenbach, *Handbuch der Naturgeschichte, Zweyter Theil* (Göttingen: Dieterich, 1780): continuously paginated (with an appended, unpaginated Register for both volumes); all subsequent editions would be bound as a single volume.

215. Part 1 dealt with "Objects of Nature Generally: their division into three realms, etc." Part 2 dealt with "Organized Bodies in General." Part 3 treated "Animals in General." Part 4 focused on mammals; part 5, on birds; part 6, on amphibians; part 7, on fish; part 8, on insects; and part 9, on "worms."

216. Klatt, the editor of the second volume of Blumenbach's correspondence, notes: "Not seldomly does the impression present itself that Blumenbach in these years was more of a mineralogist than a zoologist or physiologist" (introduction to *Correspondence of Blumenbach*, 2:xxvi).

217. See Kästner's report on Blumenbach's presentation to the Göttingen Academy of a classification scheme for mammals: *GGA* 2 (December 9, 1775): 1257–59. In part4, §54, Blumenbach elaborated: "We have attempted . . . to develop a natural system of mammals according to which, in accordance with our concepts of the natural method (§7), we have looked, not at a few, abstract characteristics, but at all external ones, the total habitus of animals" (Blumenbach, *Handbuch der Naturgeschichte* [1779], §54, 56–57). In citations of the *Handbuch*, section numbers are followed by page numbers when relevant.

218. Blumenbach, *Handbuch der Naturgeschichte* (1779), §6. Blumenbach described his research on moss in an article published separately: "Über eine ungemein einfache Fortpflanzungsart," *Göttingischen Magazin der Wissenschaften und Litteratur* 2, no. 1 (1781): 80–89.

219. Ibid., §7, 10–14.

220. Ibid., 13.

221. Ibid., §35–38, esp. §35, 39–41.

222. Ibid., §54, 57.

223. Ibid., 62.

224. Ibid., 63.

225. Ibid., 64.

226. Ibid., §10, 17.

227. Ibid., 18.

228. Ibid., §11, 19.

229. Ibid., §12, 20.

230. Georg Lichtenberg and Georg Forster, eds., *Göttingisches Magazin der Wissenschaften und Litteratur*, 1780–85.

231. J. F. Blumenbach, "Über den Bildungstrieb (*Nisus formativus*) und seinen Einfluß auf die Generation und Reproduction," *Göttingisches Magazin der Wissenschaften und Litteratur* 1, no. 5 (1780): 247–66.

232. In 1774 Blumenbach, still a medical student, made a presentation to the Göttingen Academy about his discovery of a new species of freshwater polyp in the environs of Göttingen. See Kästner's verbatim report of Blumenbach's presentation to the academy in *GGA* 2 (December 1774): 1009–11. This was the basis for experiments he then undertook, directly inspired by Trembley's, in the spring of 1778, leading to his discovery of the *Bildungstrieb*, announced in another essay for that journal in 1780. The same year, Blumenbach published a more extensive version of this report: "Von der Federbusch-Polypen in den Göttingener Gewässern," *Göttingisches Magazin der Wissenschaften und Litteratur* 1, no. 4 (1780): 117–27.

233. Blumenbach, "Über den Bildungstrieb (*Nisus formativus*)," 250–51.

234. Rudolf Vierhaus, "Bildung," in *Geschichtliche Grundbegriffe* (Stuttgart: Klett-Cotta, 1997), 1:509–51; Walter Bruford, *Culture and Society in Classical Weimar, 1775–1806* (Cambridge: Cambridge University Press, 1962).

235. See Peter Hanns Reill, "*Bildung, Urtyp* and Polarity: Goethe and Eighteenth-Century Physiology," *Goethe Yearbook* 3 (1986): 139–48.

236. Blumenbach, "Über den Bildungstrieb (*Nisus formativus*)," 251n.

237. Blumenbach, *Über den Bildungstrieb und das Zeugungsgeschäfte*, 6. All citations will be taken from the 1971 reprint edition, and page numbers will follow section numbers, where section numbers are given.

238. Ibid., 7.

239. Ibid., §4, 14–15.

240. Ibid., §5, 17–18.

241. Ibid., §16, 27–28; §17, 28–30.

242. Ibid., §20, 33.

243. Joseph Gottlieb Kölreuter, *Vorläufige Nachricht von einigen das Geschlecht der Pflanzen betreffenden Versuchen und Beobachtungen, nebst Fortsetzungen 1, 2 und 3*, Ostwald's Klassiker der exakten Wissenschaften, no. 41 (Leipzig, 1893).

244. Blumenbach, *Über den Bildungstrieb und das Zeugungsgeschäfte*, §35, 61.

245. See Stuart Gillespie and Donald McKenzie, "Lucretius and the Moderns," and Eric Baker, "Lucretius in the European Enlightenment," both in *The Cambridge Companion to Lucretius*, ed. Stuart Gillespie and Philip Hardie (Cambridge: Cambridge University Press, 2009), 306–24 and 274–88, respectively.

246. Blumenbach, *Über den Bildungstrieb und das Zeugungsgeschäfte*, §21, 38.

247. Ibid., §54, 62.

248. Ibid., 64.

249. Blumenbach, *Handbuch der Naturgeschichte*, 2nd ed. (Göttingen: Dieterich, 1782), §11, 15.

250. Ibid., §10, 14.

251. Ibid., 15.

252. Ibid., §14, 17–18.

253. Ibid., §214, 471.

CHAPTER EIGHT

1. Kant, *Kritik der Urteilskraft*, §80, AA 5:418.

2. Johann Friedrich Blumenbach, *Handbuch der Naturgeschichte* (Göttingen: Dieterich, 1779), §39, 43–44. In citations of the *Handbuch*, section numbers are followed by page numbers when relevant.

3. Johann Friedrich Blumenbach, *Handbuch der Naturgeschichte, Zweyter Theil* (Göttingen: Dieterich, 1780), pt. 16, §§243–51, 541–59. See also Blumenbach, *Handbuch der Naturgeschichte*, 2nd ed. (Göttingen: Dieterich, 1782), §§245–51, 543–61. Blumenbach differentiated various states of decomposition and metamorphosis of organic remains in §248.

4. Blumenbach, *Handbuch der Naturgeschichte, Zweyter Theil* (1780), §221, 473–74. (One sentence in the original has been divided, for English sensibilities, into two in my translation.) This passage is replicated verbatim in the second edition (1782), 476–77. It remained the basis for all subsequent versions up through 1803.

5. Blumenbach, *Handbuch der Naturgeschichte, Zweyter Teil* (1780), §222, 474 (1782 ed., 477).

6. Ibid.

7. Johann Friedrich Blumenbach, review of *Lettres physiques et morales sur l'histoire de la terre et de l'homme: Addressées à la reine de la Grande-Bretagne* (1779/80), by Jean-André Deluc, *GGA*, Zugabe to vol. 49 (December 2, 1780): 769–76.

8. Thus, the famous line from his *Beyträge zur Naturgeschichte* in 1790: "Every paving stone in Göttingen is a proof that species, or rather whole genera, of creation have disappeared." Johann Friedrich Blumenbach, *Beyträge zur Naturgeschichte, Erster Theil* (Göttingen: Dietrich, 1790). Trans. Thomas Bendyshe in *The Anthropological Treatises of Johann Friedrich Blumenbach* (1865; repr., Boston: Elibron Classics, 2005). I will cite the English translation, amending it when appropriate; here, 283.

9. Blumenbach, review of *Lettres physiques et morales*, 770.

10. Ibid., 770–71.

11. Ibid., 771.

12. Ibid., 772.

13. Ibid., 775.

14. Johann Friedrich Blumenbach, *Handbuch der Naturgeschichte*, 5th ed. (Göttingen: Dieterich, 1797), 511n. See Jean André Deluc, *Letters on the Physical History of the Earth addressed to Professor Blumenbach containing geological and historical proofs of the divine mission of Moses* (1796; London: Rivington, 1831). From start to finish, then, there are strong reasons to place Deluc's "geohistory" at the fountainhead of Blumenbach's paleontology. But this has not gone uncontested. See Frank Dougherty, "Der Begriff der Naturgeschichte nach J. F. Blumenbach anhand seiner Korrespon-

denz mit Jean-André DeLuc—ein Beigtrag zur Wissnschaftsgeschichte bei der Entdeckung der Geschichtlichkeit *ihres* Gegenstandes," in *Gesammelte Aufsätze zu Themen der klassischen Periode der Naturgeschichte* (Göttingen: Klatt, 1996), 148–59.

15. Blumenbach, *Handbuch der Naturgeschichte, Zweyter Teil* (1780), §222, 475 (1782 ed., 478).

16. On the notion of the preadamite, see Richard Popkin, *Isaac La Peyrère (1596–1676): His Life, Work and Influence* (Leiden: Brill, 1987), 115–65.

17. Blumenbach, *Handbuch der Naturgeschichte, Zweyter Teil* (1780), §223, 475 (1782 ed., 478).

18. Blumenbach, *Handbuch der Naturgeschichte*, 3rd ed. (Göttingen: Dieterich, 1788).

19. Blumenbach, *Handbuch der Naturgeschichte*, 4th ed. (Göttingen: Dieterich, 1791).

20. Johann Friedrich Blumenbach, *Beyträge zur Naturgeschichte, Erster Theil* (Göttingen: Dieterich, 1790).

21. Ibid., Vorrede, unpaginated.

22. Johann Friedrich Blumenbach, "Beyträge zur Naturgeschichte der Vorwelt," *Magazin für den Neueste aus der Physik und Naturgeschichte* 6, no. 4 (1790): 1–17.

23. Johann Friedrich Blumenbach, *Beyträge zur Naturgeschichte, Zweyter Theil* (Göttingen: Dieterich, 1806).

24. Blumenbach, "Beyträge zur Naturgeschichte der Vorwelt." Notable in the essay is Blumenbach's sharp criticism of fossil collectors who were oblivious to the location from which they extracted fossils and who did not undertake careful enough comparisons with living forms. This permitted many writers to indulge in serious errors concerning extinction, among whom, tacitly, he placed Hutton. See the immediately following article, "Dr Hutton's Theorie der Erde," *Magazin für den Neueste aus der Physik und Naturgeschichte* 6, no. 4 (1790): 17–27, with its *Vorerrinnerung* by Blumenbach making explicit reference to Deluc's critique of Hutton.

25. Johann Friedrich Blumenbach, "[Report on] 'Specimen Archaeologiae Telluris,'" *GGA* 199 (December 12, 1801): 1977–84; Johann Friedrich Blumenbach, *Specimen Archaeologiae Telluris Terrarumque Inprimis Hannoverarum* (Göttingen: Dieterich, 1803).

26. Johann Friedrich Blumenbach, "Appendix: On the Succession of Different Earth-Catastrophes," in *Beyträge zur Naturgeschichte, Erste Theil*, 2nd ed. (Göttingen: Dieterich, 1806), trans. in Bendyshe, *Anthropological Treatises*, 317–21.

27. Blumenbach, *Specimen Archaeologiae Telluris Terrarumque Inprimis Hannoverarum*, §7: "Specimina Aliquot Huius Classis Fossilium," 16–18; §10: "Recensus Nonnullorum Huius Classis Fossilium," 21–26.

28. Blumenbach, *Beyträge zur Naturgeschichte, Erster Theil* (1790), trans. 283n, amended.

29. Blumenbach, "Appendix: On the Succession of Different Earth-Catastrophes," 317–18.

30. Ibid., 318.

31. Blumenbach, *Beyträge zur Naturgeschichte, Erster Theil*, 2nd ed. (1806), trans. 285n.

32. Blumenbach, "Appendix: On the Succession of Different Earth-Catastrophes," 318.

33. Martin Rudwick, ed., *Georges Cuvier, Fossil Bones, and Geological Catastrophes: New Translations and Interpretations of the Primary Texts* (Chicago: University of Chicago Press, 1997).

34. Blumenbach, "Appendix: On the Succession of Different Earth-Catastrophes," 320–21.

35. Blumenbach, *Beyträge zur Naturgeschichte, Erster Theil* (1790), §3, trans. 285.

36. Ibid.

37. Ibid., 286.

38. Blumenbach, "Appendix: On the Succession of Different Earth-Catastrophes," 319.

39. Blumenbach, *Beyträge zur Naturgeschichte, Erster Theil* (1790), §3, trans. 285.

40. Blumenbach, "Appendix: On the Succession of Different Earth-Catastrophes," 321.

41. Ibid.

42. Walter Baron, "Die Anschauung Johann Friedrich Blumenbachs über die Geschichtlichkeit der Natur," pt. 1 of Walter Baron and Bernhard Sticker, "Ansätze zur historischen Denkweise in der Naturforschung an der Wende vom 18. zum 19. Jahrhundert," *Sudhofs Archiv* 47 (1963): 19–26, citing 23.

43. Ibid., 22. We do not know exactly who Blumenbach had in mind, but Buffon must surely have been one key figure here, and probably Camper as well.

44. Ibid., 25.

45. Ibid., 26.

46. Ibid., 21.

47. In his report in *GGA* concerning his lecture of 1801, Blumenbach stressed this as the final point to which his exposition had led: the "most sriking [*merkwürdiges*] problem" of species persistence versus extinction. Blumenbach, "Report on 'Specimen Archaeologiae Telluris,'" 1983.

48. Martin Rudwick, *Bursting the Limits of Time: The Reconstruction of Geohistory in the Age of Revolution* (Chicago: University of Chicago Press, 2005), 299.

49. Blumenbach, *Beyträge zur Naturgeschichte, Erster Theil* (1790), §5, trans. 288.

50. Ibid., 290.

51. Ibid., 289.

52. Ibid., §4: "Umschaffung der Vorwelt," 24 (my trans.; cf. Bendyshe, *Anthropological Treatises*, 287).

53. Ibid., 24 (my trans.; cf. Bendyshe, *Anthropological Treatises*, 287).

54. Ibid., 25 (modifying Bendyshe, *Anthropological Treatises*, 287).

55. Ibid., 26 (my trans.; cf. Bendyshe, *Anthropological Treatises*, 287).

56. Christoph Girtanner, *Über das Kantische Prinzip für die Naturgeschichte* (Göttingen: Vandenhoeck und Ruprecht, 1796).

57. In the rest of Blumenbach's exposition of his *Beyträge*, the topic of degenerations, or *Ausartungen*, would carry him into matters more remote to paleontology, especially his notions of racial variety.

58. Blumenbach, *Beyträge zur Naturgeschichte, Erster Theil* (1790), §5, trans. 288.

59. Blumenbach, "Report on 'Specimen Archaeologiae Telluris,'" 1983–84.

60. Blumenbach, *Beyträge zur Naturgeschichte, Erster Theil*, 2nd ed. (1806), trans. 287n.

61. On Buffon and the indestructibility of "organic molecules" and the perennial possibilities of "internal molds," see John H. Eddy Jr., "Buffon's *Histoire naturelle*: History? A Critique of Recent Interpretations," *Isis* 85 (1994): 644–61; John H. Eddy Jr., "Buffon, Organic Alterations, and Man," in *Studies in History of Biology* 7 (Baltimore: Johns Hopkins University Press, 1984), 1–46.

62. Blumenbach, "Report on 'Specimen Archaeologiae Telluris,'" 1983–84.

63. Dougherty develops this aspect of Blumenbach's thought extensively in "Der Begriff der Naturgeschichte nach J. F. Blumenbach anhand seiner Korrespondenz mit Jean-André DeLuc."

64. This would result in the conflict between Cuvier and Lamarck, then the famous controversy with Saint-Hilaire. See Toby Appel, *The Cuvier-Geoffroy Debate: French Biology in the Decades before Darwin* (Oxford: Oxford University Press, 1987).

65. On Kant as armchair natural scientist, see Erich Adickes, *Kant als Naturforscher* (Berlin: Walter de Gruyter, 1924).

66. Kant, *Kritik der Urteilskraft*, AA 5:400.

67. Raphael Lagier, *Les races humaines selon Kant* (Paris: Presses Universitaires de France, 2004), 140, 142, 47.

68. Phillip Sloan, "Kant on the History of Nature: The Ambiguous Heritage of the Critical Philosophy for Natural History," *Studies in History and Philosophy of the Biological and Biomedical Sciences* 37 (2006): 627–48, citing 636.

69. Ibid., 638.

70. Ibid.

71. Kant, "Recension von J. G. Herders *Ideen zur Philosophie der Geschichte der Menschheit*," pt. 2 (1785), AA 8:54.

72. Ibid.

73. Ibid.

74. Arno Seifert, *Cognitio Historica: Die Geschichte als Namengeberin der frühneuzeitlichen Empirie* (Berlin: Duncker und Humblot, 1976)

75. Kant, *Metaphysische Anfangsgründe der Naturwissenschaft*, AA 4:468.

76. Kant, "Bestimmung des Begriffs einer Menschenrace," AA 8:100n.

77. Ibid., 102.

78. Georg Forster, "Noch etwas über die Menschenraßen," *Teutsche Merkur* (1786): 57–86; 150–66. See my "History of Philosophy vs. History of Science: Blindness and Insight of Vantage Points on the Kant-Forster Controversy," in *Klopffechtereien—Missverständnisse—Widerspruche?*, ed. Rainer Godel and Gideon Stiening (Paderhorn: Fink, 2011 [actually 2012]), 225–44.

79. Kant, "Über den Gebrauch teleologischer Principien in der Philosophie," AA 8:157–84, citing 162.

80. Ibid., 161–62.

81. Ibid., 162.

82. Ibid., 161.

83. Ibid., 163–64.

84. Ibid., 163–64n.

85. Ibid., 178.

86. Ibid., 169.

87. Ibid.

88. Ibid., 179.

89. Sloan, "Kant on the History of Nature," 640.

90. Kant, "Recension von J. G. Herders *Ideen zur Philosophie der Geschichte der Menschheit*," pt. 2 (1785), AA 8:54.

91. These were the main claims of Kant, "Von den verschiedenen Racen der Menschen," (1775/77), AA 2:427–44.

92. Kant, *Kritik der Urtheilskraft*, §82, AA 5:428n.

93. Ibid., 428.

94. Kant, *Physische Geographie*, AA 9:162. On the basis of an internal reference to a work published in 1796, we can presume that this text from Kant is later than anything in the third *Critique*.

95. Kant, "Lectures on Physical Geography" (Barth MSS, 1786), cited in Adickes, *Kant als Naturforscher*, 394–95n. I thank Robert Bernasconi for the information about the precise dating of this text.

96. Kant, *Kritik der Urtheilskraft*, §79, AA 5:417.

97. Ibid., 418.

98. Ibid., 417. I have stressed the importance of the notion of "inscrutability" in Kant. See my "'This Inscrutable *Principle* of an Original *Organization*': Epigenesis and 'Looseness of Fit' in Kant's Philosophy of Science," *Studies in History and Philosophy of Science* 34, (2003): 73–109.

99. Maupertuis was the early Kant's paradigmatic instance of a modern hylozoist. Kant's *Dreams of a Spirit-Seer* treats him in exactly that context: "*Hylozoism* invests everything with life, while *materialism*, when carefully considered, deprives everything of life. Maupertuis ascribes the lowest degree of life to the organic particles of nourishment consumed by animals; other philosophers regard such particles as nothing but dead masses, merely serving to magnify the power of the levers of animal machines" (AA 2:330).

100. Lagier, *Les races humaines selon Kant*, 89.

101. Kant, *Kritik der reinen Vernunft*, A832–33/B860–61.

102. Lagier, *Les races humaines selon Kant*, 190.

103. Thus, the last segment of the *Kritik der Urtheilskraft* considered "moral teleology." See Kant's repudiation of physiological anthropology in the published version of his anthropology lectures, *Anthropologie in pragmatischer Hinsicht* (AA 7:117–334); this turn took place already in his early revision of the anthropology course between 1772 and 1776.

104. Kant, *Kritik der Urtheilskraft*, AA 5:400.

105. Clark Zumbach, *The Transcendent Science: Kant's Conception of Biological Methodology* (The Hague: Nijhoff, 1984).

106. See my "Teleology Then and Now: The Question of Kant's Relevance for Contemporary

Controversies over Function in Biology," *Studies in History and Philosophy of Biological and Biomedical Sciences* 37 (2006): 748–70; "Kant's Persistent Ambivalence toward Epigenesis, 1764–1790," in *Understanding Purpose: Collected Essays on Kant and Philosophy of Biology*, ed. Philippe Hunemann, North American Kant Society Studies in Philosophy (Rochester, NY: University of Rochester Press, 2007), 51–74; "Kant et la téléologie," in *L'année 1790: Kant, Critique de la faculté de juger; Beauté, vie, liberté*, ed. Christophe Bouton, Fabienne Brugère, and Claudie Levaud (Paris: Vrin, 2008), 45–54; "Should Kant Have Abandoned the 'Daring Adventure of Reason'?—the Interest of Contemporary Naturalism in the Historicization of Nature in Kant and Idealist *Naturphilosophie*," in *International Yearbook of German Idealism*, vol. 8 (Berlin: De Gruyter, 2010 [actually 2012]), 130–64; "Organism: Objective Purposiveness," in *Kant: Key Concepts*, ed. Will Dudley and Kristina Engelhard (London: Acumen, 2011), 170–83; "Kant's Notion of Intrinsic Purposiveness in the *Critique of Judgment*: A Review Essay (and an Inversion) of Zuckert's *Kant on Beauty and Biology*," *Kant Yearbook*, vol. 1, *Teleology* (Berlin: De Gruyter, 2009), 223–47; "Epigenesis in Kant: Recent Reconsiderations," in "Kant and the Empirical Sciences," ed. Dalia Nassar and Stephen Gaukroger, special issue, *Studies in History and Philosophy of Science* 58 (2016): 85–97.

107. Kant, *Kritik der Urtheilskraft*, AA 5:417.

108. Ibid., 418.

109. This was already the burden of his argumentation in "On the Use of Teleological Principles in Philosophy" in 1788 (AA 8).

110. Kant, *Kritik der Urtheilskraft*, AA 5:418–19.

111. Ibid., 419.

112. Kant's commitment to species fixity is crucial. See "Bestimmung des Begriffs einer Menschenrace," AA 8:97.

113. Kant, *Kritik der Urtheilskraft*, AA 5:419.

114. Ibid., 423–24.

115. Ibid., 424.

116. Ibid., 419n.

117. Ibid., 419–20n. But Kölreuter had already demonstrated the empirical falsity of this view. Joseph Gottlieb Kölreuter, *Vorläufige Nachricht von einigen das Geschlecht der Pflanzen betreffenden Versuchen und Beobachtungen, nebst Fortsetzungen 1, 2 und 3*, Ostwald's Klassiker der exakten Wissenschaften, no. 41 (Leipzig: Engelmann, 1893). His views might have been disregarded because he employed an obsolete language to couch his revolutionary findings, but Blumenbach had no trouble discerning the significance.

118. Kant, *Metaphysische Anfangsgründe der Naturwissenschaft*, AA 4:469–70.

119. In Blumenbach, indeed, historians have located the German impetus leading to the crystallization of biology. See John Gascoigne, "Blumenbach, Banks, and the Beginnings of Anthropology," in *Göttingen and the Development of the Natural Sciences*, ed. N. Rupke (Göttingen: Wallstein, 2002), 86–98, citing 94.

120. Kant, "Über den Gebrauch teleologischer Principien in der Philosophie," AA 8:180n, invoking the first edition of the *Handbuch der Naturgeschichte* (1779), which he owned. This was the first mention of Blumenbach in Kant's body of work. For the "daring adventure of reason," see Kant, *Kritik der Urtheilskraft*, AA 5:419n.

121. Kant, "Über den Gebrauch teleologischer Principien in der Philosophie," AA 8:180n.

122. Ibid.

123. Kant, *Kritik der Urtheilskraft*, §81, AA 5:424.

124. Kant, "Bestimmung des Begriffs einer Menschenrace," AA 8:89–106.

125. Blumenbach, *Handbuch der Naturgeschichte*, 3rd ed. (Göttingen: Dieterich, 1788).

126. Blumenbach to Kant, 1789; Kant's acknowledgment to Blumenbach, August 5, 1790; AA 11:176–77.

127. Blumenbach, *Über den Bildungstrieb und das Zeugungsgeschäfte*, 2nd ed. (Göttingen: Dieterich, 1789), 79.

128. McLaughlin, "Blumenbach und der Bildungstrieb," passim (see n. 5 above).

129. Blumenbach, *Handbuch der Naturgeschichte* (1788), 12–13.

130. Blumenbach, *Handbuch der Naturgeschichte* (1791), 14.

131. Blumenbach, *Handbuch der Naturgeschichte* (1797), 18.

132. Ibid., 80.

133. Kant to Blumenbach, August 5, 1790, AA 11, 176–77.

134. Kant, *Kritik der Urtheilskraft*, AA 5:424.

135. The issue is what to make of vitalism in emergent life science (which is quite different from Kant's position that vitalism *excludes* life from any valid *science*). Christoph Girtanner would pick this up explicitly in his *Über das Kantische Prinzip für die Naturgeschichte*. See C. Wegelin, "Dr. Med. Christoph Girtanner (1760–1800)," *Gesnerus* 14 (1957): 141–63; Hans Querner, "Christoph Girtanner und die Anwendung des Kantischen Prinzips in der Bestimmung des Menschen," in *Die Natur des Menschen: Probleme der Physischen Anthropologie und Rassenkunde (1750–1850)*, ed. Gunter Mann and Franz Dumont, Soemmerring Forschungen 6 (Stuttgart: G. Fischer, 1990), 123–36.

136. Kant, *Kritik der Urtheilskraft*, AA 5:424.

137. Robert Richards, "Kant and Blumenbach on the *Bildungstrieb*: A Historical Misunderstanding," *Studies in History and Philosophy of the Biological and Biomedical Sciences* 31 (2000): 11–32.

138. Timothy Lenoir, "The Göttingen School and the Development of Transcendental *Naturphilosophie* in the Romantic Era," *Studies in History of Biology* 5 (1981): 111–205; Timothy Lenoir, "Kant, Blumenbach, and Vital Materialism in German Biology," *Isis* 71 (1980): 77–108; Timothy Lenoir, *The Strategy of Life: Teleology and Mechanism in Nineteenth-Century Biology*, 2nd ed. (Chicago: University of Chicago Press, 1989). See my critique: "The Lenoir Thesis Revisited: Blumenbach and Kant," in *Studies in History and Philosophy of Science*, pt. C, *Studies in History and Philosophy of Biological and Biomedical Sciences* 43 (2012): 120–32.

139. Perhaps the most direct articulation of this point is James Larson, "Vital Forces: Regulative Principles or Constitutive Agents? A Strategy in German Physiology, 1786–1802," *Isis* 70 (1979): 235–49.

140. See my *Kant, Herder, and the Birth of Anthropology* (Chicago: University of Chicago Press, 2002). See also Robert Richards, *The Romantic Conception of Life: Science and Philosophy in the Age of Goethe* (Chicago: University of Chicago Press, 2002); Frederick Beiser, *German Idealism: The Struggle against Subjectivism, 1781–1801* (Cambridge, MA: Harvard University Press, 2002).

141. As Adler puts it, "Nature, the anthropological, and the history of humanity belong together for Herder." Hans Adler, "Johann Gottfried Herder's Concept of Humanity," *Studies in Eighteenth-Century Culture* 23 (1994): 55–74, citing 63. As Temkin puts it, we find in Herder "a threefold parallelism between (1) ontogeny and the ages of man, (2) successive creation of species, and (3) the history of mankind through successive civilizations." Owsei Temkin, "German Concepts of Ontogeny and History Around 1800," *Bulletin of the History of Medicine* 24 (1950): 227–46, citing 244.

142. Despite Manfred Wenzel's presentation of criticism of Herder from scientists ("Die Anthropologie Johann Gottfried Herders und das klassische Humanitätsideal," in Mann and Dumont, *Die Natur des Menschen*, 137–67), I believe that a fair appraisal of a wider array of available evidence suggests that natural scientists of the day, notwithstanding particular reservations, admired what Herder was attempting and felt it betokened the direction that inquiry in the field needed to pursue. Temkin ("German Concepts of Ontogeny," 241n) even goes so far as to contend that "Herder's *Ideen* were the starting point for the whole biological movement around 1800 including not only Kielmeyer but also Goethe, Cuvier, and Pfaff." He adds that Schelling made emphatic reference to Kielmeyer's connection with Herder even as he celebrated the former for his "new epoch of natural history," and that Hegel made reference to Schelling's derivation of *Naturphilosophie* from Kielmeyer and Herder (241 and n). Coleman writes of the Kielmeyer-Herder connection as follows: "Kielmeyer spoke of geology and biology in the distinctive context of late eighteenth-century and early nineteenth-century German reflections on world-historical development. . . . Johann Gottfried von Herder . . . extended contemporary cosmogonies beyond attention to the first

formation of the earth and planetary system to a wide-ranging concern for the preparation of the earth's surface as a suitable abode for man.... His conjectures exerted a profound influence upon Kielmeyer's developing views." William Coleman, "Limits of Recapitulation Theory: Carl Friedrich Kielmeyer's Critique of the Presumed Parallelism of Earth History, Ontogeny, and the Present Order of Organisms," *Isis* 64 (1973): 341–50, citing 342.

143. See my "'Method' vs 'Manner'?—Kant's Critique of Herder's *Ideen* in Light of the Epoch of Science, 1790–1820," *Herder Yearbook*, 1998, 1–25.

144. Johann Wolfgang von Goethe, *Scientific Studies*, ed. and trans. Douglas Miller (New York: Suhrkamp, 1986), 31.

145. Kant, *Kritik der Urtheilskraft*, AA 5:419n. See Beiser, *German Idealism*, 409, 377.

146. Girtanner's extension of Kant's work followed just the vein that Kant himself had indicated his theory of race would require were it to become a serious scientific research program. In a letter responding to the publisher Breitkopf's invitation to submit a more extended work on race in 1778, Kant, declining the invitation, explained: "my frame of reference would need to be widely expanded and I would need to take fully into consideration the place of race among animal and plant species, which would occupy me too much and carry me into extensive new reading which in a measure lies outside my field, because natural history is not my study but only my game" (Kant to Breitkopf, April 1, 1778, AA 10:227–30). The project of extending consideration of race to animals and plants took up the bulk of Girtanner's study.

147. Though Lenoir sought to exonerate Girtanner of "sinn[ing] against a ... sacred Kantian principle" [!] and to rescue him for authentic Kantian "regulative" thinking, he had to admit that "Girtanner defended a view concerning Kant's *Stammgattung* which seems to run directly counter to the regulative function attributed to it in Kant's own works.... Girtanner argued that the task of natural history was to delineate the original form (*Urbild*) of each *Stammgattung* and show how the present species were degenerated from these originals." Timothy Lenoir, "Generational Factors in the Origin of *Romantische Naturphilosophie*," *Journal of the History of Biology* 11 (1978): 57–100, citing 74.

148. Girtanner, *Über das Kantische Prinzip für die Naturgeschichte*, 2.

149. Ibid., 6.

150. Ibid., 4.

151. Ibid.

152. The term "degeneration" came to be used in very disparate ways in eighteenth-century natural science; the way Girtanner employed it signified mutation of species. It is not clear that Kant was so careful in his own usage of this term. But see Jennifer Mensch, *Kant's Organicism: Epigenesis and the Development of the Critical Philosophy* (Chicago: University of Chicago Press, 2013), 201n238.

153. Girtanner, *Über das Kantische Prinzip für die Naturgeschichte*, 11.

154. Ibid.

155. Ibid., 12.

156. Ibid., 27.

157. The clearest sign of Kant's influence on Blumenbach's theory of race is the latter's conversion to this irreversibility thesis in his significantly revised third edition of the book on human varieties in 1795. Robert Bernasconi, "Who Invented the Concept of Race? Kant's Role in the Enlightenment Construction of Race," in *Race*, ed. Robert Bernasconi (Oxford: Blackwell, 2001), 11–36.

158. Girtanner, *Über das Kantische Prinzip für die Naturgeschichte*, 15.

159. Ibid., 14–15.

160. Ibid., 17.

161. Ibid.

162. Rupke, "Study of Fossils in the Romantic Philosophy of History and Nature."

163. Exemplary is Hinrich Knittermeyer, "Gottfried Reinhold Treviranus, 1776–1837: Aus Anlaß der hundertsten Wiederkehr seines Todestages," *Bremer Beiträge zur Naturwissenschaft* 4 (1937): 3–23.

164. Brigitte Hoppe, "Le concept de biologie chez G. R. Treviranus," in *Colloque internationale "Lamarck" tenu au Museum national d'histoire naturelle, Paris, les 1–2 et 3 juillet 1971*, ed. Joseph Schiller (Paris: Hermann, 1971), 199–237. See also the earlier French essay Raoul Mourgue, "La conception de la biologie chez Gottfried Reinhold Treviranus," *Journal of Comparative Neurology* 56 (1932): 503–10, where Mourgue sees Treviranus as a predecessor for his and C. V. Monakov's theories in neurology.

165. Lenoir, "Generational Factors in the Origin of *Romantische Naturphilosophie*," 95–98.

166. Trevor DeJager, "G. R. Treviranus (1776–1837) and the Biology of a World in Transition" (PhD diss., University of Toronto, 1991); Jörg Nitzsche, "Leben und Werk des Bremer Arztes und Naturforscher Gottfried Reinhold Treviranus (1776–1837): Ein Beitrag zur Sozial- und Ideengeschichte der Medizin des frühen 19. Jahrhunderts" (MD diss., University of Lübeck, 1989).

167. Recently, there has been a new stirring of interest. See Joan Steigerwald, "Treviranus' *Biology*: Generation, Degeneration, and the Boundaries of Life," in *Race and Gender in Early Modern Philosophy*, ed. Susanne Lettow (Albany: State University of New York Press, 2014), 105–27.

168. Treviranus, curriculum vitae, cited in DeJager, "Treviranus," 31.

169. Ibid.

170. Ibid., 43.

171. Biographical information is taken largely from Knittermeyer, Nitzsche, and DeJager.

172. Knittermeyer, "Treviranus," 5; DeJager, "Treviranus," 29–30.

173. Nitzsche, "Leben und Werk," 38.

174. Ibid., 39.

175. Ibid., 58.

176. Ibid., 53–54. Bouterwek claimed to be the first to have taught Kantian philosophy at Göttingen, which was notoriously hostile to the critical philosophy. Nitzsche notes that only *Privatdozenten* taught Kant at Göttingen, not the ordinary professors, and that they had trouble getting enrollment (and hence salary). The course Treviranus took with Bouterwek concentrated on Kant's *Metaphysical Foundations of Natural Science*, *Critique of Judgment*, and elements from the *Critique of Pure Reason* dealing with knowledge in natural science.

177. Both DeJager, "Treviranus," 30–31, and Nitzsche, "Leben und Werk," 50–51, note that Richter was adamantly antitheoretical in his teaching and practice. Though a brilliant surgeon and clinician, he could not serve Treviranus in terms of the latter's philosophical and theoretical ambitions as a research scientist. For that the student would need to look elsewhere—that is, to Blumenbach at Göttingen and, in the wider context, to Carl Friedrich Kielmeyer at Tübingen and Johann Christian Reil at Halle.

178. DeJager, "Treviranus," 33; Nitzsche, "Leben und Werk," 55.

179. He would, he wrote in his curriculum vitae, "utilize my leisure time for the promotion of physiology" (cited in DeJager, "Treviranus," 31). By 1799 he managed to publish a two-volume contribution: *Physiologische Fragmente* (1797–99), en route to his magnum opus, *Biologie*, which began to appear in 1802.

180. Anonymous [Treviranus], "Über Nervenkraft und ihre Wirkungsart," *Archiv für die Physiologie* 1, no. 2 (1795): 3–20. The article was noted favorably in the reviews.

181. Treviranus, curriculum vitae (1796), cited in DeJager, "Treviranus," 31.

182. Gottfried Reinhold Treviranus, *De Emendanda Physiologia* (Göttingen, 1796).

183. Nitzsche, "Leben und Werk," 38–58.

184. Ibid., 75.

185. Ibid.

186. Knittermeyer, "Treviranus," 8, 12–14.

187. Gottfried Reinhold Treviranus, *Biologie oder Philosophie der lebenden Natur für Naturforscher und Aerzte*, 6 vols. (Göttingen: Röwer, 1802–22).

188. This was the dismissive phrase Mayr used for the achievements of 1800 in biology. See Ernst Mayr, *The Growth of Biological Thought: Diversity, Evolution, and Inheritance* (Cambridge, MA: Belknap Press, 1982), 108.

189. Treviranus, *Biologie*, vol. 1 (1802). As Knittermeyer puts it: "it is the essence of the living form, of the organism, to be able to preserve a given organized whole in the same form of existence under the most divergent external situations [*das jeweils organisierte Ganze in seiner gleichförmigen Existenz under den verschiedenartigsten äußeren Verhältnissen aufrechterhalten zu können*]" ("Treviranus," 11).

190. Thus, typically, Knittermeyer opines: "Not Schelling but Kant was his true philosophical teacher" ("Treviranus," 10). And Hoppe writes: "Treviranus follows Kant. . . . He differs from his German contemporaries on one essential point: he considers experience the point of departure for empirical science and accordingly for biology" ("Le concept de biologie," 233). But Kant was not the only philosopher of empirical inquiry to stress experience, and it is not clear that this stress on experience was not seriously complicated—even compromised—by Kant's insistence on the a priori construction of all valid knowledge.

191. DeJager, "Treviranus," 363.

192. Ibid., 329.

193. Ibid., 18.

194. Ibid., 25.

195. Ibid., 275.

196. Ibid., 338.

197. Ibid., 41.

198. Treviranus, *Biologie*, vol. 2 (1803), 264.

199. DeJager, "Treviranus," 14.

200. Nitzsche, "Leben und Werk," 193–95.

201. DeJager, "Treviranus," 78.

202. Ibid., 171. Those were questions Kant had proscribed, but they were posed with scientific hope by Herder.

203. Ibid., 245–46.

204. On Treviranus as biogeographical pioneer, see Gundolf Wingler, "Die Tiergeographie des Gottfried Reinhold Treviranus und ihre Vorgänger" (PhD diss, University of Frankfurt am Main, 1958).

205. See E. A. W. Zimmermann, *Geographische Geschichte des Menschen, und der vierfüßigen Thiere*, 3 vols. (Leipzig: Weygand, 1779–83).

206. Wingler, "Die Tiergeographie des Gottfried Reinhold Treviranus und ihre Vorgänger."

207. Treviranus was "an explicit defender of real descent . . . via adaptation to environmental influences" (Nitzsche, "Leben und Werk," 191). This establishes both that von Engelhardt is wrong to conceive of life science at this time as exclusively "ideal" in its conception of development, and that Mayr is wrong to think that there could only be evolution if it embraced Darwinian principles.

208. Treviranus, *Biologie*, vol. 3 (1805), 225–26.

209. Hoppe, "Le concept de biologie," 226.

CHAPTER NINE

1. Friedrich Schelling, *Von der Weltseele: Eine Hypothese der höheren Physik zur Erklärung des allgemeinen Organismus* (Hamburg, 1798), 298n.

2. "Whilst the extensionalist mathematical Newtonian approach offers the potential for (mathematical) a priori processing of physical nature, the price which this pays is that since forces do not have in this scheme any basic or 'essential' place, they have (because of the conceptual doubt attaching to them) to be introduced *ad hoc* (from 'without'), by way of hypothesis only. The objection to this, of course, . . . [is] that such a basic and powerful notion as force (let alone the force of attraction) ought not to be surrounded with the suspicion that—particularly during the seventeenth and eighteenth centuries—surrounded anything 'hypothetical' in science." Gerd

Buchdahl, "Kant's 'Special Metaphysics' and the *Metaphysical Foundations of Natural Science*," in *Kant's Philosophy of Physical Science*, ed. R. E. Butts (Dordrecht: Reidel, 1986), 150–51.

3. Robert Schofield, *Mechanism and Materialism: British Natural Philosophy in an Age of Reason* (Princeton, NJ: Princeton University Press, 1970); P. M. Heimann and J. E. McGuire, "Newtonian Forces and Lockian Powers: Concepts of Matter in 18th-Century Thought," *Historical Studies in the Physical Sciences* 8 (1971): 233–306.

4. Eve-Marie Engels, "Die Lebenskraft—metaphysische Konstrukt oder methodologisches Instrument? Überlegungen zum Status von Lebenskräften in Biologie und Medizin im Deutschland des 18. Jahrhunderts," in *Philosophie des Organischen in der Goethezeit*, ed. Kai Torsten Kanz (Stuttgart: Steiner, 1994), 127–52, citing 127. This is crucial for a proper understanding of what has been called the "vitalism" or "vital materialism" of the eighteenth century. These concepts must be distinguished from divine teleology and from some metaphysical mystification. The question is one, rather, of emergent properties at different orders of complexity in the natural world and the scientific understanding appropriate to each distinct level or order of emergent complexity.

5. Rudolf Stichweh, *Zur Entstehung des modernen Systems wissenschaftlicher Disziplinen* (Frankfurt: Suhrkamp, 1984).

6. Brigitte Lohff, *Die Suche nach der Wissenschaftlichkeit der Physiologie in der Zeit der Romantik* (Stuttgart: Fischer, 1990), 25.

7. Albrecht von Haller, *A Dissertation on the Sensible and Irritable Parts of Animals* (1752), in *The Natural Philosophy of Albrecht von Haller*, ed. Shirley Roe (New York: Arno, 1981), 651–91. "In France, Britain, and Germany Haller's formulations of irritability and sensibility became the starting point for a new generation of physiologists," Hansen LeRoy has aptly observed. Lee Ann Hansen LeRoy, "Johann Christian Reil and *Naturphilosophie* in Physiology" (PhD diss., University of California–Los Angeles, 1985), 42. "Haller's concepts defined the terms of physiological debate in the late eighteenth century" (8). See Erich Mende, "Die Entwicklungsgeschichte der Faktoren Irritabilität und Sensibilität in den Einfluß auf Schellings 'Prinzip' als Ursache des Lebens," *Philosophia Naturalis* 18 (1979): 327–48.

8. Caspar Friedrich Wolff, *Theoria Generationis* (Halle, 1759); Johann Friedrich Blumenbach, *Über den Bildungstrieb* (Göttingen, 1781).

9. Johann Friedrich Blumenbach, *Institutiones Physiologiae* (Göttingen: Dieterich, 1787).

10. Carl Friedrich Kielmeyer, *Über die Verhältniße der organischen Kräfte unter einander in der Reihe der verschiedenen Organisationen, die Gesetze und Folgen dieser Verhältniße* (1793), facsimile reproduction, ed. Kai Torsten Kanz (Marburg: Basilisken, 1993).

11. Stephan Schmitt, *Les forces vitales et leur distribution dans la nature: Un essai de "systématique physiologique"* (Turnhout, Belgium: Brepols, 2006); Eve-Marie Engels, *Die Teleologie des Lebendigen* (Berlin: Duncker und Humblot, 1982); Engels, "Die Lebenskraft"; Jörg Jantzen, "Physiologische Theorien," in *Schelling: Ergänzungsband zu Werke, Band 5 bis 9: Wissenschaftshistorischer Bericht zu Schellings naturphilosophischen Schriften, 1797–1800* (Stuttgart: Frommann-Holzboog, 1994), 373–668; Brigitte Lohff, "Die Entwicklung des Experimentes im Bereich der Nervenphysiologie: Gedanken und Arbeiten zum Begriff der Irritabilität und der Lebenskraft," *Sudhoffs Archiv* 64 (1980): 105–29.

12. Gottfried Reinhold Treviranus, *Biologie oder Philosophie der lebenden Natur für Naturforscher und Aerzte*, 6 vols. (Göttingen: Röwer, 1802–22), 1:4; Thomas Bach, *Biologie und Philosophie bei C. F. Kielmeyer und F. W. J. Schelling* (Stuttgart: Frommann-Holzboog, 2001), 82.

13. Friedrich Casimir Medicus, *Von der Lebenskraft* (Mannheim, 1774), 5.

14. Baron d'Holbach, *System of Nature* (Amsterdam, 1770).

15. Haller's criticisms of Caspar Friedrich Wolff derived at least in part from his own obsessive anxiety about "materialists," as we have seen.

16. Medicus, *Lebenskraft*, 5. This was the starting point for Stahl at the outset of the century and it would be the foundation of much of Kant's thought in life science by its close.

17. Ibid., 13.
18. Ibid., 6–7
19. Ibid., 6.
20. Ibid., 7.
21. Ibid.
22. Ibid., 7–8.
23. Ibid., 8.
24. Ibid., 8–12.
25. Ibid., 12.
26. This idea of a divine source of the force in the physical universe, I would contend, was the (deist) convention of the vast majority of "moderate Enlightenment" natural philosophers, from Newton onward.
27. Ibid., 15.
28. Ibid., 13.
29. Ibid., 14–16. Leibniz's notion of *petits perceptions* offered a far more subtle conception of actual consciousness, but no more efficacious *causal* connection, hence the turn to *influxus physicus* in German school philosophy by the middle of the eighteenth century. See Eric Watkins, "Development of Physical Influx in Early Eighteenth-Century Germany: Gottsched, Knutzen, and Crusius," *Review of Metaphysics* 49 (1995): 295–339; Eric Watkins, "Kant's Theory of Physical Influx," *Archiv für Geschichte der Philosophie* 77 (1995): 286–324.
30. Medicus, *Lebenskraft*, 17.
31. Ibid., 17–19.
32. Ibid., 19, 24. Thus, Medicus was conscious of the centrality of the "ape debate" as well.
33. Ibid., 23–24.
34. Ernst Plattner, *Anthropologie für Ärzte und Weltweise* (1772; repr.; Hildesheim: Olms, 2000); Johann Unzer, *Erste Gründe einer Physiologie der eigentlichen thierischen Natur thierischer Körper* (Leipzig: Weidmann, 1771).
35. Schmitt, *Les forces vitales*.
36. Jantzen, "Physiologische Theorien"; Lohff, "Die Entwicklung des Experimentes"; Engels; "Die Lebenskraft."
37. Lohff, "Die Entwicklung des Experimentes," 113. Thus, she summarizes the controversy in a striking manner: after and against Haller, "the [entire] organism [as against strictly muscle fibers] was bit by bit constructed under the irritability concept" (114).
38. The term *Bildungstrieb* whirled into ubiquitous circulation, much the way "paradigm" did in the wake of Thomas Kuhn's *Structure of Scientific Revolutions* in the 1960s and 1970s. Blumenbach was not so responsible for its polysemic character as Kuhn was for "paradigm," but the result was just as much out of his control.
39. Schelling, *Von der Weltseele*, 298n. See Nicholas Jardine, "The Significance of Schelling's 'Epoch of a Wholly New Natural History': An Essay on the Realization of Questions," in *Metaphysics and Philosophy of Science in the Seventeenth and Eighteenth Centuries*, ed. R. S. Woodhouse (Dordrecht: Kluwer, 1988), 327–50.
40. Around 1800, Roose, Burdach, Lamarck, and Treviranus all independently articulated "biology" as a term for the new science. Kristian Köchy, "Die Gesetze der grossen Maschine der organischen Welt," *Journal for General Philosophy of Science* 26 (1995): 191–98.
41. Friedrich Schelling, *First Outline of a System of the Philosophy of Nature*, trans. Keith Peterson (Albany: State University of New York Press, 2004), 141n. Bach (*Biologie und Philosophie bei Kielmeyer und Schelling*) has suggested that Schelling's detailed account of this filiation of ideas—down to the page numbers in Herder—was intended to minimize the importance of Kielmeyer, so as to create space for his own appropriation of the field. That is a noteworthy consideration, and one utterly consistent with the character and aspirations of the young Schelling.
42. J. G. Herder, *Ideen zur Philosophie der Geschichte der Menschheit*, in J. G. Herder, *Werke*, ed.

Wolfgang Pross (Munich: Hanser, 2002), vol. 3, bk. 3, chap. 2, 86. In a note, Herder mentions Alexander Monro as one such anatomist. He also probably thought of Daubenton and Camper in this light.

43. This connection has been noted by William Coleman, "Limits of the Recapitulation Theory: Carl Friedrich Kielmeyer's Critique of the Presumed Parallelism of Earth History, Ontogeny, and the Present Order of Organisms," *Isis* 64 (1973): 341–50, citing 342; and by Kai Torsten Kanz in the facsimile edition of Kielmeyer, *Über die Verhältniße*, 40.

44. Shortly after Kielmeyer's address was published, Goethe wrote a letter (1793 or 1794) to Herder indicating that he would be pleased to be aware of this publication, since Kielmeyer was somebody Herder "like[d] a lot [*vornimmt*]."

45. Schelling, *First Outline*, 141n.

46. As a result, this connection featured in standard histories in the nineteenth century, e.g., Johann Eduard Erdmann, *Grundriß der Geschichte der Philosophie*, vol. 2, *Philosophie der Neuzeit* (Berlin, 1866), 494: "Kielmeyer, inspired [*angeregt*] by Herder . . ." (cited by Bach, *Biologie und Philosophie bei Kielmeyer und Schelling*, 63n).

47. Owsei Temkin, "German Concepts of Ontogeny and History Around 1800," *Bulletin of the History of Medicine* 24 (1950): 227–46; Coleman, "Limits of the Recapitulation Theory"; Wolfgang Pross, "Herders Konzept der organischen Kräfte und die Wirkung der *Ideen zur Philosophie der Geschichte der Menschheit* auf Carl Friedrich Kielmeyer," in Kanz, *Philosophie des Organischen in der Goethezeit*, 81–99; Bach, *Biologie und Philosophie bei Kielmeyer und Schelling*.

48. See Jacob Friedrich Abel, *Einleitung in die Seelenlehre* (Stuttgart, 1786). On Abel, see Ingrid Schumacher, "Karl Friedrich Kielmeyer, ein Wegbereiter neuer Ideen: Der Einfluß seiner Method des Vergleichens auf die Biologie der Zeit," *Medizinhistorisches Journal* 14 (1979): 81–99, citing 84; Dorothea Kuhn, "Die naturwissenschaftlich Unterricht an der Hohen Karlsschule," *Medizinhistorisches Journal* 11 (1976): 319–34; Bach, *Biologie und Philosophie bei Kielmeyer und Schelling*, 130, esp. 130n.

49. A vigorous correspondence about Herder's *Ideen* circulated among the leaders of German life science in these years, including Blumenbach, Forster, Sömmerring, Camper, and others.

50. Kai Torsten Kanz, "Die Naturgeschichte (Botanik, Zoologie, Mineralogie) an der Hohen Karlsschule in Stuttgart (1772–1794)," *Jahreshefte der Gesellschaft für Naturkunde in Württemberg* 148 (1993): 5–23, citing 19.

51. Ibid.

52. Bach, *Biologie und Philosophie bei Kielmeyer und Schelling*, 90.

53. Kanz, "Einführung" to Kielmeyer, *Über die Verhältniße*, cited in Bach, *Biologie und Philosophie bei Kielmeyer und Schelling*, 90.

54. Pross, "Herders Konzept der organischen Kräfte."

55. Bach, *Biologie und Philosophie bei Kielmeyer und Schelling*, 151.

56. "By transforming the hierarchy of immutable creatures into a dynamic gradation of changing, developing, and self-regulating life-forms, [Kielmeyer] lent concrete scientific shape to Herder's vague notion of a 'gradation of organizations' to sketch a bold scheme of natural development that envisioned the gradual mutation of species." Gabrielle Bersier, "Visualizing Carl Friedrich Kielmeyer's Organic Forces: Goethe's Morphology on the Threshold of Evolution," *Monatshefte* 97 (2005): 18–32, citing 21. As Bach puts it, "Kielmeyer's task accordingly consisted in overcoming the indeterminacy in the array [*Gliederung*] of organic forces via the formulation of a more consistent conceptualization [*Begrifflichkeit*] of organic forces" (*Biologie und Philosophie bei Kielmeyer und Schelling*, 161).

57. Gustav Wilhelm Münter, *Allgemeine Zoologie oder Physik der organischen Körper* (Halle, 1840). See Ilse Jahn, "War Gustav Wilhelm Münter (1804–1870) ein 'Plagiator' Kielmeyers? Zur Autorschaft der *Allgemeinen Zoologie oder Physik der organischen Körper* (Halle, 1840); Mit einem Anhang: Die Münter-Rexzension von Friedrich Stein," in Kanz, *Philosophie des Organischen in der Goethezeit*, 174–210. Jahn establishes that Münter indeed plagiarized from an 1807 version of

Kielmeyer's lectures. While Balss offered a charitable suggestion that Münter might have had Kielmeyer's tacit consent, Jahn's research has made that highly unlikely. Still, the direct plagiarism allowed Balss to use Münter's published text to reconstruct Kielmeyer's original arguments. See Heinrich Balss, "Kielmeyer als Biolog," *Sudhofs Archiv* 23 (1930): 268–88: the charitable suggestion is at 273, and the balance of the essay is a summary of the ideas in that text.

58. Kai Torsten Kanz, "Carl Friedrich Kielmeyer (1765–1844)—Leben, Werk, Wirkung: Perspektiven der Forschung und Edition," in Kanz, *Philosophie des Organischen in der Goethezeit*, 13–32. But Gabrielle Bersier has noted that "after several inconclusive attempts, an interdisciplinary research project sponsored by the Deutschen Forschungsgemeinschaft is currently underway at the University of Stuttgart to produce the first critical edition of Kielmeyer's works and letters" ("Visualizing Carl Friedrich Kielmeyer's Organic Forces," 29n).

59. Kanz, "Die Naturgeschichte," 5–7. The Karlsschule has a historiography dating to the mid-nineteenth century in which the natural sciences and medicine have a significant place. See Arthur Moll, "Die medizinische Fakultät der Carlsakademie: Eine historische Studie bei Schiller's 100jähriger Geburtsfeier" (1859); Otto Krimmel, "Die hohe Karlsschule und die Naturwissenschaften" (1895); Gustav Hauber, *Die Hohe Karlsschule* (Eßlingen, 1909)—all cited in Kanz, "Die Naturgeschichte," 21–22. For more recent research, see Robert Uhland, *Geschichte der Hohen Karlsschule in Stuttgart* (Stuttgart: Kohlhammer, 1953); Wilhelm Theopold, *Der Herzog und die Heilkunst: Die Medizin an der Hohen Carlsschule zu Stuttgart* (Cologne: Deutscher Aerzte-Verlag, 1967).

60. Kuhn, "Die naturwissenschaftlich Unterricht an der Hohen Karlsschule," 323.

61. Ibid., 325.

62. Ibid., 325–26. This has a bearing on Kielmeyer's alleged Kantianism.

63. Ibid., 326–27; Friedrich Schiller, *Versuch über den Zusammenhang der tierischen Natur des Menschen mit seiner geistigen* (Stuttgart, 1780).

64. On Schiller's medical dissertations and their context, see G. Rudolph, "Schiller und Albrecht von Haller," *Annales Universitatis Saraviensis: Medizin* 8 (1961): 1–11; Kenneth Dewhurst and Nigel Reeves, *Friedrich Schiller: Medicine, Psychology, and Literature* (Berkeley: University of California Press, 1978); Wolfgang Riedel, *Die Anthropologie des jungen Schiller: Zur Ideengeschichte der medizinischen Schriften und der "Philosophische Briefe"* (Würzburg: Königshausen und Neumann, 1985).

65. Kanz, "Die Naturgeschichte," 6–7. Kanz points out that the university at Tübingen had no such dedicated chair (8). See also Irmtraut Scheele, "Grundzüge der institutionellen Entwicklung der biologischen Disziplinen an den deutschen Hochschulel seit dem 18. Jahrhundert," in *"Einsamkeit und Freiheit" neu besichtigt: Universitätsreformen und Disziplinenbildung in Preußen als Modell für Wissenschaftspolitik im Europa des 19. Jahrhunderts* (Stuttgart: Steiner, 1991), 144–54, citing 149.

66. Kanz, "Die Naturgeschichte," 15.

67. Kuhn, "Die naturwissenschaftlich Unterricht an der Hohen Karlsschule," 334.

68. Ibid., 330.

69. Kanz, "Die Naturgeschichte," 8.

70. Ibid., 19.

71. "What I learned at [the Karlsschule], apart from the most basic foundations, I owe . . . very little to my teachers but mostly to myself; as I can illustrate with chemistry, which I learned all on my own." Kielmeyer to his parents, cited in Kai Torsten Kanz, "Carl Friedrich Kielmeyer, Lichtenberg und Göttingen, 1786–1796," *Lichtenberg Jahrbuch* 2 (1989): 140–60, citing 144.

72. Reinhard Löw, *Pflanzenchemie zwischen Lavoisier und Liebig* (Straubings: Donau Verlag, 1979), draws extensively from two unpublished sources from Kielmeyer's years as professor of chemistry at the University of Tübingen: Kielmeyer's *Antrittsvorlesung* of 1801, *Exposito Discrepantiarum Quarundam quo Corpora Organica et Anorganica quoad Mixtionem Intercedere Videntur*; and the lectures for his chemistry course of 1802–05, both preserved in the Württemberg State Library, Stuttgart.

73. Kielmeyer to his parents, cited in Kanz, "Carl Friedrich Kielmeyer, Lichtenberg und Göttingen," 144.

74. Kielmeyer to his parents, February 7, 1787, in ibid., 142.

75. Kielmeyer to his parents, May 13, 1787, in ibid.

76. Ibid., 142–43.

77. John Zammito, "The Lenoir Thesis Revisited: Blumenbach and Kant," in *Studies in History and Philosophy of Science*, Part C, *Studies in History and Philosophy of Biological and Biomedical Sciences* 43 (2012): 120–32.

78. Dorothea Kuhn, "Uhrwerk und Organismus," third, revised version in Kanz, *Philosophie des Organischen in der Goethezeit*, 33–49; Kanz, "Carl Friedrich Kielmeyer, Lichtenberg und Göttingen"; Kanz, "Carl Friedrich Kielmeyer (1765–1844)—Leben, Werk, Wirkung"; Frank Dougherty, "Über den Einfluß Johann Friedrich Blumenbachs auf Kielmeyers feierliche Rede von 1793: Mit einer Anhang über Kielmeyers Göttinger Lektüre," in Kanz, *Philosophie des Organismus in der Goethezeit*, 50–80.

79. Dougherty, "Über den Einfluß Johann Friedrich Blumenbachs auf Kielmeyers feierliche Rede von 1793," 54n.

80. Ibid., 60–65.

81. Ibid., 52n.

82. Kanz, "Die Naturgeschichte," 16.

83. Ibid., 9; on Storr's courses and textbooks, see ibid., 7–8.

84. Carl Friedrich Kielmeyer, "Über Naturgeschichte," in Kielmeyer, *Gesammelte Schriften*, ed. F. Holler (Berlin: Keiper, 1938), 228, cited in Kanz, "Die Naturgeschichte," 10–11.

85. In this sense, already, Kielmeyer was prefiguring the development of the research field.

86. Kanz, "Die Naturgeschichte," 16.

87. Schumacher, "Karl Friedrich Kielmeyer," 81–82n.

88. Jahn, "War Gustav Wilhelm Münter (1804–1870) ein 'Plagiator' Kielmeyers?"

89. Schumacher, "Karl Friedrich Kielmeyer," 82n.

90. See Christoph Heinrich Pfaff, *George Cuvier's Briefe an C. H. Pfaff aus den Jahren 1788 bis 1792, naturhistorischen, politischen und literarischen Gehalts: Nebst biographischen Notizen über G. Cuvier*, ed. W. F. G. Behn (Kiel, 1845).

91. Karl Friedrich Kielmeyer, "Ideen zu einer allgemeineren Geschichte und Theorie der Entwicklungserscheinungen der Organisationen 1793–94," in Kielmeyer, *Gesammelte Schriften*, 102–94.

92. We can note simply Karl Ernst von Baer's great work of that same title, *Über Entwickelungsgeschichte der Thiere: Beobachtung und Reflexion* (1828).

93. Schumacher, "Karl Friedrich Kielmeyer," 87.

94. Ibid., 90.

95. Ibid., 92.

96. Ibid., 88. The resonances with Goethe's thinking in these years concerning the *Urpflanze* arise immediately, with Herder as a key mediation.

97. Ibid., 86.

98. Kielmeyer, *Über die Verhältniße*.

99. Schumacher, "Karl Friedrich Kielmeyer," 86.

100. Ibid.

101. Felix Buttersack, "C. V. Kielmeyer: Ein vergessenes Genie," *Sudhofs Archiv* 23 (1930): 236–46, citing 243.

102. In a note to his published address, Kielmeyer promised that he would soon provide a more extensively documented and argued version of his ideas in a "Geschichte und Theorie der Entwicklung der Organisationen" (History and theory of the organic forces of living things) (Kielmeyer, *Über die Verhältniße*, 8n).

103. In the 1930s compilation of Kielmeyer's *Gesammelte Schriften* we have the outline and

prospectus Kielmeyer himself eventually prepared—in 1814 (!)—for his promised synthetic presentation: "Entwurf zu einer vergleichende Zoologie" (Draft for a comparative zoology). We are told that a first segment was already in proofs before Kielmeyer withdrew it. Not only did his longstanding dread of publication appear a factor, but by then he was also retiring from teaching, and his lecture notes had been circulating throughout German academia for two decades and had entered quite thoroughly (if also without the authoritative rigor of a published text) into the discourse of the field. At least one scholar has gone so far as to suggest that by 1814—indeed, 1804—Kielmeyer may no longer have believed in the general synthetic framework he had developed in the early 1790s: William Coleman comes very close to such an interpretation in his influential essay, "Limits of the Recapitulation Theory." While I am very grateful for his contribution, I think Coleman is wrong about this and about much else in this pathbreaking essay.

104. T. Ballauf, *Die Wissenschaft vom Leben*, vol. 1 (Freiburg: Alber, 1954), 345.

105. Reinhard Mocek, *Johann Christian Reil (1759–1813): Das Problem des Übergangs von der Spätaufklärung zur Romantik in Biologie und Medizin in Deutschland* (Frankfurt: Peter Lang, 1995), 92.

106. Kristian Köchy, "Die Gesetze der grossen Maschine der organischen Welt," *Journal for General Philosophy of Science* 26 (1995), 191–98, citing 192, 193, 194. This is a book review of Kanz's facsimile edition of Kielmeyer's address.

107. Bach, *Biologie und Philosophie bei Kielmeyer und Schelling*, 87.

108. Ibid., 89.

109. Kielmeyer, *Gesammelte Schriften*, 111.

110. Bach, *Biologie und Philosophie bei Kielmeyer und Schelling*, 137.

111. Ibid., 135.

112. Kielmeyer to Cuvier, December 1807, in Kielmeyer, *Gesammelte Schriften*, 251.

113. As Bach puts it in apt summation: in Kielmeyer "the forces of the soul become the forces of the body" (*Biologie und Philosophie bei Kielmeyer und Schelling*, 198).

114. Ibid., 141.

115. Ibid., 123.

116. Kielmeyer used the concept of a "parabola" as a two-dimensional figure that, in the third dimension of temporality, would assume the features of a spiral: it "expressed thus the temporal developmental path[,] . . . a regular but not periodic continual change" (ibid., 101).

117. A. O. Lovejoy, "The Temporalization of the Chain of Being," in *The Great Chain of Being* (New York: Harper, 1960), chap. 9.

118. Bach, *Biologie und Philosophie bei Kielmeyer und Schelling*, 124.

119. Kielmeyer, *Gesammelte Schriften*, 29.

120. Kielmeyer, *Über die Verhältniße*, 3–4. Bach emphasizes that "Kielmeyer bases himself in these considerations not, as Leibniz had done, on metaphysics but rather on empirical findings" (*Biologie und Philosophie bei Kielmeyer und Schelling*, 97).

121. Kielmeyer, *Über die Verhältniße*, 3–4. See Louis Bourguet, *Lettres philosophiques sur la formation des sels et des crystaux, et sur la generation et le mechanisme organique des plantes et des animaux* (1729).

122. Kielmeyer, *Über die Verhältniße*, 3–4.

123. Ibid., 6.

124. Ibid., 5.

125. Ibid., 6.

126. Ibid., 7.

127. Ibid., 16–17.

128. Ibid., 18.

129. Ibid., 17.

130. Ibid., 19.

131. Ibid., 20–23.

132. Ibid., 24.

133. Ibid., 31, 34.

134. Ibid., 28–30.

135. Ibid., 35.

136. Ibid., 36, 38–39.

137. A. W. Meyer, "Some Historical Aspects of the Recapitulation Idea," *Quarterly Review of Biology* 10 (1935): 379–96.

138. Kielmeyer, *Über die Verhältniße*, 39.

139. Kant, *Kritik der Urteilskraft*, AA 5:428n.

140. Bersier, "Visualizing Carl Friedrich Kielmeyer's Organic Forces," 21.

141. Kielmeyer, *Über die Verhältniße*, 37.

142. Notably, Herder had articulated this view earlier, and Kielmeyer's key contemporary, J. C. Reil, made exactly the same argument.

143. Kielmeyer to Cuvier, March 9, 1801, cited in J. H. F. Kohlbrugge, "Cuvier und Kielmeyer," *Biologische Centralblatt* 32 (1912): 291–95, citing 293–95.

144. Kielmeyer to Windischmann, 1804, in Kielmeyer, *Gesammelte Schriften*, 205–6.

145. Ibid., 206.

146. Ibid., 207.

147. Ibid.

148. Ibid., 209.

149. Ibid.

150. Ibid., 210.

151. Ibid.

152. Kielmeyer, "Ideen zu einer allgemeineren Geschichte und Theorie der Entwicklungserscheinungen der Organisationen 1793–94," cited in Schumacher, "Karl Friedrich Kielmeyer," 96.

153. Karl Eschenmayer, *Grundriss der Naturphilosophie* (1832), cited in Schumacher, "Karl Friedrich Kielmeyer," 98.

154. Löw (*Pflanzenchemie*) gives a detailed account of this inaugural lecture and its impact.

155. Alexander von Humboldt, cited in Balss, "Kielmeyer als Biolog," 270.

156. Johann Meckel, *Abhandlungen aus der menschlichen und organischen Anatomie und Physiologie* (1806).

157. Johannes Müller, *Vergleichende Physiologie des Gesichtssinnes* (Leipzig, 1826), 29, cited in Schumacher, "Karl Friedrich Kielmeyer," 88.

158. Rudolf Virchow, "Göthe als Naturforscher" (1861), 126, cited in Schumacher, "Karl Friedrich Kielmeyer," 90.

159. "Kielmeyer was no *Naturphilosoph*, but as one of his bright students put it to me, he was a philosopher of nature." Karl Friedrich Philipp von Martius, *Denkrede auf Carl Friedrich von Kielmeyer* (1845), cited in Schumacher, "Karl Friedrich Kielmeyer," 99. I suspect that "bright student" was Christoph Heinrich Pfaff.

160. Pfaff, *George Cuvier's Briefe*, 18.

161. Ibid. In historical appraisals, Kielmeyer has been called the decisive link between Blumenbach and Cuvier (Kanz, "Carl Friedrich Kielmeyer (1765–1844)—Leben, Werk, Wirkung," 17). Kielmeyer was also, and more notoriously, called the "Father of German *Naturphilosophie*," in large measure because of his presumed influence on Schelling. The phrase can be traced to Cuvier; it had wide currency especially in France in the first decades of the nineteenth century, though Kielmeyer explicitly disowned it to Cuvier himself. See Thomas Bach, "Kielmeyer als 'Vater der Naturphilosophie'? Anmerkungen zu seiner Rezeption im deutschen Idealismus," in Kanz, *Philosophie des Organischen in der Goethezeit*, 232–51.

162. Pfaff, *George Cuvier's Briefe*, 18.

163. See David Cahan, ed., *From Natural Philosophy to the Sciences* (Chicago: University of Chicago Press, 2003).

164. "Kant, zu dessen Schule Kielmeyer gehörte . . ." (Pfaff, *George Cuvier's Briefe*, 18).

165. See Cahan, *From Natural Philosophy to the Sciences*.

166. See my *A Nice Derangement of Epistemes: Post-positivism in the Study of Science from Quine to Latour* (Chicago: University of Chicago Press, 2004); Joseph Rouse, *Engaging Science: How to Understand Its Practices Philosophically* (Ithaca, NY: Cornell University Press, 1996); Joseph Rouse, *How Scientific Practices Matter: Reclaiming Philosophical Naturalism* (Chicago: University of Chicago Press, 2002).

167. Carl Friedrich Kielmeyer, "Versuche über die sogenannte animalische Elektrizität," *Grens Journal der Physik* 8 (1794): 65–77. As it happens, Kielmeyer never intended to publish this. It was put together from letters sent to a friend regarding experiments in "animal electricity" that Kielmeyer had conducted in 1792, immediately upon the circulation of Galvani's key paper. That friend submitted them to Gren's journal for publication. Kielmeyer was offended by this, at first, as indicated in an open letter about the impropriety of the publication. Somewhat later, he relented. In any event, he stood out as a pioneer in the German scientific reception of galvanism. That impetus was transmitted to his student, Christoph Pfaff, who produced a dissertation on galvanism in 1793, under Kielmeyer's supervision.

168. Luigi Galvani, *De Viribus Electricitatis in Motu Musculari Commentarius* (1791). Trans. Margaret Foley as *Commentary on the Effects of Electricity on Muscular Motion* (Norwalk: Burndy Library, 1953).

169. This was a highly conflictual element in the German chemical community. One of the earliest converts to the "chemical revolution" was Christoph Girtanner. Another was Heinrich Friedrich Link. These were two key figures in the so-called "Göttingen School," which makes the question of its role all the more salient.

170. Stuart Strickland, "Galvanic Disciplines: The Boundaries, Objects, and Identities of Experimental Science in the Era of Romanticism," *History of Science* 33 (1995): 449–68, citing 457.

171. Ibid., 458.

172. Christoph Girtanner, "Abhandlung über die Irritabilität, als Lebensprincip in der organisierten Natur," *Journal der Physik* 3 (1791): 317–51, 507–37.

173. Johann Christian Reil, *Von der Lebenskraft*, in *Archiv für die Physiologie* 1 (1795): 8–162.

174. Marcello Pera, *The Ambiguous Frog: The Galvani-Volta Controversy on Animal Electricity*, trans. Johnathan Mandelbaum (Princeton, NJ: Princeton University Press, 1992), xxiii–xxv. Originally published in Italian in 1986.

175. Ibid., 85–86. Kipnis agrees: "Galvani was concerned with a discovery of what makes life *different* from inanimate nature, whereas Volta was interested in finding their *common* features." Naum Kipnis, "Luigi Galvani and the Debate on Animal Electricity, 1791–1800," *Annals of Science* 44 (1987): 107–42, citing 116.

176. Pera, *Ambiguous Frog*, 98. "Galvani saw the frog's contractions and sought their explanation in an *electrobiologial* and not only *electrophysical* context[;] . . . for him the primacy went to biology and physiology, not to physics or even to chemistry" (Pera, *Ambiguous Frog*, 77). Strickland argues that this "disciplinary" reconstruction by Pera, "a contest between physicists and physicians," is "difficult to maintain" in light of the rise of *chemistry* as a dominant force in *Naturlehre* and its appropriation of Volta's results ("Galvanic Disciplines," 461). Certainly, Strickland makes a good case against Pera's too-ready acceptance of physics as the dominant discipline, hence Pera's characterization of physiology as "underdog," but even he concedes that "galvanism was something entirely different within the discourse familiar to Volta than it was to Galvani or . . . to German chemists, physiologists, or physicians" ("Galvanic Disciplines," 462). That deserves more consideration than Strickland allows, especially in that specific German context.

177. Kipnis, "Luigi Galvani," 118–19. Muscle reaction was key; hence, the focus was not generally *organic*; plants played only a very marginal role in the episode.

178. Pera, *Ambiguous Frog*, 160. "Volta managed to pass off an instrument, the pile, as something it was not—the living confutation of Galvani's theory" (*Ambiguous Frog*, 178). "It was the

discovery of the pile that swung prevailing opinion in favor of Volta's theory," Kipnis recognizes ("Luigi Galvani," 136). The best indicator came in Delamétherie's annual reports on progress in the sciences, which by 1801 and 1802 clearly discerned Volta as victor ("Luigi Galvani," 136–37). Volta made his case first in a series of letters to the German physicist Gren in 1796. That made this particularly salient for the German research community. But Kipnis argues that this does not account for the curious phenomenon that Galvani's "theory was neither refuted nor fully abandoned" ("Luigi Galvani," 109).

179. One of the finest features of Kipnis's account is his careful exposition of the multiplicity and the inconclusiveness of Volta's experimental practices and claims. It was not just Galvani who committed errors of experimentation or interpretation. In fact, Kipnis makes it clear that this was inevitable for all concerned. That shifts the question from some abstract form of "correctness" to the more interesting issue of contextual impact.

180. Kipnis, "Luigi Galvani," 107–42. Kipnis urges us to "judge the overall significance of Galvani's contribution to physiology," taking note that his work fell "at the junction of physics and physiology" and would be received and construed differently depending on the disciplinary concern (138, 116).

181. Ibid., 134.

182. As Kipnis notes, the debate about the electric eel (and the Torpedo) had been going on for some time before Galvani presented his results, and the interest would not abate for some time thereafter. And "animal electricity" retained interest well into the nineteenth century.

183. Ibid., 139, 142.

184. Ibid., 142.

185. The turning point in the German enthusiasm may be dated as early as 1796, when Volta sent letters to Gren about his new experiments and the implications he drew from them.

186. Brigitte Lohff puts it effectively: "What arises in the discussion concerning the fundamental forces of the organism, then, with the new argument about bioelectricity? Perhaps some titles give us enough of an indication: Thus, for example, Johann Wilhelm Ritter, *Beweis, daß ein beständiger Galvanismus den Lebensprozeß im Tierreich begleitet* (1798); Christoph Heinrich Pfaff, *Über thierische Elektricität und Reizbarkeit* (1795); Alexander von Humboldt, *Versuche über die gereizte Muskelfaser, nebst Vermutungen über den chemischen Prozeß des Lebens in der Thier- und Pflanzenwelt* (1797)" ("Die Entwicklung des Experimentes," 120).

187. In Scotland, Alexander Monro II made that clear. Insisting on the autonomy of the organismic, he claimed that galvanic fluid was "not the nervous fluid but only a stimulus to it" (Monro cited in Kipnis, "Luigi Galvani," 118). Likewise, Erasmus Darwin argued that the vital functions of organic life could not be reduced to mechanics, chemistry, or electricity. (Erasmus Darwin, *Zoonomia* [1794], 2:65–66, cited in Kipnis, "Luigi Galvani," 130). Nonetheless, for all these figures vital force was still an element of the *physical* world, not something hyperphysical. Hence, Engels ("Die Lebenskraft") notes the important distinction between a *methodological* vitalism and a *metaphysical* one. In the former instance, the term stood for an unknown to be investigated empirically. Only in the latter case was a transcendent intervention invoked.

188. Johann Ritter, *Beweis, daß ein beständiger Galvanismus den Lebensprozeß im Tierreich begleitet* (1798), cited in Kipnis, "Luigi Galvani," 129.

189. Kipnis ("Luigi Galvani," 116–17) makes a list of the physicists and the physiologists, then explicitly recognizes the imprecision of those categories for the self-conception or the practices of individual figures. Yet the disciplinary *orientation* or *interest* did prove meaningfully distinct in the period.

190. Blumenbach praised Kant for attentiveness to the importance of proper boundaries among disciplines of inquiry, writing: "However, as not so long ago Mr Kant noted so properly, 'it is not the advancement but rather the disruption of the sciences if one allows their boundaries to cross over into each other.'" Blumenbach, *Medizinische Bibliothek* 3, no. 1 (1788): 85, cited in Jantzen, "Physiologische Theorien," 497n.

191. Kipnis, "Luigi Galvani," 129, referring to Christoph Pfaff, "Ein Beytrag zu Alexander von Humboldts 2ten Bande der Versuche über die gereizte Muskel- und Nervenfaser" (1796).

192. Joan Steigerwald, "Goethe's Morphology: *Urphänomene* and Aesthetic Appraisal," *Journal of the History of Biology* 35 (2002): 291–328.

193. Christoph Girtanner, "Mémoires sur l'irritabilité considerée comme principe de vie dans la nature organisée" (1790). Trans. T. Beddoes in *Observations on the nature and cure of calculus, sea scurvy, consumption, catarrh, and fever. Together with conjectures upon several other subjects of physiology and pathology* (Philadelphia: Webster, 1815).

194. Girtanner, "Abhandlung über die Irritabilität, als Lebensprincip in der organisierten Natur."

195. Christoph Girtanner, *Anfangsgründe der antiphlogistischen Chemie* (Berlin: Unzer, 1791; 2nd ed., 1795; 3rd ed. (posthumous), 1801); Christoph Girtanner, *Neue chemische Nomenklatur für die deutsche Sprache* (Berlin: Unzer, 1791).

196. Abbé Bertholon, *L'electricité du corps humain* (1786).

197. Girtanner, "Mémoires sur l'irritabilité."

198. Andreas Röschlaub exposed Girtanner's blatant borrowings from John Brown.

199. Johann Ulrich Gottlieb Schäffer, *Über Sensibilität als Lebensprinzip in der organisierte Natur* (Frankfurt am Main, 1793).

200. Christoph Pfaff, *Über thierische Elektricität und Reizbarkeit* (Leipzig: Crusius, 1795), 243. This verdict against Schäffer is confirmed by Jantzen, "Physiologische Theorien," 502–5, esp. 502n.

201. See John Heilbron, *Elements of Early Modern Physics* (Berkeley/L.A.: U. of California P., 1982).

202. Christoph Pfaff, "Abhandlung über die sogennante thierische Elektricität," *Neues Journal der Physik*: 8 (1794): 196–280; Pfaff, *Über thierische Elektricität und Reizbarkeit.*

203. In *Neues Journal der Physik* 2 (1795): 116–22; and 3 (1796): 165–84, respectively.

204. Pfaff, *Über thierische Elektricität und Reizbarkeit*, esp. pt. 2:1 "On Sensibility and Irritability, the Forces Which Form the Basis of These Properties, the Interactive Relationship of These Forms to One Another, and the Principles Pertaining" (1795); Heinrich Friedrich Link, *Beyträge zur Naturgeschichte*, vol. 2, *Über die Lebenskräfte in naturhistorischer Rücksicht . . .* (Rostock: Stiller, 1795); Alexander von Humboldt, "Die Lebenskraft oder der rhodische Genius," *Die Horen*, 1795, reprinted in *Biologie der Goethezeit*, ed. A. Meyer-Abisch (Stuttgart: Hippokrates, 1949), 184–88; Joachim Brandis, *Versuch über die Lebenskraft* (1795).

205. Ignaz Döllinger, "Über den jetzigen Zustand der Physiologie," *Jahrbücher der Medizin als Wissenschaft* 1 (1805): 119–42, citing 124. This body of work provoked an equally important critical response from Johann Christian Reil in Halle: *Von der Lebenskraft* (1795). Theodor Georg August Roose in turn penned a lengthy rejoinder to Reil two years later: *Grundzüge der Lehre von der Lebenskraft* (1797).

206. It is of great historical significance that Pfaff would prove to be one of Schelling's mentors in the natural sciences in his crucial years at Leipzig, triggering his adventures with *Naturphilosophie*. See Manfred Durner, "Schellings Begegnung mit den Naturwissenschaften in Leipzig," *Archiv für Geschichte der Philosophie* 72 (1990): 220–36.

207. Pfaff, *Über thierische Elektricität und Reizbarkeit*, 235. It should be remembered that it was Pfaff who dragged his teacher Kielmeyer's address into print; the manuscript written in Pfaff's hand remains as archival proof of his crucial intervention.

208. Ibid., 236.

209. Ibid., 236–37.

210. Ibid., 238.

211. Ibid., 240–43.

212. Ibid., 239.

213. Ibid., 240.

214. Ibid., 241.

215. Ibid., 242. It is noteworthy how important Plattner's text of 1771, *Anthropologie für Ärzte und Weltweise*, appears in these contexts. Plattner added an important supplement in 1796: "Schwierigkeiten des Hallerischen Systems," in his *Vermischte Aufsätze über medizinische Gegenstände* (Leipzig, 1796), 142–61. See Jantzen, "Physiologische Theorien," 431.

216. Pfaff, *Über thierische Elektricität und Reizbarkeit*, 243.

217. Much of Pfaff's commentary on nervous force derives directly from Unzer's pathbreaking work of 1771, e.g., ibid., 253, etc.

218. Ibid., 244.

219. Ibid., 253. Pfaff proved quite ambivalent about Humboldt, but he recognized his formidable presence.

220. Ibid., 257–93. The chapter was entitled "On Irritability or Contractility, and on Oxygen as the Principle of Irritability."

221. Ibid., 278–79, 282n, 283n. Here was the most explicit acknowledgment of Pfaff's indebtedness to the physiology of his teacher.

222. Ibid., 293.

223. Ibid., 291. See Alexander von Humboldt, *Aphorismen über die chemische Physiologie der Pflanzen* (1794).

224. See Philippe Huneman, *Bichat, la vie et la mort* (Paris: Presses Universitaires de France, 1998).

225. Cited in Meyer-Abisch's editorial introduction to Alexander von Humboldt, "Die Lebenskraft oder der rhodische Genius," in *Biologie der Goethezeit*, ed. A. Meyer-Abisch (Stuttgart: Hippokrates, 1949), 182.

226. See Ilse Jahn, ed., *Die Jugendbriefe Alexander von Humboldts, 1787–1799* (Berlin: Akademie Verlag, 1973), 454–56, 465–72.

227. "Already from 1792 on he had become involved with the phenomenon of Galvanism." K. E. Rothschuh, "Alexander von Humboldt und die Physiologie seiner Zeit," *Sudhoffs Archiv* 43 (1959): 97–113, citing 99.

228. Maria Jean Trumpler, "Questioning Nature: Experimental Investigations of Animal Electricity in Germany, 1791–1810" (PhD diss., Yale University, 1992), 128–58, citing 130.

229. Ibid., 136.

230. See Jahn, *Die Jugendbriefe Alexander von Humboldts, 1787–1799*, 454–56, 465–72.

231. Alexander von Humboldt, "Über die gereizte Muskelfasser," *Neues Journal der Physik* 2 (1795): 115–29; Alexander von Humboldt, "Aus einem Brief des Herrn Oberbergraths von Humboldt an Herrn Hofrath Blumenbach," *Neues Journal der Physik* 2 (1795): 471–73; Alexander von Humboldt, "Neue Versuche über den Metallreitz, besonders in Hinsicht auf die verschiedenartige Empfänglichkeit der thiereischen Organe (Aus einem Briefe an den Herrn Hofrath Blumenbach)," *Neues Journal der Physik* 3 (1796): 165–84.

232. Ilse Jahn, "Die Anatomischen Studien der Brüder Humboldt unter Justus Christian Loder in Jena," *Beiträge zur Geschichte der Universität Erfurt (1392–1816)* 14 (1956): 91–97.

233. Humboldt, "Die Lebenskraft oder der rhodische Genius," reprinted in Meyer-Abisch, *Biologie der Goethezeit*, 184–88.

234. Alexander von Humboldt, *Versuche über die gereizte Muskel- und Nervenfaser nebst Vermuthungen über den chemischen Process des Lebens in der Thier- und Pflannzenwelt*, 2 vols. (Posen: Decker und Rottmann, 1797–99).

235. Meyer-Abisch, editor's introduction to Humboldt, "Die Lebenskraft oder der rhodische Genius," 181–84.

236. Rothschuh ("Alexander von Humboldt und die Physiologie seiner Zeit," 98) claims they despised each other; what is clear is that they were extremely attentive to each other's research and findings, a level of respect they did not bestow on many of their contemporaries.

237. Ibid., 98n.

238. "In the intervening five years, many of the initial hopes for medical application of Gal-

vanism had proven illusory" (Trumpler, "Questioning Nature," 142–43). Humboldt thus tried to begin his volume with what hopes remained for medical applications, and this won him some positive reception among the more clinically oriented commentators in the medical community, including Christoph Hufeland (ibid., 156). In a review for the *Medicinisch-chirugische Zeitung*, Humboldt was praised for "provid[ing] invaluable material for the founding of physiology as a science [*Wissenschaft*], for the explanation of the most important phenomena of organic nature which until now were enveloped in impenetrable darkness, and for the enrichment of medical science in general" (cited in ibid., 157).

239. "The lengthy work of Humboldt remained, together with that by Pfaff, the standard compendium on the phenomenology of galvanic appearance in the literature of the early nineteenth century" (Rothschuh, "Alexander von Humboldt und die Physiologie seiner Zeit," 102).

240. "It was not possible for Humboldt to construct from a few elements an unequivocal experimental coordination, so that he very often was unable to achieve any clear results" (ibid., 110).

241. Ibid., 97.

242. Ibid., 99. "In a time when almost all researchers were inclined with Volta to deny animal electricity, he did not allow himself to be misled from the conviction that there was a specific galvanic fluid which was of physiological origin" (ibid., 109).

243. Trumpler, "Questioning Nature," 152.

244. This did not mean his ideas were not intensely discussed, especially by Pfaff and Johann Ritter.

245. H. F. Link was an extremely versatile scholar with wide-ranging interests from the outset. One of his first publications was a treatment of the mineralogical theories of the Freiberg School of Mines: *Versuch einer Anleitung zur geologischen Kenntnis der Mineralien* (Göttingen: Dieterich, 1790). In 1790 he published one of the earliest essays in support of Lavoisier ("Über die chemische Verwandschaft," *Crells Annalen* 1 [1790]: 484–90), and then he went on to translate volumes 4 and 5 of Lavoisier's collected writings: *Herrn Lavosier . . . Physikalisch-chemische Schriften, 4, 5* (Greinfswald: Röse, 1792, 1794). Notably, he tried to reconcile Lavoisier's new chemistry with the long-established phlogiston theory of the German chemical community. See H. F. Link, *Beyträge zur Physik und Chemie: 1.1, Über einige Grundlehren der Physik und Chemie* (Rostock: Stiller, 1795); *1:2, Beobachtungen über den Wärmestoff* (Rostock: Stiller, 1796). Early on, Link published an important defense of Blumenbach's notion of *Bildungstrieb*. He published a three-volume series of essays, *Beyträge zur Naturgeschichte* (Rostock: Stiller, 1794–97), in clear emulation of his mentor.

246. Link, *Beyträge zur Naturgeschichte*, vol. 2, *Über die Lebenskräfte in naturhistorischer Rücksicht* (1795), §1, "Various Kinds of Life-Forces," 1–6. Thus, it would appear that the crucial tension between Kielmeyer's scheme of forces and those of Blumenbach revolved around this notion. Kielmeyer never used the term *Bildungstrieb* in his work; conversely, *Reproduktionskraft* was always only a *part* of the function Blumenbach associated with his own notion of *Bildungstrieb*. This terminological difference may serve as a thread by which to trace the affiliations of scholars in the 1790s and thereafter.

247. Ibid., 6.

248. Ibid., 13n. Link wrote that in 1791, in an essay for *Annalen der Naturgeschichte*, he had not sufficiently discriminated between force and drive in discussing Blumenbach's *Bildungstrieb*. Here, in his 1795 essay in *Beyträge zur Naturgeschichte*, vol. 2, he proposed to "properly distinguish among force, drive, capacity." A great deal of his contribution lay in the careful conceptual discrimination of these matters with respect to each specific life-force.

249. Ibid., 3.

250. Ibid., 7–9. That is, physiology was in no position to resolve the mind-body problem of metaphysics.

251. Ibid., 4, 10.

252. Ibid., 13–14.

253. Ibid., 15.

254. Ibid., §3, "On the Life-Forces in Plants," 16ff. For Link's later work in plant physiology, see his *Grundlagen der Physiologie der Pflanzen*, 2 vols. (Göttingen, 1807–9).

255. Perhaps most notable in this regard was Reimarus, as we have seen.

256. Link, *Beyträge zur Naturgeschichte*, "On the Life-Forces in Plants," 2:16ff. Blumenbach clearly assigned *Bildungstrieb* to plants. Kielmeyer did as well. And, for all his equivocations about *life*, so too did Kant.

257. Ibid., 17.

258. Ibid., 18.

259. Ibid., 19–20.

260. Ibid., 22.

261. Link's review of Kielmeyer's address appeared in *Neues Allgemeine deutsche Bibliothek* 7 (1794): 363–64.

262. Link, *Beyträge zur Naturgeschichte*, 2:40.

263. Ibid.

264. Ibid., 38.

265. Brandis, *Versuch über die Lebenskraft* (1795), 71.

266. Ibid., preface, xiv.

267. Ibid.

268. Ibid., xvi.

269. Ibid., 4.

270. Ibid., xviii.

271. Ibid., 13.

272. Ibid., 26–27.

273. Ibid., 15.

274. This corroborates Engels's thesis (in "Die Lebenskraft") that these naturalists were working with a "methodological," rather than a "metaphysical," orientation.

275. Brandis, *Versuch über die Lebenskraft*, 15.

276. Ibid., 15–16.

277. Ibid., 17. This is one of the most important conceptualizations of life-force from the era: namely, that a life-force was necessary not as a hyperphysical intervention but as a force at a different level of organization, which modified the prevailing physicochemical laws, without contradicting them. This is still the fundamental issue for a special science of biology that is not simply reducible to physics or chemistry.

278. Ibid., 2, 18. Alexander von Humboldt, who became far more familiar with the French intellectual scene, clearly correlated his own similar view (in "Die Lebenskraft oder der rhodische Genius") and his early work on plant physiology with Bichat's theory. Later, Humboldt backed away from the notion of life-force altogether. In 1797, at the conclusion of the second volume of his *Versuche über die gereizte Muskel- und Nervenfaser*, he pronounced that there was no empirical foundation for the notion of life-force. In still-later work, he chose to distinguish between living and nonliving entities, but with no reference to a life-force.

279. Brandis, *Versuch über die Lebenskraft*, 18.

280. Ibid., 26. *This* sounds more metaphysical than methodological, it must be said!

281. Ibid., 14n.

282. Ibid.

283. Ibid.

284. Brandis found tremendous reinforcement for his approach in Erasmus Darwin's *Zoonomia*, which he had just read as he was completing his own monograph. He would eventually translate that text into German. Between his efforts and those of Girtanner, who published *Ausführliche Darstellung des Darwinschen Systemes der praktischen Heilkunde, nebst einer Kritik desselben* (Göttingen: Rosenbüsch, 1799), *Zoonomia* became a significant force for German thinking about life science by 1800.

285. Brandis, *Versuch über die Lebenskraft*, 29–33.

286. Ibid., 36. This was one of the most problematic elements in his formulation.

287. Ibid., 6n–7n.

288. Ibid., 52. It is always important to recognize that the first sense of *Reproduktion* in this epoch was not propagation of species, as it would be for us, but rather the regeneration of tissue or tissue growth more generally.

289. Ibid., 54–62.

290. Ibid., 71. This is one of the most striking phrases in Brandis's monograph.

291. Ibid., 73.

292. Ibid., 83ff. This second segment was entitled: "Further Grounds for the Doctrine That the Phlogistic Process Takes Place in the Fiber Itself."

293. Ibid., 72.

294. Ibid., 74–75.

295. Ibid., 83ff.

296. Ibid., 81.

297. Johann Christian Reil, review of *Versuch über die Lebenskraft*, by Joachim Brandis, *Archiv für Physiologie* 1, no. 2 (1795): 178–92, citing 190.

298. Ibid., 192.

299. Ibid., 184.

300. Ibid., 182.

301. Ibid., 180.

302. Reil began his university studies at Göttingen in 1779 but he stayed only a year and a half before shifting to Halle. Why Reil disliked Göttingen is an unanswered question: "In the biographical literature on Reil one finds only the terse observation that he swiftly left Göttingen because of the 'stultifying dogmatism' that prevailed there and transferred to 'democratic Halle.' But Göttingen University was in those very years at its height, and especially also in medicine it had much that was new and exciting to offer" (Mocek, *Johann Christian Reil*, 30). Indeed, this is baffling; we have not had any good explanation. Without suggesting this explained the matter, Hansen Le Roy ("Johann Christian Reil," 108) notes that Reil's father died during this year. After his move to Halle, Reil became the protégé of the practice-oriented Johann Friedrich Gottlieb Goldhagen (1742–88). His academic research for his dissertation picked up the threads of the crucial Halle "philosophical physicians" of midcentury: neurophysiology, mind-body relations, physiological psychology, and mental illness. The most important intellectual development for Reil came in the year after he completed his studies at Halle, 1782–83, when he moved to Berlin and came into contact with the eminent Jewish physician-philosopher Marcus Herz. Herz had a direct connection to Kant, first as his student and then as his key correspondent over the "silent decade" of the composition of the *Kritik der reinen Vernunft*. He was prominent in propagating Kantianism in Berlin over the 1770s and '80s. Martin Davies, *Identity or History? Marcus Herz and the End of the Enlightenment* (Detroit: Wayne State University Press, 1995). Hansen observes that Herz shared all Reil's interests: "Herz's concerns centered on the relationship of mind and body, most particularly as revealed by disorders of the nervous system, both mental and physiological. He wanted to construct theories of nerve function and of mind-body interaction that had an empirical basis in clinical medicine and a philosophical foundation in the *Critique of Pure Reason*." Lee Ann Hansen, "From Enlightenment to *Naturphilosophie*: Marcus Herz, Johann Christian Reil, and the Problem of Border Crossings," *Journal of the History of Biology* 26 (1993): 39–64, citing 45.

303. Hansen, "From Enlightenment to *Naturphilosophie*," 62.

304. Ibid., 63.

305. Mocek, *Johann Christian Reil*, 84.

306. "The *Bildungstrieb* became a name for the laws of polar dynamics which produced ever more complex metamorphoses of phenomena" (Hansen Le Roy, "Johann Christian Reil," 203).

307. Mocek cites the introduction to Reil's multivolume study on *The Cures of Fever* from 1799: "I write at a time when nerve and humoral pathologists, Brownians and anti-Brownians, go to

battle with one another, where the going theories of medicine have all come apart. . . . [T]here are regions of medicine where it is darkest night" (*Johann Christian Reil*, 87).

308. Johann Christian Reil, "An die Professoren Herrn Gren und Herrn Jakob in Halle," *Archiv für die Physiologie* 1 (1795): 5.

309. Lohff, *Die Suche nach der Wissenschaftlichkeit*, 28. See also Erna Lesky, "Cabanis und die Gewißheit der Heilkunst," *Gesnerus* 11 (1954): 152–84. Broman writes that "Kant's critical philosophy had by the 1790s come to define the epistemological conditions of academic *Wissenschaft* in Germany." Thomas Broman, *The Transformation of German Academic Medicine* (Cambridge: Cambridge University Press, 1996), 136. But Broman also recognizes that not everyone followed Kant's definitions of proper science, and he sees two alternative discourses emerging in the controversy. One sought to place medicine under the umbrella of an alternative model of science: history. The other was *Naturphilosophie* (*Academic Medicine*, 138ff.).

310. "You, my friends, can contribute *a great deal* to the perfection of physiology through your philosophical, physical, and chemical expertise" (Reil, "An die Professoren Herrn Gren und Herrn Jakob in Halle," 6).

311. Ibid., 4.

312. Ibid. "Supersensible substrate" is a distinctly Kantian construct, from the *Kritik der Urteilskraft*, a speculative basis through which to conceive of the unity of the subjective purposiveness of human agency with the order of the phenomenal world. *Soul* was the key concept of Stahlian (and other traditional) theories of life. The "universal world spirit" was a conception Reil identified with Herder's *Ideen* (Reil, *Von der Lebenskraft*). Mocek makes an important observation on this score: "To be sure, in the *Lebenskraft* essay [Reil] had expressed more skepticism toward than agreement with Herder's world-spirit interpretation of Spinoza; only his own Spinozistic point of departure corresponded entirely with [Herder's] interpretation" (*Johann Christian Reil*, 146). Finally, *Lebenskraft* was, of course, the term on everyone's lips in the 1790s, the occasion for Reil's extended essay. According to Mocek, "The concept of 'life-force' belonged among the questionable conceptual resources of biology and medicine at the close of the eighteenth century" (69). That was certainly what moved Reil.

313. Reil, *Von der Lebenskraft*, 6.

314. "The theoretical insufficiency of biology and medicine played its part in landing the search for orientation in the lap of philosophy—but where else?" (Mocek, *Johann Christian Reil*, 102). "With Kant the world of science begins anew" (ibid., 93).

315. Hansen, "From Enlightenment to *Naturphilosophie*," 40. Hansen is emphatic that the project of Herz and Reil had nothing to do with Lenoir's "teleomechanical school of German physiology." She distinguishes "two Kantian strands in the organic sciences of late eighteenth- and early nineteenth-century Germany." One, associated with Blumenbach, took up embryology and natural history, as Lenoir elaborated. "But there is also a second Kantian strand," concentrated in "clinical medicine, anatomy, and physiology" and focused on "the relationships of mind, emotions, and the body" (45; see also 41, 44). This was the line that Herz and Reil pursued.

316. Ibid., 45.

317. Ibid., 41.

318. Mocek, *Johann Christian Reil*, 32.

319. In the opening pages of his *Anthropology*, Kant dismissed the ambitions of physiological psychology. In his discussion of "empirical psychology" on the basis of the "critical philosophy," he eschewed utterly any bodily significance for cognition. Kant made the problem of self-consciousness one of the deepest abysses of his critical philosophy, and in *Metaphysical Foundations of Natural Science* he denied that psychology could ever achieve any scientific results.

320. Hansen, "From Enlightenment to *Naturphilosophie*," 53. "Herz was not content to rest with the Kantian limit on understanding the mind-body connection" (56).

321. Ibid., 49.

322. How could Kant's transcendental idealism encode Reil's empirical realism? That became

the problem. Hansen notes: "Although Reil made extensive use of Kantian concepts [in *Von der Lebenskraft*], he was by no means a Kantian even here" (Hansen Le Roy, "Johann Christian Reil," 7). Reil, she affirms, "was a dubious Kantian from the very beginning despite the lip service he gave to the critical philosophy," and "he undercut the Kantian agenda even as he espoused it" (Hansen, "From Enlightenment to *Naturphilosophie*," 41). Whatever Kantian vocabulary Reil had appropriated or even believed that he had accepted was undercut by his reliance on chemistry as the fundamental reality and by the way in which he subverted the Kantian limits of reason.

323. Reil, *Von der Lebenskraft*, 9.

324. Hansen Le Roy, "Johann Christian Reil," 96, citing Reil, *Von der Lebenskraft*, 25. "Reil thus rejected the entire physiological tradition of an immaterial vital force responsible for animal functions" (Hansen Le Roy, "Johann Christian Reil," 115).

325. Mocek writes of his having "landed in an empirical materialism" (*Johann Christian Reil*, 139).

326. Reil, *Von der Lebenskraft*, 9. The blurring of the distinction between "physics" and "chemistry" had to do with the ambiguity of *Naturlehre* as the generic term for physical science. Thus, Gren's journal was entitled *Journal der Physik*, but it was the central journal for German chemistry. Reil's work demonstrated the preponderance of chemistry in his notion of the physical sciences, and this was becoming widespread among German naturalists at that moment, as Stuart Strickland has argued ("Galvanic Disciplines," 453).

327. Reil, *Von der Lebenskraft*, 11.

328. Ibid., 23.

329. Reil, review of *Versuch über die Lebenskraft* by Brandis, *Archiv für die Physiologie* 1, no. 2 (1795): 180.

330. Reil, *Von der Lebenskraft*, 19.

331. Ibid., 49, 53.

332. Ibid., 50.

333. Hansen, "From Enlightenment to *Naturphilosophie*," 63.

334. Johannes Köllner, "Prüfung der neuesten Bemühungen und Untersuchungen in der Bestimmung der organischen Kräfte, nach Grundsätze der kritischen Philosophie," *Archiv für die Physiologie* 2 (1797): 240–396. There is a thorough analysis of this essay in Hansen Le Roy, "Johann Christian Reil," 130–48.

335. Schelling, *Von der Weltseele* (1798), cited in Hansen Le Roy, "Johann Christian Reil," 97.

336. Kant was perfectly happy to "think"—and, I would urge, *believe in*—the "supersensible substrate"; he just would not permit any *knowledge claims* about it. It was a "regulative ideal."

337. "Above all, the much-discussed question regarding how much Schelling's philosophy really advanced the natural sciences and medicine of this era is to be answered concretely in the works of Reil" (Mocek, *Johann Christian Reil*, 111).

338. Hansen, "From Enlightenment to *Naturphilosophie*," 64.

339. "The conception formulated in the text on *Lebenskraft* remained the foundation of Reil's work, in the natural-theoretical sense and also in the narrower physiological sense, at least until 1804 and, I believe I can establish, essentially until 1806" (Mocek, *Johann Christian Reil*, 36). Mocek makes some good points (137–38) about an interpretation of a text from 1802 that Hansen Le Roy offered in her "Johann Christian Reil" (176–78). Hansen Le Roy took Reil's warnings about the danger that "transcendental philosophy" posed to empirical medical science (in Reil's important review essay of T. G. A. Roose's *Grundzüge der Lehre von der Lebenskraft* in *Archiv für Physiologie* 5, no. 2 [1802]: 319–21) as a repudiation of Kantianism, but the phrase "transcendental philosophy" in that context, as Mocek persuasively contends, had a far more likely connection to Schelling's recent *System of Transcendental Idealism* (1800). It should be recalled that Schelling, in the name of Kant, had been particularly scathing in his criticism of Reil in 1798 and had subsequently embarked upon the conquest of medicine by *Naturphilosophie*—something Reil was not yet ready to accept. It is thus highly plausible that Reil meant to disparage Schelling in the cited

passage of 1802. Still, Mocek makes the crucial error of believing that Reil's insistence upon empiricism and observation in his review could only mean that he was an orthodox Kantian. Of this, alas, we have no reason to be so confident. In general I believe that Hansen is correct about the disintegration of Kantianism in Reil's thought.

340. Broman, *Academic Medicine*, 86.

341. "A continually recurring area of interest for Reil was the reform of medical education" (Mocek, *Johann Christian Reil*, 39).

342. Broman, *Academic Medicine*, 87–88.

343. Thomas Broman, "J. C. Reil and the 'Journalization' of Physiology," in *The Literary Structure of Scientific Argument: Historical Studies*, ed. Peter Dear (Philadelphia: University of Pennsylvania Press, 1991), 13–45, citing 23.

344. Ibid., 22.

345. Ibid.

346. Ibid., 26.

347. D. Von Madai, "Über die Wirkungen der Reize und der thierischen Organe," *Archiv für die Physiologie* 1, no. 3 (1795–96): 64–148. And see Johann Christian Reil, "Veränderte Mischung und Form der thierischen Materie, als Krankheit oder nächste Ursache der Krankheitszufälle betrachtet," *Archiv für die Physiologie* 3, no. 3 (1799): 424–61.

348. Broman, "J. C. Reil," 29.

349. Ibid.

350. Ibid., 35.

351. Ibid., 31.

352. Ibid., 33.

353. Ibid., 31. Broman finds this less than salutary: "From a science of argumentation about causes, physiology in the thrall [!] of *Naturphilosophie* became a science of description" (ibid., 35). That betrays all the old positivist misapprehensions of *Naturphilosophie* and misses the historical significance of Reil's conversion and the subsequent developments of his journal and of the field of physiology in Germany.

354. Broman, *Academic Medicine*, 186.

355. Ibid., 188; Broman, "J. C. Reil," 36–40.

356. Broman, *Academic Medicine*, 201.

357. Broman, "J. C. Reil," 40. I dispute that "withering" is the proper term for this disciplinary speciation.

358. Lynn Nyhart, *Biology Takes Form: Animal Morphology and the German Universities, 1800–1900* (Chicago: University of Chicago Press, 1995), 193.

CHAPTER TEN

1. Friedrich Schelling, *Von der Weltseele: Eine Hypothese der höheren Physik zur Erklärung des allgemeinen Organismus* (Hamburg, 1798), vi.

2. Here I dissent from Stephane Schmitt, "Type et métamorphose dans la morphologie de Goethe, entre classicisme et romanticisme," *Revue d'histoire des sciences* 54 (2001): 495–521.

3. Timothy Lenoir, "The Eternal Laws of Form: Morphotypes and the Conditions of Existence in Goethe's Biological Thought," in *Goethe and the Sciences: A Reappraisal*, ed. F. Amrine, J. Zucker, and H. Wheeler (Dordrecht: Reidel, 1987), 17–28, citing 17.

4. "In spite of the less than enthusiastic reception of his scientific work, Goethe gave up neither his interest in morphology nor his attempts to find a forum for his scientific insights. . . . Instead, he kept searching for and experimenting with different ways of making them public." Dorothea von Mücke, "Goethe's Metamorphosis: Changing Forms in Nature, the Life Sciences, and Authorship," *Representations* 95 (2006): 27–53, citing 29.

5. On Goethe and the Humboldt brothers, see Nicholas Boyle, *Goethe: The Poet and the Age*, vol. 2, *Revolution and Renunciation* (Oxford: Clarendon Press, 2000). On the scientific endeavors of the Humboldt brothers at Jena, see Ilse Jahn, "Die anatomischen Studien der Brüder Humboldt unter Justus Christian Loder in Jena," *Beiträge zur Geschichte der Universität Erfurt (1392–1816)* 14 (1956): 91–97.

6. Johann Wolfgang von Goethe, "Toward a General Comparative Theory," in *Scientific Studies*, ed. and trans. Douglas Miller (New York: Suhrkamp, 1988), 53–56.

7. Gabrielle Bersier, "Visualizing Carl Friedrich Kielmeyer's Organic Forces: Goethe's Morphology on the Threshold of Evolution," *Monatshefte* 97 (2005): 18–32, citing 25.

8. Boyle, *Goethe*, 2:482.

9. Ibid., 483. "The dualism we find in Kant and Humboldt is an extended and radicalized form of Leibniz's original distinction between the monads and the 'well-founded phenomena' to which they gave rise" (ibid., 53).

10. Ibid., 53.

11. Ibid., 486.

12. Johann Wolfgang von Goethe, "A Commentary on the Aphoristic Essay 'Nature' (Goethe to Chancellor von Müller)," in *Scientific Studies*, 6.

13. Ibid.

14. Ibid.

15. "In the three years after his return from Venice Goethe wrote even less of literary merit than between 1782 and 1786," his biographer Nicholas Boyle writes. "Far from being released into new creativity by his Italian experiences, he seems . . . to have reverted to a state of latency like that which preceded his self-discovery as a writer in 1769" (*Goethe*, 2:170).

16. Julia Gauss, "Goethe und die Prinzipien der Naturforschung bei Kant," *Studia Philosophica* 29 (1970): 54–71, citing 54–55.

17. Boyle, *Goethe*, 2:94.

18. Goethe to von Stein, cited in Heinrich Henel, "Type and Proto-phenomenon in Goethe's Science," *PMLA* 71 (1956): 651–68, citng 655.

19. Ibid., 655n.

20. Ibid.

21. Schmitt, "Type et métamorphose"; Dorothea Kuhn, "Goethe's Relationship to the Theories of Development of His Time," in Amrine, Zucker, and Wheeler, *Goethe and the Sciences*, 3–15.

22. There is a curious feedback loop here. Goethe worked out these ideas, perhaps without the specific terms yet, in his *Versuch die Metamorphose der Pflanzen zu erklären* in 1790, then Schelling took them up and developed them in his works of 1798–99, and Goethe, his ideas confirmed by Schelling's adoption, reaffirmed them as central in his publications starting in 1817. "'Polarity' was to become one of the controlling ideas in Goethe's later understanding both of Nature and of human life but it was more important to him as a way of evoking the ungraspably ambiguous processes of struggle and change, than as a means of identifying specific pairs of permanently opposed forces" (Boyle, *Goethe*, 2:679).

23. Goethe to Reichardt, October 25, 1790, in Gauss, "Goethe und die Prinzipien der Naturforschung bei Kant," 55n.

24. It appears the talk had dealt with Linnaean classification in biology. By this point, Goethe was thoroughly disenchanted with Linnaeus. See James Larson, "Goethe and Linnaeus," *Journal of the History of Ideas* 28 (1967): 590–96.

25. Johann Wolfgang von Goethe, "Fortunate Encounter," in *Scientific Studies*, 20. This holistic approach to causation immediately calls to mind Kant's discussion of the *intellectus archetypus* from §§76–77 of the *Kritik der Urteilskraft*, especially in the light of Förster's brilliant exposition of these sections and their impact on post-Kantian thought. Eckart Förster, "Die Bedeutung von §§76–77 der *Kritik der Urteilskraft* für die Entwicklung der nachkantichen Philosophie," *Zeitschrift für philosophische Forschung* 56 (2002): pt. 1, 169–90, and pt. 2, 321–45; Eckart Förster,

"Goethe and the 'Auge des Geistes,'" *Deutsche Vierteljahrsschrift für Litteraturwissenschaft und Geistesgeschichte* 75 (2001): 87–101.

26. Johann Wolfgang von Goethe, "A Study Based on Spinoza," in *Scientific Studies*, 8. Goethe's notion of such an "intellectual intuition" had unquestionable roots in his study of Spinoza. Neubauer brings this out clearly by referring back to Goethe's correspondence with F. H. Jacobi in the context of the *Pantheismusstreit*. On May 6, 1786, Goethe wrote to Jacobi: "when Spinoza speaks of the *scientia intuitiva* . . . these few words give me courage to devote all my life to the contemplation of things I can reach [*reichen*] and of whose essential form I can hope to develop an adequate idea." John Neubauer, "Goethe and the Language of Science," in *The Third Culture: Literature and Science*, ed. Elinor Schaffer (Berlin: De Gruyter, 1998), 51–65, citing 52–53n.

27. On Goethe learning about Peter Camper's techniques from his son in Rome, see Robert Paul Willem Visser, *The Zoological Work of Petrus Camper (1722–1789)* (Amsterdam: Rodopi, 1995).

28. See Lorraine Daston and Peter Galison, *Objectivity* (New York: Zone, 2007), 67–75.

29. Goethe, "Fortunate Encounter," 20. The sense that Goethe was after a simple perception, without the mediation of reflection, is a misconstrual of the notion of "visualization" at the heart of his thinking, as he articulated already in a text dated to September 27, 1786: "But what is seeing without thinking?" We cannot be sure, however, that this was not an interjection by Goethe at the time he composed his *Italian Journey* rather than an actual note from that date. See Joan Steigerwald, "Goethe's Morphology: *Urphänomene* and Aesthetic Appraisal," *Journal of the History of Biology* 35 (2002): 291–328, for strong reservations along these lines. I take it to be a thought from the Italian years themselves, and I believe this crucial rhetorical question indicates that Goethe was never someone who believed in unmediated perception of facts of nature but rather from the outset grasped the intuitive configuration of human cognition.

30. See Schiller's key letter to Goethe, August 1794, in *Der Briefwechsel zwischen Schiller und Goethe*, ed. Emil Staiger (Frankfurt: Insel, 1966), taking up and elaborating on their initial conversation.

31. "Indeed it is Schiller's unforgettable contribution that he brought [Goethe] into contact with the brothers Humboldt, with Reinhold and the University of Jena, and thus introduced him into a decisive circle of Kantians and linked him with it for years" (Gauss, "Goethe und die Prinzipien der Naturforschung bei Kant," 65).

32. See the essays in Helmut Brandt, ed., *Goethe und die Wissenschaften* (Jena: Friedrich Schiller Universität, 1984). "It was Goethe's position as an administrator at the Weimar court, his involvement through this position in the institutions at Jena and his frequent contact with the intellectual figures who gravitated to Jena, that enabled him to generate interest in his science of morphology and the methods for its study" (Steigerwald, "Goethe's Morphology," 317). Goethe himself wrote, later: "Among the scholars who offered me some solidarity from their side, I can count anatomists, chemists, literary scholars, philosophers—like Loder, Soemmerring, Göttling, Wolf, Forster, Schelling—but by contrast no physicists." Goethe cited in Wilhelm Raimund Beyer, "Natur und Kunst: Goethes Interesse am Jenenser Schelling," *Goethe Jahrbuch* 92 (1975): 9–28, citing 13.

33. Goethe, "Fortunate Encounter," 20–21.

34. Gauss writes, fairly enough, that for all his reservations "it is nevertheless not proper to speak of a real rejection [*Lossagung*] of Kant." But even she acknowledges the essential point, citing Goethe's own words: Kant "made me more attentive to myself [*mich auf mich selber aufmerksam gemacht*]" ("Goethe und die Prinzipien der Naturforschung bei Kant," 71).

35. Johann Wolfgang von Goethe, "The Influence of Modern Philosophy," in *Scientific Studies*, 28–32, citing 28–29.

36. Ibid., 29.

37. Ibid.

38. Immanuel Kant, *Kritik der reinen Vernunft*, A175–78/B222.

39. Goethe, "Influence of Modern Philosophy," 29.

40. Ibid., 29–30.

41. Cited in Förster, "Goethe and the 'Auge des Geistes,'" 91n.

42. Goethe, "Influence of Modern Philosophy," 30.

43. Goethe, "Doubt and Resignation," in *Scientific Studies*, 33.

44. Ibid. "This Kantian phase was gradually to stifle the progress of Goethe's morphological speculations—which, with their dependence on the notion of a unified scale of being, were essentially Leibnizian—and in the new area of 'comparative anatomy' that he was so hopefully entering none of his more substantial projects would advance beyond the fragmentary state of the *Essay on the Form of Animals*" (Boyle, *Goethe*, 2:82). Goethe needed to be liberated from Kantianism to conduct his empirical science. This comes through clearly in a letter Goethe sent to Soemmerring concerning his controversial *Organ der Seele* and the involvement of Kant. See Peter McLaughlin, "Soemmerring und Kant: Über das Organ der Seele und den Streit der Fakultäten," in *Samuel Thomas Soemmerring und die Gelehrten der Goethezeit*, ed. Gunter Mann and Franz Dumont (Stuttgart: Fischer, 1985), 191–202. That was a ubiquitous experience in that moment. It was what made Schelling seem to be a liberator.

45. Johann Wolfgang von Goethe, "Judgment through Intuitive Perception," 31.

46. Ibid. See Kant, *Kritik der Urteilskraft*, AA 5:405–10.

47. Goethe, "Judgment through Intuitive Perception," 31.

48. Ibid., 32.

49. With no little disapproval, Boyle writes: "Goethe sketches a remarkable anticipation of that nineteenth century materialist biology which accepted Kant's critical conclusions while ignoring his concern for their rational context" (*Goethe*, 2:82). That is, they disregarded Kant's cardinal distinction between the regulative and the constitutive. As it happens, I believe, that was a *progressive* step for science. See my "Should Kant Have Abandoned the 'Daring Adventure of Reason'?—the Interest of Contemporary Naturalism in the Historicization of Nature in Kant and Idealist *Naturphilosophie*," in *International Yearbook of German Idealism*, vol. 8 (Berlin: De Gruyter, 2010 [actually 2012]), 130–64.

50. Goethe, "Influence of Modern Philosophy," 30.

51. For Goethe, *type*—most famously, the leaf—was, in the terms of Förster, "neither any existing leaf, nor is it a generalization from all existing leaves, like a Lockean 'abstract idea.' Nor is it a primitive natural ancestor of the present plant from which the latter has evolved in a Darwinian manner. Rather, it is (comparable to Hegel's Idea) a concrete universal" (Förster, "Goethe and the 'Auge des Geistes,'" 97–98).

52. Daston and Galison, *Objectivity*, 64–75.

53. Johann Wolfgang von Goethe, "Empirical Observation and Science," in *Scientific Studies*, 24.

54. In the words of Overbeck, "The external phenomenon is at the same time the exterior of the law and the law is the interior of the phenomenon." Gertrud Overbeck, "Goethes Lehre von der Metamorphose der Pflanzen und ihre Widerspiegelung in seiner Dichtung," *Publications of the English Goethe Society* 31 (1961): 38–59, citing 48.

55. Goethe, "Empirical Observation and Science," 24.

56. Kant, *Kritik der Urteilskraft*, AA 5:233–35.

57. Kant defined a rational idea at *Kritik der reinen Vernunft*, A326; cf. Schiller's invocation of it in July 1794.

58. Pfau is right to urge us to investigate "the close affinities between the life-sciences and aesthetic models of autopoiesis around 1800." Thomas Pfau, "'All Is Leaf': Difference, Metamorphosis, and Goethe's Phenomenology of Knowledge," *Studies in Romanticism* 49 (2010): 3–41, citing 25, 28. "Goethe defines 'seeing' as recognizing within a given phenomenon the law governing its existence, which cannot appear *per se* but nonetheless conditions the object's specific mode of appearance" (ibid., 37). For a brilliant exploration of this issue, see Förster, "Goethe and the 'Auge des Geistes.'"

59. I submit that this was Goethe's stance all along. The real tension is between Goethe's *figural* sense of language and Kant's *discursive* sense. The latter, as Goethe clearly put it in a late text,

held apart intuition and understanding in a manner that was destructive for scientific insight. Goethe himself, however, never accepted this dichotomy and even under the maximal pressure of Schiller to adopt a Kantian vantage point, Goethe's resistance was not what Schiller had earlier scorned as a kind of infatuation on Goethe's part with his own powers of observation but rather a coherent sense of meaning in language that was far more integral than Kant's discursivism could accommodate.

60. Pfau, "'All Is Leaf,' 37.

61. Johann Wolfgang von Goethe, *Maxims and Reflections*, trans. Elizabeth Stopp (Harmondsworth: Penguin, 1998), 75.

62. Goethe, "Doubt and Resignation," 33.

63. Olaf Breidbach, *Goethes Metamorphosenlehre* (Munich: Fink, 2006).

64. Goethe wrote: "The German has a word, *Gestalt*, for the complex existence of an actual being. He abstracts, with this expression, from the moving and assumes a congruous whole to be determined, completed, and fixed in its character. . . . But . . . we find that independence, rest, or termination nowhere appear, but everything fluctuates rather in continuous motion. Our speech is therefore accustomed to use the word *Bildung* pertaining to both what has been brought forth and the process of bringing forth." Johann Wolfgang von Goethe, *Bildung und Umbildung* (1817; Frankfurt: Deutscher Klassiker Verlag, 1999), 24:392. See also Astride Orle Tantrillo, "Goethe's 'Classical' Science," in *The Literature of Weimar Classicism*, ed. Simon Richter (Rochester, NY: Camden House, 2005), 323–45, citing 330.

65. Ronald Brady, "Form and Cause in Goethe's Morphology," in Amrine, Zucker, and Wheeler, *Goethe and the Sciences*, 257–300, citing 275–77. "The turn from *Gestalt* to *Bildung* is a shift of focus from the static product to the transformation which leads to and from the product, and thus eventually to a consideration, not of the products, but of the generative field of movement" (280–81).

66. Dalia Nassar, "From a Philosophy of Self to a Philosophy of Nature: Goethe and the Development of Schelling's *Naturphilosophie*," *Archiv für Geschichte der Philosophie* 92 (2010): 304–21, citing 309. Here, Goethe proposed to construe a principle that constituted unity across *temporal* variation of a specific form. That takes us back to a classical problem in philosophy that preoccupied Leibniz: namely, the problem of *identity* across change in manifestations.

67. Fergus Henderson, "Goethe's 'Naturphilosophie'—Essay Review of R. H. Stephenson, *Goethe's Conception of Knowledge and Science* (Edinburgh University Press, 1995)," *Studies in History and Philosophy of Science* 29 (1998): 143–53, citing 146.

68. Förster, "Goethe and the 'Auge des Geistes,'" 92.

69. Johann Wolfgang von Goethe, "Analysis and Synthesis," in *Scientific Studies*, 49.

70. Förster, "Goethe and the 'Auge des Geistes,'" 93.

71. Goethe, cited in ibid.

72. Goethe wrote: "the phenomenon is not detached from the observer, but intertwined and involved with him" (*Maxims and Reflections*, 155). Pfau draws a conclusion that is as presentist as it is historicist: "The challenge for science thus is to understand that it is only ever involved with 'objects' at the level of 'phenomenon' and that, consequently, it must at all times stay focused on the dynamic process whereby phenomena make their 'appearance' and so show themselves to be constitutively entwined with an observing intelligence" ("'All Is Leaf,'" 35). Pfau elaborates: "engaging the world as a practical, sensory, and intelligent being means to understand mind and world as ontologically entwined" (40).

73. Goethe, "Bildung und Umbildung" (1817), cited in Brady, "Form and Cause in Goethe's Morphology," 274.

74. Johann Wolfgang von Goethe, "Problems," in *Scientific Studies*, 44.

75. Goethe, "Influence of Modern Philosophy," 28.

76. On this notion of mechanism in Kant, especially as articulated in the *Kritik der Urteilskraft*, see esp. Peter McLaughlin, *Kant's Critique of Teleology in Biological Explanation: Antinomy and Teleology* (Lewiston, NY: Mellen, 1990); Peter McLaughlin, "Newtonian Biology and Kant's Mecha-

nistic Causality," in *Kant's Critique of the Power of Judgment: Critical Essays*, ed. Paul Guyer (London: Rowman and Littlefield, 2003), 209–18; Peter McLaughlin, "Mechanical Explanation in the 'Critique of the Teleological Power of Judgment,'" in *Kant's Theory of Biology*, ed. Ina Goy and Eric Watkins (Berlin: De Gruyter, 2014), 149–66. For a critique of McLaughlin's view, see Hannah Ginsburg, "Two Kinds of Mechanical Inexplicability in Kant and Aristotle," *Journal of the History of Philosophy* 42 (2004): 33–65; Hannah Ginsburg, "Kant on Understanding Organisms as Natural Purposes," in *Kant and the Sciences*, ed. Eric Watkins (Oxford: Oxford University Press, 2001), 231–58.

77. Amrine writes that morphology defined the proper study of nature for Goethe: "to understand forms, to see and then conjecture as to their processes of modification and adaptation . . ." (Frederick Amrine, "Editorial Preface," in Amrine, Zucker, and Wheeler, *Goethe and the Sciences*, vii).

78. The "archetype is constantly evolving and transforming itself through the process of reproduction," and so, even as "the underlying archetype or structure defines the constraints within which changes of form occur," those very constraints change. The result, for Goethe, is that change becomes "an open-ended process." There is no overarching teleology to natural process as a whole, no singular goal fostered by an intelligent design, and certainly not mankind. Instead, "Goethe . . . adopts a modified version of functionalism, one that betrays a considerable proximity to Aristotle's locally specific teleology[;] . . . each animal is its own purpose" (von Mücke, "Goethe's Metamorphosis," 33–35).

79. "Throughout his scientific corpus, Goethe speaks of the economy of nature and how it can through small variations and limited principles take the most basic of elements and create the most infinite and varied forms" (Tantrillo, "Goethe's 'Classical' Science," 329).

80. Johann Wolfgang von Goethe, "The Formative Impulse," in *Scientific Studies*, 35.

81. Ibid.

82. "If we . . . reconsider evolution and epigenesis, they will strike us as terms which only avoid the issue" (ibid., 36).

83. Ibid.

84. Goethe, cited by George Wells, "Goethe and Evolution," *Journal of the History of Ideas* 28 (1967): 537–50, citing 537–38n.

85. Dorothea Kuhn notes: "Throughout his life Goethe was intrigued by Buffon's sketch of a self-creating nature. . . . He saw in Buffon a kind of precursor of his own typology" ("Goethe Relationship to the Theories of Development," 6).

86. On idealization versus transformism in morphology, see Dietrich von Engelhardt, "Die Naturwissenschaft der Aufklärung und die romantisch-idealistische Naturphilosophie," in *Idealismus und Aufklärung*, ed. Christoph Jamme and Gerhard Kunz (Stuttgart: Klett-Cotta, 1988), 80–96.

87. For one effort to sort all this out, see Manfred Wenzel, "Verzeitlichungstendenze im Vorfeld des Evolutionismus," *Natur und Museum* 112 (1982): 15–25. For a more extensive consideration, see Dietrich von Engelhardt, *Historisches Bewußtsein in der Naturwissenschaft von der Aufklärung bis zum Positivismus* (Freiburg im Breisgau: Alber, 1979).

88. Kant, *Kritik der Urteilskraft*, AA 5:419n.

89. Carl Friedrich Kielmeyer, *Gesammelte Schriften*, ed. F. Holler (Berlin: Keiper, 1938), 299.

90. "A second reading of 'Kielmeyers Rede' was recorded on November 9, 1806, in conjunction with plans to publish the morphological essays written in the nineties." Gabrielle Bersier, "Visualizing Carl Friedrich Kielmeyer's Organic Forces: Goethe's Morphology on the Threshold of Evolution," *Monatshefte* 97 (2005): 18–32, citing 19. Wenzel suggests that Goethe's work on morphology proceeded in spurts. The first came around 1795, followed by a lull. Then Goethe picked up matters again in the mid-1800s, falling off again until 1816–17, when he began to put together the series of publications that would comprise *Vom Morphologie*. Manfred Wenzel, "Goethes Morphologie in ihrer Beziehung zum Darwinischen Evolutionsdenken," *Medizinhistorisches Journal* 18 (1983): 52–68, citing 54–55.

91. Bersier, "Visualizing Carl Friedrich Kielmeyer's Organic Forces," 19.

92. Ibid., 18.

93. Ibid., 23.

94. Ibid.

95. Goethe, *Tag- und Jahres-Hefte*, cited in ibid., 24.

96. Nisbet explicitly conceives of Goethe as aiming to "write a neo-Lucretian epic of nature for the modern age." H. B. Nisbet, "Lucretius in Eighteenth-Century Germany, with a Commentary on Goethe's 'Metamorphose der Tiere,'" *Modern Language Review* 81 (1986): 97–115, citing 99.

97. The first *prose* translation of Lucretius into German came only in 1784. A poetic translation became one of the objects of key interest in the 1790s. See ibid., 98.

98. Immanuel Kant, *Allgemeine Naturgeschichte*, AA 1:221–28.

99. Kant, *Kritik der Urteilskraft*, AA 5:393.

100. "Herder made copious notes from the *De rerum natura* in 1766, and his posthumous papers suggest that he was seriously preoccupied with philosophical materialism around this time" (Nisbet, "Lucretius in Eighteenth-Century Germany," 101). See also H. B. Nisbet, *Herder and the Philosophy and History of Science* (Cambridge: Modern Humanities Research Association, 1970), 126–27.

101. Nisbet draws attention to August von Einsiedel, whose explicit materialism proved very important to Herder and presumably to Goethe in this context. His "unpublished reflections on atoms and the struggle for existence Herder copied out for his private use" ("Lucretius in Eighteenth-Century Germany," 102). See August von Einsiedel, *Ideen* (Berlin, Akademie, 1957). But Boyle suggests that by this point in the 1790s, "the days were past in which he would wish to call himself, in any sense, a Lucretian materialist" (*Goethe*, 2:145).

102. While Knebel began in the 1790s, when "Goethe and Herder were intensively involved," he did not publish a complete version until 1821 (Nisbet, "Lucretius in Eighteenth-Century Germany," 99).

103. Ibid.

104. Schiller to Goethe, January 30, 1798, in Staiger, *Briefwechsel*, 559–61.

105. Goethe to Schiller, January 26, 1798, in ibid., 555–59.

106. Goethe had achieved great success with his *Roman Elegies*. Johann Wolfgang von Goethe, *Roman Elegies and Other Poems*, trans. Michael Hamburger (London: Anvil Press Poetry, 1996).

107. A good analysis of the formal issues in the composition is Gertud Overbeck, "Goethes Lehre von der Metamorphose der Pflanzen und ihre Widerspiegelung in seiner Dichtung," *Publications of the English Goethe Society* 31 (1961): 38–59.

108. The most powerful piece of writing associated with this project was Goethe's *Über den Granit* (1784).

109. "Romance" was a pejorative term for speculation in science in the eighteenth century, invoked by Bonnet, Haller, and even Kant.

110. See Wolf von Engelhardt, "Goethes Beschäftigung mit Gesteinen und Erdgeschichte im ersten Weimarer Jahrzehnt," in *Genio Huius Loci: Dank a Leiva Petersen*, ed. Leiva Petersen, Dorothea Kuhn, and Bernhard Zeller (Vienna: Böhlau, 1982), 169–204.

111. Nisbet, "Lucretius in Eighteenth-Century Germany," 106. "Goethe modeled his poem . . . very closely indeed on Lucretius—so much so that only such ideas and formulations as are compatible with the *De rerum natura* are included, and the rest either omitted or expressed in the most general of terms" (ibid., 108). As to dating, Nisbet is supported by Boyle: "It is . . . likely that the sixty-one hexameter lines of *Metamorphosis of Animals* were written in 1799 or 1800, probably as several discrete fragments of this project, which were only later collected together" (*Goethe*, 2:677).

112. "The poems condense a lifetime's study of nature" in a "lyrical cosmology" (Jeremy Adler, "The Aesthetics of Magnetism: Science, Philosophy and Poetry in the Dialogue between Goethe and Schelling," in Schaffer, *Third Culture*, 66–102, citing 67).

113. Beyer, "Natur und Kunst," 9.

114. See Goethe to Schiller, January 3, 6, 13, 1798, in Staiger, *Briefwechsel*, 532–44.

115. Goethe to Voigt, May 29, 1798, cited in Beyer, "Natur und Kunst," 9.

116. Adler, "Aesthetics of Magnetism," 72.

117. Goethe issued the official call to Schelling for the Jena position on July 5, 1798. Schelling would arrive to take up the position in October of that year. Hans Behring, "Goethe und Schelling," *Festschrift für Berthold Litzmann* (Bonn: Literarhistorische Gesellschaft Bonn, 1920), 77–104, citing 78.

118. Goethe to Voigt, June 21, 1798, cited in Nassar, "From a Philosophy of Self," 311.

119. Adler, "Aesthetics of Magnetism," 74.

120. Ibid.

121. Nassar, "From a Philosophy of Self," 304

122. Schelling, letter of November 9, 1799, cited in Nassar, "From a Philosophy of Self," 311.

123. "Goethe adopted central concepts of Schelling's, such as 'Polarität' and the 'Weltseele'; he also redeployed Schelling's redaction of his own ideas. Both Goethe's science and his poetry were enriched by Schelling's thought; correlatively, both Schelling's philosophy and his poetry engaged with Goethe's ideas" (Adler, "Aesthetics of Magnetism," 62). The upshot for Adler is that "the young Schelling adopted central Goethean ideas" while "the mature Goethe adopted categories from Schelling's thought" such that "Goethe's science around 1805–6, his mature epistemology, and his cosmology were decisively shaped by the encounter with Schelling" (68).

124. Nassar, "From a Philosophy of Self," 305, 307. Boyle takes a similar view, though with a twist: "In an important respect [Schelling's] *First Draft* showed the effect of Goethe's criticism of his earlier thought. It took a Spinozistic turn and even went further in that direction than Goethe himself was prepared to go" (*Goethe*, 2:621).

125. Nassar, "From a Philosophy of Self," 315. Nassar notes that research has established that the concept of productivity begins to appear in Schelling's texts only with the *Einleitung* and in later revisions of the *Erster Entwurf* (314n, referring to the work of Paul Ziche).

126. For a crucial attestation, see Kielmeyer's letter to Cuvier of 1807, discussed above.

127. Nassar, "From a Philosophy of Self," 319.

128. "In 1800, Goethe gave up his plan for a great epic of nature, and made it over to the young philosopher Schelling" (Nisbet, "Lucretius in Eighteenth-Century Germany," 110). This occurred in a personal meeting between the two figures in October 1800. Boyle extrapolates: "Goethe seems to have felt that the time had come to hand on the torch of morphology to the Nature philosophers. Certainly, after delivering this impetus to Schelling's development, his own morphological studies made no serious progress for many years" (*Goethe*, 2:622).

129. Margarethe Plath, "Der Goethe-Schellingische Plan eines philosophischen Naturgedichts: Eine Stuide zu Goethe's 'Gott und Welt,'" *Preussissche Jahrbücher* 106 (1901): 44–74, citing 48.

130. Ibid., 49.

131. Haller: "Ins innre der Natur dringt kein erschaffner Geist," from *Falschheit menschlicher Tugenden* (1730), cited in Richard Toellner, *Albrecht von Haller: Über die Einheit im Denken des letzten Universalgelehrten*, Sudhoffs Archiv, Beiheft 10 (Wiesbaden: Steiner, 1971), 65.

132. Johann Wolfgang von Goethe, "Allerdings," in *Gott und Welt* (1820), cited in Toellner, *Albrecht von Haller*, 64. Toellner argues that Goethe was not targeting Haller so much as the use of Haller's lines by Nicolai and others (64n). I am not convinced.

133. Plath, "Der Goethe-Schellingische Plan," 56.

134. Boyle, *Goethe*, 2:595.

135. Immanuel Kant, "Recension von J. G. Herders *Ideen zur Philosophie der Geschichte der Menschheit*," pt. 2 (1785), AA 8:43–66.

136. If I have stressed Herder's role, it is because the historiography of both orthodox Kantianism and the Modern Synthesis in biology have overemphasized Kant to the exclusion of other influences. Indeed, I agree with Huneman not only about the indispensability of Kant but even about the explicit identification with Kant by the key proponents of *Naturphilosophie* (often suppressing Herder's role)—and here Schelling proved paramount. Schelling was eager to affiliate

himself with Kant and hesitant to acknowledge the influence of Herder. This has constituted a very important hermeneutic problem in the reconstruction of his intellectual development. The crucial intermediary for the clarification of the relations is Kielmeyer. The irony, which Huneman recognizes, is that Kant wanted *none* of this affiliation. Philippe Huneman, "From the *Critique of Judgment* to the Hermeneutics of Nature," *Continental Philosophy Review* 39 (2006): 1–34, citing 9. "It might appear that in the end Kant was quite alone in insisting on the sharp distinction between regulative and constitutive, and mainly for metaphysical reasons" (ibid., 8). I could not agree more.

137. "In both *Weltseele* and *Erster Entwurf*, Schelling clearly goes beyond Kielmeyer: firstly, by suggesting magnetic duality as a mutual principle of both the inorganic and the organic world; secondly, by viewing sunlight as the cause of earthly duality; thirdly, by treating the metamorphosis of insects and plants as analogous to the process of the creation of the world; and fourthly, by linking this metamorphosis to the sexual reproduction of the human race" (Adler, "Aesthetics of Magnetism," 76).

138. Goethe to Knebel, June 16, 1798, cited in Adler, "Aesthetics of Magnetism," 75.

139. Johann Wolfgang von Goethe, *Maximen und Reflexionen*, cited in Adler, "Aesthetics of Magnetism," 91.

140. Goethe to Schiller, June 27, 1798, in Staiger, *Briefwechsel*, 641.

141. Hence the famous final line of the introduction Schelling gave to his *Ideen*: "Nature should be Mind made visible, Mind the invisible Nature." Friedrich Schelling, *Ideas for a Philosophy of Nature*, trans. Errol E. Harris and Peter Heath (New York: Cambridge University Press, 1988), 30.

142. Here, Kant's dynamic theory of attraction and repulsion, as articulated in *Metaphysical Foundations of Natural Science* (1785), proved influential for both Goethe and Schelling. Schelling's endeavor in his *Ideas* was the reformulation of these two forces into the principles of polarity and intensification, which would become the cornerstones of Goethe's mature philosophy of science.

143. This study makes no pretense to a comprehensive treatment of Schelling; its concern is strictly with *Naturphilosophie* and its connection with the emergent life sciences.

144. On the later Schelling, see, e.g., Andrew Bowie, *Schelling and Modern European Philosophy: An Introduction* (London: Routledge, 1993).

145. Schelling, *Ideas for a Philosophy of Nature*, preface to the 1797 ed. Already at the outset, Schelling clearly aims to dispute Kant's approach to a philosophy of natural science even as he affirms his allegiance to transcendental philosophy.

146. Ibid., supplement to preface (2nd ed., 1803).

147. Ibid., 23.

148. Kant, *Kritik der reinen Vernunft*, A832/B860; Kant, *Kritik der Urteilskraft*, AA 5:186.

149. This tonal shift is not without significance; there are also substantive differences between the "regulative" idea of the transcendental dialectic and the "reflective" idea of the *Kritik der Urteilskraft*. See Rudolf Makkreel, *Imagination and Interpretation in Kant: The Hermeneutical Import of the "Critique of Judgment"* (Chicago: University of Chicago Press, 1990).

150. The question of what the proper relation between philosophy and natural science ought to be—the grounding question for philosophy of science—turns out to be far murkier in Kant's critical philosophy and a fortiori in its legacy than this brief characterization can evoke. See my "Should Kant Have Abandoned the 'Daring Adventure of Reason'?"

151. Olaf Breidbach, "Schelling und die Erfahrungswissenschaft," *Sudhofs Archiv* 88 (2004): 153–74.

152. Schelling, *Ideas for a Philosophy of Nature*, 22.

153. Peter McLaughlin, "Kants Organismusbegriff in der *Kritik der Urteilskraft*," in *Philosophie des Organischen in der Goethezeit*, ed. Kai Torsten Kanz (Stuttgart: Steiner, 1994), 100–110, citing 100.

154. Adler, "Aesthetics of Magnetism."

155. Sibille Mischler, *Der verschlungen Zug der Seele: Natur, Organismus und Entwicklung bei Schelling, Steffens und Oken* (Würburg: Königshausen und Neumann, 1997).

156. Schelling, *Ideas for a Philosophy of Nature*, 9.

157. Ibid., 18.

158. Thus, there was a strong anti-Baconian leaning in this posture, which became doctrinaire by 1803, when Schelling denounced "the blind and mindless type of natural research . . . since the corruption of philosophy by Bacon and of physics by Boyle and Newton" (ibid., 52).

159. This went against the grain of the seventeenth- and especially eighteenth-century suspicions of "hypothesis" in physical science, most famously uttered by Newton but buttressed by a far wider phalanx of thinkers, including Locke in England and Réaumur and his school in France. It also drew the scorn of nineteenth-century (positivist) philosophy of science for "speculation." In terms of contemporary philosophy of science, however, this might be construed more liberally in terms of the relation between *theory* and *experimental practice*, models and target objects of inquiry.

160. This reading of Kant is made more plausible by a close consideration of the arguments in the two versions of the introduction to Kant's *Kritik der Urteilskraft* (1790), over against the position in the transcendental analytic of the *Kritik der reinen Vernunft* (1781), and also by a consideration of the extensive efforts now collected as Kant's *Opus Postumum*, of which, of course, Schelling could have had no knowledge.

161. Schelling, *Ideas for a Philosophy of Nature*, 23.

162. Ibid., 15.

163. Ibid., 32.

164. Indeed, this is the crucial point for any revisionism regarding Kant's philosophy of biology. See my "Should Kant Have Abandoned the 'Daring Adventure of Reason'?"

165. On the aporias in Kant's philosophy of science, see my "'This Inscrutable *Principle* of an Original *Organization*': Epigenesis and 'Looseness of Fit' in Kant's Philosophy of Science," *Studies in History and Philosophy of Science* 34 (2003): 73–109.

166. Schelling, *Ideas for a Philosophy of Nature*, 35.

167. Ibid., 31.

168. Ibid., 35.

169. Ibid., 16.

170. Ibid., 30.

171. Ibid., 31.

172. Kant, *Kritik der Urteilskraft*, AA 5:185–86.

173. Schelling, *Ideas for a Philosophy of Nature*, 42.

174. Ibid., 30.

175. Frederick Beiser, *German Idealism: The Struggle against Subjectivism, 1781–1801* (Cambridge, MA: Harvard University Press, 2002), 559.

176. Ibid., 466.

177. Ibid., 550.

178. Ibid., 530.

179. Ibid., 367.

180. Ibid. Thus, "Schelling, like Hölderlin, Novalis, and Schlegel, is less a strict Spinozist than a follower of Herder's vitalist reinterpretation of Spinoza" (ibid., 530).

181. Schelling, *Ideas for a Philosophy of Nature*, 30.

182. Ibid., 87.

183. Ibid., 37.

184. Ibid., 18.

185. Karl Hufbauer, *The Formation of the German Chemical Community (1720–1795)* (Berkeley: University of California Press, 1982).

186. Schelling, *Ideas for a Philosophy of Nature*, 59.

187. Ibid., 62–63.

188. Ibid., 62.

189. Ibid., 4.

190. Others helped him to advance. "Schelling's *Naturphilosophie* has to be placed within a broader tradition. . . . There can be no doubt that Schelling was strongly influenced by the work of Eschenmayer and Kielmeyer" (Beiser, *German Idealism*, 514).

191. Schelling, *Ideas for a Philosophy of Nature*, 35.

192. Friedrich Schelling, *Von der Weltseele: Eine Hypothese der höheren Physik zur Erklärung des allgemeinen Organismus* (Hamburg, 1798). Amazingly, this text has never been translated into English, and it has even been left out of a number of the more accessible collections of Schelling's works.

193. Joachim Brandis, *Versuch über die Lebenskraft* (1795); Johann Christian Reil, *Von der Lebenskraft* (1795).

194. Schelling specifically refers to Christoph Girtanner's *Über das Kantische Prinzip für die Naturgeschichte* (1796) in *Von der Weltseele*, 250; it would become even more important for his thinking in the *Erste Entwurf*, as I will argue below.

195. Schelling, *Von der Weltseele*, vi.

196. Ibid., x.

197. Friedrich Schelling, *First Outline of a System of the Philosophy of Nature*, trans. Keith Peterson (Albany: State University of New York Press, 2004), 6.

198. Ibid., 154.

199. Ibid., 17.

200. Schelling, *Von der Weltseele*, 4.

201. Ibid., 5.

202. Ibid., 46.

203. Ibid., 26.

204. Ibid., iv.

205. Ibid., 128.

206. Ibid., 9.

207. Ibid., 117.

208. Ibid., xi.

209. Schelling, *First Outline*, 57.

210. "It is the height of unphilosophy to maintain that life is a *property* of matter" (Schelling, *Von der Weltseele*, 185). Here, Schelling is posing the essential Kantian objections against hylozoism.

211. Schelling, *First Outline*, 57. Schelling (*Von der Weltseele*, 186) explicitly identified the notion of "spontaneous animal matter" with Reil.

212. Schelling, *Von der Weltseele*, 227.

213. Christoph Girtanner, "Abhandlung über die Irritabilität, als Lebensprincip in der organisierten Natur," *Journal der Physik* 3 (1791): 317–51, 507–37.

214. Schelling, *Von der Weltseele*, 259.

215. Ibid., 262. See Christoph Pfaff, *Über thierische Elektricität und Reizbarkeit* (Leipzig: Crusius, 1795). It was Pfaff who was Schelling's principal guide to the new life sciences during his stay in Leipzig. See Manfred Durner, "Schellings Begegnung mit den Naturwissenschaften in Leipzig," *Archiv für Geschichte der Philosophie* 72 (1990): 220–36.

216. Schelling, *Von der Weltseele*, 281ff. Galvanism would become a major preoccupation in Schelling's subsequent works.

217. Ibid., 265.

218. Ibid., 280.

219. Schelling, *First Outline*, 110.

220. Schelling, *Ideas for a Philosophy of Nature*, 37.

221. Ibid., 38.

222. Schelling, *Von der Weltseele*, 234.

223. Ibid., 280.

224. Schelling, *First Outline*, 111.

225. Schelling, *Von der Weltseele*, 234.
226. Ibid., 235.
227. Ibid.
228. Ibid., 215.
229. Ibid., 277.
230. Ibid., 189.
231. Ibid. Note the double invocation of "dead" in this passage; the "inert," the "passive," had become the commonplace notions of matter, but they could not make sense of life, or even of force.
232. Ibid.
233. Ibid., 226.
234. Ibid., 237.
235. Ibid., 235.
236. Ibid.
237. Ibid., 235–36.
238. Ibid., 238.
239. Ibid.
240. Ibid., 249.
241. Ibid., 299.
242. Ibid., 300.
243. Ibid., 302.
244. Ibid., 301.
245. Ibid., 298.
246. Ibid., 305.
247. Ibid., 238.
248. Ibid., 305.
249. Steigerwald, "Goethe's Morphology."
250. Schelling, *Von der Weltseele*, 223.
251. Ibid., vi–vii.
252. Ibid., 297.
253. Ibid., 298n.
254. Norbert Hinske, Erhard Lange, and Horst Schröper, eds., *Der Aufbruch in den Kantianismus: Der Frühkantianismus an der Universität Jena von 1785–1800 und seine Vorgeschichte* (Stuttgart: Frommann-Holzboog, 1995); Anthony La Vopa, *Fichte: The Self and the Calling of Philosophy, 1762–1799* (Cambridge: Cambridge University Press, 2001).
255. Schelling, *First Outline*, 5.
256. Ibid.
257. Ibid.
258. Ibid., 157.
259. Ibid., 10. As Schelling would phrase it later in his *First Outline*, it was clear that "no experience reaches to the first origin of duplicity itself" (122–23). It required an *ideal* construction.
260. Ibid., 13.
261. Ibid.
262. Ibid.
263. Ibid., 14.
264. Ibid.
265. Ibid., 15n.
266. Ibid., 16.
267. Ibid.
268. Ibid., 19.
269. Ibid.
270. Ibid., 6.

271. Ibid.
272. Ibid., 48n.
273. Ibid.
274. Ibid., 42.
275. Ibid., 40.
276. Schelling, *Introduction to the Outline of a System of the Philosophy of Nature, or, On the Concept of Speculative Physics and the Internal Organization of a System of This Science (1799)*, in Schelling, *First Outline of a System of the Philosophy of Nature*, trans. Peterson, 193–232, citing 228n.
277. Ibid., 212.
278. Schelling, *First Outline*, 43.
279. Kant, *Kritik der Urteilskraft*, AA 5:419n.
280. Schelling, *First Outline*, 43.
281. Ibid., 37n.
282. Girtanner, *Über das Kantische Prinzip für die Naturgeschichte*.
283. Schelling, *First Outline*, 44. Schelling illustrated this with an example drawn directly from Kant and replicated by Girtanner: "Kant said very truly[,] . . . 'even the character of the *Caucasian* is only the development of one of the original natural predispositions'" (ibid.). Presumably, Schelling got this from Kant's essay of 1785, but did Kant already use the term "Caucasian" there? This term has conventionally been associated only with Blumenbach's work of 1795, which may very well, via Girtanner, have been Schelling's source.
284. Ibid.
285. In *Anthropologie in pragmatischer hinsicht* (Anthropology from a pragmatic point of view; 1798) Kant specifically celebrated Girtanner for the latter's exposition of the theory of race. Under the heading "On the Character of Races" Kant wrote: "As to this subject I can refer to what Girtanner has stated so beautifully and carefully in explanation and further development [of my principles]" (*Anthropologie*, AA 7:320). In the lecture materials for his *Physische Geographie*, edited and published by Rink in 1802, Kant also referred to Girtanner repeatedly as authoritative (*Physische Geographie*, AA 9:185, 234, 313–14, 319).
286. Schelling, *First Outline*, 44.
287. Ibid., 45.
288. Ibid., 47.
289. Ibid.
290. Ibid., 47n.
291. Ibid., 48.
292. Ibid., 43, translation modified.
293. Ibid., 48.
294. Ibid. That is, what he proposed fits the classical dialectical sense of *Aufheben*.
295. Ibid., 48n.
296. Ibid., 49.
297. Ibid.
298. Ibid., 218.
299. Ibid., 50.
300. Ibid., 53.
301. Ibid.
302. Ibid.
303. Schelling, *Von der Weltseele*, vi.
304. Ibid., vii.
305. Schelling, *First Outline*, 146n.
306. Ibid., 147.
307. Schelling, *Von der Weltseele*, 238.
308. Schelling, *First Outline*, 148n.

309. Ibid., 149.

310. Robert Richards, *The Romantic Conception of Life: Science and Philosophy in the Age of Goethe* (Chicago: University of Chicago Press, 2002), 11.

311. Ibid, 490.

CHAPTER ELEVEN

1. Frederick Beiser, *German Idealism: The Struggle against Subjectivism, 1781–1801* (Cambridge, MA: Harvard University Press, 2002), 508.

2. "Intellectually speaking, Jena, in early 1797, was the most exciting place in the world," Boyle writes. Nicholas Boyle, *Goethe: The Poet and the Age*, vol. 2, *Revolution and Renunciation* (Oxford: Clarendon Press, 2000), 468. After the explosive arrival of Johann Fichte, it had become the pacesetter of philosophy in Germany.

3. Henrik Steffens, "Recension der neueren philosophischen Schriften des Herausgebers," *Zeitschrift für spekulative Physik* 1, no. 1 (1800): 3–48; 1, no. 2 (1800): 88–121.

4. Adolph Karl August von Eschenmayer, "Spontaneität = Weltseele," *Zeitschrift für spekulative Physik* 2, no. 1 (1801), 2–63.

5. This endeavor has been pioneered by Olaf Breidbach and Thomas Bach. See Olaf Breidbach, "Jenaer Naturphilosophie un 1800," *Sudhofs Archiv* 84 (2000): 19–49; Thomas Bach, "Zur Institutionalisierung der Naturphilosophie in Jena," in *Christian Gottfried Nees von Eschenbeck—Politik und Naturwissenschaften in der ersten Hälfte des 19. Jahrhunderts*, Acta Leopoldina 43 (Leipzig: Barth, 2004), 167–84; Thomas Bach and Olaf Breidbach, eds., *Naturphilosophie nach Schelling* (Stuttgart: Frommann-Holzboog, 2005). See also Olaf Breidbach and Paul Ziche, eds., *Naturwissenschaften um 1800: Wissenschaftskultur in Jena-Weimar* (Weimar: Böhlaus, 2001); Olaf Breidbach and Roswitha Burwick, eds., *The Transformation of Science in Germany at the Beginning of the Nineteenth Century: Physics, Mathematics, Poetry, and Philosophy* (Lewiston, NY: Edwin Mellen, 2013).

6. Paul Diepgen, *Geschichte der Medizin: Die historische Entwicklung der Heilkunde und des ärztlichen Lebens*, vol. 2.1, *Von der Medizin der Aufklärung bis zur Begründung der Zellularpathologie*, 2nd ed. (Berlin: De Gruyter, 1959), 17ff.

7. I borrow the phrase from W. B. Gallie, "Essentially Contested Concepts," in *Philosophy and the Historical Understanding*, 2nd ed. (New York: Schocken, 1968), 157–91. As Reill aptly notes, Kant "considered such adventures highly suspect." By contrast, "the *Naturphilosophen* elevated this 'daring adventure of reason' into a fundamental imperative." Peter Hanns Reill, *Vitalizing Nature in the Enlightenment* (Berkeley: University of California Press, 2005), 201.

8. Kant, *Kritik der Urteilskraft*, AA 5:419n.

9. Richards puts it succinctly: "The impact of Kant's *Kritik der Urteilskraft* on the disciplines of biology has, I believe, been radically misunderstood by many contemporary historians. . . . Those biologists who found something congenial in Kant's third *Critique* either misunderstood his project (Blumenbach and Goethe) or reconstructed certain ideas to have very different consequences from those Kant originally intended (Kielmeyer and Schelling)." Robert Richards, *The Romantic Conception of Life: Science and Philosophy in the Age of Goethe* (Chicago: University of Chicago Press, 2002), 229.

10. Phillip Sloan, "Kant on the History of Nature: The Ambiguous Heritage of the Critical Philosophy for Natural History," *Studies in History and Philosophy of the Biological and Biomedical Sciences* 37 (2006): 627–48, citing 643.

11. Goethe described his project in this manner in retrospect. See Johann Wolfgang von Goethe, "Judgment through Intuitive Perception," in Goethe, *Scientific Studies*, ed. and trans. Douglas Miller (New York: Suhrkamp, 1988), 32.

12. Sloan, "Kant on the History of Nature," 643.

13. Immanuel Kant, preface to *Metaphysische Anfangsgründe der Naturwissenschaft*, AA 4:467–

79. "His critical philosophy was very quickly regarded as the decisive breakthrough in the project of setting philosophy on a secure foundation and therewith providing the proof of the proper science of its pronouncements." Olaf Breidbach and Paul Ziche, "Einführung: Naturwissen und Naturwissenschaften—zur Wissenschaftskultur in Weimar/Jena," in Breidbach and Ziche, *Naturwissenschaften um 1800*, 7–24, citing 16.

14. Soemmerring to Kant, August 10, 1795; see Peter McLaughlin, "Soemmerring und Kant: Über das Organ der Seele und den Streit der Fakultäten," in *Samuel Thomas Soemmerring und die Gelehtren der Goethezeit*, ed. Gunter Mann and Franz Dumont (Stuttgart: Fischer, 1985), 191–201. On Soemmerring, see also Franz Dumont and Gunter Mann, eds., *Gehirn—Nerven—Seele: Anatomie und Psychologie im Umfeld S. Th. Soemmerrings* (Stuttgart: G. Fischer, 1988).

15. On the neurophysiology of Soemmerring's essay, see Walther Riese, "The 150th Anniversary of S. T. Soemmerring's *Organ of the Soul*: The Reaction of His Contemporaries and Its Significance Today," *Bulletin of the History of Medicine* 20 (1946): 310–21.

16. Samuel Thomas Soemmerring, *Über das Organ der Seele* (Königsberg, 1786; repr., Amsterdam: Bonset, 1966).

17. Ibid., §§1–50, 1–56.

18. Kant, comment on Soemmerring, AA 12:31–35.

19. Kant, first draft, AA 13:398–412. See Arthur Warda, "Zwei Entwürfe Kants zu seinem Nachwort für Soemmerrings Werk 'Ueber das Organ der Seele,'" *Altpreussische Monatsschrift* 40 (1903): 84–120; Luigi Marino, "Soemmerring, Kant and the Organ of the Soul," in *Romanticism in Science*, ed. S. Poggi and M. Bossi, 127–42; Werner Euler, "Die Suche nach dem 'Seelenorgan'—Kants philosophisehe Analyse einer anatomishcen Entdeckung Soemmerrings," *Kant Studien* 93 (2002): 453–80; Alexander Rueger, "Brain Water, the Ether, and the Art of Constructing Systems," *Kant-Studien* 86 (1995): 26–40.

20. See my "History of Philosophy vs. History of Science: Blindness and Insight of Vantage Points on the Kant-Forster Controversy," in *Klopffechtereien—Missverständnisse—Widerspruche?*, ed. Rainer Godel and Gideon Stiening (Paderhorn: Fink, 2011 [actually 2012]), 225–44.

21. In his ultimate response, Kant wrote: "two faculties could fall into conflict over their respective jurisdictions (their *forum competens*)—the *medical* faculty, in its anatomical-physiological expertise, and the *philosophical* faculty, in its psychological-metaphysical expertise" (AA 12:31). In his first draft, Kant took a far more defensive stance, fearing that Soemmerring was setting a trap for metaphysicians: "to dare to set foot beyond their borders into the field of physiology and thus to expose their own weakness." Kant acknowledged his temptation to cross that border and speculate about life-forces, but he recognized that this should properly be left to the experts in physiology. Still, since he proclaimed himself "not entirely unfamiliar with *empirically constrained* natural science" (AA 13:405; note that he dropped this modifier in the ultimate version), he felt entitled to offer hypotheses, even if he risked the scorn that Maupertuis had drawn from Voltaire for similar adventures.

22. Kant, comment on Soemmerring, AA 12:31–35; AA 13:398–412.

23. McLaughlin, "Soemmerring und Kant," 200.

24. Brigitte Lohff, *Die Suche nach der Wissenschaftlichkeit der Physiologie in der Zeit der Romantik* (Stuttgart: Fischer, 1990), 14, 42, 44.

25. Immanuel Kant, *Logic*, trans. Robert S. Hartman and Wolfgang Schwarz (New York: Dover, 1974).

26. Stanley L. Jaki, introduction to *Universal Natural History*, by Immanuel Kant (Edinburgh: Scottish Academic Press, 1981).

27. It is no surprise that this was Herder's favorite work by Kant. See my "'Method' vs 'Manner'?—Kant's Critique of Herder's *Ideen* in Light of the Epoch of Science, 1790–1820," *Herder Yearbook*, 1998, 1–25.

28. Wilhelm von Humboldt to Schiller, cited in McLaughlin, "Soemmerring und Kant," 201.

29. "Kant's contemporaries . . . knew from the obscurity of such doctrines as that of the 'in-

ner sense' or the 'a priori schematism' and from the ambiguity of Kant's position on the 'thing-in-itself' that a reform of his premises was needed. Moreover, Kant's official position was complex (and confused) enough to make each believe, at least in the beginning, that in adopting his respective line of thought he was only making explicit the *real* position of the master." George Di Giovanni, "Kant's Metaphysics of Nature and Schelling's Ideas for a Philosophy of Nature," *Journal of the History of Philosophy* 17 (1979): 197–215, citing 206.

30. Peter McLaughlin, "Kants Organismusbegriff in der *Kritik der Urteilskraft*," in *Philosophie des Organischen in der Goethezeit*, ed. Kai Torsten Kanz (Stuttgart: Steiner, 1994), 100–110.

31. Ibid., 100.

32. Ibid., 101.

33. "How does one explain the explosive development of those six to seven years from 1789 to 1795/6—a development initially away from Kant, which at once led to the situation that Kant was virtually totally dropped from discussion within it, and then in reaction against Fichte as well, which began already in the year 1794?" Dieter Henrich, *Konstellationen: Probleme und Debatten am Ursprung der idealistischen Philosophie (1789–1795)* (Stuttgart: Klett-Cotta, 1991), 223.

34. Here I take issue with Dietrich von Engelhardt, particularly his "Die Naturwissenschaft der Aufklärung und die romantisch-idealistische Naturphilosophie," in *Idealismus und Aufklärung*, ed. Christoph Jamme and Gerhard Kunz (Stuttgart: Klett-Cotta, 1988), 80–96; and his "Zur Naturwissenschaft und Naturphilosophie um 1780 und 1830," in *Hegel und die Chemie: Studie zur Philosophie und Wissenschaft der Natur um 1800* (Wiesbaden: G. Pressler, 1975), 5–30.

35. Goethe argued that the experiment was "merely a stage in the process of knowledge [*Zwischenstufe im Erkenntnisprozeß*]." Johann Wolfgang von Goethe, "The Experiment as Mediator between Object and Subject" (1792), in *Scientific Studies*, 11–17.

36. McLaughlin, "Kants Organismusbegriff in der *Kritik der Urteilskraft*."

37. August Friedrich Hecker (1763–1811), cited by Lohff, *Die Suche nach der Wissenschaftlichkeit*, 171.

38. Nelly Tsouyopoulos, "Der Streit zwischen Friedrich Wilhelm Joseph Schelling und Andreas Röschlaub über die Grundlagen der Medizin," *Medizinhistorisches Journal* 13 (1978): 229–46, citing 245; see also 230 and Nelly Tsouyopous, "The Influence of John Brown's Ideas in Germany," in *Brunonianism in Britain and Europe*, ed. W. F. Bynum and R. Porter, Medical History Supplement no. 8 (London: Wellcome Institute for the History of Medicine, 1988), 63–74, citing 68–70.

39. Lohff, *Die Suche nach der Wissenschaftlichkeit*, 42, 44.

40. Thus, Carl Christian Schmid appeared to claim that only the a priori could warrant knowledge, and he "dismissed every value of knowledge from experience," whereas Johann Heinrich Autenrieth (1772–1835) maintained that only concrete experience constituted knowledge while everything else was mere speculation (Lohff, *Die Suche nach der Wissenschaftlichkeit*, 70–71). The "critical philosophy" was very hard to digest as a whole (as is no less the case today), and as Lohff very aptly observes, the philosophically troubled physiological community was not necessarily sophisticated in coming to terms with Kant's system. In fact, Lohff makes clear, "never was the complete philosophical system of doctrine assimilated." Instead, only "philosophical slants [*Tendenzen*] and turns of phrase" got taken up (36). Similarly, Mocek observes: "Reil and his medical colleagues of the time had by no means studied this philosophy in all its ramifications, that is, in a sense, developed themselves into philosophical competitors." Instead, they oriented themselves to a few lines of thought and emphases, "categories, high-profile definitions, fundamental determinations." Reinhard Mocek, *Johann Christian Reil (1759–1813): Das Problem des Übergangs von der Spätaufklärung zur Romantik in Biologie und Medizin in Deutschland* (Frankfurt: Peter Lang, 1995), 101. Kant's very terminology was daunting, presenting, in Lohff's delicate phrasing, "difficulties for comprehension [*Verständnisschwierigkeiten*]" so substantial that lexica were required to render his technical terms accessible for ordinary readers (Lohff, *Die Suche nach der Wissenschaftlichkeit*, 42). The most prominent such lexicon of the day was published by Carl Christian Schmid: *Wörterbuch zum leichtern Gebrauch der Kantischen Schriften* (Darmstadt: Wissenschaftliche Buchgesellschaft, 1976: Reprograf. Nachdr. d. 4., verm. Ausg. [*sic*]; Jena: Cröker, 1798).

41. Carl Christian Schmid, *Physiologie philosophische bearbeitet*, 3 vols. (Jena: Akademische Buchhandlung, 1798). See Hanns-Peter Gosau, "Über den vergeblichen Versuch Carl Christian Schmidts, die spekulative Naturphilosophie Fichtes und Schellings aus dem medizinischen Denkens des ausgehenden 18. Jahrhunderts zu verdrängen," *Hippokrates: Zeitschrift für praktischen Heilkunde* 35 (1964): 968–70.

42. For a subtle and effective problematizing of these matters, see Nicholas Jardine, "The Significance of Schelling's 'Epoch of a Wholly New Natural History': An Essay on the Realization of Questions," in *Metaphysics and Philosophy of Science in the Seventeenth and Eighteenth Centuries*, ed. R. S. Woodhouse (Dordrecht: Kluwer, 1988), 327–50. For the most explicit consideration of this question, see Frederick Gregory, "Kant's Influence on Natural Scientists in the German Romantic Period," in *New Trends in the History of Science*, ed. R. P. W. Visser, H. J. M. Bos, L. C. Palm, and H. A. M. Snelders (Amsterdam: Rodopi, 1989), 53–72.

43. For a drastic emphasis on the charismatic role of Schelling, see K. Rothschuh, "Ansteckende Ideen in der Wissenschaftsgeschichte, gezeigt an der Entstehung und Ausbreitung der romantischen Physiologie," in *Physiologie im Werden* (Stuttgart: Fischer, 1969), 45–58.

44. Here, I find myself in agreement, on perhaps slightly different grounds, with L. Pearce Williams, "Kant, Naturphilosophie and Scientific Method," in *Foundations of Scientific Method: The Nineteenth Century*, ed. R. Giere and R. Westfall (Bloomington: Indiana University Press, 1973), 3–22.

45. Kielmeyer to Cuvier, December 1807, in Carl Friedrich Kielmeyer, *Gesammelte Schriften*, ed. F. Holler (Berlin: Keiper, 1938), 234–54.

46. Gregory, "Kant's Influence on Natural Scientists in the German Romantic Period," 60. See also Frederick Gregory, "Romantic Kantianism and the End of the Newtonian Dream in Chemistry," *Archives internationales d'histoire des sciences* 34 (1984): 108–23; Frederick Gregory, "Kant, Schelling, and the Administration of Science in the Romantic Era," *Osiris*, 2nd ser., 5 (1989): 17–35.

47. Lohff, *Die Suche nach der Wissenschaftlichkeit*.

48. Ibid. Thus, "in every publication on theoretical medicine one can find the author setting forth his epistemological position" (ibid., 21).

49. Ibid., 28. See Erna Lesky, "Cabanis und die Gewißheit der Heilkunst," *Gesnerus* 11 (1954): 152–84.

50. Thomas Broman, *The Transformation of German Academic Medicine* (Cambridge: Cambridge University Press, 1996), 131.

51. Ibid., 131–32.

52. "Kant's critical philosophy had by the 1790s come to define the epistemological conditions of academic *Wissenschaft* in Germany," Broman writes (ibid., 136), but he also recognizes that not everyone followed Kant's definitions of "proper" science. Two alternative discourses emerged in the controversy. One sought to place medicine under the umbrella of an alternative model of science: history (ibid., 138ff.). The other was *Naturphilosophie*.

53. Ibid., 133ff.

54. On the simultaneity of Kant and Schelling, see Philippe Huneman, "From the *Critique of Judgment* to the Hermeneutics of Nature," *Continental Philosophy Review* 39 (2006): 1–34; Philippe Huneman, "Naturalising Purpose: From Comparative Anatomy to the 'Adventure of Reason,'" *Studies in History and Philosophy of the Biological and Biomedical Sciences* 37 (2006): 649–74.

55. Friedrich Schelling, *Von der Weltseele: Eine Hypothese der höheren Physik zur Erklärung des allgemeinen Organismus* (Hamburg, 1798), vi. See Marie Luise Heuser-Kessler, "Schellings Organismusbegriff und seine Kritik des Mechanismus und Vitalismus," *Allgemeine Zeitschrift für Philosophie* 14 (1989): 17–36, citing 31.

56. Lee Ann Hansen Le Roy, "Johann Christian Reil and *Naturphilosophie* in Physiology" (PhD diss., University of California–Los Angeles, 1985), 101.

57. On all this background, see Schelling's extensive commentary: "Anhang zu den voraufstehenden Aufsatz, betreffend zwei naturphilosophischen Rezensionen und die Jenaische Allgemeine Literaturzeitung," *Zeitschrift für spekulative Physik* 1, no. 1 (1800): 51–99.

58. The *ALZ* had been the first journal to embrace Kantianism. By 1800, it regarded itself as the authoritative guardian of Kantianism in Germany, and as such it excoriated any deviations. Schelling lambasted it as a dead Kantianism, a Kantianism of the letter, lacking in all spirit. For the general context, see Frederick Beiser, *The Fate of Reason: German Philosophy from Kant to Fichte* (Cambridge, MA: Harvard University Press, 1987).

59. Henrik Steffens, *Was ich erlebte* (Breslau: Mar, 1841), 4:149.

60. "The main question at stake between [Kant and *Naturphilosophie*] is whether rationality is something that we create and impose on the world, or whether it is something that exists within the world itself and is reflected in our own activity" (Beiser, *German Idealism*, 556). *Naturphilosophie* took the position that "rather than beginning with self-consciousness, the philosopher should start with the *natura naturans* and derive self-consciousness from it" (ibid., 471). In Richards's view, "the task of *Naturphilosophie* (as distinct from transcendental philosophy) is to begin with a refined understanding of nature, a nature articulated with the help of the latest empirical theories, and to show how its various phenomenal relationships can be regressively chased back [to] their only possible source," an *absolute* "modeled . . . on Spinoza's *Deus sive Natura* and Plato's *Good*" (Richards, *Romantic Conception of Life*, 133, 184).

61. The new generation sought a grasp of "the universe as a whole, nature in itself, the *natura naturans*, which subsists apart from consciousness and explains its very possibility according to necessary laws" (Beiser, *German Idealism*, 557).

62. Steffens, *Was ich erlebte*, 4:76.

63. R. C. Stauffer, "Persistent Errors Regarding Oersted's Discovery of Electromagnetism," *Isis* 44 (1953): 307–10; R. C. Stauffer, "Speculation and Experiment in the Background of Oersted's Discovery of Electromagnetism," *Isis* 48 (1957): 33–50; Owsei Temkin, "Basic Science, Medicine, and the Romantic Era," *Bulletin of the History of Medicine* 37 (1963): 97–129; H. A. M. Snelders, "Romanticism and *Naturphilosophie* in the Inorganic Sciences, 1797–1840," *Studies in Romanticism* 9 (1970): 193–215.

64. Walter Wetzel, *Johann Wilhelm Ritter: Physik im Wirkungsfeld der deutschen Romantik* (PhD diss., Princeton University, 1968); Jocelyn Holland, *German Romanticism and Science: The Procreative Poetics of Goethe, Novalis, and Ritter* (London: Routledge, 2009).

65. Ralph Marks, *Konzeption einer dynamischen Naturphilosophie bei Schelling und Eschenmayer* (Munich: Holler, 1982).

66. Schelling, *Ideas for a Philosophy of Nature*.

67. Eschenmayer, "Spontaneität = Weltseele"; Friedrich Schelling, "Anhang zu dem Aufsatz des Herrn Eschenmayer betreffend den wahren Begriff der Natyurphilosophie und die richtige Art ihre Probleme aufzuhölen," *Zeitschrift für spekulative Physik* 2, no. 1 (1801): 111–46.

68. Adolph Karl August von Eschenmayer, *Sätze aus der Naturmetaphysik auf chemische Gegenstände angewandt* (Tübingen, 1797), 62–96.

69. Tsouyopoulos, "Streit zwischen Friedrich Wilhelm Joseph Schelling und Andreas Röschlaub," 231–32.

70. See Ernst Hirschfeld, *Romantische Medizin*, vol. 1 (Leipzig: Thieme, 1930); George Rosen, "Romantic Medicine: A Problem in Historical Periodization," *Bulletin of the History of Medicine* 25 (1951): 149–58; Ugo D'Orazio, "'Romantische Medizin': Entstehung eines medizin-historischen Epochenbegriffs," *Medizinhistorisches Journal* 32 (1997): 179–217.

71. On Steffens, see Dietrich von Engelhardt, "Henrik Steffens," in Bach and Breidbach, *Naturphilosophie nach Schelling*, 701–36; Fritz Paul, *Henrich Steffens* (Munich: Fink, 1973).

72. Henrik Steffens, "Über den Oxydations- und Desoxydations-Proceßs der Erde: Eine Abhandlung vorgelesen in der naturforschenden Gesellschaft zu Jena," *Zeitschrift für spekulative Physik* 1 (1800): 139–68.

73. Henrik Steffens, *Beyträge zur inneren Naturgeschichte der Erde* (Freyberg, 1801).

74. "In such a science all the divisions of natural inquiry into physics, chemistry, physiology, etc. as sciences set apart from one another will fall away, for its purpose would be the unification

of all these branches under higher principles" (Steffens, "Recension der neueren philosophischen Schriften des Herausgebers," 6).

75. Ibid., 7. But Steffens reiterated Schelling's essential insistence that what speculative physics postulated theoretically had to be confirmed by empirical research. If it failed, the theory would need to be reformulated. And, in fact, Schelling modulated his system from version to version of his *Naturphilosophie*, and often not in small ways.

76. Ibid.

77. Ibid., 8.

78. Ibid., 9, 26.

79. Ibid., 36.

80. Ibid., 10–11.

81. Ibid., 20–23.

82. Ibid., 23.

83. Ibid., 101.

84. Ibid.

85. Ibid., 34.

86. Ibid., 115.

87. Because Steffens believed that there was no general theoretical formulation of Werner's theory available, he offered an extended summary of Werner's geognosy at the outset of his *Beyträge zur inneren Geschichte der Erde*, 15–23.

88. Ibid., 14.

89. Ibid.

90. Ibid., 28.

91. Ibid., 40–41.

92. Ibid., 58.

93. Ibid., 88.

94. Ibid., 91.

95. Ibid., 98.

96. Ibid., 99.

97. Ibid., 279.

98. Ibid., 305.

99. Ibid., 310.

100. Henrik Steffens, *Grundzüge der philosophischen Naturwissenschaft* (Berlin, 1806).

101. Mocek, *Johann Christian Reil*, 170–72. Steffens owed Reil a great deal for his appointment; they became close friends, and Steffens even wrote a significant memorial about Reil after the latter's sudden death during a typhus epidemic: *Johann Christian Reil: Eine Denkschrift* (Halle, 1815).

102. On Schelver, see Thomas Bach, "Franz Joseph Schelver," in Bach and Breidbach, *Naturphilosophie nach Schelling*, 596–626; Klaus-Dieter Müller, *F. J. Schelver, 1778–1832: Romantischer Naturphilosoph, Botaniker und Magnetiseur im Zeitalter Goethes* (Stuttgart: Wissenschaftliche Verlagsgesellschaft, 1992).

103. Thomas Bach, "'Für wen das hier gesagte nicht gesagt ist, der wird es nicht für überflüssig halten': Franz Joseph Schelvers Beitrag zur Naturphilosophie um 1800," in Breidbach and Ziche, *Naturwissenschaften um 1800*, 65–82, citing 68n.

104. Ibid., 68.

105. Ibid., 75.

106. Guenter Risse, "Schelling, 'Naturphilosophie' and John Brown's System of Medicine," *Bulletin of the History of Medicen* 80 (1976): 321–34; Guenter Risse, "'Philosophical' Medicine in Nineteenth-Century Germany: An Episode in the Relations between Philosophy and Medicine," *Journal of Medicine and Philosophy* 1 (1976): 72–92; Guenter Risse, "Kant, Schelling, and the Early Search for a Philosophical 'Science' of Medicine in Germany," *Journal of the History of Medicine and Allied Sciences* 27 (1972): 145–58.

107. See Bynum and Porter, *Brunonianism in Britain and Europe*, esp. Guenter Risse, "Brunonian Therapeutics: New Wine in Old Bottles?," 46–62; Nelly Tsouyopoulos, "The Influence of John Brown's Ideas in Germany," 63–74. See also Guenter Risse, "The Brownian System of Medicine: Its Theoretical and Practical Implications," *Clio Medica* 5 (1970): 45–51; Guenter Risse, "Scottish Medicine on the Continent: John Brown's Medical System in Germany, 1796–1806," *Proceedings of the XXIII International Congress of the History of Medicine, London, 2–9 September 1972*, 2 vols. (London: Wellcome Institute of the History of Medicine, 1974), 1:682–87; John Neubauer, "Dr. John Brown (1735–1788) and Early German Romanticism," *Journal of the History of Ideas* 28 (1967): 367–82.

108. Christoph Girtanner, "Mémoires sur l'irritabilité considerée comme principe de vie dans la nature organisée" (1790). Trans. T. Beddoes, in *Observations on the nature and cure of calculus, sea scurvy, consumption, catarrh, and fever. Together with conjectures upon several other subjects of physiology and pathology* (Philadelphia: Webster, 1815).

109. Tsouyopoulos, "Influence of John Brown's Ideas in Germany," 67. See Nelly Tsouyopoulos, *Andreas Röschlaub und die Romantische Medizin* (Stuttgart: G. Fischer, 1982).

110. Ibid., 65.

111. Ibid., 64. See C. Pfaff, *John Brown: System der Heilkunde begleitet von einer neuen kritischen Abhandlung über die Brownschen Grundsätze* (Copenhagen, 1798); Christoph Girtanner, *Ausführliche Darstellung des Brownischen Systemes der praktischen Heilkunde*, 2 vols. (Göttingen, 1797–98).

112. Risse, "'Philosophical' Medicine," 75–78. Kant's philosophy of science proved a crucial source of anxiety in this regard. See Risse, "Kant, Schelling," 147–49; Tsouyopoulos, "Influence of John Brown's Ideas in Germany," 69–70.

113. Neubauer, "Dr. John Brown and Early German Romanticism," 381.

114. Lohff, *Die Suche nach der Wissenschaftlichkeit*, 171.

115. Manfred Durner, "Schellings Begegnung mit den Naturwissenschaften in Leipzig," *Archiv für Geschichte der Philosophie* 72 (1990): 220–36.

116. Risse, "Schelling, 'Naturphilosophie'," 323–27.

117. Risse, "Kant, Schelling," 149. See also Andreas Röschlaub, *Untersuchungen über Pathogenie*, 3 vols. (1798–1800).

118. "C. A. Eschenmayer proposed to make empirical sciences, such as chemistry and medicine, philosophical by amalgamating philosophy and the applied sciences in 'Natur-Metaphysik.' The second half of [his] book was devoted to medicine and adopted Dr. Brown's concepts of stimulation" (Neubauer, "Dr. John Brown and Early German Romanticism," 371). "Eschenmayer attempted to give transcendental foundations to medicine, and Schelling incorporated it in his *Naturphilosophie*, which aimed at the philosophical construction of nature" (ibid., 381). Eschenmayer's work *Sätze aus der Naturmetaphysik auf chemische und medizinishce Gegenstände angewandt* was published in Tübingen in 1797.

119. "The relationship of Schelling to Brown went through . . . at least three phases—skeptical distance in the world soul text of 1798, agreement in the *First Draft of a System of Natural Philosophy* in 1799, and discriminating critique in *Preliminary Sketch of the Status of Medicine according to the Principles of Natural Philosophy* in 1805" (Mocek, *Johann Christian Reil*, 91).

120. Tsouyopoulos, "Der Streit zwischen Friedrich Wilhelm Joseph Schelling und Andreas Röschlaub," 231–32.

121. Marcus had reported on their tests in *Prufung des Brownschen System der Heilkunde* (1797).

122. See Wolfgang Kretschmer, "Die Bedeutung Friedrich W. Schellings für die Medizin," *Deutsche medizinische Wochenschrift* 79 (1954): 1488–91; Erich Mende, "Der Einfluß von Schellings 'Princip' auf Biologie und Physik der Romantik," *Philosophia Naturalis* 15 (1975): 461–85.

123. Richards, *Romantic Conception of Life*, 179.

124. Thus, the last time Schelling actually delivered lectures on *Naturphilosophie* in Jena was 1801. The field was perpetuated by a series of epigones over the next years, largely elaborating his philosophical ideas but without his intense interest in integrating these with the results of the empirical sciences. Only with the arrival of Lorenz Oken in Jena in 1807 did a regular program of

Naturphilosophie integrating philosophy with natural science get back on a regular footing, but this would be Oken's *Naturphilosophie*, not exactly the same thing as Schelling's, though clearly identifying with his innovation.

125. Michael Vater and David Wood, eds., *The Philosophical Rupture between Fichte and Schelling: Selected Texts and Correspondence (1800–1802)* (Albany: State University of New York Press, 2012).

126. Andreas Röschlaub, *Lehrbuch der Nosologie: Zu seinen Vorlesungen entworfen* (Bamberg : T. Göbhardt, 1801).

127. Tsouyopoulos, "Der Streit zwischen Friedrich Wilhelm Joseph Schelling und Andreas Röschlaub," 232.

128. Ibid.

129. It was entitled *Über die Afteranwendung des neuen Systems der Philosophie auf die Medizin: Eine Rede vogetragen von Professor Andreas Röschlaub, herausgegeben von einen seiner Freunde*. The likely author, according to Nelly Tsouyopoulos, was Joseph Reubeln, a camp follower of the movement at Bamberg. Tsouyopoulous notes that this publication was ascribed by one commentator writing in 1831 to Ignaz Döllinger, and she suspects that Döllinger and even Marcus participated in this send-up (Tsouyopoulos, "Der Streit zwischen Friedrich Wilhelm Joseph Schelling und Andreas Röschlaub," 232). Given Döllinger's personality and practice, it is unlikely that he was the author, though it is not impossible that he and Marcus felt some resentment toward Röschlaub's stature and sought to win Schelling over to their side.

130. Tsouyopoulos, "Der Streit zwischen Friedrich Wilhelm Joseph Schelling und Andreas Röschlaub," 235.

131. Ibid.

132. Werner Gerabek, *Friedrich Wilhelm Joseph Schelling und die Medizin der Romantik: Studien zu Schellings Wurzburger Periode* (Frankfurt am Main: Peter Lang, 1995).

133. Schelling, foreword to *Jahrbücher der Medizin als Wissenschaft* 1 (1805): vi.

134. Ibid.

135. Ibid., ix.

136. Ibid., xiii.

137. Eschenmayer, "Spontaneität = Weltseele."

138. Friedrich Schelling, "Vorläufige Bezeichnung des Standpunktes der Medicin nach Grundsätzen der Naturphilosophie," *Jahrbücher der Medizin als Wissenschaft* 1, no. 1 (1805): 165–207.

139. Ibid., 171.

140. Ibid., 173.

141. Ibid., 176.

142. Ibid., 183. See Ignaz Troxler, *Ideen zur Grundlage der Nosologie und Therapie* (1803). On Troxler, see Stefan Büttner, "Ignaz Paul Vital Troxler," in Bach and Breidbach, *Naturphilosophie nach Schelling*, 775–802.

143. Tsouyopoulos, "Influence of John Brown's Ideas in Germany," 65.

144. Tsouyopoulos, "Der Streit zwischen Friedrich Wilhelm Joseph Schelling und Andreas Röschlaub," 240–41.

145. Hansen Le Roy, "Johann Christian Reil," 73–74.

146. Tsouyopoulos, "Der Streit zwischen Friedrich Wilhelm Joseph Schelling und Andreas Röschlaub," 240.

147. Lorenz Oken, *Frühe Schriften zur Naturphilosophie*, vol. 1 of Oken, *Gesammelte Werke*, ed. Thomas Bach, Olaf Breidbach, and Dietrich von Engelhardt (Weimar: Böhlaus, 2007), most notably "Abriß des Systems der Biologie" (1805) and "Die Zeugung" (1805), published in Bamberg and Würzburg, when Oken was close to the Schelling circle. His first journal project was the *Beiträge zur vergleichenden Zoologie, Anatomie und Physiologie* (1806–7).

148. See Bernhard Milt, "Lorenz Oken und seine Naturphilosophie," *Notizen zur schweizerischen Kulturgeschichte, Vierteljahrsshcirft der Naturforschende Gesellschaft in Zürich* (1951): 181–202; Michael Ghiselin, "Lorenz Oken," in Bach and Breidbach, *Naturphilosophie nach Schelling*, 433–58; Olaf Breidbach and Michael Ghiselin, "Lorenz Oken and *Naturphilosophie* in Jena, Paris

and London," *History and Philosophy of the Life Sciences* 24 (2002): 219–47. Peter Hanns Reill's critique of *Naturphilosophie* centers on an examination of Oken's writings (Reill, *Vitalizing Nature in the Enlightenment*, 229–32), including the memorable sentence—quite true in this case—that "there are few historians who do not wince at some of the *naturphilosophic* scientific explanations" (ibid., 214).

149. Breidbach and Ghiselin, "Lorenz Oken and *Naturphilosophie* in Jena, Paris and London," 219.

150. See H. F. Link, *Über Naturphilosophie* (Leipzig: Stiller, 1806). On Fries, see Frederick Gregory, "Die Kritik von J.F. Fries an Schellings Naturphilosophie." *Sudhoffs Archiv* 67 (1983), 145–57. For a very famous repudiation of *Naturphilosophie*—targeting Schelling and Hegel, not Oken, however—see Matthias Schleiden, *Schelling's und Hegel's Verhältnis zur Naturwissenschaft: Zum Verhältnis der physikalistischen Naturwissenschaft zur spekulativen Naturphilosophie* (1844; repr., with introduction by Olaf Breidbach, Weinheim: VCH Verlagsgesellschaft, 1988).

151. Kai Torsten Kanz, "'. . . die Biologie als die Krone order der höchste Stehepunct aller Wissenschaften': Zur Rezeption des Biologiebegriffs in der romantischen Naturforschung (Lorenz Oken, Ernst Bartels, Carl Gustav Carus)," *NTM: Zeitschrift für Geschichte der Naturwissenschaften, Technik und Medizin* 14 (2006): 77–92; Ilse Jahn, "Biologische Konzeptionen der Goethezeit," in *Geschichte der Biologie: Theorien, Methoden, Institutionen, Kurzbiographien*, ed. Ilse Jahn, 3rd ed. (Stuttgart: Fischer, 1998), 275–89.

152. See Jane M. Oppenheimer, *Autobiography of Dr. Karl Ernst von Baer*, trans. H. Schneider (Canton, MA: Science History Publications, 1986) for the details of the crucial role played by Döllinger in setting him on his career path.

153. Lynn Nyhart, *Biology Takes Form: Animal Morphology and the German Universities, 1800–1900* (Chicago: University of Chicago Press, 1995).

154. Philipp Franz von Walther, *Rede zum Andenken an Ignaz Döllinger* (Munich: Bavarian Academy of Sciences, 1841), 34.

155. Ignaz Döllinger, *Von den Fortschritten, welche die Physiologie seit Haller gemacht hat* (Munich: Lindauer, 1824).

156. Ignaz Döllinger, *Über den Werth und die Bedeutung der vergleichenden Anatomie* (Würzburg: Nitribitt, 1814).

157. Ignaz Döllinger, *Grundriss der Naturlehre des menschlichen Organismus* (Bamberg: Goebhardt, 1805).

158. See Arthur William Meyer, *Human Generation: Conclusions of Burdach, Döllinger and von Baer* (Stanford: Stanford University Press, 1956).

159. Eckhard Struck, "Ignaz Döllinger (1770–1841): Ein Physiologe der Goethe-Zeit und der Entwicklungsgedanke in seinem Leben und Werk" (diss., Munich University, 1977), 53.

160. Heinrich Haeser, *Lehrbuch der Geschichte der Medecin und epidemischen Krankheiten*, 3rd ed. (Jena, 1875–82), cited in Meyer, *Human Generation*, 29.

161. Neuburger and Pagel, *Handbuch der Geschichte der Medezin* (Jena, 1902–5), cited in Meyer, *Human Generation*, 29.

162. Haeser, *Lehrbuch*, cited in ibid.

163. Albrecht Kölliker, *Zur Geschichte der medezinischen Fakultät an der Universität Würzburg* (Würzburg, 1871), cited in ibid.

164. Döllinger, preface to *Grundriss der Naturlehre des menschlichen Organismus*, unpaginated (v).

165. For biographical information, see Struck, "Ignaz Döllinger."

166. Döllinger, preface to *Über die Metamorphose der Erd- und Steinarten aus der Kieselreihe* (Erlangen: n.p., 1803).

167. "[A]fter his call to Würzburg he withdrew himself . . . entirely from medical practice and followed therewith the example of the majority of German professors of anatomy" (Walther, *Rede zum Andenken an Ignaz Döllinger*, 76).

168. Hans Heinz Eulner, *Die Entwicklung der medizinischen Spezialfächer an den Universitäten des deutschen Sprachgebietes* (Stuttgart: Enke, 1970), 48.

169. Broman, *Academic Medicine*, 180.

170. The reorganization of the University of Würzburg led to a restructuring of the medical faculty—the "Sektion für Heilskunde," as it was called—into four divisions: physiology, considered as "general natural science of the organic"; anatomy, both human and comparative; general medical theory, subdivided into pathology and therapy (including pharmacology); and, finally practical medicine (treatment of wounds, surgery, and obstetrics) (Eulner, *Die Entwicklung der medizinischen Spezialfächer*, 52). Döllinger was ordinary professor for the first two fields.

171. Walther, *Rede zum Andenken an Ignaz Döllinger*.

172. This was the heyday of materialistic experimental physiology. See Everett Mendelsohn, "The Biological Sciences in the Nineteenth Century: Some Problems and Sources," *History of Science* 3 (1964), 39–59; Everett Mendelsohn, "Physical Models and Physiological Concepts: Explanation in Nineteenth-Century Biology," *British Journal of the History of Science* 2 (1965), 201–19.

173. See Philipp Franz von Walther, *Physiologie des Menschen mit durchgängiger Rücksicht auf die comparative Physiologie der Thiere* (Landshut, 1807).

174. Walther, *Rede zum Andenken an Ignaz Döllinger*, cited in Struck, "Ignaz Döllinger," 28–29.

175. Struck, "Ignaz Döllinger," 42.

176. Ibid., 15.

177. Ibid., 161–65.

178. Döllinger, *Über die Metamorphose der Erd- und Steinarten aus der Kieselreihe*.

179. Struck, "Ignaz Döllinger," 54.

180. [Carl Windischmann], review of *Grundriss der Naturlehre des menschlichen Organismus*, by Ignaz Döllinger, *Jahrbücher der Medizin als Wissenschaft* 2, no. 1 (1806): 94–115. See Gerabek, *Friedrich Wilhelm Joseph Schelling und die Medizin der Romantik*, 251–52.

181. For Schelling's instigation, see Schelling to Windischmann, September 5, 1805, cited in Gerabek, *Friedrich Wilhelm Joseph Schelling und die Medizin der Romantik*, 251n.

182. Cited in Gerabek, *Friedrich Wilhelm Joseph Schelling und die Medizin der Romantik*, 372n.

183. Ignaz Döllinger, "Über den jetzigen Zustand der Physiologie," *Jahrbücher der Medicin als Wissenschaft* 1, no. 1 (1805): 119–42. In the sequel, "Tubingae, *Cogitata Nonnulla de Idea Vitae, Hujusque Formis Praecipuis: Autore Carolo Eberh. Schelling*, 1803, 38.S., 8.," *Jahrbücher der Medicin als Wissenschaft* 1, no. 2 (1805): 136–60, a review of Karl Eberhard Schelling's dissertation under Kielmeyer at the University of Tübingen, Döllinger repeatedly referred to the first part as "Über den Standpunkt der Medicin."

184. Döllinger, "Über den jetzigen Zustand der Physiologie," 119.

185. Ibid., 120.

186. Ibid., 121–22.

187. Ibid., 122.

188. Ibid., 126ff.

189. Stuart Strickland, "Galvanic Disciplines: The Boundaries, Objects, and Identities of Experimental Science in the Era of Romanticism," *History of Science* 33 (1995): 449–68, citing 456–57.

190. Reil recognized the inadequacy of "Mischung und Form" for physiology by 1805.

191. Döllinger, "Über den jetzigen Zustand der Physiologie," 127.

192. Ibid., 129.

193. Ibid., 130.

194. Ibid., 125.

195. Ibid., 137.

196. Ibid.

197. Ibid., 139.

198. Ibid., 140.

199. Ibid., 141.
200. Döllinger, "Tubingae," 134.
201. Ibid., 136.
202. Döllinger, *Grundriss der Naturlehre des menschlichen Organismus*, 1 (§2).
203. Ibid., 3 (§8).
204. Ibid., 3 (§6). In the preface (unpaginated [ii]), Döllinger noted: "in my eyes comparative anatomy has too great a value for one to skip it in lecturing on the natural science of the human organism."
205. Ibid., 7 (§15).
206. Ibid., 8 (§17).
207. Ibid., 9 (§19).
208. Ibid., 15 (§31).
209. Ibid., 15 (§32); 16 (§35).
210. Ibid., 16 (§35).
211. Ibid., 29 (§61).
212. Ibid., 25 (§51). "The formative drive is either the basis for the emergence or, in what has already emerged, the basis for the maintenance of the organism" (29 [§62]).
213. Ibid., 239 (§440).
214. Ibid., 241 (§445).
215. Ibid., 25 (§55).
216. Ibid., 31 (§§64–65).
217. Ibid., 31 (§66).
218. Ibid., 26 (§§57–58).
219. Ibid., 240 (§442).
220. Ibid., 258 (§479).
221. Ibid., 258–59 (§480).
222. Ibid., 260 (§483).
223. Döllinger, *Über den Werth und die Bedeutung der vergleichenden Anatomie*, 12–13, on the spirit of 1813 in the Germanies.
224. Ibid., 25.
225. Ibid., 27.
226. Ibid., 24.
227. Ibid., 23.
228. Ibid., 29.
229. Ibid., 33–34.
230. Döllinger, *Von den Fortschritten, welche die Physiologie seit Haller gemacht hat* (1824), 5.
231. Ibid.
232. Ibid., 6.
233. Ibid., 9.
234. Ibid., 12.
235. On Pander and embryology, see Meyer, *Human Generation*.
236. Döllinger, *Von den Fortschritten, welche die Physiologie seit Haller gemacht hat* (1824), 12.
237. Ibid., 13–14.
238. Ibid., 16.
239. Ibid., 17.
240. Ibid., 21.
241. Ibid.
242. Ibid.
243. Ibid.
244. Ibid., 22.

Index of Names

Index of Subjects